Systems Engineering and Analysis

PRENTICE HALL INTERNATIONAL SERIES
IN INDUSTRIAL AND SYSTEMS ENGINEERING

W. J. Fabrycky and J. H. Mize, Editors

second edition

Systems Engineering and Analysis

BENJAMIN S. BLANCHARD
Virginia Polytechnic Institute and State University

WOLTER J. FABRYCKY
Virginia Polytechnic Institute and State University

PRENTICE HALL, Englewood Cliffs, New Jersey 07632

Library of Congress Cataloging-in-Publication Data

BLANCHARD, BENJAMIN S.
   Systems engineering and analysis / Benjamin S. Blanchard, Wolter
J. Fabrycky.—2nd ed.
   p.   cm.—(Prentice-Hall international series in industrial
and systems engineering)
   Includes bibliographical references.
   ISBN 0-13-880758-2
   1. Systems engineering.   2. System analysis.   I. Fabrycky, W. J.
(Wolter J.).   II. Title.   III. Series.
   TA168.B58   1990
   620'.001'1—dc20   89-23096
                                                              CIP

Editorial/production supervision
   and interior design: Rob DeGeorge
Manufacturing buyer: Denise Duggan

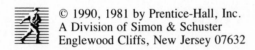 © 1990, 1981 by Prentice-Hall, Inc.
A Division of Simon & Schuster
Englewood Cliffs, New Jersey 07632

Printed in the United States of America

10   9   8   7   6   5

ISBN 0-13-880758-2

PRENTICE-HALL INTERNATIONAL (UK) LIMITED, *London*
PRENTICE-HALL OF AUSTRALIA PTY. LIMITED, *Sydney*
PRENTICE-HALL CANADA INC., *Toronto*
PRENTICE-HALL HISPANOAMERICANA, S.A., *Mexico*
PRENTICE-HALL OF INDIA PRIVATE LIMITED, *New Delhi*
PRENTICE-HALL OF JAPAN, INC., *Tokyo*
SIMON & SCHUSTER ASIA PTE. LTD., *Singapore*
EDITORA PRENTICE-HALL DO BRASIL, LTDA., *Rio de Janeiro*

Contents

PART TWO: THE SYSTEM DESIGN PROCESS

PART THREE: TOOLS FOR SYSTEMS ANALYSIS

7 ALTERNATIVES AND MODELS IN DECISION MAKING   122

8 MODELS FOR ECONOMIC EVALUATION   146

9 OPTIMIZATION IN DESIGN AND OPERATIONS   182

10 PROBABILITY AND STATISTICAL METHODS   238

Preface

This book is about systems. It focuses on the engineering of systems and upon systems analysis. In the first case, the concern is with the process of bringing systems into being. In the latter, the concern is with improving systems already in existence. Additionally, systems analysis applied through modeling is presented as an integral part of the systems engineering process.

Systems may be classified as either "natural" or "man-made." Natural systems are those which came into existence by natural processes. Man-made systems are those where humans have intervened through system components, attributes, and relationships. Only man-made systems are treated in this book.

The experience of recent decades indicates that properly coordinated and functioning man-made systems, with a minimum number of undesirable side effects, require the application of a well-integrated "systems" approach. Systems engineering has been recently recognized as the process by which the orderly evolution of man-made systems can be achieved. The objective of this book is to provide the systems engineer and systems analyst with the knowledge and tools needed for the implementation of this process.

Topics presented in this book have been developed and organized into six parts. Part One presents an introduction to systems and systems engineering in the context of the system life cycle. Part Two addresses the system design process as a series of evolutionary steps leading from the identification of a need through conceptual design, preliminary design, detail design and development, system test and evaluation, and system operation and sustaining support. Part Three develops some of the mathematical tools and techniques commonly used in systems analysis. The application of analysis and modeling techniques in the context of the systems engineering process is introduced. Part Four introduces design for operational feasibility and covers those characteristics of system design considered to be most significant in the systems engineering process. Reliability, maintainability, manability (human factors), supportability, and economic feasibility are highlighted. Part Five presents an overview of systems engineering management. Part Six introduces three different examples illustrating design applications of selected concepts and techniques presented in the earlier chapters.

This book is intended for use in the classroom at either the undergraduate or graduate level or by the practicing professional in industry, business, or government. The concepts and techniques presented are applicable to any type of system and the topics discussed may be "tailored" for both large- and small-scale systems. Many practical problems are introduced to illustrate concepts. The text material is arranged in such a manner as to guide the practicing engineer or analyst through the entire system life cycle. Appropriate references are included at the end of each chapter.

Five other Prentice-Hall books by the authors provided some of the raw material from which this text was fashioned. These books cover the subject areas of Applied Operations Research and Management Science, Economic Decision Analysis, Engineering Economy, Engineering Organization and Management, and Logistics Engineering and Management.

BENJAMIN S. BLANCHARD
WOLTER J. FABRYCKY

PART I: INTRODUCTION TO SYSTEMS

1

System Definitions and Concepts

Systems are as pervasive as the universe in which we live. At one extreme, they are as grand as the universe itself. At the other, they are as infinitesimal as the atom. Systems appeared first in natural forms, but with the appearance of human beings, a variety of man-made systems have come into existence. Only recently have we come to understand the underlying structure and characteristics of natural and man-made systems in a scientific sense.

In this chapter some system definitions and system science concepts are presented that provide a basis for the study of systems engineering and analysis. This will include definitions of system characteristics, a classification of systems into various types, a discussion of the current state of system science, and a description of the transition to the systems age now underway. Finally, this chapter outlines the nature of engineering requirements in the systems age.

1.1. SOME SYSTEM DEFINITIONS

A *system* is an assemblage or combination of elements or parts forming a complex or unitary whole such as a river system or a transportation system; any assemblage or set of correlated members such as a system of currency; an ordered and comprehensive assemblage of facts, principles, or doctrines in a particular field of knowledge or thought, such as a system of philosophy; a coordinated body of methods or a complex scheme or plan of procedure, such as a system of organization and management; any regular or special method of plan of procedure, such as a system of marking,

1

numbering, or measuring.[1] Not every set of items, facts, methods, or procedures is a system. A random group of items lying on a table would constitute a set with definite relationships between the items, but they would not qualify as a system because of the absence of unity, functional relationship, and useful purpose.

The Elements of a System

Systems are composed of components, attributes, and relationships. These are described as follows:

1. *Components* are the operating parts of a system consisting of input, process, and output. Each system component may assume a variety of values to describe a system state as set by control action and one or more restrictions.
2. *Attributes* are the properties or discernible manifestations of the components of a system. These attributes characterize the system.
3. *Relationships* are the links between components and attributes.

A system is a set of interrelated components working together toward some common objective. The set of components has the following properties:

1. The properties and behavior of each component of the set has an effect on the properties and behavior of the set as a whole.
2. The properties and behavior of each component of the set depends upon the properties and behavior of at least one other component in the set.
3. Each possible subset of components has the two properties listed above; the components cannot be divided into independent subsets.

The properties given above ensure that the set of components comprising a system always has some characteristic or behavior pattern that cannot be exhibited by any of its subsets. A system is more than the sum of its component parts. However, the components of a system may themselves be systems, and every system may be part of a larger system in a hierarchy.

The objective or purpose of a system must be explicitly defined and understood so that system components may provide the desired output for each given set of inputs. Once defined, the objective or purpose makes it possible to establish a measure of effectiveness indicating how well the system performs. Establishing the purpose of a man-made system and defining its measure of effectiveness is often a most challenging task.

The purposeful action performed by a system is its *function*. A common system function is that of altering material, energy, or information. This alteration embraces input, process, and output. Some examples are the materials processing in a manufacturing system or a digestive system, the conversion of coal to electricity in a power plant system, and the information processing in a computer system.

[1] This definition was adapted from J. Stein, ed., *The Random House Dictionary of the English Language.* (New York: Random House, Inc., 1966).

Systems that alter material, energy, or information are composed of structural components, operating components, and flow components. *Structural components* are the static parts, *operating components* are the parts that perform the processing, and *flow components* are the material, energy, or information being altered. Of course, a motive force must be present to provide the alteration within the restrictions set by structural and operating components.

Structural, operating, and flow components have various attributes that affect their influence on the system. The attributes of an electrical system may be described in terms of inductance, capacitance, impedance, and so on. The system may change its condition over time in only certain ways, as in the *on* or *off* state of an electrical system. A system, condition, situation, or state is set forth to describe a set of components, attributes, and relationships.

A systems view is only one way of understanding complexity. Another is that of a *relational view*. Three major differences exist between a relation and a system. First, a relation exists between two and only two components, while a system is described by the interaction between many components. Second, a relation is formed out of the imminent qualities of the components, whereas a system is created by the particular position and spatial distribution of its components. The components of a relation are separated spatially, while a system is made up of the interacting distribution of its components. Third, the connection between the components of a relation is direct, whereas the connection in a system depends upon a common reference to the entire set of components making up the system.

Relationships that are functionally necessary to each other may be characterized as *first-order*. An example is symbiosis, the necessary relationship of dissimilar organisms such as an animal and a parasite. *Second-order* relationships, called *synergistic,* are those which are complementary and add to system performance. *Redundance* in a system exists when duplicate components are present for the purpose of assuring continuation of the system function.

Systems and Subsystems

The definition of a system is not complete without consideration for its position in the hierarchy of systems. Every system is made up of *components,* and any component can be broken down into smaller components. If two hierarchial levels are involved in a given system, the lower is conveniently called a *subsystem.* For example, in an air transportation system, the aircraft, terminals, ground support equipment, and controls are subsystems. Equipment items, people, and information are components. Clearly, the designations of system, subsystem, and component are relative, since the system at one level in the hierarchy is the component at another.

In any particular situation it is important to define the system under consideration by specifying its limits or boundaries. Everything that remains outside the boundaries of the system is considered to be the *environment*. However, no system is completely isolated from its environment. Material, energy, and/or information must often pass through the boundaries as *input* to the system. In reverse, material, energy, and/or information that passes from the system to the environment is called *output*. That which enters the system in one form and leaves the system in another form is usually called *throughput*.

The total system, at whatever level in the hierarchy, consists of all components, attributes, and relationships needed to accomplish an objective. Each system has an objective, providing a purpose for which all system components, attributes, and relationships have been organized. Constraints placed on the system limit its operation and define the boundary within which it is intended to operate. Similarly, the system places boundaries and constraints on its subsystems.

An example of a total system is a fire department. The components of this "fire control system" are the building, the fire engines, the firefighters and small equipment, the communication equipment, and the maintenance facilities. These components are actually the major subsystems of the fire department. Each of these subsystems has several contributing subsystems or components. At each level in the hierarchy the description must include all components, all attributes of these components, and all relationships.

The systems viewpoint looks at a system from the top down rather than from the bottom up. Attention is first directed to the system as a black box that interacts with its environment. Next, attention is focused on how the smaller black boxes (subsystems) combine to achieve the system objective. The lowest level of concern is then with individual components.

The process of bringing systems into being, and of improving systems already in existence, in a wholistic sense is receiving continued attention. By bounding the total system for study purposes, the systems engineer or systems analyst will be more likely to achieve a satisfactory result. Focusing on systems, subsystems, and components in a hierarchy forces consideration of all pertinent functional relationships. Components and attributes are important, but only to the end that the purpose of the whole system is achieved through the functional relationships linking them.

1.2. A CLASSIFICATION OF SYSTEMS

Systems may be classified for convenience and to provide insight into their wide range. This will be accomplished by several dichotomies conceptually illustrating system similarities and dissimilarities. In this section descriptions are given of natural and man-made systems, physical and conceptual systems, static and dynamic systems, and closed and open systems.[2]

Natural and Man-Made Systems

The origin of systems gives a most important classification opportunity. *Natural systems* are those which came into being by natural processes. *Man-made systems* are those in which human beings have intervened through components, attributes, or relationships.

---

[2] The classifications in this section are only a few that could be presented. All system types have embedded information flow components to a greater or lesser degree.

All man-made systems, when brought into being, are embedded into the natural world. Important interfaces often exist between man-made systems and natural systems. Each affects the other in some way. The effect of man-made systems on the natural world has only recently become a keen subject for study by concerned people, especially in those instances where the effect is undesirable.

Natural systems exhibit a high degree of order and equilibrium. This is evidenced in the seasons, the food chain, the water cycle, and so on. Organisms and plant life adapt themselves to maintain an equilibrium with the environment. Every event in nature is accompanied by an appropriate adaptation, one of the most important being that material flows are cyclic. In the natural environment there are no dead ends, no wastes, only continual recirculation.

Only recently have significant man-made systems appeared. These systems make up the man-made world, their chief engineer being human. The rapid evolution of human beings is not adequately understood, but their coming upon the scene has significantly affected the natural world, often in undesirable ways. Primitive beings had little impact on the natural world, for they had not yet developed a potent and pervasive technology.

A good example of the impact of man-made systems on natural systems is the set of problems that arose from building the Aswan Dam on the Nile River. Construction of this massive dam ensures that the Nile will never flood again, solving an age-old problem. However, several new problems arose. The food chain was broken in the eastern Mediterranean, thereby reducing the fishing industry. Rapid erosion of the Nile Delta took place, introducing soil salinity into upper Egypt. No longer limited by periodic dryness, the population of bilharzia (a water-borne snail parasite) has produced an epidemic of intestinal disease along the Nile. These side effects were not adequately considered by those responsible for the project. A systems view encompassing both natural and man-made elements might have led to a better solution to the problem of flooding.

Physical and Conceptual Systems

Physical systems are those which manifest themselves in physical terms. They are composed of real components and may be contrasted with *conceptual systems,* where symbols represent the attributes of components. Ideas, plans, concepts, and hypotheses are examples of conceptual systems.

A physical system consumes physical space, whereas conceptual systems are organizations of ideas. One type of conceptual system is the set of plans and specifications for a physical system before it is actually brought into being. A proposed physical system may be simulated in the abstract by a mathematical or other conceptual model. Conceptual systems often play an essential role in the operation of physical systems in the real world.

The totality of elements encompassed by all components, attributes, and relationships focused on a given result employ a process in guiding the state of a system. A process may be mental (thinking, planning, learning), mental-motor (writing, drawing, testing), or mechanical (operating, functioning, producing). Processes exist equally in physical and conceptual systems.

Process occurs at many different levels within both systems. The subordinate process essential to the operation of a total system is provided by the subsystem. The subsystem may, in turn, be dependent upon more detailed subsystems. System complexity is the feature that defines the number of subsystems present and, consequently, the number of processes involved. A system may be bounded for the purpose of study at any process or subsystem level.

Static and Dynamic Systems

Another system dichotomy is the distinction of static and dynamic systems. A *static system* is one having structure without activity, as exemplified by a bridge. A *dynamic system* combines structural components with activity. An example is a school, combining a building, students, teachers, books, and curricula.

For centuries people have viewed the universe of phenomena as unchanging. A mental habit of dealing with certainties and constants developed. The substitution of a process-oriented description for the static description of the world is one of the major characteristics separating modern science from earlier thinking.

A dynamic conception of the world has become a necessity. Yet, a general definition of a system as an ongoing process is incomplete. Many systems would not be included under this broad definition because they lack motion in the usual sense. A highway system is static, yet contains the system elements of components, attributes, and relationships.

It is recognized that a system is static only in a limited frame of reference. A bridge is constructed over a period of time, and this is a dynamic process. It is then maintained and perhaps altered to serve its intended purpose more fully. Even a crystal passes through a number of forms during its period of growth.

Systems may be characterized as having random properties. In almost all systems in both the natural and man-made categories, the inputs, process, and output can only be described in statistical terms. Uncertainty often occurs in both the number of inputs and the distribution of these inputs over time. For example, it is difficult to predict exactly the number of passengers that will check in for a flight, or the exact time they will arrive at the airport. However, each of these factors can be described in terms of probability distributions, and system operation is said to be *probabilistic*.

Closed and Open Systems

A *closed system* is one that does not interact significantly with its environment. The environment only provides a context for the system. Closed systems exhibit the characteristic of equilibrium resulting from internal rigidity that maintains the system in spite of influences from the environment. An example is the chemical equilibrium eventually reached in a closed vessel when various reactants are mixed together. The reaction can be predicted from a set of initial conditions. Closed systems involve deterministic interactions, with a one-to-one correspondence between initial and final states.

An *open system* allows information, energy, and matter to cross its boundaries. Open systems interact with their environment, examples being plants, ecological systems, and business organizations. They exhibit the characteristics of *steady state,* wherein a dynamic interaction of system elements adjusts to changes in the environment. Because of this steady state, open systems are self-regulatory and often self-adaptive.

It is not always easy to classify a system as either open or closed. Open systems are typical of those which have come into being by natural processes. Man-made systems have characteristics of open and closed systems. They may reproduce natural conditions not manageable in the natural world. They are closed when designed for invariant input and statistically predictable output, as in the case of an aircraft in flight.

Both closed and open systems exhibit the property of entropy. *Entropy* is defined here as the degree of disorganization in a system, and is analogous to the use of the term in thermodynamics. In the thermodynamic usage, entropy is the energy unavailable for work resulting from energy transformation from one form to another.

In systems, increased entropy means increased disorganization. A decrease in entropy takes place as order occurs. Life represents a transition from disorder to order. Atoms of carbon, hydrogen, oxygen, and other elements become arranged in a complex and orderly fashion to produce a living organism. A conscious decrease in entropy must occur to create a man-made system. All man-made systems, from the most primitive to the most complex, consume entropy—the creation of more orderly states from less orderly states.

1.3. SCIENCE AND SYSTEMS SCIENCE

The significant accumulation of scientific knowledge, which began in the eighteenth century and rapidly expanded in the twentieth, made it necessary to classify what was discovered into scientific disciplines. Science began its separation from philosophy over a century ago. It then proliferated into more than 100 distinct disciplines. A relatively recent unifying development is the idea that systems have general characteristics independent of the area of science to which they belong. In this section, the evolution of a science of systems is presented through an examination of cybernetics, general systems theory, and systemology.

Cybernetics

The word *cybernetics* was first used in 1947 by Norbert Wiener, but it is not explicitly defined in his classical book.[3] Cybernetics comes from the Greek word meaning "steersman" and is a cognate of "governor." In its narrow view, cybernetics is

---

[3] N. Wiener, *Cybernetics,* (New York: John Wiley & Sons, Inc., 1948).

equivalent to servo theory in engineering. In its broad view, it may encompass much of natural science. Cybernetics has to do with self-regulation, whether mechanical, electromechanical, electrical, or biological.

The concept of feedback is central to cybernetic theory. All goal-seeking behavior is controlled by the feedback of corrective information regarding deviation from a desired state. The best known and most easily explained illustration of feedback is the action of a thermostat. The thermometer component of a thermostat senses temperature, and when the actual temperature falls below that set into the thermostat, an internal contact is made, activating the heating system. When the temperature rises above that set into the thermostat, the contact is broken, shutting the heating system off.

Biological organisms are endowed with the capacity for self-regulation, called *homeostasis*. The biological organism and the physical world are both very complex. Instructive analogies exist between them and man-made systems. Through these analogies humans have learned some things about their properties which might have not been learned from the study of natural systems alone. As we develop even more complex systems, we will gain a better understanding of how to control them and our environment.

The science of cybernetics has made three important contributions to the area of regulation and control. First, it stresses the concept of information flow as a distinct system component and clarified the distinction between the activating power and the information signal. Second, it recognizes that similarities in the action of control mechanisms involve principles which are fundamentally identical. Third, the basic principles of feedback control are given mathematical treatment.

A practical application of cybernetics has been the tremendous development of automatic equipment and processes, many controlled by mini- and microcomputers. However, its significance is greater than this technological contribution. The science of cybernetics is important not only for the control engineer but also for the purest of scientists. Cybernetics is a new science of purposeful and optimal control applicable to complex processes in nature, society, and business organizations.

General Systems Theory

An even broader unifying concept than cybernetics took shape during the late 1940s. It was the idea that basic principles common to all systems could be found which went beyond the concept of control and self-regulation. A unifying principle for science and a common ground for interdisciplinary relationships needed in the study of complex systems was being sought. Ludwig von Bertalanffy used the phrase *general systems theory* around 1950 to describe this endeavor.[4]

General systems theory is concerned with developing a systematic framework for describing general relationships in the natural and the man-made world. The

[4]L. von Bertalanffy, "General System Theory: A New Approach to Unity of Science," *Human Biology*, December 1951. A related contribution was by K. Boulding, "General Systems Theory: The Skeleton of Science," *Management Science*, April 1956.

need for a general theory of systems arises out of the problem of communication among the various disciplines. Although the scientific method brings similarity between the methods of approach, the results are often difficult to communicate across disciplinary boundaries. Concepts and hypotheses formulated in one area seldom carry over to another where they could lead to significant forward progress. The difficulties are greatest among the various disciplines, the physical and life sciences, the social and behavioral sciences, and the humanities.

One approach to an orderly framework is the structuring of a hierarchy of levels of complexity for basic units of behavior in the various fields of inquiry. A *hierarchy of levels* can lead to a systematic approach to systems which has broad application. Such a hierarchy was formulated by Boulding approximately as follows:[5]

1. The level of static structure or *frameworks*, encompassing the geography and anatomy of the universe.
2. The level of the simple dynamic system of *clockworks*, encompassing a significant segment of chemistry, physics, and engineering science.
3. The level of the *thermostat* or cybernetic system, encompassing the transmission and interpretation of information.
4. The level of the *cell*, the self-maintaining structure or open system where life begins to be evident.
5. The level of the *plant*, with genetic-societal structure making up the world of botany.
6. The level of the *animal*, encompassing mobility, teleological behavior, and self-awareness.
7. The level of the *human*, encompassing self-consciousness and the ability to produce, absorb, and interpret symbols.
8. The level of *social organization*, where the content and meaning of messages, value systems, transciption of images into historical record, art, music, and poetry, and complex human emotion are of concern.
9. The level of the *unknowables*, where structure and relationship may be postulated but where answers are not yet available.

The first level in Boulding's hierarchy is the most pervasive. Static systems are everywhere, and this category provides a basis for analysis and synthesis of systems at higher levels. Dynamic systems with predetermined outcomes are predominant in the natural sciences. At higher levels cybernetic models are available, mostly in closed-loop form. Open-loop systems are currently receiving scientific attention, but modeling difficulties arise regarding their self-regulating properties. Beyond this level there is little systematic knowledge available. However, general systems theory provides science with a useful framework within which each specialized discipline

[5] Boulding, "General Systems Theory: The Skeleton of Science," *Management Science*, April 1956.

may contribute. It allows scientists to compare concepts and similar findings, its greatest benefit being that of communication across disciplines.

Systemology

The science of systems or their formation is called *systemology*. As system science is pushed forward by the formation of interdisciplines, humankind will benefit from a more appropriate application of the discipline. The problems and problem complexes faced by human beings are not organized along disciplinary lines. It is only through a new organization of scientific and professional effort based on the common attributes and characteristics of problems that beneficial progress will be made.

Disciplines in science and the humanities developed largely by what society permitted scientists and humanists to investigate. Areas that provided the least challenge to cultural, social, and moral beliefs were given priority. The survival of science was also of concern in the progress of the discipline, and this is still so. However, recent developments have added to the respectability of most areas. Much credit for this can be given to the rise of interdisciplinary inquiry.

During the 1940s, scientists of established reputation accepted the challenge of attempting to understand a host of common processes in military operations. Their team effort was called *operations research* and the focus of their attention was the science of military systems. After the war this interdisciplinary area began to take on the attributes of a discipline and a profession. Today a body of systematic knowledge exists for military and commercial operations. But operations research is not the only science of systems available today. Cybernetics, general systems research, organizational and policy sciences, management science, and the communication sciences are others.

Formation of interdisciplines in the past 50 years has brought about an evolutionary synthesis of knowledge. This has occurred not only within science, but between science and technology and between science and the humanities. The forward progress of systemology in the study of large-scale complex systems requires a synthesis of science and the humanities as well as a synthesis of science and technology. One of the most important contributions of systemology is that it offers a single vocabulary and a unified set of concepts applicable to many types of systems.

1.4. TRANSITION TO THE SYSTEMS AGE[6]

There is considerable evidence to suggest that the advanced nations of the world are leaving one technological age and entering another. It appears that this transition is bringing about a change in the conception of the world in which we live. This conception is both a realization of the complexity of natural and man-made systems and a basis for improvement in the human's position relative to these systems.

---

[6] This section was adapted from R. L. Ackoff, *Redesigning the Future* (New York: John Wiley, 1974).

The Machine Age

Two ideas have been dominant in the way we seek to understand the world around us. The first is called *reductionism*. It consists of the belief that everything can be reduced, decomposed, or disassembled to simple indivisible parts. These were taken to be atoms in physics, simple substances in chemistry, cells in biology, and monads, instincts, drives, motives, and needs in psychology.

Reductionism gives rise to an analytical way of thinking about the world, a way of seeking explanations and understanding. Analysis consists, first, of taking apart what is to be explained, disassembling it, if possible, down to the independent and indivisible parts of which it is composed; second, of explaining the behavior of these parts; and, finally, of aggregating these partial explanations into an explanation of the whole. For example, the analysis of a problem consists of breaking it down into a set of as simple problems as possible, solving each, and assembling their solutions into a solution of the whole. If the analyst succeeds in decomposing a problem into simpler problems that are independent of each other, aggregating the partial solutions is not required, because the solution to the whole is the sum of the solutions to its independent parts. In the *Machine Age,* understanding the world was taking to be the sum, or result, of an understanding of its parts, which were conceptualized as independently of each other as was possible.

The second basic idea was that of *mechanism.* All phenomena were believed to be explainable by using only one ultimately simple relation, cause and effect. One thing or event was taken to be the *cause* of another (its *effect*) if it was both necessary and sufficient for the other. Because a cause was taken to be sufficient for its effect, nothing was required to explain the effect other than the cause. Consequently, the quest for causes was environment-free. It employed what is now called "closed-system" thinking. Laws such as that of freely falling bodies were formulated so as to exclude environmental effects. Specially designed environments, called *laboratories,* were used so as to exclude environmental effects on phenomena under study.

Causal laws permit no exceptions. Effects are completely determined by causes. Hence, the prevailing view of the world was deterministic. It was also mechanistic, because science found no need for teleological concepts (such as functions, goals, purposes, choice, and free will) in explaining any natural phenomenon; they considered such concepts to be unnecessary, illusory, or meaningless. The commitment to causal thinking yielded a conception of the world as a machine; it was taken to be like a hermetically sealed clock—a self-contained mechanism whose behavior was completely determined by its own structure.

The *Industrial Revolution* brought about *mechanization,* the substitution of machines for people as a source of physical work. This process affected the nature of work left for people to do. They no longer did all the things necessary to make a product; they repeatedly performed a simple operation in the production process. Consequently, the more machines were used as a substitute for people at work, the more workers were made to behave like machines. The dehumanization of human work was the irony of the Industrial Revolution.

The Systems Age

Although eras do not have precise beginnings and endings, the 1940s can be said to have contained the beginning of the end of the Machine Age and the beginning of the *Systems Age*. This new age is the product of a new intellectual framework in which the doctrines of reductionism and mechanism and the analytical mode of thought are being supplemented by the doctrines of expansionism, teleology, and a new synthetic (or systems) mode of thought.

Expansionism is a doctrine which maintains that all objects and events, and all experiences of them, are parts of larger wholes. It does not deny that they have parts, but it focuses on the wholes of which they are part. It provides another way of viewing things, a way that is different from, but compatible with, reductionism. It turns attention from ultimate elements to a whole with interrelated parts—to systems.

Preoccupation with systems brings with it the synthetic mode of thought. In the *analytic* mode, an explanation of the whole was derived from explanations of its parts. In *synthetic* thinking, something to be explained is viewed as part of a larger system and is explained in terms of its role in that larger system. The Systems Age is more interested in putting things together than in taking them apart.

Analytic thinking is outside-in thinking; synthetic thinking is inside-out thinking. Neither negates the value of the other, but by synthetic thinking we can gain understanding that we cannot obtain through analysis, particularly of collective phenomena.

The synthetic mode of thought, when applied to systems problems is called the *systems aproach*. This way of thinking is based on the observation that, when each part of a system performs as well as possible, the system as a whole may not perform as well as possible. This follows from the fact that the sum of the functioning of the parts is seldom equal to the functioning of the whole. Accordingly, the synthetic mode seeks to overcome the often observed predisposition to perfect details and ignore system outcomes.

Because the Systems Age is *teleologically oriented,* it is preoccupied with systems that are goal-seeking or purposeful, that is, systems that can display choice of either means or ends, or both. It is interested in purely mechanical systems only insofar as they can be used as instruments of purposeful systems. Furthermore, the Systems Age is most concerned with purposeful systems, some of whose parts are purposeful; these are called *social groups*. The most important class of social groups is the one containing systems whose parts perform different functions, that have a division of functional labor; these are called *organizations*. In the Systems Age interest is focused on groups and organizations that are themselves parts of larger purposeful systems.

1.5. ENGINEERING IN THE SYSTEMS AGE

Engineering activities of analysis and design are not an end in themselves but are a means for satisfying consumer wants. Thus, engineering has two aspects. One aspect concerns itself with the materials and forces of nature; the other is concerned with the needs of people.

In the Systems Age, successful accomplishment of engineering objectives requires a combination of technical specialties and expertise. Engineering in the Systems Age must be a team activity where various individuals involved are cognizant of the important relationships between specialties and between economic factors, ecological factors, political factors, and societal factors. Engineering decisions of today require consideration of these factors in the early stage of system design and development, and the results of such decisions have a definite impact on these factors. Conversely, these factors usually impose constraints on the design process. Thus, technical expertise must include not only the basic knowledge of individual specialty fields of engineering, but a knowledge of the context of the system being brought into being.

System Complexity and Scope

While relatively small products, such as a portable radio, an electrical household appliance, or even an automobile, may employ a limited amount of direct engineering resources, there are many large-scale systems that require the combined input of specialists representing a wide variety of engineering disciplines. An example is that of a ground mass-transit system.

Civil engineers are required for the layout and/or design of railroad tracks, tunnels, bridges, cables, and facilities. Electrical engineers are involved in the design of automatic train control provisions, traction power, substations for power distribution, automatic fare collection, digital data systems, and so on. Mechanical engineers are necessary in the design of passenger vehicles and related mechanical equipment. Architectural engineers provide support in the construction of passenger terminals. Reliability and maintainability engineers are involved in the design for system availability and the incorporation of supportability characteristics. Industrial engineers deal with the production aspects of passenger vehicles and vehicle components. Test engineers evaluate the system to ensure that all performace, effectiveness, and system support requirements are met. Engineers in the planning and marketing areas are required to keep the public informed and to promote the technical aspects of the system (that is, to keep the politicians and local citizens informed). General systems engineers are required to ensure that all aspects of the system are properly integrated and function as a single entity.

Although the example described above is not all-inclusive, it is apparent that many different engineering disciplines are directly involved. In fact, there are some large projects, such as the development of a new airplane, where the number of engineers assigned to perform engineering functions varies from 5,000 to 10,000. In addition, the quantity of different engineering types often ranges in the hundreds. These engineers, forming a part of a large organization, must not only be able to communicate with each other but must be conversant with such interface areas as purchasing, accounting, personnel, and legal.

Large projects of this type will usually fluctuate in terms of manpower loading. Depending on the functions to be performed on a project phase-by-phase basis (the system life cycle), some engineers will be assigned to a project until system development and production are completed, and others will be brought in for a short term to perform specific tasks. The requirements in terms of project personnel loading will also vary from phase to phase since the emphasis changes as system development progresses. During product planning, individuals with a broad systems engineering background are needed in greater quantities than detail design specialists, whereas during the product design phase, the reverse may be true. In any event, the need for engineering will change as a system evolves through its life cycle.

Another major factor associated with large projects is that much system development, production, evaluation, and support is often accomplished at supplier (sometimes known as *subcontractor*) facilities located throughout the world. Often there is a prime producer or contractor who is ultimately responsible for the development and production of the total system as an entity, and there are numerous suppliers providing different system components. Thus, much of the project work and many of the associated engineering functions may be accomplished at dispersed locations.

Technological Growth and Change

Technological growth and change is occurring continuously and is stimulated by an attempt to respond to some unmet current need and/or by attempting to perform ongoing activities in a more effective and efficient manner. In addition, changes in this area are being stimulated by social factors, political objectives, and ecological constraints.

Technological and economic feasibility can no longer be the main determinants of what engineers do. Ecological, political, social, cultural, and even psychological influences are equally important considerations. The number of variables in any given engineering project has multiplied. Because of the shifts in social attitudes toward moral responsibility, the ethics of personal decisions are becoming a major professional concern. Engineering is, of course, not alone in facing up to these changes.

Many examples of concern may be cited. For instance, the concern for ecology has resulted in certain legislation which creates additional needs for developing new methods of reducing air and water pollution and the handling and disposal of solid waste. More recent concerns involving the energy shortage have caused a shift in emphasis toward the further development of resources (solar energy, nuclear power, etc.) to supplement current oil and gas supplies. These and comparable situations, created either through properly planned programs or as a result of political or societal panic reactions, have definitely stimulated technological growth.

The response in fulfilling technology requirements is highly dependent on the scientists and engineers available in the needed fields of expertise and whether they are up to date and creative in their respective specialty areas. In some fields, such as electronics and the medical profession (from the standpoint of engineering innovations), technological growth is rapid. Engineers in these fields have an extremely difficult time maintaining their skills.

The continuing trend in technological advances has created an increasing demand for engineers in many fields. There will always be a demand for engineers who can synthesize and adapt. Certain technical specialties will become obsolete with time. The astute engineer should be able to detect trends and plan accordingly for satisfactory job transition by acquiring knowledge to broaden his or her horizons. To help in this process is one aim of this book.

SELECTED REFERENCES

(1) Ackoff, R. L., *Redesigning the Future*. New York: John Wiley, 1974.

(2) Boulding, K., "General Systems Theory: The Skeleton of Science," *Management Science,* April 1956.

(3) Johnson, R. A., F. W. Kast, and J. E. Rosenzweig, *The Theory and Management of Systems* (3rd ed.). New York: McGraw-Hill, 1973.

(4) Machol, R. E. (Ed.), *System Engineering Handbook,* McGraw-Hill Book Co., 1965.

(5) Optner, S. L., *Systems Analysis for Business and Industrial Problem Solving*. Englewood Cliffs, N.J.: Prentice-Hall, 1965.

(6) Rubinstein, M. F., *Patterns of Problem Solving,* Englewood Cliffs, N.J.: Prentice-Hall, 1975.

(7) Sandquist, G. M., *Introduction to System Science*. Englewood Cliffs, N.J.: Prentice-Hall, 1985.

(8) Von Bertalanffy, L., *General Systems Theory*. New York: George Braziller, 1968.

(9) Von Bertalanffy, L., "General System Theory: A New Approach to Unity of Science," *Human Biology,* December 1951.

(10) Wiener, N., *Cybernetics*. New York: John Wiley, 1948.

QUESTIONS AND PROBLEMS

1. Pick a system that alters material and identify its structural components, operating components, and flow components.
2. Select a complex system and discuss it in terms of its hierarchy of subsystems.
3. Pick a natural system and describe it in terms of components, attributes, and relationships; repeat the description for a man-made system.
4. Identify and contrast a physical and a conceptual system.
5. Identify and contrast a static and a dynamic system.
6. Identify and contrast a closed and an open system.
7. Describe cybernetics through an example of your choice.
8. Give a system example at each level in Boulding's hierarchy.
9. Give an example of a problem requiring an interdisciplinary approach and identify the needed disciplines.
10. Identify the attributes of the Machine Age; the Systems Age.
11. What are the special engineering requirements in the Systems Age?

2
Bringing Systems Into Being

The world in which we live may be divided into the natural world and the man-made world. Included in the former are all elements of the world that came into being by natural processes. The man-made world is made up of all systems, structures, and products made by people for the use of people. But we are not satisfied with the impact of the man-made world on the natural world and upon ourselves. Since the physical sciences and engineering are largely responsible for bringing the man-made world into being, it is not surprising that there is some dissatisfaction with these fields of endeavor.

Emerging technologies are revealing unprecedented opportunities for bringing new and improved systems and products into being that will be more competitive in the private and/or public sectors worldwide. These technologies are acting to expand physically realizable design options and to enhance capabilities for developing more cost-effective entities for consumer use. This chapter introduces a technologically based process, including the extension of engineering through all phases of the life cycle of a system—design and development, production or construction, operational use, and support.

2.1. ENGINEERING FOR ECONOMIC COMPETITIVENESS

In these times of intensifying international competition, producers are searching for ways to gain a sustainable competitive advantage in the marketplace. Acquisitions, mergers, and/or extensive advertising campaigns seem unable to create the intrinsic wealth so essential for long term corporate health. On the other hand, economic competitiveness is essential. Engineering, with an emphasis on economic competitiveness, must become coequal with concerns for advertising, finance, production, and the like. Further, engineering viewed in this context, implies a life-cycle approach. It is, through a life-cycle approach to engineering, that economic competitiveness can be realized.

The purpose of engineering activities of design and analysis is to determine how physical factors may be altered to create the most utility for the least cost, in terms of product cost, product service cost, and social cost. This economic concern is clear in the definition of engineering adopted by the Accreditation Board for Engineering and Technology (ABET).

> "Engineering is the profession in which a knowledge of the mathematical and natural sciences gained by study, experience, and practice is applied with judgement to develop ways to utilize, economically, the materials and forces of nature for the benefit of mankind."

The Consumer-to-Consumer Process

Fundamental to the application of engineering, and the practice of systems engineering, is an understanding of the consumer-to-consumer process illustrated in Figure 2.1.[1] This process begins with the identification of a need and extends through planning, research, design, production or construction, evaluation, consumer use, maintenance and support, and ultimate retirement (phaseout). The process is generic in nature and represents the life-cycle activities of large-scale systems, as described in Section 1.5. Although these activities may vary somewhat from one system to the next, they do reflect a *process* common to all.

For the purposes of illustration, Figure 2.2 is presented to show the system life cycle in simple form. The program activities identified in Figure 2.1 have been classified in two basic phases, the system *acquisition phase* and the system *utilization phase*. Referring to the figure, systems progress from the identified need through conceptual and preliminary design, detail design and development, and so on.

In general, engineers have focused mainly on the acquisition phase of the life cycle and have been involved in early design and analysis activities alone. Product performance has been a main objective, rather than the development of an overall system with economic factors in mind. However, experience in recent decades indicates that a properly functioning system which is competitive in the marketplace cannot be achieved through efforts applied largely after it comes into being. Accordingly, it is essential that engineers be sensitive to operational outcomes during the early stages of system development, and that they assume the responsibility for *life-cycle engineering* which has been largely neglected in the past.

---

[1] Each phase of the system life cycle is given detailed treatment in Parts Two and Four.

CONSUMER-TO-CONSUMER PROCESS			
	CONSUMER	Identification of Need	Wants or desires for systems because of obvious deficiencies or problems are made evident through basic research results.
	PRODUCER	System Planning Function	Marketing analysis; feasibility study; advanced system planning (system selection, specifications and plans, acquisition plan, research, design and production, evaluation plan, system use and logistic support plan); planning review; proposal.
		System Research Function	Basic research; applied research (need-oriented); research methods; results of research; evolution from basic research to system design and development.
		System Design Function	Design requirements; conceptual design; preliminary system design; detail design; design support; engineering model/prototype development; engineering test; transition from design to production.
		Production and/or Construction Function	Production and/or construction requirements; industrial engineering and operations analysis (plant engineering, manufacturing engineering, methods engineering, production control); quality control; production operations.
	CONSUMER	System Evaluation Function	Evaluation requirements; categories of test and evaluation; test preparation phase (planning, resource requirements, etc); formal operational test and evaluation; data collection, analysis, reporting, and corrective action; retesting.
		System Use and Logistic Support Function	System distribution and operational use; elements of logistics and life cycle maintenance support; system evaluation; modifications; product phaseout; material disposal, reclamation, and/or recycling.

Figure 2.1   The consumer-to-consumer process.

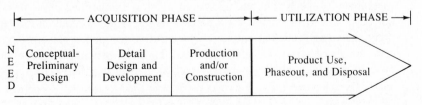

Figure 2.2   The system life cycle.

Designing for the Life Cycle

The life-cycle or concurrent engineering design approach for bringing competitive products into being goes beyond consideration of the life cycle of the product itself.[2] It must simultaneously embrace the life cycle of the manufacturing process as well as the life cycle of the product service system. In effect, there are three concurrent life cycles progressing in parallel as is illustrated in Figure 2.3.

The need for the product comes into focus first. This recognition initiates conceptual design to meet the need. Then, during conceptual or preliminary design of the product, consideration should simultaneously be given to its production. This gives rise to a parallel life cycle for bringing a manufacturing capability into being. It requires many production-related activities to become ready for manufacturing.

Also shown in Figure 2.3 is another life cycle of great importance which is often neglected until product and production design is completed. This is the life cycle for the logistic support activities needed to service the product during use and to support the manufacturing capability during its duty cycle. Logistic and maintenance requirements planning should begin during product conceptual design in a coordinated manner.

The communication and coordination needed to develop the product, the process, and the support capability in a coordinated manner is not easy to achieve. Progress in this will probably be facilitated by new technologies that make more timely acquisition and use of design information possible. CAD/CAM technology is only one of these. Others are being developed which can integrate relevant design and development activities over the entire life cycle of the system and its important phases shown in Figure 2.3.

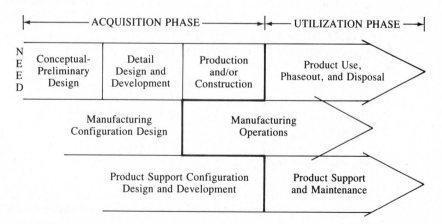

Figure 2.3   Product, process, and support life cycles.

[2] A *product*, for example, may be a television set, an automobile, or an appliance. The product cannot function properly without an operator, a manufacturing capability, a support capability, and so on. Accordingly, in dealing with systems, we must not only consider the product, but its manufacturing process, consumer use, and life-cycle maintenance and support.

The objective behind engineering for the life cycle in a concurrent manner is to ensure that the entire life of a system is considered from inception. An engineering design should not only transform a need into a definitive product configuration for customer use, but should ensure the design's compatability with related physical and functional requirements. Further, it should take into account life-cycle outcomes as measured by performance, effectiveness, producibility, reliability, maintainability, and cost.

Concern for the entire *life cycle* is very strong within the Department of Defense (DOD). The concepts dealing with "Concurrent Engineering" and "Simultaneous Engineering" stress a life-cycle approach. This may be attributed to the fact that acquired defense systems are owned, operated, and maintained by the DOD. This is unlike the situation most often encountered in the private sector, where the consumer or user is usually not the producer. Those private firms serving as defense contractors are obliged to design and develop in accordance with DOD directives, specifications, and standards. Since the DOD is the customer and also the user of the resulting system, considerable intervention takes place during the acquisition phase. This intervention is guided by DOD Directive 5000.1 and a host of subordinate directives and documentation.[3]

Many firms that produce for private sector markets have chosen to design with the life cycle in mind. For example, design for energy efficiency is now quite common in appliances like water heaters and air conditioners. Fuel efficiency is a required design characteristic for automobiles. Some truck manufacturers promise that life-cycle maintenance requirements will be within stated limits. These developments are commendable, but they do not go far enough. When the producer is not the consumer, it is less likely that potential operational problems will be addressed during development. Undesirable outcomes too often end up as problems of the user of the product instead of the producer.

All other factors being equal, people will meet their needs by buying goods and services which offer the highest value-cost ratio, subjectively evaluated. This ratio can be increased by giving more attention to the resource-constrained world within which engineering is practiced. To ensure economic competitiveness with regard to the end item, engineering must become more closely associated with economics and economic feasibility. This is best accomplished through a life-cycle approach to engineering design.

Systems Engineering

Systems engineering relates primarily to the design and development activities depicted in Figures 2.1, 2.2, and 2.3. More specifically, systems engineering is a process that has only recently been recognized to be essential in the orderly evolution of man-made systems. It involves the application of efforts to

---

[3] DODD 5000.1, "Major System Acquisitions," Department of Defense, Washington, DC. *Concurrent Engineering* is defined as a systematic approach to creating a product design that considers all elements of the product life cycle from conception through disposal to include consideration of manufacturing processes, transportation processes, maintenance processes, and so on.

1. Transform an operational need into a description of system performance parameters and a preferred system configuration through the use of an iterative process of functional analysis, synthesis, optimization, definition, design, test, and evaluation;

2. Incorporate related technical parameters and assure compatability of all physical, functional, and program interfaces in a manner that optimizes the total system definition and design; and

3. Integrate performance, producibility, reliability, maintainability, manability, supportability, and other specialties into the overall engineering effort.

Systems engineering per se is not considered to be an engineering discipline in the same context as the technical specialties it represents. Actually, systems engineering is a process employed in the evolution of systems from the point when a need is identified through production and/or construction and ultimate deployment of that system for consumer use. This process involves a series of steps accomplished in a logical manner and directed toward the development of an effective and efficient product or system. The requirement for systems engineering is brought about because many of the engineering specialists in one or more of the conventional engineering areas (such as aeronautical engineering, civil engineering, electrical-electronic engineering) are not sufficiently experienced to ensure that *all* elements of the system are considered in a proper and timely manner.

2.2. THE SYSTEM ENGINEERING PROCESS

An understanding of the system engineering process requires some knowledge of the functions necessary for bringing systems into being. There are variations in the application of engineering functions to the system life cycle, depending on the scope and complexity of the system and the extent of new design and development required. The role of the systems engineer will usually be different from one situation to the next. However, in spite of these differences, it can be stated that engineering functions of one type or another are performed in each applicable phase in the system life cycle.

The process of design evolution is illustrated in Figure 2.4. It is *tailored* to meet a specific requirement. Tailoring refers to the application of the proper level of engineering effort to the system being developed. The application of too much or too little effort could be quite costly. Thus, the steps presented in Figure 2.4 should be considered as a *thought* process, with each step being addressed to the extent and depth necessary to fulfill the requirement.

Regardless of the system type and size, one begins with an identified need and a completed feasibility study for the purposes of establishing a set of requirements, constraints, and design criteria. Based on the results, functional analyses and allocations are generated to apportion the appropriate system-level requirements down to the subsystem, unit, and lower levels of the system.

System analyses are accomplished to evaluate the various alternative ap-

22

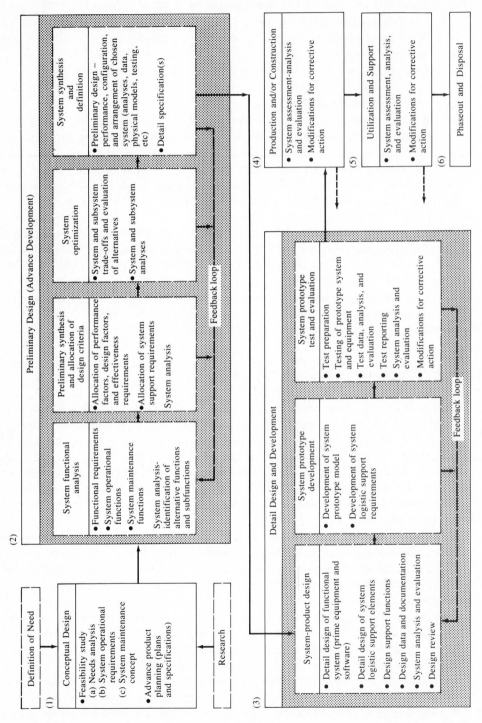

Figure 2.4 The system life-cycle process.

proaches that are considered feasible in meeting the identified need. The output reflects a preferred system configuration. Inherent in this activity of requirements identification, analysis, and system definition are the feedback provisions illustrated in Figure 2.5. The systems engineering process is continuous, iterative, and incorporates the feedback actions necessary to ensure convergence.

Definition of System Requirements

The system life-cycle approach stems from the identification of a need that develops as a result of a problem or deficiency and the subsequent want or desire for a system of some type. From the identification of a given need, one must define the basic requirements for the system in terms of input criteria for design. To facilitate this process, the following questions should be addressed:

1. What is the system to accomplish in terms of operations and functional performance characteristics (range, accuracy, speed of performance, power output, widgets produced, liquid flow in gallons per hour, items processed per month, etc.)?

2. When is the system needed? What are the consumer requirements? What is the expected operational life of the system?

3. How is the system to be used in terms of hours of operation per day, number of on-off cycles per month, and so on?

4. How is the system to be distributed and deployed? Where are the various elements of the system to be located, and for how long?

5. What effectiveness requirements should the system exhibit? Effectiveness figures-of-merit may include factors for cost effectiveness, system effectiveness, availability, dependability, reliability, maintainability, and supportability.

6. What are the environmental requirements for the system (temperature, humidity, shock and vibration, etc.)? Will the system be operated in arctic, tropical areas, or mountainous or flat terrain, and what are the anticipated transportation, handling, and/or storage modes?

7. How is the system to be supported throughout its life cycle? This includes a definition of levels of maintenance, functions at each level, and anticipated logistic support requirements (test and support equipment, supply support and spare/repair parts, personnel and training, transportation and handling requirements, facilities, software, and technical data).

8. When the system becomes obsolete and/or when items are removed from the inventory, what are the requirements for disposal? Can specific items be reclaimed and recycled? What are the effects on the environment?

Regardless of the size and type of system (large or small, mechanical, electrical or electronic, chemical, commercial or defense), these questions apply in varying degrees and need to be addressed. The answers generally evolve from feasibility

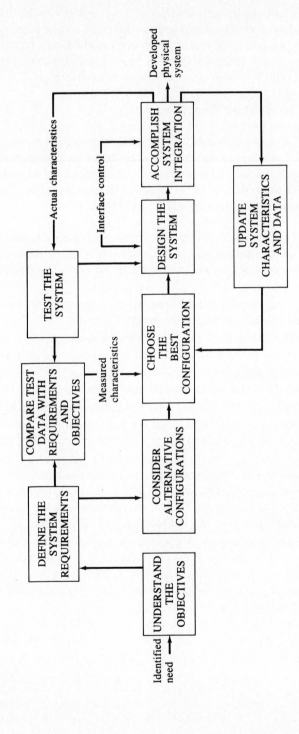

Figure 2.5 Feedback in the systems engineering process.

studies, the development of operational requirements and the maintenance concepts, and the preparation of the system specifications. This activity, constituting the baseline upon which the systems engineering process depends, is reflected in Figure 2.4, Block 1, and is covered in depth in Part Two of this text.

System Design and Development

The design process follows from a set of stated requirements for a given system and evolves through (1) conceptual design (the establishment of performance parameters, operational requirements, and support policies), (2) preliminary systems design (sometimes called *advanced development*), and (3) detail design. This process generally begins with a visualization of what is required and extends through the development, test, and evaluation of an engineering or prototype model of the system. The output constitutes a configuration that can be directly produced or constructed from specifications, a set of drawings, and supporting documents.

The engineer's role in this phase involves a variety of functions that depend on the type of system and the extent of new development necessary. These functions may include all or any combination of the following:

1. Accomplishing functional analyses and allocations to identify the major operational and maintenance support functions that the system is to perform.
2. Establishing criteria (qualitative and quantitative technical parameters, bounds, and constraints) for system design.
3. Evaluating different alternative design approaches through the accomplishment of system/cost effectiveness analyses and trade-off studies.
4. Preparing system development, process, and material specifications.
5. Selecting components for the system and recommending supplier sources.
6. Assisting the purchasing and contracting functions in the preparation of supplier specifications and contractual documentation.
7. Preparing functional design layouts, engineering drawings, parts and material lists, standards, and so on, with the objective of thoroughly defining the product or process through documentation. This includes preliminary drawings, interface drawings, installation drawings, manufacturing drawings, and associated data bases.
8. Assessing the design through predictions, analyses, and the performance of periodic design reviews. This assessment is basically accomplished through a review of the engineering documentation.
9. Developing breadboards, engineering models, and prototypes for system test and evaluation purposes.
10. Developing system software (computer programs), associated data bases, and related documentation required to define, design, test, produce, operate, and maintain the system.
11. Developing system and component test specifications and procedures, and accomplishing specific tests to ensure that all design requirements are met.

12. Performing design modifications as necessary to correct deficiencies and/or to improve the system design.

These activities are inherent within the context of Bocks 2 and 3 in Figure 2.4, and they must address all facets of the system as depicted in Figure 2.3. Of particular significance to systems engineering are

1. The proper evolution from system-level definition to subsystem design, component design, and system integration and test, and
2. The proper integration of the various design activities into a total effective engineering effort. Design is a team approach, involving many different disciplines, and it is essential that an effective integration effort be maintained.

Specifically, the role of systems engineering is to provide the necessary input to the design process through specifications, establish the appropriate checks and balances throughout the design process by conducting selected design reviews, implement an effective integrated test and evaluation effort, and provide the necessary feedback for corrective action as required.

Production and/or Construction

Production and/or construction may constitute (1) the production of a multiple quantity of like items (mass production), (2) the production of small quantities of a wide variety of different items (a job-shop type of operation), and/or (3) the construction of a single item, such as a large structure of some type. In production and/or construction operations, material and personnel resources must be combined in such a manner as to provide the necessary product-system output in an effective and efficient manner. Production actually begins from the point where system design is considered fixed and includes the total flow of materials, from the acquisition of raw materials to delivery of the finished product for evaluation and ultimate consumer use. The production flow process, regardless of the product type, involves inventories, material acquisition and control provisions, tooling and test equipment, transportation and handling methods, facilities, personnel, and data.

Engineering is directly required in the design and development of a production capability and for defining the resources necessary for a large construction project. These engineering functions may entail the following:

1. Design of facilities for product fabrication, assembly, and test functions. This includes determining the capacity and location of both manufacturing and storage facilities, utility requirements, capital equipment needs, and material handling provisions.
2. Design of manufacturing processes (such as sequencing of tasks, human-machine operations, and process specifications).
3. Selection of materials and the determination of inventory requirements.

4. Design of special tools, test equipment, transportation and handling equipment, and fixtures used in production-construction operations.

5. Establishment of work methods and processes, time and cost standards, and the subsequent evaluation of production/construction operations in terms of the established parameters.

6. Evaluation of production/construction operations to ensure that product performance, quality, reliability, maintainability, safety, and other desired features are maintained throughout the production-construction process.

These activities are represented by Block 4 in Figure 2.4. Emphasis at this point is to ensure that those characteristics which have been designed into the system during the earlier phases of a program are indeed maintained throughout the production-construction process.

System Use and Sustaining Support

Functions during the utilization phase constitute (1) consumer use of the system throughout its intended life cycle, (2) incorporation of product or system modifications for improvement, (3) the logistic support requirements necessary to ensure that the product or system is deployed and operationally available when needed, and (4) the ultimate phaseout and disposal of the product or system due to obsolescence or wear. These functions may be accomplished by the consumer alone, by the consumer with the support of the producer in certain areas, or by the consumer with the support of an outside organization (other than the producer) performing specific tasks.

The engineering role, regardless of the organizations represented, may encompass the following:

1. Providing engineering assistance in the initial deployment, installation, and checkout of the system in preparation for consumer operational use.

2. Providing field service or customer service engineers at strategic geographical locations to assist the consumer in the day-to-day operation and maintenance support of the system.

3. Providing engineering support in the installation of system modifications and in the subsequent checkout of the system to ensure satisfactory operation.

4. Providing engineering assistance in the collection and analysis of data covering system operations in the field, special tests, and in the actual assessment of system operations.

5. Providing engineering support in the phase-out and disposal of the system at the end of the life cycle and in the subsequent reclamation and recycling of system components as appropriate.

Referring to Figure 2.4, Block 5, systems engineering activities primarily involve customer service and the ongoing overall assessment of the system in the user environment.

2.3. SYSTEMS ENGINEERING MANAGEMENT

A basic objective of systems engineering is to provide an entity that will satisfactorily meet some identified need. Not only must the system perform in a specified manner, but it must do so both effectively and efficiently. That is, the development and production of a system must consider the appropriate combination of

1. System performance and physical parameters such as capacity, energy output, delivery rate, range, accuracy, speed, volume, weight shape, and so on.
2. System operational and support factors such as system effectiveness, availability, dependability, operational readiness, reliability, maintainability, supportability, transportability, and so on.
3. System economic factors such as initial cost, operating and support cost, life cycle cost, and so on.

Good system design reflects an optimum balance among performance, support, and economic factors which is attained through a trade-off and analysis effort accomplished in the early stages of system development. The results of this effort lead to engineering decisions that have a significant impact on what will occur during system production or construction, operational use and support, and ultimate phaseout. The proper attention given to certain factors early in the life cycle may avoid problems later, while ignoring others may prove to be quite costly.

Research and Development in the Life Cycle

System life-cycle engineering is a professional research and development (R & D) activity which holds great promise for bringing scientific and engineering talent systematically to bear on an entire spectrum of technological activities embraced by the life cycle. Life-cycle engineering has its greatest impact during the early phases of system development. These early phases have a strong R & D orientation. Accordingly, it is the R & D sector to which strategies for improving the life cycle outcome can be most profitably directed.

The National Science Foundation (NSF) uses the following definitions of R & D in its resource surveys:[4]

1. Basic research has as its objective "a fuller knowledge or understanding of the subject under study, rather than a practical application thereof."

---

[4] National Science Foundation, *National Patterns of R & D Resources*, Final Report NSF 89-308, Washington, D. C.: 1989.

2. Applied research is directed toward gaining "knowledge or understanding necessary for determining the means by which a recognized and specific need may be met."

3. Development is the "systematic use of the knowledge or understanding gained from research directed toward the production of useful materials, devices, systems or methods, including design and development of prototypes and processes."

Based on the above definitions, it is clear that system life-cycle considerations are more directly dependent upon applied research and development than upon basic research. For this reason, life-cycle engineering must include those individuals, activities, and organizations engaged in applied research and development.

Early life-cycle activities are the concern of a significant percentage of all scientists and engineers in government and industry. Forty-one (41%) percent of all scientists and engineers are engaged in applied research, development, and the management of research and development. This numbers more than one million individuals and does not include those performing R & D related functions through consulting, teaching, professional services, and others.

National expenditures for R & D by character of work reveal that the greatest portion is related to the early life-cycle activities of applied research and development. NSF's Science Indicators for 1989 give estimated expenditures (in millions) for applied research to be $18,570 and for development to be $113,780. This is 86% of the total expended, with the remaining 14% for basic research. Small improvements in the effectiveness of these activities will yield very beneficial results over the system life cycle.

Managing Systems Engineering

As a system progresses through its life cycle, engineering decisions are made that could have a significant feedback effect on what has already been accomplished. For instance, if a system is used in a manner different from what was intended in the design process, causing unanticipated stresses on the system, the results may be damaging. Engineering decisions are made at various stages in the life cycle, and the consequences, in most instances, will ultimately determine the success or failure of the system in meeting its objectives.

For most small- and large-scale systems, the accomplishment of engineering activities is the result of team effort, involving a combination of

1. Personnel with the technical expertise, individual skills, and the proper attitudinal characteristics necessary to perform required engineering functions.

2. Personnel trained as technicians, draftsmen, computer programmers, model builders, and other comparable tradespeople who directly support the professional engineering specialist in the performance of engineering functions.

3. Personnel representing nontechnical fields whose skills are necessary in

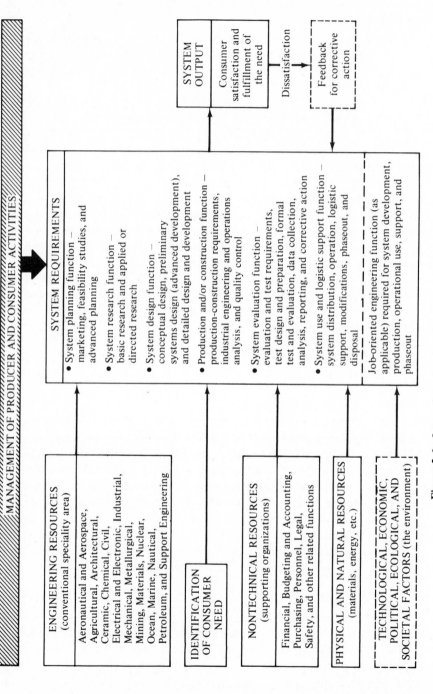

Figure 2.6 Integration of engineering and support functions.

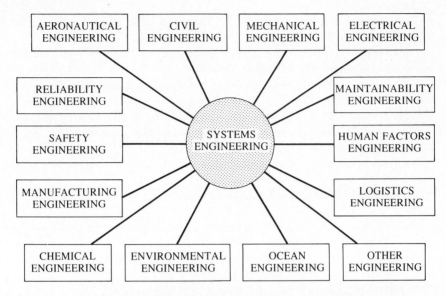

Figure 2.7 Major system engineering interfaces (example).

providing direct support to ensure that engineering functions are accomplished in an effective and efficient manner. This includes financial management, budgeting and accounting, purchasing, personnel, legal, safety, and other related functions.

4. The physical and natural resources (e.g., energy, materials) required for support in the accomplishment of manual tasks and in the production-construction of both hardware and software components of the system.

These activities must be properly integrated through effective organization and management in order to meet the requirements of the system life cycle as discussed in Section 2.2. Figure 2.6 illustrates this integration process, and Figure 2.7 shows some of the major relationships that exist on a given project. The various aspects of systems engineering management (like planning, organization, implementation, and control of a systems engineering program) are covered further in Chapter 18.

SELECTED REFERENCES

(1) Beakley, G. C., D. L. Evans, and J. B. Keats, *Engineering: An Introduction to a Creative Profession*. New York: Macmillan, 1986.

(2) Blanchard, B. S., *Engineering Organization and Management*. Englewood Cliffs, N.J.: Prentice-Hall, 1976.

(3) Chase, W. P., *Management of System Engineering*. New York: John Wiley, 1974.

(4) Chestnut, H., *Systems Engineering Methods*. New York: John Wiley, 1967.

(5) Chestnut, H., *System Engineering Tools*. New York: John Wiley, 1965.

(6) David, E. E., E. J. Piel, and J. G. Truxal, eds., *The Man-Made World*. New York: McGraw-Hill, 1971.

(7) Defense Systems Management College (DSMC), *Systems Engineering Management Guide*. Fort Belvoir, Virginia, December 1986.

(8) Dieter, G. E., *Engineering Design: A Materials and Processing Approach*. New York: McGraw-Hill, 1983.

(9) Drew, D. R., and C. H. Hsieh, *A Systems View of Development: Methodology of Systems Engineering and Management*. Cheng Yang Publishing Co., No. 4, Lane 20, Gong-Yuan Road, Taipei, ROC, 1984.

(10) Fabrycky, W. J., "Designing for the Life Cycle," *Mechanical Engineering,* January 1987.

(11) Hall, A. D., *A Methodology for Systems Engineering*. Princeton, N.J.: Van Nostrand, 1962.

(12) Karger, D. W., and R. G. Murdick, *Managing Engineering and Research*. New York: Industrial Press, 1969.

(13) Meredith, D. D., and others, *Design and Planning of Engineering Systems*. Englewood Cliffs, N.J.: Prentice-Hall, 1973.

(14) MIL-STD-499A, Military Standard, "Engineering Management," Headquarters, U. S. Air Force, Air Force Systems Command, Attention SDDE, Andrews Air Force Base, MD 20331.

(15) National Science Foundation, *National Patterns of R & D Resources,* Final Report NSF 89-308, Washington, D. C.: 1989.

(16) Ostrofsky, B., *Design, Planning and Development Methodology*. Englewood Cliffs, N.J.: Prentice-Hall, 1977.

(17) Sage, A. P., *Economic System Analysis: Microeconomics for Systems Engineering, Engineering Management, and Project Selection*. New York: Elsevier Science Publishing Co., 1983.

(18) Sage, A. P., *Methodology for Large Scale Systems*. New York: McGraw-Hill, 1977.

(19) Singh, M. G., ed., *Systems And Control Encyclopedia: Theory, Technology, Applications*. Elmsford, N.Y.: Pergamon Press, 1989.

QUESTIONS AND PROBLEMS

1. The various phases of the system life cycle shown in Figure 2.1 are applicable to *all* systems. True or false? Please explain.

2. Select a system of your choice and define the system life cycle. Construct a detailed flow diagram.

3. Define *systems engineering*. What is included? Why is it important? How does systems engineering differ from *system science* and *system analysis*?

4. Refer to Figure 2.3. Describe the interrelationships between the three illustrated life cycles.

5. One of the first steps in the system engineering process constitutes the definition of system requirements. How is this accomplished and what information is included?

6. What is meant by the *checks and balances* in the systems engineering process?

7. What are some of the basic objectives of systems engineering (identify at least three)?

8. How is systems engineering impacted as a result of political factors? Economic factors? Technological factors? Please explain.

9. How does the consumer become involved in the design process?

10. What are the major systems engineering functions in conceptual design? Preliminary system design? Detail design and development? Production-construction? System use and life-cycle support?

11. What is the significance of the feedback process in Figure 2.5?

12. What is meant by *tailoring* to meet a specific need? Why is it important?

13. Identify some of the major management challenges in implementing a systems engineering program.

14. Why is the R & D sector such a fruitful area for the use of system engineering concepts?

PART II: THE SYSTEM DESIGN PROCESS

3

Conceptual System Design

Conceptual design is the first step in system design and development which, in turn, is accomplished through application of the systems engineering process. The systems engineering process evolves functional detail and design requirements with the goal of achieving the proper balance among operational, economic, and logistic factors. It employs a sequential and iterative methodology to reach cost-effective solutions to design alternatives. Systems engineering is directly concerned with the transition from requirements identification to a fully defined system configuration ready for production and consumer use. Information developed through the systems engineering process is used to plan and integrate the engineering effort for the system as a whole, as was described in Chapter 2.

Several essential elements constitute conceptual design. A needs analysis must be made. Feasibility studies are accomplished to support technology applications. System operational requirements and a maintenance concept are defined. Trade-off analyses are completed, and a system specification is developed. At the same time, advanced systems planning is initiated. These conceptual design activities, reflected in Block 1 of Figure 2.4, are addressed in this chapter. They provide a basis for all subsequent design activities to be presented in the chapters that follow.

3.1. DEFINITION OF SYSTEM REQUIREMENTS

Section 2.2 introduces the subject of system requirements with a number of key questions that must be addressed at program inception. Response to these questions, viewed on a collective basis, leads to defining a baseline from which system design and development evolves. Activities in this area are critical when considering the impact on the results downstream in the system life cycle. While this phase of a program is particularly difficult because of the usual lack of system definition, it is essential that a baseline of some sort be established. Changes will occur from here on; however, these changes must be related to a reference and must be controlled in their implementation.

The systems engineering process begins in conceptual design, and the major activities inherent in this phase of the life cycle are depicted in Figure 3.1. This figure constitutes an expansion of Block 1 in Figure 2.4. The role of systems engineering is to control these functions to ensure that the overall design and development effort is integrated in an effective manner.

Definition of Need

The system life cycle and the system engineering process begin with the identification of a need based on a *want* or *desire* for something arising from a deficiency (perceived or real). An individual and/or organization identifies a need for an item or for function to be performed, and a new or modified system (or product) is procured to fulfill the requirement.

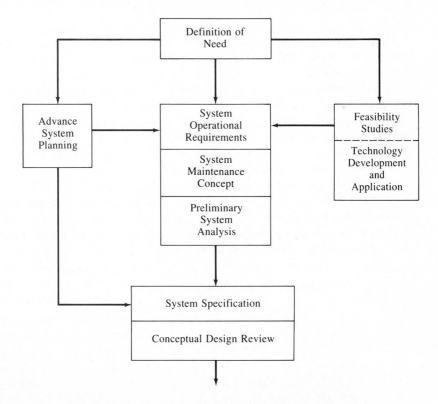

Figure 3.1 System requirements definition process (example).

Given that there may be multiple needs, combined with limited available resources, it is important that

1. The nature of the existing deficiency be well defined (inadequate performance characteristics, inadequate system support capability, excessive ownership cost, and so on).
2. The date by which the new system must be installed and operational be established.
3. The magnitude of the resources for investing in the new system capability be identified.
4. The relative priority for the new system capability be established.

At this point, problem solutions should not be identified, nor should hardware or software configurations be proposed. The purpose is to establish, beyond any doubt, that there is an *actual identified need*. This may appear to be rather basic; however, one often becomes involved in the design of some item without first having justified the need for such.

Feasibility Studies

A feasibility analysis, comprised of a series of technology-oriented studies, is often conducted as an extension of the early system definition task. Basically, the need for a system is established, followed by the evaluation of various technical approaches that can be applied in system development. For example, in defining a given system architecture, there may be a requirement to consider possible applications of various communication approaches, energy sources, software configurations (like artificial intelligence or expert systems), processor methods, and the like. This evaluation process may lead to a direct application of some existing technology or may identify an area where some further research is required.

While on one hand we are fostering a *top-down* systems engineering approach, we are also identifying possible technologies that can be applied in resolving specific design-related problems (a *bottom-up* approach). At this stage in the system life cycle, our task is to propose feasible technology applications and to integrate these into the system requirements definition process.

System Operational Requirements

The technical parameters to be established for system design evolve from an analysis of the need, combined with the feasibility of various technology applications. More specifically, the following questions must be asked. What are the anticipated quantities of equipment and personnel in the field and where are they to be located? How will the system be used? What is the expected environment at each location? The answer to these and comparable questions leads to the definition of system operating characteristics, the maintenance support concept for the system, and the identification of specific design criteria.[1] The operational concept includes the following:

1. *Mission definition*—identification of the prime operating mission of the system along with alternative or secondary missions. What is the system to accomplish? How will the system accomplish its objectives? The mission may be defined through one or a set of scenarios or system operational profiles. For example, in aircraft design there may be a number of flight profiles that should be addressed, like those illustrated in Figure 3.2. Although all possibilities can not be covered, a few representative approaches should be addressed. The objective is to cover the *dynamic aspects* of the scenario, as compared to addressing only the upper and lower bounds.

2. *Performance and physical parameters*—definition of the operating characteristics or functions of the system (e.g., size, weight, speed, accuracy, output rate, capacity). What are the critical system performance parameters?

3. *Use requirements*—anticipated use of the system and its elements (e.g., hours of operation per day, on-off sequences, operational cycles per month). How is the system to be used in the field? The objective is to relate specific performance and use characteristics to the operational profiles identified in Figure

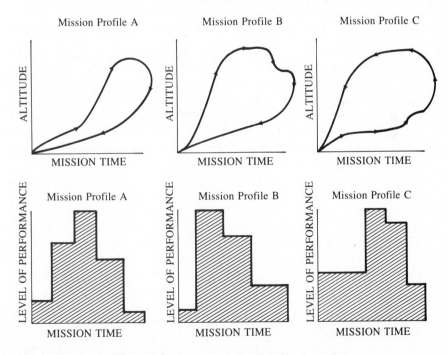

Figure 3.2   System operational profiles (examples).

[1] *Design criteria* refers to design standards, constraints, guidelines, and the like which constitute an *input* to the system design process.

3.2. This leads to the identification of system effectiveness requirements and the development of operational functions (discussed in Chapter 4).

4. *Operational deployment or distribution*—identification of the quantity of equipment, personnel, facilities, and so on, and the expected geographical location to include transportation and mobility requirements. How much equipment and associated software is distributed, and where is it to be located? When are they required? When does the system become fully operational? Figure 3.3 presents an illustrated example of the distribution of system components, leading into the identification of system support locations and the development of the system maintenance concept.

5. *Operational life cycle (horizon)*—anticipated time that the system will be in operational use. What is the total inventory profile throughout the system life cycle? Who will be operating the system and for what period of time? This establishes the basis for the identification of life-cycle activities and the development of life-cycle cost data.

6. *Effectiveness factors*—system requirements specified as figures-of-merit for cost-system effectiveness, operational availability, dependability, logistic support effectiveness, mean time between maintenance (MTBM), failure rate (λ) maintenance downtime (MDT), facility use (in percent), operator skill levels and tasks, personnel efficiency, and so on. Given that the system will perform, how effective or efficient is it? It should be possible to relate system effectiveness requirements directly to performance factors as they are applied to the mission scenario.[2]

7. *Environment*—definition of the environment in which the system is expected to operate (e.g., temperature, humidity, arctic or tropics, mountainous or flat terrain, airborne, ground, shipboard). This should include a range of values as applicable and should cover all transportation, handling, and storage modes. How will the system be handled in transit? What will the system be subjected to during operational use, and for how long? A complete environmental profile should be developed.

In essence, the producer (management and engineering) must project how the system will be deployed and utilized in the field by the consumer. Engineering must design the system to do the job under the conditions stated, and management must ensure that this is accomplished effectively, efficiently, and in a timely manner.

When reviewing operational requirements in total, many of the specific factors identified above are based on the projected distribution of equipment and personnel. This not only influences system operation on a day-to-day basis, since each organizational entity operating the system usually functions somewhat differently, but the equipment site locations intuitively specify environmental conditions, transportation and handling methods, and storage requirements. Thus, one of the first steps in defining operational requirements is to determine consumer locations. The example illustrated in Figure 3.3 shows that 592 pieces of equipment are located at 15 operat-

---

[2] Effectiveness factors such as availability, reliability, maintainability, supportability, life-cycle cost, and so on, are presented in Part Four.

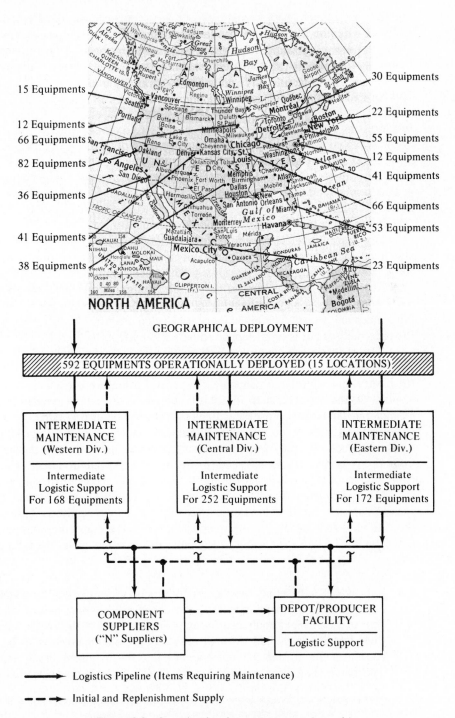

GEOGRAPHICAL DEPLOYMENT

15 Equipments

12 Equipments
66 Equipments

82 Equipments

36 Equipments

41 Equipments

38 Equipments

30 Equipments

22 Equipments

55 Equipments

12 Equipments

41 Equipments

66 Equipments

53 Equipments

23 Equipments

//// 592 EQUIPMENTS OPERATIONALLY DEPLOYED (15 LOCATIONS) ////

INTERMEDIATE MAINTENANCE (Western Div.)	INTERMEDIATE MAINTENANCE (Central Div.)	INTERMEDIATE MAINTENANCE (Eastern Div.)
Intermediate Logistic Support For 168 Equipments	Intermediate Logistic Support For 252 Equipments	Intermediate Logistic Support For 172 Equipments

COMPONENT SUPPLIERS ("N" Suppliers)	DEPOT/PRODUCER FACILITY
	Logistic Support

→ Logistics Pipeline (Items Requiring Maintenance)

---→ Initial and Replenishment Supply

Figure 3.3   Operational-maintenance concept (example).

ing sites. Consumer operations at each site are estimated, and requirements are established based on the demand for the equipment. This, in turn, determines expected utilization and effectiveness factors to which the system must be responsive. The figure also includes the first step in illustrating the level of support anticipated.

System Maintenance Concept

In addressing system requirements, the tendency is to cover primarily those elements of the system that relate to performance; that is, prime mission-oriented equipment, operational software and associated data, operating personnel, and so on. At the same time, there is usually very little attention given to system support, particularly at the early stages of design and development. In essence, the emphasis has been placed on only part of the system, not the entire system. This practice of not addressing the system in its entirely can be rather costly.

In covering overall system requirements, it is essential that all aspects of the system be addressed. This includes not only prime equipment and similar factors, but the system support capability as well. System support must be considered on an integrated basis from the beginning if the ultimate output is to be cost-effective. This is initially accomplished through the definition of the system maintenance concept during conceptual design. The objective is to develop a before-the-fact concept on how the proposed system is to be supported on a life-cycle basis.[3]

The maintenance concept, which evolves from the definition of system operational requirements, delineates (1) the anticipated levels of maintenance support, (2) the basic responsibilities for support, (3) general overall repair policies and/or constraints, (4) the major elements of logistic support as they apply to a new system (for example, a standard item of test equipment that must be incorporated or a given test philosophy), (5) the effectiveness requirements associated with the system support capability (such as supply responsiveness, test equipment reliability, facility utilization rate, personnel efficiency), and (6) the maintenance environment. The maintenance concept describes, in general terms, the overall support environment in which the system is to exist. Further, it constitutes the baseline for the determination of specific support requirements through the logistic support analysis. Although there are different approaches for illustration, Figure 3.4 presents an overview of the scope entailed. More specifically, the maintenance concept serves the following purposes.

1. It provides a baseline for the establishment of supportability requirements (e.g., reliability, maintainability, and usability or human factors characteristics) in system and equipment design. For instance, if the repair policy indicates that no scheduled or detail unscheduled maintenance is to be accomplished at the consumer's operational site, the equipment design should not incorporate diagnostic test provisions, special packaging considerations for the

---

[3] The maintenance concept is an input to the system design process, while the detailed maintenance plan depicts actual support requirements based on an assumed fixed design configuration, after-the-fact and an output from design. The maintenance plan is discussed further in Chapter 16.

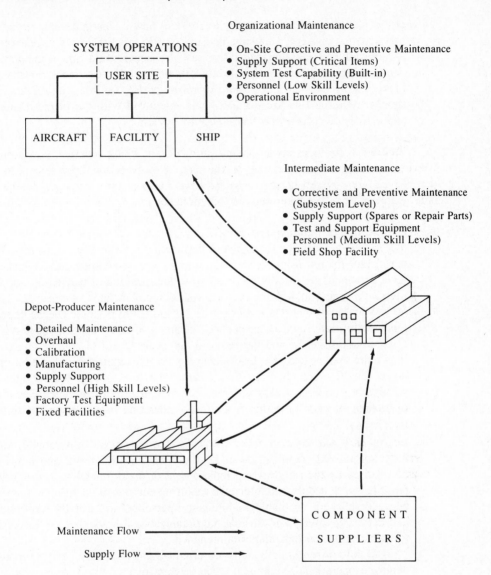

Figure 3.4   System operation and support flow.

benefit of the consumer, and so on. On the other hand, if it is decided that a certain amount of maintenance should be performed on site, then built-in self-test provisions, accessibility, readily removable functional packages, and related characteristics should be incorporated in the design. The maintenance concept aids in establishing design criteria related to system support.

2. It provides that basis for the establishment of requirements for total logistic support. Given an assumed design configuration of the prime mission equipment (i.e., that part of the total system which is directly related to the mission or operational requirement, such as airplane, electronic equipment, or nuclear

power plant), it is then necessary to consider how it should be supported.[4] The maintenance concept, supplemented by logistic support analyses, leads to the identification of maintenance tasks, task frequencies and times, maintenance personnel quantities and skill levels, training needs, test and support equipment, supply support (e.g., spare or repair parts), facilities, and data. These support requirements are evaluated and integrated with the prime equipment and associated software to form the total system.

Figure 3.5 provides some general guidelines as to the levels of maintenance, which may vary from one system to the next. However, for discussion purposes, three levels are covered herein: *organizational* maintenance, *intermediate* maintenance, and *depot* or producer-supplier maintenance.

1. *Organizational maintenance* is performed at the consumer's operational site (e.g., the 15 locations identified in Figure 3.3). Generally, it includes tasks performed by the using organization on its own equipment. Organizational-level personnel are usually involved with the operation of equipment and have minimum time available for detailed system maintenance.

   Maintenance at this level normally is limited to periodic checks of equipment performance, visual inspection, cleaning of equipment, some servicing, external adjustments, and the removal and replacement of some components. Personnel assigned to this level generally do not repair the removed components, but forward them to the intermediate level. From the maintenance standpoint, the least skilled personnel are assigned to this function. The design of equipment must take this fact into consideration (e.g., design for simplicity).

2. *Intermediate maintenance* tasks are performed by mobile, semimobile, and/or fixed specialized organizations and installations. At this level, end items may be repaired by the removal and replacement of major modules, assemblies, or piece parts. Scheduled maintenance requiring equipment disassembly may also be accomplished. Available maintenance personnel are usually more skilled and better equipped than those at the organizational level, and are responsible for performing more detailed maintenance.

   Mobile or semimobile units are often assigned to provide close support to deployed operational equipments. These units may comprise vans, trucks, or portable facilities containing some test and support equipment and spares. The objective is to provide on-site maintenance (beyond that accomplished by organizational-level personnel) to facilitate the return of the system to its full operational status on an expedited basis. A mobile unit may be used to support more than one operational site. A good example is the maintenance vehicle that is deployed from the airport hangar to an airplane needing extended maintenance while parked at a commercial airline terminal gate.

---

[4] This does not mean to imply that the design is fixed. The engineer must start with some baseline assumption, then synthesize it, evaluate it, modify it, synthesize and evaluate the modified configuration, and so on. Logistic support requirements can be determined for each configuration being evaluated.

CRITERIA	ORGANIZATIONAL MAINTENANCE	INTERMEDIATE MAINTENANCE		DEPOT/PRODUCER MAINTENANCE
Done where?	At the system operating site or wherever the prime equipment is located	Mobile or semimobile units	Fixed units	Depot or producer facility
		Truck, van, portable shop, or equivalent	Fixed field shop	Specialized repair activity, or producer's manufacturing plant
Done by whom?	System/equipment operating personnel (low maintenance skills)	Personnel assigned to mobile, semimobile, or fixed units (intermediate maintenance skills)		Depot facility personnel or producer's production personnel (mix of intermediate fabrication skills and high maintenance skills)
On whose equipment?	Using organization's equipment	Equipment owned by using organization		
Type of work accomplished?	Visual inspection Operational checkout Minor servicing External adjustments Removal and replacement of some components	Detailed inspection and system checkout Major servicing Major equipment repair and modifications Complicated adjustments Limited calibration Overload from organizational level of maintenance		Complicated factory adjustments Complex equipment repairs and modifications Overhaul and rebuild Detailed calibration Supply support Overload from intermediate level of maintenance

Figure 3.5   Major levels of maintenance.

Fixed installations (permanent shops) are generally established to support both the organizational level tasks and the mobile or semimobile units. Maintenance tasks that cannot be performed by the lower levels, because of limited personnel skills and test equipment, are performed here. High personnel skills, additional test and support equipment, more spares, and better facilities often enable equipment repair to the module and piece-part level. Permanent shops are usually located within specified geographical areas. The three intermediate facilities supporting the equipment illustrated in Figure 3.3 are examples. Rapid maintenance turnaround times are not as imperative here as at the lower levels of maintenance (i.e., organizational level).

3. *Depot or producer maintenance* constitutes the highest type of maintenance and supports the accomplishment of tasks above and beyond the capabilities available at the intermediate level. Physically, the depot may be a specialized repair facility supporting a number of systems or equipment in the inventory or may be the equipment supplier or producer's plant. Depot facilities are fixed and mobility is not a problem. Complex and bulky equipment, large quantities of spares, environmental control provisions, and so on, can be provided if re-

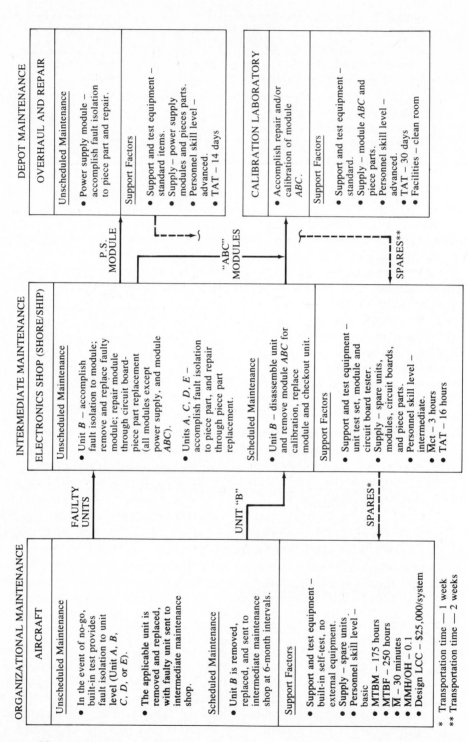

Figure 3.6 System *XYZ* maintenance concept flow (projected example).

* Transportation time — 1 week
** Transportation time — 2 weeks

quired. The high-volume potential in depot facilities fosters the use of assembly-line techniques which, in turn, permit the use of relatively unskilled labor for a large portion of the workload with a concentration of highly skilled specialists in such key areas as fault diagnosis and quality control.

Depot-level maintenance includes the complete overhauling, rebuilding, and calibration of equipment, as well as the performance of highly complex maintenance actions. In addition, the depot serves as a supply facility, and may be a special capability established for a specific need or the producer's main plant. The depot facilities are generally remotely located to support specific geographical areas or designated product lines.

The foregoing definitions basically cover the maintenance aspects of system support as reflected by the solid lines in Figure 3.4 (the flow of items from operational sites to the intermediate facilities and to the depot if the requirement exists). The reverse flow, indicated by the dashed lines, represents supply support or the flow of spares and repair parts to support maintenance operations.

Within the scope of the identified levels of maintenance support, the producer should attempt to define a basic repair policy. The repair policy may vary from discarding the entire system when a failure occurs (i.e., the system is replaced as an entity, and no maintenance is accomplished on the faulty item), to complete repair of the system, or to removal and replacement of a component part. There are many combinations of repair policies involving mixed approaches, an example of which is illustrated in Figure 3.6.

Repair policies at this stage in the life cycle are by no means fixed. Assumptions must be made in order to provide an input for advanced product planning and for system design. These assumptions may be based on past experience with comparable systems and should be evaluated in terms of total cost effectiveness. For instance, it may appear to be too costly to discard or throw away the entire system if a failure occurs as a result of the high cost of replacement spares, and it may be too costly to accomplish repair down to the Nth item because of the cost of test and support equipment, personnel, facilities, and other required resources. The best approach may represent a policy somewhere in between, as illustrated in Figure 3.6. In any event, a policy must be assumed (the most likely approach envisioned at this time) with the intent of updating it as system development progresses.

3.2. PRELIMINARY SYSTEMS ANALYSIS

The definition of system operational and maintenance requirements is a starting point in the engineering development process. The responsible engineering activity must know how the system is to be utilized and supported (and in what environment) if the output is to effectively meet the need.

Operational and maintenance support requirements specified for the system provide goals for the development of prime mission equipment, software, the elements of logistic support, and other segments of the system. These goals may be stated both qualitatively and quantitatively with quantitative factors presented in the form of discrete values and/or a range of values. Sometimes, the specified quantitative goals are in the form of minimum or maximum constraints (e.g., the range shall be at least 200 miles, the size shall not be greater than 2 ft by 4 ft, the operational reliability shall be at least 98%, the plant capacity shall be at least 300 units per month, the spare parts availability shall be 90%, the shop turnaround time for maintenance shall be less than 16 hrs, the mean time between maintenance shall be greater than 500 hrs, the equivalent life-cycle cost on a unit-item basis shall not exceed $1,500, etc.). The specification of these and comparable factors is based on a preliminary evaluation and interpretation of the need.

Given a set of guidelines for system development, the systems engineer (through a review of current and predicted future technology identified from the results of the feasibility studies) investigates and determines different technical means by which the system requirements can be met. At this point, the various alternatives are defined in general terms. However, decisions must be made concerning the feasibility of proceeding further with the project, and engineers must do their best with what is available. Once feasible alternative approaches have been identified, a preliminary analysis will be performed to determine the specific approach most appropriate for the occasion. Inherent in this is the systems analysis process illustrated in Figure 3.7 and briefly described below.

1. *Define the problem*—The initial step in any analysis effort begins with the clarification of objectives, defining the issues of concern, and limiting the problem so that it can be studied in an efficient and timely manner. In many cases, the nature of the problem appears to be obvious, whereas the precise definition of the problem may be the most difficult part of the entire process. However, unless the problem is clearly and precisely defined, it is doubtful whether an analysis of any type will be meaningful.

2. *Identify feasible alternatives*—The next step is to identify alternative solutions to the problem. All possible candidates must be initially considered, and yet the more alternatives that are considered, the more complex the analysis process becomes. Thus, it is desirable to list all possible candidates to ensure against inadvertent omissions, and then to eliminate those candidates which are clearly unattractive, leaving only the most promising for evaluation.

3. *Select the evaluation criteria*—The criteria used in the evaluation process may vary considerably, depending on the stated problem and the level and complexity of the analysis. For instance, at the system level, parameters of primary importance include system performance, cost effectiveness, logistics effectiveness, and operational availability. The parameters selected as evaluation criteria should relate directly to the problem statement.

4. *Apply modeling techniques*—The next step in the analysis process involves the application of analytical techniques in the form of a model or a series of models. The model may be simple or complex, highly mathematical or not mathematical, computerized or manually implemented, and so on. The extensiveness of the model will depend on the nature of the problem relative to the number of variables, input-parameter relationships, number of alternatives being evaluated, and the complexity of operation.

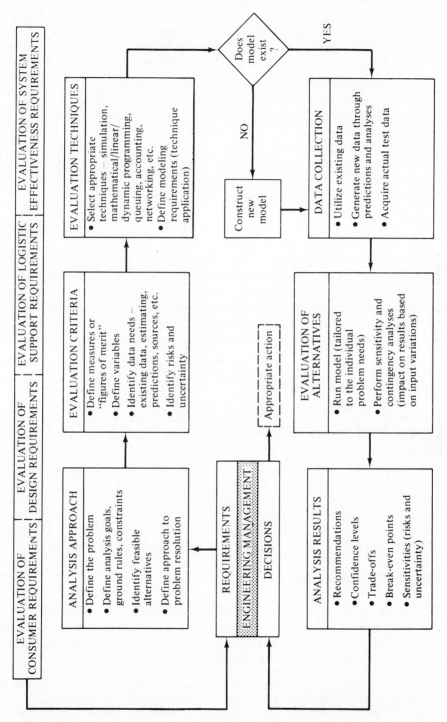

Figure 3.7 The systems analysis process.

EVALUATION OF CONSUMER REQUIREMENTS | EVALUATION OF DESIGN REQUIREMENTS | EVALUATION OF LOGISTIC SUPPORT REQUIREMENTS | EVALUATION OF SYSTEM EFFECTIVENESS REQUIREMENTS

ANALYSIS APPROACH
• Define the problem
• Define analysis goals, ground rules, constraints
• Identify feasible alternatives
• Define approach to problem resolution

EVALUATION CRITERIA
• Define measures or "figures of merit"
• Define variables
• Identify data needs — existing data, estimating, predictions, sources, etc.
• Identify risks and uncertainty

EVALUATION TECHNIQUES
• Select appropriate techniques — simulation, mathematical/linear/dynamic programming, queuing, accounting, networking, etc.
• Define modeling requirements (technique application)

Does model exist ?

NO

YES

Construct new model

DATA COLLECTION
• Utilize existing data
• Generate new data through predictions and analyses
• Acquire actual test data

EVALUATION OF ALTERNATIVES
• Run model (tailored to the individual problem needs)
• Perform sensitivity and contingency analyses (impact on results based on input variations)

ANALYSIS RESULTS
• Recommendations
• Confidence levels
• Trade-offs
• Break-even points
• Sensitivities (risks and uncertainty)

REQUIREMENTS
ENGINEERING MANAGEMENT
DECISIONS

Appropriate action

5. *Generate input data*—A most important step in the analysis process is to specify the requirements for appropriate input data. The right type of data must be collected in a timely manner and presented in the proper format. Specific data requirements are identified from the evaluation criteria and from the input requirements of the model used for evaluation purposes.

6. *Manipulate the model*—When data are gathered and inputed, the model may be exercised. The analysis results will lead to a recommendation for some type of action. However, the performance of a sensitivity analysis should be accomplished prior to making a final recommendation. In a given analysis, there may be a few key input parameters about which the analyst is very uncertain, and the sensitivity of the analysis results to variations in these uncertain parameters should be determined. This will lead to a recommendation, along with the identification of areas of potential risk.

The preliminary system analysis, identified in Figure 3.1, is iterative in nature and covers the various trade-off studies that must be accomplished in the early definition of system requirements. Major decisions must be made through the evaluation of alternative operational concepts and system use profiles, alternative maintenance and support policies, alternative technology applications, and so on. Regardless of the problem, the basic approach conveyed in Figure 3.7 is applicable. This approach is also employed in the accomplishment of lower-level trade-off studies as one progresses through the system life cycle.[5]

3.3. ADVANCE SYSTEM PLANNING

Referring to Figure 3.1, early system planning commences with the identification of a need and evolves throughout the conceptual design phase. The results from such planning may be classified in terms of (1) *technical* requirements included in specifications and (2) *management* requirements included in a program management plan. The documentation associated with these requirements is highlighted in Figure 3.8. Of particular significance for systems engineering are the System Specification (Type A) and the System Engineering Management Plan (SEMP). It is these documents that initiate the implementation of top-level requirements for a given program, as well as providing the appropriate checks and balances throughout.

System Specification

The system specification is the top-level technical document and includes information derived from the definition of operational requirements and the maintenance concept and the results of the feasibility analysis. Specific specification content usually covers the following general areas:

[5] Inherent in the systems analysis process in Figure 3.7 is the application of various analytical methods for evaluation purposes. While it is appropriate to introduce the overall process here, it is also essential that one adapt the proper technique to the specific problem at hand. General coverage of some of these analytical techniques is included in Part Three.

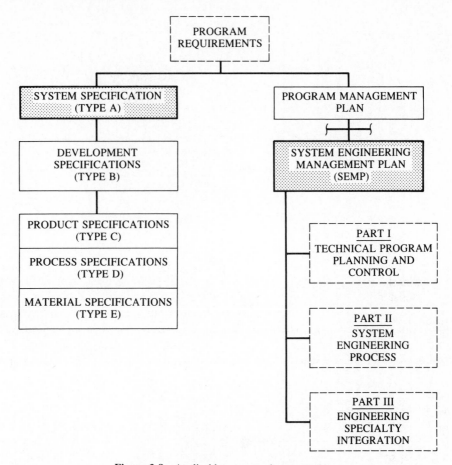

Figure 3.8 Applicable program documentation.

1. General description of the system and its function.
2. Operational requirements.
3. Maintenance concept definition.
4. System functional diagram and functional interfaces.
5. Performance characteristics.
6. Physical characteristics.
7. Effectiveness characteristics.
8. Design characteristics—reliability, maintainability, manability, supportability, transportability, producibility, interchangeability, and others.
9. Construction—materials, processes, components, workmanship, and so on.
10. Logistic support—test equipment, supply support, publications, personnel, training, facilities, and maintenance planning.
11. Design documentation.
12. Quality assurance provisions—test and evaluation.

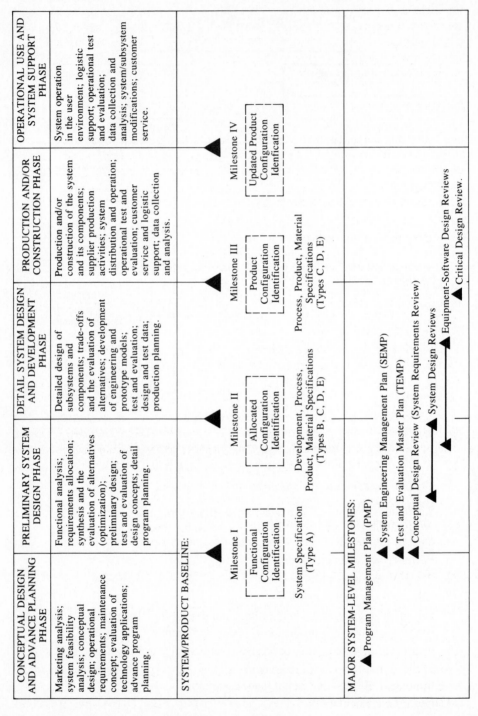

Figure 3.9 The system acquisition process and major milestones (example).

50

The preparation of specifications is an engineering function. The system specification is prepared initially at program inception and updated to reflect the output of the conceptual design activities represented in Figure 3.9. As the system design progresses, the system specification forms the basis for the preparation of development specifications, procurement specifications, process specifications, material specifications, and so on.

System Engineering Management Plan (SEMP)

In viewing the overall system engineering objectives defined in Chapter 2, it is important that the appropriate program management documentation be developed to ensure that these objectives are met. This is accomplished through the System Engineering Management Plan (SEMP) prepared in the conceptual design phase (refer to Figure 3.9). The plan may differ in format from one program to the next, but a general approach is presented in Figure 3.8. More specifically, the plan may be divided into three major sections:

1. *Technical program planning and control* describes the program tasks that must be planned and developed to ensure system engineering objectives (statement of work, work breakdown structure, organization, task schedules and cost, technical performance measurement, program-design reviews, supplier interfaces, risk management, and so on).

2. *System engineering process* describes the system engineering process as it applies to program requirements (the functions identified in Figure 2.4 to include system operational requirements and maintenance concept, functional analysis and allocation, system synthesis, system analysis and trade-off studies, system design, system test and evaluation, and so on).

3. *Engineering specialty integration* describes the major system-level requirements in the engineering speciality areas (reliability, maintainability, human factors, supportability or logistic support, producibility, quality assurance, and so on).

The details associated with the preparation and implementation of the System Engineering Management Plan (SEMP) are covered further in Chapter 18.

3.4. CONCEPTUAL DESIGN REVIEW

Design progresses from an abstract notion to something that has form and function, is fixed, and can ultimately be reproduced in designated quantities to satisfy a need. Initially, a need is identified. From this point a design evolves through a series of stages (i.e., conceptual design, preliminary system design, detail design and development). In each major stage of the design process, an evaluative function is accomplished to ensure that the design is correct at that point prior to proceeding with the next stage. The evaluative function includes both the informal day-to-day project coordination and data review, and the formal design review. Design information is released and reviewed for compliance with the basic system-equipment requirements (i.e., performance, reliability, maintainability, manability, etc., as defined by the system specification). If the requirements are satisfied, the design is approved as is. If not, recommendations for corrective action are initiated and discussed as part of the formal design review.

The formal design review constitutes a coordinated activity (including a meeting or series of meetings) directed to satisfy the interests of the design engineer and the technical discipline support areas (reliability, maintainability, human factors, logistics, manufacturing, industrial engineering, quality assurance, program management). The purpose of the design review is to formally and logically cover the proposed design from the total system standpoint in the most effective and economical manner through a combined integrated review effort. The formal design review serves a number of purposes.

1. It provides a formalized check (audit) of the proposed system/subsystem design with respect to specification requirements. Major problem areas are discussed and corrective action is taken.

2. It provides a common baseline for all project personnel. The design engineer is provided the opportunity to explain and justify his or her design approach, and representatives from the various supporting organizations (e.g., maintainability, logistic support) are provided the opportunity to hear the design engineer's problems. This serves as an excellent communication medium and creates a better understanding among design and support personnel.

3. It provides a means for solving interface problems and promotes the assurance that all system elements will be compatible.

4. It provides a formalized record of what design decisions were made and the reasons for making them. Analyses, predictions, and trade-off study reports are noted and are available to support design decisions. Compromises to performance, reliability, maintainability, human factors, cost, and logistic support are documented and included in the trade-off study reports.

5. It promotes a higher probability of mature design, as well as the incorporation of the latest techniques (where appropriate). Group review may identify new ideas, possibly resulting in simplified processes and ultimate cost savings.

The formal design review, when appropriately scheduled and conducted in an effective manner, causes a reduction in the producer's risk relative to meeting specification requirements and results in improvement of the producer's methods of operation. Also, the consumer often benefits through receipt of a better product.

Design reviews are generally scheduled prior to each major evolutionary step in the design process, as illustrated in Figure 3.9. In some instances, this may entail a single review toward the end of each stage (i.e., conceptual, preliminary system design, detail design and development). For other projects, where a large system is involved and the amount of new design is extensive, a series of formal reviews may be conducted on designated elements of the system. This may be desirable to allow for the early processing of some items while concentrating on the more complex, high-risk items.

Although the number and type of design reviews scheduled may vary from program to program, four basic types are readily identifiable and are common to most programs. They include the conceptual design review (i.e., system requirements review), the system design review, the equipment design review, and the critical design review. Referring to Figure 2.4, the conceptual design review generally covers the results of the activities specified in block 1, system and equipment design reviews relate to the activities in block 2, and equipment and critical design reviews primarily pertain to block 3 activities.

Of particular interest relative to the activities discussed in this chapter is the *conceptual design review*. The conceptual design review may be scheduled during the early part of a program (preferably not more than 4 to 8 weeks after program start) when operational requirements and the maintenance concept have been defined. Feasibility studies justifying preliminary design concepts should be reviewed. Logistic support requirements at this point are generally included in the maintenance-concept definition.

SELECTED REFERENCES

(1) Blanchard, B. S., *Engineering Organization and Management*. Englewood Cliffs, N.J.: Prentice-Hall, 1976.

(2) Blanchard, B. S., *Logistics Engineering and Management* (3rd ed.). Englewood Cliffs, N.J.: Prentice-Hall, 1986.

(3) Chase, W. P., *Management of System Engineering*. New York: John Wiley, 1974.

(4) Chestnut, H., *Systems Engineering Methods*. New York: John Wiley, 1967.

(5) Defense Systems Management College (DSMC), *Systems Engineering Management Guide*. Fort Belvoir, Virginia, December 1986.

(6) Karger, D. W., and R. G. Murdick, *Managing Engineering and Research* (2nd ed.). New York: Industrial Press, 1969.

(7) Miels, R., ed., *Systems Concepts*. New York: John Wiley, 1973.

(8) Ostrofsky, B., *Design, Planning, and Development Methodology*. Englewood Cliffs, N.J.: Prentice-Hall, 1977.

QUESTIONS AND PROBLEMS

1. Why is the definition of system operational requirements important? What is included? Why is the definition of the mission profile important?

2. Select a system of your choice and define the basic operational requirements for that system.

3. For the system selected in Question 2, develop a detailed maintenance concept flow. Identify alternative repair policies where appropriate.

4. Refer to Figure 3.7. How does the engineering analysis process relate to the system design process in Figure 2.4?

5. Why is it important to define the system's operational requirements and the maintenance concept at this stage in the life cycle?

6. What information is included in the maintenance concept? How does it differ from the maintenance plan?

7. What is the purpose of the feasibility analysis? Where does research fit in?

8. What is most significant about the feedback provision in the systems engineering approach?

9. What program documentation is important in systems engineering? Provide some examples.

10. How do the following factors affect logistic support: operational scenario or profile, operational deployment, operational life cycle, system use, system MTBM, and operational environment?

11. What is the impact on logistic support if system use is increased? System reliability is decreased? The life cycle of the system is extended? Environmental factors became more constraining than initially anticipated?

12. Refer to Figure 3.6. What is the impact of supply transportation times on logistics? What impact does the mean active maintenance time at the organizational level have on spares? What impact does TAT at the intermediate level have on spares? What is the impact of test thoroughness on spares?

13. When developing a maintenance concept, all applicable levels of maintenance must be considered on an integrated basis. Why?

14. What factors should be considered in identifying which maintenance functions should be accomplished at the organizational level, at the intermediate level, and at the depot level?

15. Select a system of your choice and develop a system specification.

16. What is meant by design criteria? How are design criteria developed? How are criteria applied to the design process?

17. What is the purpose of formal design review? Name at least three major benefits. What is covered specifically in a conceptual design review?

4

Preliminary System Design

Preliminary system design, sometimes referred to as advance development, begins with the technical baseline for the system as defined in the previous chapter and extends through the translation of established system-level requirements into detailed qualitative and quantitative design requirements. Preliminary design includes the process of functional analysis and requirement allocation, the accomplishment of trade-off studies and optimization, system synthesis, and configuration definition in the form of detailed specifications. These activities are reflected in Figure 2.4, Block 2.

A requirement is initially defined, feasible design approaches are considered, the results are evaluated in terms of specified effectiveness criteria, and a preferred configuration is described through subsystem specifications. If the results are not satisfactory, alternatives are evaluated and changes are implemented as required. This is a continuous and iterative process, incorporating the necessary feedback provisions allowing for corrective action. The activities described in this chapter reflect a continuation of a top-down approach to overall system development.

4.1. SYSTEM FUNCTIONAL ANALYSIS

The next major activity is the system's functional analysis, illustrated in Figure 4.1. An essential element of preliminary design is the employment of a functional approach as a basis for the identification of design requirements for each hierarchical level of the system. A function constitutes a specific or discrete action required to

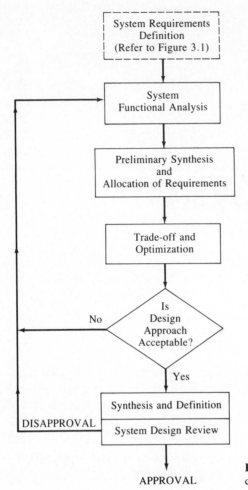

Figure 4.1 System development process (partial).

achieve a given objective (e.g., an operation that the system must perform to accomplish its mission, a maintenance action that is required to restore the system to operational use). Such actions may be accomplished through the use of equipment, personnel, facilities, firmware and software, data, or a combination thereof. The functional approach helps to assure

1. That all facets of system development, operation, and support are covered. This includes design, production/construction, test, deployment, transportation, training, operation, and maintenance.
2. That all elements of the system (e.g., prime equipment, test and support equipment, facilities, personnel, data, software, etc.) are fully recognized and defined.
3. That a means of relating equipment packaging concepts and support requirements to given functions is provided. This identifies the relationship between the need and the resources required to support that need.

4. That the proper sequences and design relationships are established, along with critical design interfaces.

Functional analysis is a logical and systematic approach to system design and development. It constitutes the process of translating system operational and support requirements into specific qualitative and quantitative design requirements. This process is iterative, and is accomplished through the development of functional flow diagrams.

The functional analysis (and the generation of functional flow diagrams) is intended to facilitate the design, development, and system definition process in a complete and logical manner. The functional analysis is based on the definition of system operational requirements and the system maintenance concept, and is subsequently used as the basis for detail design. There are a number of interrelated detail design tools which must track the top-level functional analysis (e.g., operational and maintenance functional block diagrams). An objective is to (1) identify system/subsystem functions, (2) identify the method for accomplishing the various functions—manually, automatically, or a combination thereof, and (3) identify the resources required to accomplish the function. Both the *operational* and *maintenance support* aspects, as related to anticipated system life-cycle use in the consumer environment, must be addressed.

Functional Flow Diagrams[1]

Functional flow diagrams are developed for the primary purpose of structuring system requirements into *functional terms*. They are developed to indicate basic system organization, and to identify functional interfaces. Functional blocks are concerned with *what* is to be accomplished, versus the realization of *how* something should be done. It is relatively easy to evolve prematurely into equipment block diagrams without having first established functional requirements. The decision concerning which functions should be performed by a piece of equipment, or by an element of software, or by a human being, or by a combination of each should not be made until the complete scope of functional requirements has been clearly defined. In other words, *not one piece of equipment should be defined or acquired without first justifying its need through the functional requirements definition process.*

Functional flow diagrams are employed as a mechanism for portraying system design requirements in a pictorial manner, illustrating series and parallel relationships, the hierarchy of system functions, and functional interfaces. Functional flow diagrams are designated as top level, first level, second level, and so on. The top-level diagram shows gross operational functions. The first-level and second-level diagrams represent progressive expansions of the individual functions of the preceding level. Functional flow diagrams are prepared down to the level necessary to establish the needs (hardware, software, facilities, personnel, data) of the system. The inden-

---

[1] The development of functional block diagrams is amplified further in Appendix A. The details associated with block identification and numbering, along with some sample applications, are included.

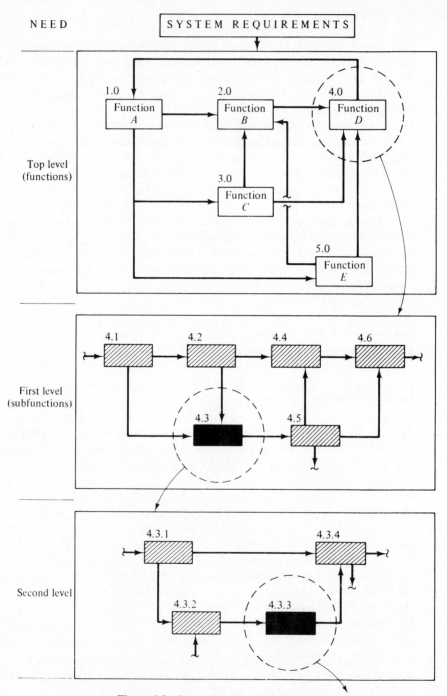

Figure 4.2 System functional indenture levels.

ture relationships of functions by level are illustrated in Figure 4.2, and the major steps involved in developing functional flows are described in Appendix A.

The functions identified should not be limited to only those necessary for operation of the system, but must consider the possible impact of maintenance on system design. Maintenance functional flow diagrams will, in most instances, evolve from operational functional flow diagrams as illustrated in the evolutionary sequence presented in Figure 4.3. Maintenance requirements (evolving from the maintenance concept) should be addressed to preclude the possibility of developing a technically feasible system from an operational viewpoint, without first determining whether or not the system can be effectively and economically supported throughout its planned life cycle. Experience has indicated that the costs associated with system maintenance and support often far exceed the cost of system acquisition. The objective is to attain the proper balance of performance, effectiveness, support, and economic factors.

The benefits associated with the generation of functional flows are many. First, the process enables the systems engineer to approach design from a logical and systematic standpoint. The proper sequences and design relationships are readily established. Second, the preparation of functional flows forces the integration of the numerous interfaces that exist in system development and operation. Both internal and external interface problems are quickly identified at an early stage in the life cycle. Sometimes these benefits are difficult to visualize unless one has actually had some experience in functional analysis. However, it has been shown that many operational problems which occur later in the life cycle could have been avoided had this approach been followed initially.

Operational Functions

The functional description represents an overall protrayal of the functions that are necessary to describe total system activities. Gross operational activities are defined in terms of mission activities (i.e., the system operational requirements described in Chapter 3). This may constitute a description of the various modes of system operation and utilization. For instance, typical gross operating functions might be (1) "prepare aircraft for flight," (2) "fly aircraft from point A to point B," and (3) "recycle aircraft for the next flight." In the case of communications system, typical operation functions might include (1) "develop a communications system for an urban area of a given geographical size and population," (2) "produce and install a communications network," and (3) "accomplish the communication of certain designated information throughout the urban area through seven days per week, six hours per day." System functions necessary to support the identified modes of operation are then described. Ultimately, the functional description is completed to the level where initial packaging concepts are visualized.[2]

---

[2] It should be noted that equipment design is by no means formulated at this time. A gross-level configuration is developed for the purpose of allocating requirements. This configuration, which serves as a starting point, may be verified or may change as a result of subsequent analyses.

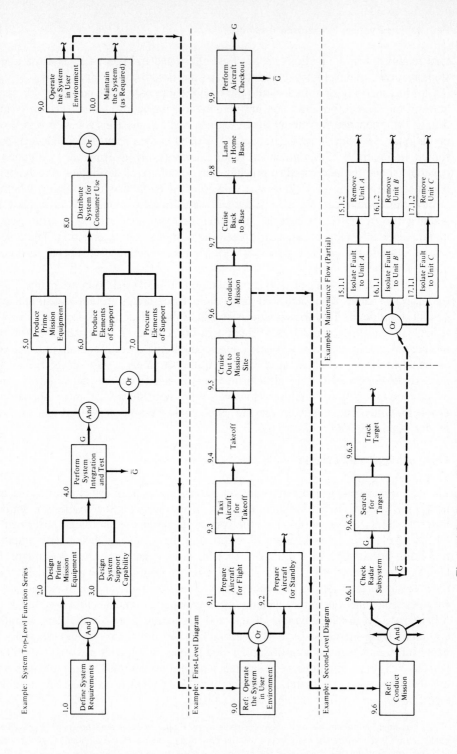

Figure 4.3 Series of flow diagrams (evolutionary development).

60

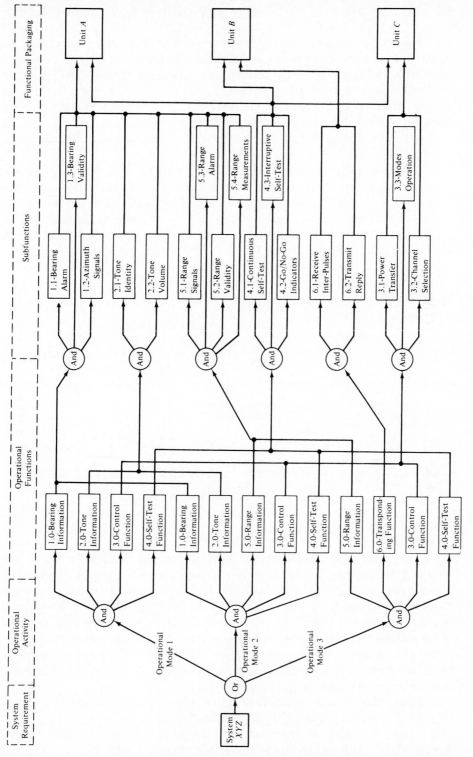

Figure 4.4 System *XYZ* operational functional flow.

Figure 4.4 illustrates an abbreviated application of a system operational functional flow diagram, showing the overall process leading from the various modes of operational use to a proposed system physical packaging scheme. The system covered will henceforth be noted as System *XYZ*. Referring to the figure, the major opertional functions of System *XYZ* are described, and the respective blocks are numerically identified for reference purposes. Each major function, in turn, is then analyzed and expanded through the identification of subfunctions as indicated in Figure 4.5.

Considering the desirability of packaging equipment by function (consistent with system size and weight constraints), the packaging scheme for System *XYZ* results in three basic units (units *A*, *B*, and *C*). The operational functions incorporated in each of the three units are indicated in Figure 4.4. Developing the design concept further, an analysis of the functions contained within each unit will lead to the identification of major assemblies. Given a broad functional packaging scheme, it is then possible to allocate performance parameters and operational effectiveness factors to the unit level and possibly the assembly level. This allocation serves as the basis for subsequent detail design, and is described in Section 4.2.

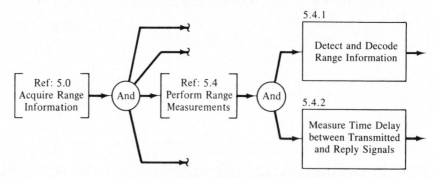

Figure 4.5   Range information functional flow diagram.

Maintenance Functions

Once operational functions are defined, the system description leads to the development of gross maintenance functions. For example, there are specified performance requirements (signal level, tolerances, accuracies, unit dimension, output capacity, effectiveness measure, etc.) for each operational function. A check of the applicable function will indicate either a go or a no-go decision. A *go* decision leads to a check of the next operational function. A *no-go* indication (constituting a symptom of malfunction) provides a starting point for the development of detailed maintenance functional flows and logic troubleshooting diagrams. A gross-level maintenance functional flow is illustrated in Figure 4.6.

Maintenance functions identified at this point will reflect consideration for the effectiveness and supportability factors specified at the system level as well as available logistic resources. In some instances, it may be possible to expand top-level functional flows in a manner similar to the example presented in Figure 4.5. In most

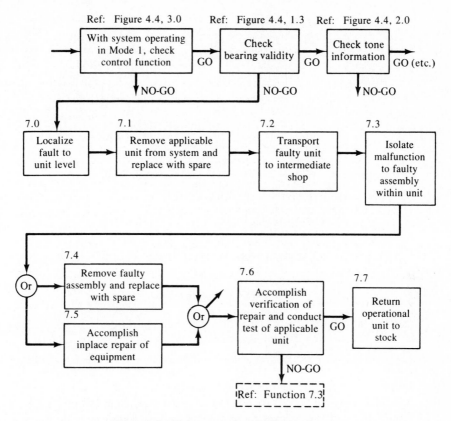

Figure 4.6   Maintenance functional flow diagram.

cases, however, only an estimate of first-level maintenance functions can be made since the availability of preliminary engineering data is limited. Maintenance functional flows can and should be prepared for corrective maintenance, preventive maintenance, transportation and handling functions, support equipment maintenance, servicing, inspections, and so on.

Maintenance functional flow diagrams are developed and used to supplement and/or update the system maintenance concept discussed in Chapter 3, in terms of functions by level and a preliminary equipment packaging scheme. The maintenance concept and functional flow development process is iterative and continues throughout the early system definition process.

Summary of Functional Analysis

The functional analysis provides an initial description of the system and, as such, its application is extensive. It serves as a baseline for the definition of later equipment requirements, personnel requirements, software requirements, maintenance and logistic support requirements, and so on.

In the past, these requirements have been derived largely on an independent

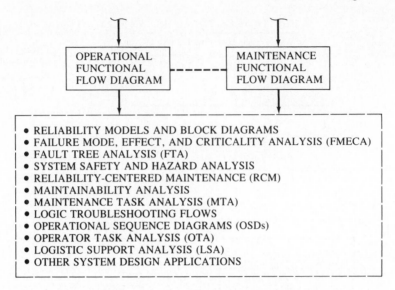

Figure 4.7   Functional requirements flow.

basis. The results have often been conflicting as different baselines have been followed. In other words, the electrical design has evolved from one baseline, the mechanical design from another, the reliability and maintainability requirements have been derived from another, logistics requirements from a different source, and so on. This, in turn, has led to some costly results.

The purpose here is to develop *one* set of requirements and define *one* baseline from which all lower-level requirements evolve. Figure 4.7 illustrates some of the areas which require a functional analysis as an input. It is the objective of systems engineering to ensure that all such activities evolve from the *same* functional definition.

4.2. ALLOCATION OF REQUIREMENTS

The preceding discussion pertains to the first step in the process of translating system operational and maintenance requirements into specific design criteria for various elements of the system. The functional analysis provides a description of major system functions and results in a preliminary *synthesis* of the overall system configuration.[3] The next step involves the allocation of top-level system factors to various subsystems and lower-level elements of the system.[4] In essence, a system can be broken down into different categories of components, as illustrated in Figure 4.8, each of which must support the overall performance and effectiveness requirements at the system level. The question is what should one include in a "design-to"

---

[3] *Synthesis* refers to the combining and structuring of parts and elements of an item in a manner that forms a functional entity. Refer to Section 4.4 for additional discussion.

[4] Allocation refers to the distribution, allotment, or apportionment of top-level requirements to lower indenture levels of the system.

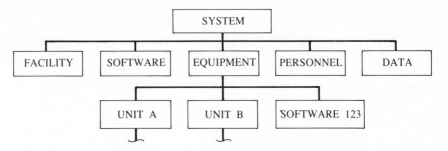

Figure 4.8   Hierarchy of system components.

specification covering a given element of the system? Hopefully, the response is meaningful in terms of meeting some overall mission objective.

Figure 4.9 illustrates the allocation process for System *XYZ*. With requirements identified at the system level, what should be specified at the *unit* level, or *assembly* level? For instance, if the allowable weight of System *XYZ* is 750 lbs, what should unit *A*, unit *B*, and unit *C* weigh? If the mission reliability mean time between failure (MTBF) specified at the system level is 450 hrs, what values should be specified at the unit level? If the system life-cycle cost goal is $50,000, what cost goals should be established for units *A*, *B*, and *C*?[5]

These and other factors must be allocated to lower indenture levels of the system to provide technical parameters and constraints, functional requirements, and design criteria where needed. Otherwise, individual engineering designers assigned to different elements of the system and working independently will establish their own goals, and the results when combined may not comply with the initially established requirements for the overall system. Thus, it is necessary to first establish requirements at the system level and then allocate these requirements to the depth necessary to provide guidance in the design process. Subsequently, as design progresses, it is necessary to check (on a continuing basis) to ensure that the design results at the lower indenture levels are in turn compatible with the overall system requirements.

When accomplishing allocation, one should consider all appropriate qualitative and quantitative criteria that will significantly influence the design process. These criteria will vary from system to system and must directly support operational requirements and the maintenance concept. In some instances the allocation of design criteria to the subsystem and unit level may be sufficient, and in other cases it may be appropriate to assign goals down to the assembly and subassembly level. The depth of coverage depends on the design controls that engineering and management may wish to establish.[6] Controls that are too stringent may inhibit the decision-making process relative to allowing for trade-offs at the lower levels in order to meet to-

---

[5] Many system-level requirements to include effectiveness characteristics, reliability and maintainability factors, logistic support factors, and cost factors are discussed further in Part Four.

[6] If an item is to be procured from an outside supplier, design criteria are established to the level needed to effect the necessary design controls on the supplier. On the other hand, the level to which criteria are established may be different if the item is to be developed within the major producer's organization.

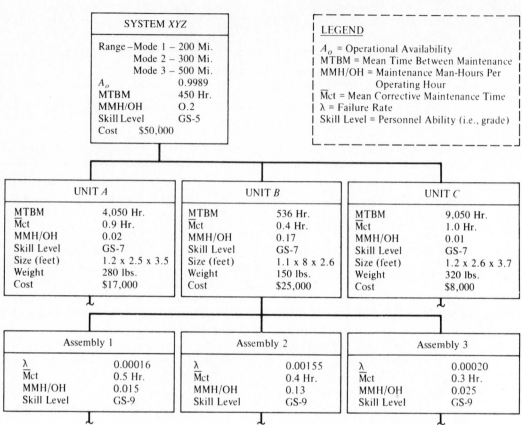

Figure 4.9   System *XYZ* functional requirements and allocation.

tal system requirements. On the other hand, insufficient design controls may not provide the desired results.

The process of allocation stems from the functional packaging scheme identified in Figure 4.4. Although still preliminary in nature because of the unavailability of necessary design data, this scheme (i.e., the identification of units *A*, *B*, and *C*) does form an initial baseline for design guidance. The next step is to attempt to subdivide the units into assemblies, the assemblies into subassemblies, and so on. This may be accomplished through a preliminary equipment breakdown such as the example presented in a partial form in Figure 4.9.

Some of the parameters normally established at the system level are allocated to the unit level and then to the assembly level. Basically, allocation should consider *all* significant system parameters stated in the form of maximum or minimum requirements (with tolerance bands where appropriate) to include

1. System effectiveness factors such as operational availability, readiness, reliability, maintainability, and supportability.
2. System performance and physical parameters such as range, accuracy, speed, capacity and output rate, power output, weight, size, and volume.

3. Factors covering the system support capability such as transportation or supply times between levels of maintenance, parts availability, test and support equipment use and availability, facility utilization, personnel effectiveness, transportation and handling rates, and maintenance turnaround times.

4. Life-cycle cost factors, which include research and development cost, investment or production cost, operation and maintenance support cost, and system retirement and disposal cost.

In summary, the purpose of allocation is to provide, as an input, some guidelines to the design engineer to assist him or her in developing a product that will be compatible with system requirements. These guidelines may be stated both qualitatively and quantitatively and will vary from system to system. At times, it may be impossible to meet the allocated parameters when considering available technology, and trade-offs must be made with other elements of the system. In such instances, the allocations are revised and the best approach is defined representing a modified baseline. In any event, a baseline of some type must be established before proceeding further.

4.3. TRADE-OFF AND OPTIMIZATION

The allocation process described in Section 4.2 establishes boundaries and constraints for system design (i.e., maximum and/or minimum values to which the design must conform). Within these bounds and constraints, the systems engineer may envision any number of design configurations that will satisfy the specified requirements. The problem is to select the best approach possible through the iterative process of system analysis using various analytical methods.

Figure 4.10 illustrates the steps involved in performing trade-offs and evaluating alternative approaches. These steps are an integral part of the systems analysis process illustrated in Figure 3.7. Alternatives are identified, evaluation criteria are established, analytical methods are selected, input data are collected, and the various alternatives are evaluated on an equivalent basis.[7]

When selecting evaluation criteria, the factors used may vary considerably, depending on the stated problem and the level and complexity of the analysis. For instance, at the system level, parameters of primary importance include cost effectiveness, system effectiveness, logistics effectiveness, life-cycle cost, operational availability, and performance. At the detailed level (e.g., subsystem, unit, assembly) the order of parameters will be different, as illustrated in Figure 4.11.

The parameters selected as evaluation criteria should relate directly to the problem statement. For instance, the problem may be to design a system that will perform a certain mission with a specified degree of effectiveness at minimum life-cycle cost. There may be several possible alternative design approaches, each of which is evaluated in terms of system effectiveness and life-cycle cost. On the other hand, the problem may entail the selection of the best among several alternative "off-the-shelf" equipment items from different suppliers, using supportability characteristics in design as criteria (e.g., accessibility, standardization of components,

[7] The role of models for determining equivalence is discussed in Chapter 7.

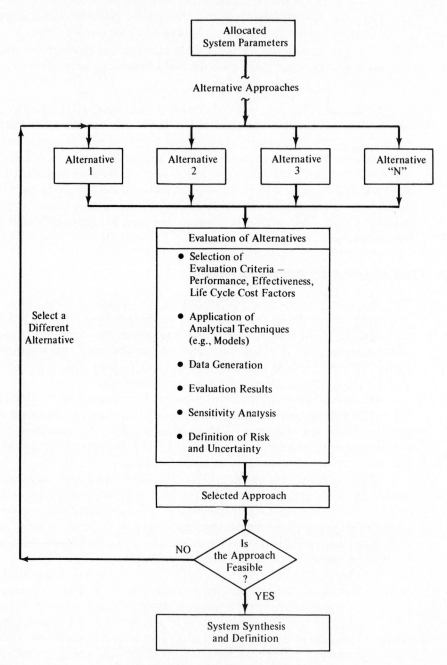

Figure 4.10   Evaluation of alternatives.

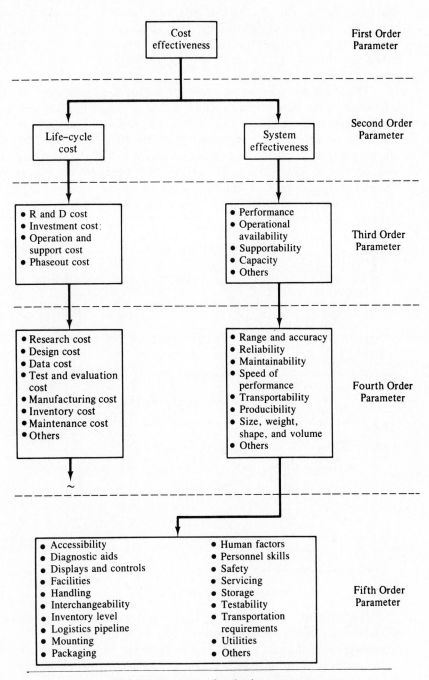

Figure 4.11   Order of evaluation parameters.

diagnostic aids, interchangeability, etc.). In this instance, there are a number of evaluation factors, and the various parameters should be reviewed from the standpoint of relevancy or degree of importance. The degree of importance may be realized by applying parameter weighting factors (the most important items receiving the heaviest weighting). The application of weighting factors will depend on the evaluation technique employed.

The evaluation process itself may involve the accomplishment of specific trade-off studies where two or more individual system parameters are reviewed independently in terms of the effects on each other. Figure 4.12 illustrates some approaches where the possible feasible solutions fall within the shaded areas. Individual trade-offs may deal with different performance parameters in terms of cost, performance as a function of reliability, weight as a function of range, capacity as a function of size and cost, and so on. Ultimately, these individual trade-offs are combined and reviewed in terms of higher-order system parameters such as system effectiveness and life-cycle cost.

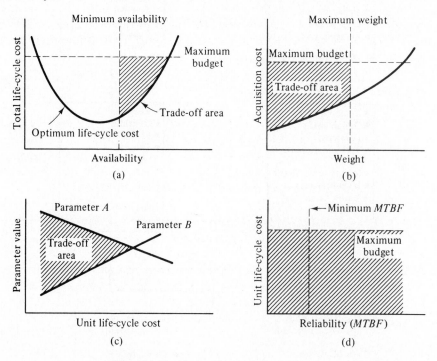

Figure 4.12 Trade-off areas for optimization.

The evaluation process may be facilitated through the use of various mathematical techniques in the form of a model or series of models. A model is a simplified representation of the real world which abstracts the features of the situation relative to the problem being analyzed. It is a tool employed by an analyst to assess the likely consequences of various alternative courses of action being examined. The model must be adapted to the problem at hand and the output must be oriented to the selected evaluation criteria. The model, in itself, is not the decision maker, but is a tool that provides the necessary data in a timely manner in support of the decision-making process.

The extensiveness of the model will depend on the nature of the problem, the number of variables, input parameter relationships, number of alternatives being evaluated, and the complexity of operation. The ultimate objective in the selection and development of a model is simplicity and usefulness. The model utilized should incorporate the following features.

1. The model should represent the dynamics of the system configuration being evaluated in a way that is simple enough to understand and manipulate, and yet close enough to the operating reality to yield successful results.
2. The model should highlight those factors that are most relevant to the problem at hand, and suppress (with discretion) those that are not as important.
3. The model should be comprehensive by including *all* relevant factors and be reliable in terms of repeatability of results.
4. Model design should be simple enough to allow for timely implementation in problem solving. Unless the tool can be utilized in a timely and efficient manner by the analyst or the manager, it is of little value. If the model is large and highly complex, it may be appropriate to develop a series of models where the output of one can be tied to the input of another. Also, it may be desirable to evaluate a specific element of the system independently from other elements.
5. Model design should incorporate provisions for ease of modification and/or expansion to permit the evaluation of additional factors as required. Successful model development often includes a series of trials before the overall objective is met. Initial attempts may suggest information gaps which are not immediately apparent and consequently may suggest beneficial changes.

The use of mathematical models offers significant benefits. In terms of system application, a number of considerations exist—operational considerations, design considerations, production/construction considerations, testing considerations, and logistic support considerations. There are many interrelated elements that must be integrated as a system and not treated on an individual basis. The mathematical model makes it possible to deal with the problem as an entity and allows consideration of all major variables of the problem on a simultaneous basis.

1. The mathematical model will uncover relations between the various aspects of a problem which are not apparent in the verbal description.
2. The mathematical model enables a comparison of *many* possible solutions and aids in selecting the best among them rapidly and efficiently.
3. The mathematical model often explains situations that have been left unexplained in the past by indicating cause-and-effect relationships.
4. The mathematical model readily indicates the type of data that should be collected to deal with the problem in a quantitative manner.
5. The mathematical model facilitates the prediction of future events such as ef-

fectiveness factors, reliability and maintainability parameters, logistics requirements, and so on.

6. The mathematical model aids in identifying areas of risk and uncertainty.

When analyzing a problem in terms of selecting a mathematical model for evaluation purposes, it is desirable to first investigate the tools that are currently available. If a model already exists and is proven, then it may be feasible to adopt that model. However, extreme care must be exercised to relate the right technique with the problem being addressed and to apply it to the depth necessary to provide the sensitivity required in arriving at a solution. Improper application may not provide the results desired, and the result may be costly.

On the other hand, it might be necessary to construct a new model. In accomplishing such, one should generate a comprehensive list of system parameters that will describe the situation being simulated. Next, it is necessary to develop a matrix showing parameter relationships, each parameter being analyzed with respect to every other parameter to determine the magnitude of relationship. Model input-output factors and parameter feedback relationships must be established. The model is constructed by combining the various factors and then testing for validity. Testing is difficult to do since the problems addressed primarily deal with actions in the future which are impossible to verify. However, it may be possible to select a known system or equipment item which has been in existence for a number of years and exercise the model using established parameters. Data and relationships are known and can be compared with historical experience. In any event, the analyst might attempt to answer the following questions.

1. Can the model describe known facts and situations sufficiently well?
2. When major input parameters are varied, do the results remain consistent and are they realistic?
3. Relative to system application, is the model sensitive to changes in operational requirements, design, production/construction, and logistic support?
4. Can cause-and-effect relationships be established?

Models and appropriate systems analysis techniques are presented further in Part Three. Trade-off and system optimization relies heavily on these techniques and their application. A fundamental knowledge of probability and statistics, economic analysis techniques, modeling and optimization, simulation, queuing theory, control techniques, and the other analytical techniques is essential to accomplishing effective systems analysis.

Finally, the results of the overall system evaluation effort will lead to the selection of a preferred approach. If the configuration selected is significantly better than other options, the decision may be clear-cut and the systems engineer may proceed toward greater system definition. However, in most cases the engineer will perform something of a self-analysis to ensure the correct decision. The following questions will usually arise.

1. How much better is the selected approach than the next-best alternative? Is there a significant difference between the results of the comparative evaluation?
2. Have all feasible alternatives been considered?
3. What are the areas of risk and uncertainty?

These and other comparable questions should be addressed, and to obtain some of the answers may require one or more iterations of the process illustrated in Figure 4.10. The objective, of course, is to arrive at a decision where the selected approach is clearly the best among the alternatives evaluated with the associated risk and uncertainty minimized.

4.4. SYNTHESIS AND DEFINITION

Synthesis refers to the combining and structuring of parts and elements in such a manner so as to form a functional entity. System synthesis has been achieved when sufficient trade-offs and preliminary design have been accomplished to confirm and assure the completeness of system performance and design requirements allocated for detail design. The performance, configuration, and arrangement of a chosen system and its elements are portrayed along with the techniques for their test, operation, and life-cycle support. These portrayals cover intra- and intersystem and item interfaces, and provide enough information forming a definitive baseline that can be presented in the form of detail specifications. Synthesis may be accomplished by analytical means or through the development and preliminary test of physical models.

When synthesizing a specific system design configuration employing analytical means, it may be feasible to accomplish a system availability analysis, an evaluation of consumer operations, a shop turnaround time analysis, a level of repair analysis, and/or a spares inventory policy evaluation as part of verifying design adequacy in terms of compliance with system requirements. One may also wish to determine the impact of one system parameter on another, or the interaction effects between two elements of logistic support. In accomplishing such, problem resolution may require the utilization of a number of different models combined in such a manner to provide a variety of output factors.

The analysis of a system (to include the prime equipment and its associated elements of logistic support) is sometimes fairly complex; however, this process can be simplified by applying the right techniques and developing the proper analytical tools. The combining of techniques to facilitate the analysis task can be accomplished by following the steps illustrated in Figure 3.7 (i.e., defining the problem, identifying feasible alternatives, selecting evaluation criteria, etc.). The output may appear as illustrated in Figure 4.13.

Referring to the figure, the analyst may wish to develop a series of individual models as illustrated. Each model may be used separately to solve a specific detailed

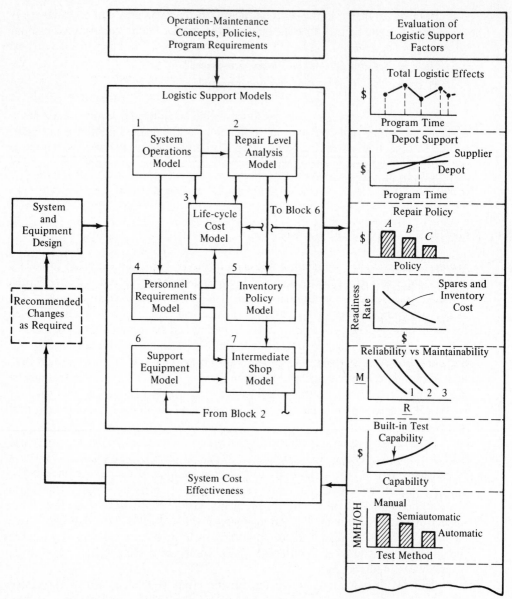

Figure 4.13 Application of models (example).

problem, or the models may be combined to solve a higher-level problem. The models may be used to varying degrees, depending on the depth of analysis required. In an event, the overall analytical task must be approached in an organized methodical manner and applied judiciously to provide the results desired.

In summary, the proposed system configuration is synthesized, and the results are included in the appropriate specifications (e.g., development, material, and/or process specification as illustrated in Figure 3.8). These specifications are used as input to the detail design of the various elements of the system. The major activities and output of the preliminary system design phase are illustrated in Figure 2.4.

4.5. SYSTEM DESIGN REVIEW

The basic objectives of formal design review were introduced in Section 3.4. In essence, design is a progression from a defined need to an entity that will perform a useful function in a satisfactory manner. Design evolves through a series of stages to include conceptual design, preliminary system design, and detail design and development. In each major stage of the design process, an evaluative function is accomplished to ensure that the design is correct at that point prior to proceeding with the next stage.

As is the case for conceptual design, the results of the preliminary system design stage must be reviewed prior to entering into detail design. Referring to Figure 2.4, the system design review generally covers the results of the activities specified in Block 2 and discussed thoughout this chapter.

SELECTED REFERENCES

(1) Blanchard, B. S., *Engineering Organization and Management*. Englewood Cliffs, N.J.: Prentice-Hall, 1976.

(2) Blanchard, B. S., *Logistics Engineering and Management* (3rd ed.). Englewood Cliffs, N.J.: Prentice-Hall, 1986.

(3) Chase, W. P., *Management of System Engineering*. New York: John Wiley, 1974.

(4) Chestnut, H., *Systems Engineering Methods*. New York: John Wiley, 1967.

(5) Chestnut, H., *Systems Engineering Tools*. New York: John Wiley, 1965.

(6) Defense Systems Management College (DSMC), *Systems Engineering Management Guide*. Fort Belvoir, Virginia, December 1986.

(7) Dieter, G. E., *Engineering Design: A Materials and Processing Approach*. New York: McGraw-Hill, 1983.

(8) Fabrycky, W. J., P. M. Ghare, and P. E. Torgersen, *Applied Operations Research and Management Science*. Englewood Cliffs, N.J.: Prentice-Hall, 1984.

(9) Ostrofsky, B., *Design, Planning, and Development Methodology*. Englewood Cliffs, N.J.: Prentice-Hall, 1977.

(10) Pressman, R. S., *Software Engineering: A Practitioner's Approach*. New York: McGraw-Hill, 1982.

(11) Vick, C. R., and C. V. Ramamoorthy, eds. *Handbook of Software Engineering*. New York: Van Nostrand Reinhold, 1984.

(12) Woodson, W. E., *Human Factors Design Handbook*. New York: McGraw-Hill, 1981.

QUESTIONS AND PROBLEMS

1. Describe in your own words what is meant by functional analysis. What purpose(s) does it serve?

2. Why is functional analysis important in the system engineering process?

3. Is functional analysis applicable to any system or equipment item? Why?

4. Select a system (or an element of a system) of your choice and develop operational and maintenance functional flow diagrams to the second level. Show or describe the relationship between operation functions and maintenance functions.

5. What is the purpose of allocation? How does it affect system design?

6. Select a system (or an element of a system) of your choice and assign top-level requirements. Allocate performance factors, effectiveness factors, reliability and maintainability factors, cost factors, and other factors as appropriate to the second level.

7. What is meant by design criteria? How are design criteria developed? How are these criteria applied to the design process?

8. What are the steps involved in the evaluation of design alternatives? Draw a flow diagram that illustrates these steps and their interrelationship(s). How would you select a design approach?

9. What is a model? List the basic desired characteristics.

10. List the benefits associated with the use of mathematical models in system analysis.

11. What are some of the concerns and/or problems associated with mathematical models and their application?

12. What is meant by synthesis?

13. What type of information is covered by the system design review?

14. What is the desired output of the preliminary system design stage?

5
Detail Design and Development

The detail design phase begins with the concept and configuration derived through the preliminary system design activities identified in Block 2, Figure 2.4. When an overall system design configuration has been established, it is necessary to progress through further definition leading to the realization of hardware, software, and items of support. The process from this point includes

1. The description of subsystems, units, assemblies, and lower-level components and parts of the prime mission equipment and the elements of logistic support (e.g., test and support equipment, facilities, personnel and training, technical data, spare and repair parts).
2. The preparation of design documentation (e.g., specifications, analysis results, trade-off study reports, predictions, detail drawings and related data bases) describing *all* elements of the system.
3. The definition and development of computer software (as applicable).
4. The development of an engineering model, a service test model, and/or a prototype model of the system and its elements for test and evaluation to verify design adequacy.
5. The test and evaluation of the system physical model that has been developed.
6. The redesign and retest of the system, or an element thereof, as necessary to correct any deficiencies noted through initial system testing.

This chapter primarily addresses the activities identified in items 1 through 4 above (i.e., the first two areas within Block 3 of Figure 2.4), while Chapter 6 covers test and evaluation.

5.1. DETAIL DESIGN REQUIREMENTS

The basic design objective(s) for the system and its elements must (1) be compatible with the operational requirements and maintenance concept covered in Chapter 3, (2) support the qualitative and quantitative allocated design criteria discussed in Chapter 4, and (3) meet all specification requirements. The specific objectives will vary depending on the type of system and the nature of its mission. However, in any case, the design function must consider the following goals:

1. *Design for functional capability or performance (functional design)*—the characteristic of design that deals with the technical performance of the system. This includes size, weight, volume, shape, accuracy, capacity, flow rate, speed of performance, power output, and all of the technical and physical characteristics that the system (when operating) must exhibit to accomplish its planned mission. This primarily constitutes the main thrust of mechanical design, electrical design, chemical design, process design, industrial design, aeronautical design, structural design, and so on.

2. *Design for reliability*—the characteristic of design and installation concerned with the successful operation of the system throughout its planned mission. Reliability (R) is often expressed as the probability of success or measured in terms of MTBF (mean time between failure). An objective is to consider maximizing operational reliability while minimizing system failures. Chapter 13 provides coverage of reliability.

3. *Design for maintainability*—the characteristic of system design and installation that is concerned with the ease, economy, safety, and accuracy in the performance of maintenance functions. The objective is to minimize maintenance times, maximize the supportability characteristics in system design (e.g., accessibility, diagnostic provisions, standardization), minimize the logistic support resources required in the performance of maintenance, and minimize maintenance support costs. Chapter 14 provides coverage of maintainability.

4. *Design for manability (human factors)*—the characteristic of system design that is directed toward the optimum human-machine interface (i.e., ensuring the compatibility between system physical and functional design features and the human element in the operation, maintenance, and support of the system or equipment). Human factors considers operability and equipment aesthetic features and results in reducing personnel skill levels, minimizing training requirements, and minimizing potential personnel error rates. Chapter 15 presents manability and human factors.

5. *Design for producibility*—the characteristic of system that allows for the effective and efficient production of one or a multiple quantity of items of a given configuration. The objective is to minimize resource requirements (e.g., hu-

man resources, materials, facilities, energy) during the production/construction process.

6. *Design for supportability*—the characteristic of system design directed toward ensuring that the system can ultimately be supported effectively and efficiently throughout its planned life cycle. An objective is to consider both the internal characteristics of equipment design (e.g., reliability, maintainability, human factors) and the design of the logistic support capability. Further coverage of supportability (and logistic support) is included in Chapter 16.

7. *Design for economic feasibility*—the characteristic of system design and installation which is directed toward maximizing the benefits and cost effectiveness of the overall system configuration. An objective is to base design decisions on *life-cycle cost,* and not just on system acquisition cost (or purchase price). Chapter 17 provides additional coverage in this area.

8. *Design for social acceptability*—the characteristic of system design directed toward ensuring that the system can become an acceptable part of the social system. The objective is to seek minimum pollutability, ease of disposability, minimum safety risk, high transportability, and many others.

Actually, there are many different considerations in system/product design that are important. The ultimate output must not only meet performance requirements but must do so in a cost-effective manner. Accomplishing this requires an optimum balance between performance features, system availability characteristics, reliability, maintainability, human factors, logistic support, life-cycle cost, and so on. This balance is often difficult to attain, since some of the stated objectives are in opposition to others. For instance, incorporating high performance and highly sophisticated functional design techniques will undoubtedly reduce reliability and increase logistic support requirements and life-cycle cost; incorporating too much reliability may require the use of costly system components and thus increase both the system acquisition cost and life-cycle cost; or the incorporation of too many provisions for maintainability and human factors may reduce reliability. In essence, the goal is to incorporate only the necessary characteristics to meet the requirement, not too many or too few. Thus, one must be careful not to overdesign or to underdesign.

5.2. TECHNICAL PERFORMANCE MEASURES

Technical performance measures refer to design-related factors, expressed quantitatively, that can be applied in the evaluation of a system and/or any of its components. With a design objective to develop a system that will perform its intended function in a cost-effective manner, one needs to recognize that there are many considerations that need to be addressed by both engineering and management alike. *All* elements of the system need to be covered on an *integrated* basis, and trade-offs are accomplished to seek a preferred approach leading to detail system design and definition (refer to Figure 4.10). Inherent in the trade-off analysis process is the use of various technical measures for evaluation.

When establishing evaluation measures, there are different applications depending on the characteristics of the system or product being evaluated. In general, however, reference can be made to Figure 4.11 as an example of a hierarchy of system parameters. Although there are many different levels of trade-offs performed, the ultimate criterion is some version of cost-effectiveness.[1]

Cost-effectiveness relates to the measure of a system in terms of mission fulfillment (system effectiveness) and total life-cycle cost and can be expressed in various terms, depending on the specific mission or system parameters that one wishes to evaluate. *True* cost-effectiveness is impossible to measure, since there are many factors that influence the operation and support of a system that cannot realistically be quantified (e.g., interaction effects as a result of other systems, political implications, certain environmental factors); thus, it is common to employ specific cost-effectiveness figures of merit (FOM), such as

$$\text{FOM} = \frac{\text{system benefits}}{\text{life-cycle cost}} \tag{5.1}$$

$$\text{FOM} = \frac{\text{system effectiveness}}{\text{life-cycle cost}} \tag{5.2}$$

$$\text{FOM} = \frac{\text{availability}}{\text{life-cycle cost}} \tag{5.3}$$

$$\text{FOM} = \frac{\text{system capacity}}{\text{life-cycle cost}} \tag{5.4}$$

$$\text{FOM} = \frac{\text{supply effectiveness}}{\text{life-cycle cost}} \tag{5.5}$$

These factors and others (which may represent a combination of system parameters) are often presented as delta values to allow the comparison of alternatives on the basis of the relative merits of each. Given two or more alternatives evaluated in a consistent manner, one can select the best among the alternatives based on these delta values.

The prime ingredients of cost effectiveness are illustrated in Figure 5.1. Although there are different ways of presenting cost effectiveness, this illustration is used for the purpose of showing the many influencing factors and their relationships. Design attributes influence both the system-effectiveness and life-cycle cost sides of the balance, and the results of design dictate the requirements for logistic support, which in turn affect both areas. For instance, the complexity of prime equipment design influences the personnel skill level and training requirements, which can in turn have a significant impact on both system effectiveness and the system operation and maintenance cost. The packaging scheme and standardization in design affect the

[1] Although industrial organizations as a whole rely primarily on an economic analysis approach for top-level management decision making (i.e., considering revenues and profit along with the cost), the cost-effectiveness concept discussed herein is more relevant in making specific design-related decisions. These decisions should then be evaluated in terms of their effects on revenues and profit with the necessary feedback and corrective action provisions incorporated to cover possible changes.

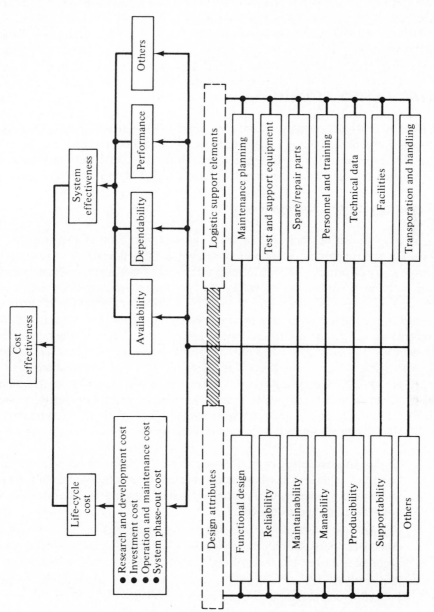

Figure 5.1 The elements of cost-effectiveness.

test and support equipment and the type and quantity of spare and repair parts and supporting inventories, which in turn affect both effectiveness and life-cycle cost. Similarly, the design of the various elements of logistic support (e.g., test and support equipment, facilities) has an impact on prime mission equipment. In essence, there are many system characteristics that interact, and the consequences of these interactions (as well as the effects of changes) must be evaluated and thoroughly understood.

5.3. DETAIL DESIGN ACTIVITIES

Once the goals and objectives have been established for detail design, it is necessary to organize a design team of engineering technical and nontechnical support personnel. From the activities of this design team will come detail design documentation and associated data bases developed with the support of various design aids.

Establishing the Design Team

It is necessary to organize and establish an integrated team of management, engineering, and supporting personnel (constituting the proper levels of expertise) for the purpose of performing the transformation functions from a defined "paper" entity in the preliminary system design stage to a prototype system configuration ready for production and/or construction.[2] In essence, a project organization must be established to accomplish detail design of equipment, software, and elements of logistic support as well as the follow-up design verification functions involving preliminary system test and evaluation. The project organization must be responsive to the system technical requirements covered in the detail plans and specifications.

As stated in Section 2.4, project organizations will vary with the system objectives. Some projects may involve thousands of engineers plus supporting personnel, while others may require only a few individuals. In any event, the project team at this stage will probably include a combination of

1. *Engineering technical expertise*—electrical engineers, mechanical engineers, computer engineers, civil engineers, nuclear engineers, system engineers, reliability and maintainability engineers, logistics engineers, and/or others as appropriate to the project needs.
2. *Engineering technical support*—draftsmen, technical publication specialists, component-parts specialists, laboratory technicians, model builders, computer programmers, test technicians, and the like.
3. *Nontechnical support*—marketing, purchasing and procurement, contracts, budgeting and accounting, legal, industrial relations, and so on.

[2] The design team effort identified here is not to be considered separate and aside from the system-level design activity required for conceptual or preliminary design. The overall design effort is continuous from one stage to the next, except that the level of emphasis shifts.

The project team, with expertise in a number of different areas (especially for large projects), must be properly integrated, and the individuals involved should be highly motivated to accomplish their respective assigned functions in a successful manner. This objective is attained through the proper organization and management of the resources made available. Organizational goals, project organizations, functions and tasks, and the associated management of project resources are covered further in Chapter 18.

Evolution of Detail Design

The evolution of design is illustrated basically in Figure 5.2 and amplified in Figure 5.3. The process shown in the figure is iterative, proceeding from system-level definition to an output that can be produced in single or multiple quantities. There are checks and balances in the form of reviews at each stage of design progression, and a feedback loop is provided for corrective action.[3] In this respect, the process is similar in concept to the system analysis, optimization, synthesis, and definition accomplished in the preliminary system design stage except that the requirements now are at a lower indenture level (i.e., subsystems, units, assemblies, subassemblies, etc.).

As detail design progresses, actual definition is accomplished through documentation in the form of specifications and plans (already discussed), procedures, drawings, material and part lists, reports and analyses, computer programs, and so on. The design configuration may be the best possible in the eyes of the designer. However, the results are practically useless unless properly documented, so that oth-

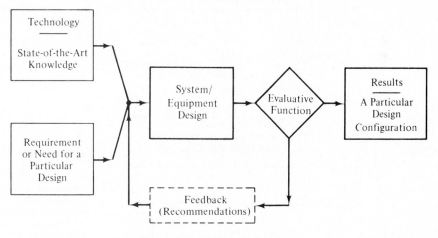

Figure 5.2   Evolution of design.

[3] The design reviews reflected in Figure 5.3 represent the day-to-day design progress in working from the system level down to the small-component level. These reviews are informal and occur continuously as compared to formal design reviews which are scheduled at specific times in the system life cycle.

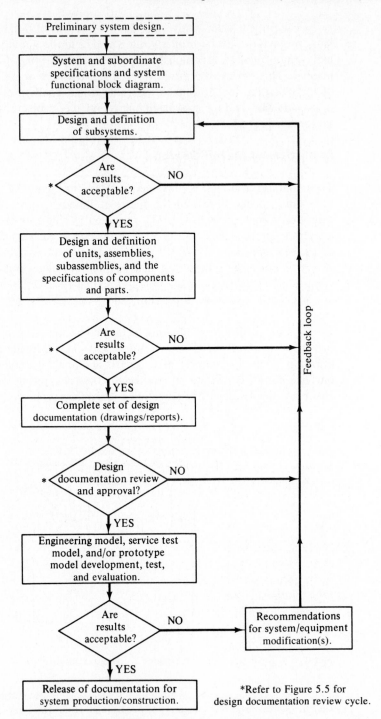

Figure 5.3 Basic design sequence.

ers can first understand what is being conveyed and then be able to translate the output into an entity that can be constructed or produced.

When addressing the aspect of design documentation, technological advances have promoted the use of electronic data bases for the purposes of information storage and retrieval. Through the use of computer-aided design techniques, information can be stored in the form of three-dimensional presentations, regular two-dimensional line drawings, in digital format, or combinations of these. By applying computer-graphics methods, word processor capabilities, video disc technology, and the like, design information can be presented faster, in greater detail, and in a more descriptive manner. The methods for documenting design are changing rapidly as a result of advances in information systems technology.

Although many advances have been made in the application of computerized methods to data acquisition, storage, and retrieval, there are still needs for some of the more conventional methods of design documentation. These include a combination of the following:

1. *Design drawings*—assembly drawings, control drawings, logic diagrams, installation drawings, schematics, and so on.
2. *Material and part lists*—part lists, material lists, long-lead-item lists, bulk-item lists, provisioning lists, and so on.
3. *Analyses and reports*—trade-off study reports supporting design decisions, reliability and maintainability analyses and predictions, human factors analyses, safety reports, logistic support analyses, configuration identification reports, computer documentation, installation and assembly procedures, and so on.[4]

Design drawings, constituting a primary source of definition, may vary in form and function depending on the design objective; that is, the type of equipment being developed, the extent of development required, whether the design is to be subcontracted, and so on. Some typical types of drawings are specified in Figure 5.4.

During the process of detail design, engineering documentation is rather preliminary, and then gradually progresses to the depth and extent of definition necessary to enable product manufacture. The responsible designer, using appropriate design aids, produces a functional diagram of the overall system. The system functions are analyzed and initial packaging concepts are assumed. With the aid of specialists representing various disciplines (i.e., electrical, mechanical, components, reliability, maintainability, etc.) and supplier data, detail design layouts are prepared for subsystems, units, assemblies, and subassemblies. The results are analyzed and evaluated in terms of functional capability, reliability, maintainability, human factors, safety, producibility, and other design parameters to assure compliance with the allocated requirements and the initially established design criteria. This review and

---

[4] Trade-offs and analyses are accomplished throughout the design process as discussed in Chapters 3 and 4 and Section 5.1. It is important that the results of these analyses be adequately documented to support design decisions. The reports identified here may be newly generated, or reports prepared during conceptual and preliminary system design that have been updated to reflect new information.

1. *Arrangement drawing* — shows in any projection or perspective, with or without controlling dimensions, the relationship of major units of the item covered.

2. *Assembly drawing* — depicts the assembled relationship of (a) two or more parts, (b) a combination of parts and subassemblies, or (c) a group of assemblies required to form the next higher indenture level of the equipment.

3. *Connection diagram* — shows the electrical connections of an installation or of its component devices or parts.

4. *Construction drawing* — delineates the design of buildings, structures, or related construction (including architectural and civil engineering operations).

5. *Control drawing* — an engineering drawing that discloses configuration and configuration limitations, performance and test requirements, weight and space limitations, access clearances, pipe and cable attachments, support requirements, etc., to the extent necessary that an item can be developed or procured on the commercial market to meet the stated requirements. Control drawings are identified as envelope control (i.e., configuration limitations), specification control, source control, interface control, and installation control.

6. *Detail drawing* — depicts complete end item requirements for the part(s) delineated on the drawing.

7. *Elevation drawing* — depicts vertical projections of buildings and structures or profiles of equipment.

8. *Engineering drawing* — an engineering document that discloses by means of pictorial or textual presentations, or a combination of both, the physical and functional end product requirements of an item.

9. *Installation drawing* — shows general configuration and complete information necessary to install an item relative to its supporting structure or to associated items.

10. *Logic diagram* — shows by means of graphic symbols the sequence and function of logic circuitry.

11. *Numerical control drawing* — depicts complete physical and functional engineering and product requirements of an item to facilitate production by tape control means.

12. *Piping diagram* — depicts the interconnection of components by piping, tubing or hose, and when desired, the sequential flow of hydraulic fluids or pneumatic air in the system.

13. *Running (wire) list* — a book-form drawing consisting of tabular data and instructions required to establish wiring connections within or between items.

14. *Schematic diagram* — shows, by means of graphical symbols, the electrical connections and functions of a specific circuit arrangement.

15. *Wiring and cable harness drawing* — shows the path of a group of wires laced together in a specified configuration, so formed to simplify installation.

Figure 5.4 Typical engineering drawing classifications.

evaluation occurs at each stage in the basic design sequence and generally follows the steps presented in Figure 5.5.

Engineering documentation is reviewed using checklist criteria developed from both program and specification requirements. If the review indicates complete compliance with these requirements, the documentation is considered approved on an

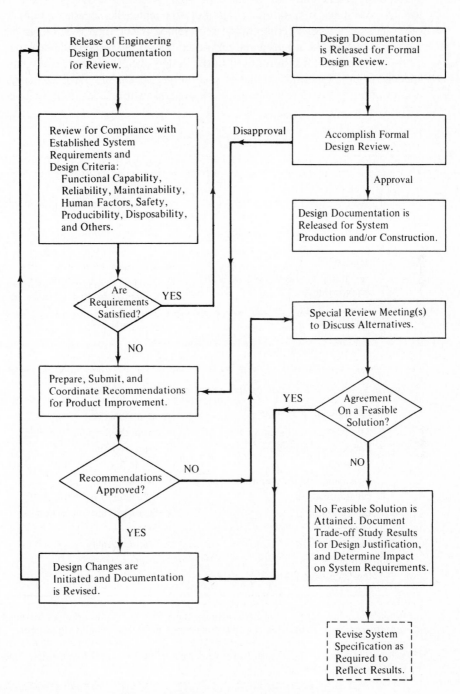

Figure 5.5 Design documentation review cycle.

initial basis and is ready for the formal design review process. Approval at this point usually constitutes signing off or initialing the applicable documentation. If, on the other hand, the design configuration represented is not approved, recommendations for corrective action are prepared and submitted to the responsible designer.

The checklist of criteria used in the design documentation review may be of a specific nature, quantitative and/or qualitative, emanating from the specifications and program plans, or may be of more general nature, constituting characteristics

---

System Design Review Checklist

General

1. System operational requirements defined
2. Effectiveness factors established
3. System maintenance concept defined
4. Functional analysis and allocation accomplished
5. System trade-off studies documented
6. System specification and supporting specifications completed
7. System engineering management plan completed
8. Design documentation completed
9. Logistic support requirements defined
10. Ecological requirements met
11. Societal requirements met
12. Economic feasibility determined

Design Features — Does the design reflect adequate consideration of

1. Accessibility
2. Adjustments and alignments
3. Cables and connectors
4. Calibration
5. Packaging and mounting
6. Disposability
7. Environment
8. Fasteners
9. Handling
10. Human factors
11. Interchangeability
12. Maintainability
13. Panel displays and controls
14. Producibility
15. Reliability
16. Safety
17. Selection of parts/materials
18. Servicing and lubrication
19. Software
20. Standardization
21. Supportability
22. Testability

When reviewing design (layouts, drawings, parts lists, computer graphics, engineering reports, program plans), this checklist may prove beneficial in covering various program functions and design features. The items listed can be supported with more detailed criteria like those included in Appendix B. The response to each item listed should be YES.

Figure 5.6   Sample design review checklist.

representative of desirable commonsense features in design. In any event, appropriate checklists are developed early in the design process to serve as an aid in engineering documentation review. Figure 5.6 constitutes an abbreviated sample checklist which can be supported by more in-depth criteria.[5]

5.4. ENGINEERING DESIGN TECHNOLOGIES

To design is to project and evaluate ideas for bringing a needed system or product into being. A design is an abstraction of what could be. For the design process to yield a good result, numerous abstractions (design alternatives) must be formulated and evaluated. This must be accomplished accurately, with relative ease, and in a limited period of time. The design engineer and/or the design team needs to assure a final design output which meets the need effectively and efficiently. A number of computer based technologies are emerging to facilitate this design assurance.

Computer-Aided Engineering Design

The designer must solve many different problems and evaluate numerous alternatives in a limited amount of time and still must define a total system or product that will meet all requirements effectively and efficiently in a highly competitive international environment. To meet this challenge requires utilization of a wide range of computer-based design aids. Generic categories of computer-aided methods include Computer-Aided Design (CAD), Computer-Aided Engineering Design (CAED), Computer-Aided Design/Computer-Aided Manufacturing (CAD/CAM), and Computer-Aided Acquisition and Logistic Support (CALS).

Computer usage in the design process is not new. It is used to generate drawings and graphic displays, to facilitate the accomplishment of analyses, to develop material lists, and so on. However, these various applications have been accomplished primarily on an individual basis. One computer program has been developed to produce design drawings; another program has been designed to aid in accomplishing a reliability analysis; a third program is used to generate component part lists; and so on. In essence, the expression relating to "islands of automation" addresses a reality. There are many different independent approaches intended to help solve a wide variety of problems. Further, the language requirements and program formats are different, and there is limited (if any) integration of these design methods. This approach can be quite costly overall.

The application of CAD, CAM, CALS, and related tools promotes the integration of these various methods into a single entity. These design-related aids must be readily available and easy to use, compatible with each other facilitating the flow of information in a timely manner, and adaptable to the system or product design configuration at hand. The overall concept illustrated in Figure 5.7 reflects some of the major interfaces that must be addressed.

[5] An example of a detail design checklist is included in Appendix B of this text. Another example is included in B. S. Blanchard, *Logistics Engineering And Management*, 3rd ed., (Englewood Cliffs, N.J.: Prentice-Hall, Inc., 1986) Appendix E.

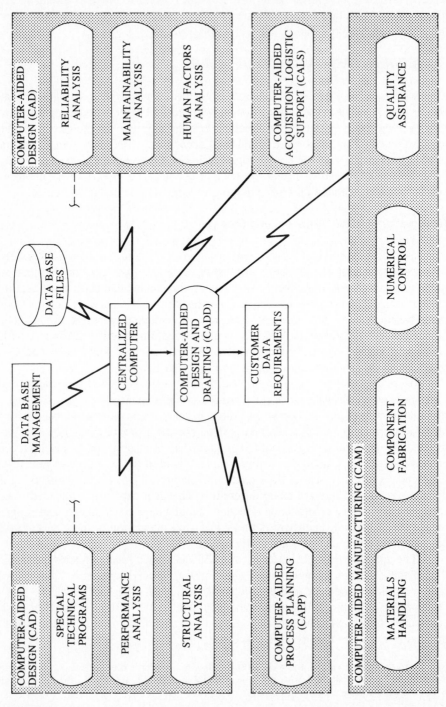

Figure 5.7 An example of CAD, CAM, and CALS in interface.

Referring to the figure, the objective of CAD is to allow for the accomplishment of design analyses, the projection of various design configurations in terms of three-dimensional graphical presentations, the accomplishment of reliability and maintainability predictions, the generation of design material lists, etc. There are many individual problems that require resolution and, through CAD, are integrated into the overall design process. The results are projected through the use of electronic data bases and, subsequently, in the form of hard drawings using CAD.

The results of CAD, in terms of components that must be manufactured, should feed directly into CAM. In other words, components that have been designed through CAD and must be produced should be addressed using CAM. Computer-Aided Process Planning (CAPP) is used in responding to the CAD output with the objective of determining specific process requirements. Again, the information flow should be continuous, providing the necessary continuity between design activities and production activities.

CALS, which primarily addresses the logistic support aspects of the system, is oriented both to system design and production. The objective of CALS is to provide a vehicle for design evaluation, as well as a tool for the processing of logistics data. CALS is addressed further in Chapter 16.

The use of CAD, CAM, and CALS methods, implemented on an integrated basis, offers many advantages. A few of these are the following:

1. The designer can address many different alternatives in a relatively short time frame. With the capability of evaluating a greater number of possible options, the risks associated with design decision making are reduced.
2. The designer is able to simulate, and verify design, for a greater number of configurations by using three-dimensional projections. Employing an electronic data base to facilitate this objective may eliminate the need to build a physical model later on, thus reducing cost.
3. The ability to incorporate design changes is enhanced, both in terms of the reduced time for accomplishing them and in the accuracy of data presentation. If the information flow is properly integrated, changes will be reflected throughout the data base as applicable. Also, the vehicle for incorporating a change is greatly simplified.
4. The quality of design data is improved, both in terms of the methods for data presentation and in the reproduction of individual data elements.
5. The availability of an improved data base earlier in the system life cycle (with better methods for presenting design configurations) facilitates the training of personnel assigned to the project. Not only is it possible to better describe the design through graphical means, but a common data base will help to ensure that all design activities are tracking the same baseline.

Morphology for Concurrent Engineering Design

Incorporating quality features into the design early in the life cycle is a multidisciplinary effort. It is critical to establish an engineering team with a representative from each functional area. The key, however, is not a continuation of the sequential approach for design, but a concurrent approach. The multi-disciplinary team should have the responsibility of analyzing and updating the algorithms and tools incorporated into the computer-aided design process. This is dependent upon are advanced computer-aided workstation environment incorporating the elements illustrated in Figure 5.8.[6]

The numbered blocks in Figure 5.8 may be briefly described as follows:

Block 1: The human element of the design system (an individual or group of individuals) links with CAD and/or CAE design tools via a human computer interface (HCI).

Block 2: The CAD and/or CAE design generator provides the framework and facilitates the process by which the designer creates a design and determines its features.

Block 3: Design features (conceptual, preliminary, or detailed) are passed to estimators and predictors where design dependent parameter (Y_d) values are determined by estimation and prediction.

Block 4: Evaluators provide the designers, via a human computer interface, an evaluation (E) of the design at its current iteration or state of development.

Block 5: A data base of economic and physical factors is maintained as a source of design independent parameters (Y_i) (interest rates, labor rates, material costs, physical properties, part lists, etc.).

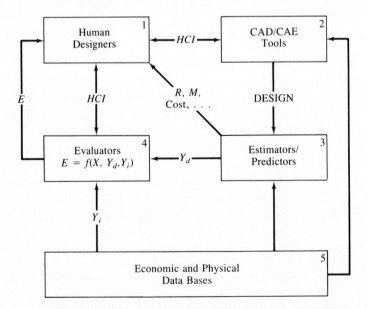

Figure 5.8   A morphology for computer-aided engineering design.

[6] Full realization of computer-aided design generation and evaluation, as structured in Figure 5.8, requires further progress in the interfacing and integration of the many tools named in Figure 5.7.

Design Evaluation Methodology

Most computer-aided design processes provide separate evaluations of effectiveness and cost.[7] It is usually assumed that performance measures are to be satisfied minimally, with attention then focused on reducing life-cycle cost as is illustrated in Figure 5.9.

Application of the above approach in design requires consideration of the following:

1. Each design alternative will have a calculated performance measure, as well as an associated life-cycle cost.
2. As design iterations are carried out, a pairwise comparison of design alternatives may be performed, and the alternative that meets the performance requirements at the lowest life-cycle cost will be best.

It is essential that engineers assume the responsibility to design systems and products that satisfy performance and quality requirements at minimum cost. A means to help them meet this responsibility are computer-aided concurrent engineer-

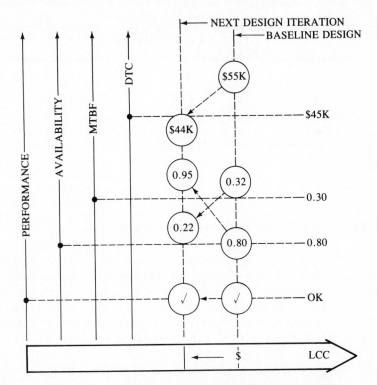

Figure 5.9   Performance requirements and LCC (example values).

---

[7] Figure 5.1 illustrates the relationship of cost and effectiveness along with performance measures.

ing design approaches incorporating multiple life-cycle considerations (refer to Figure 2.3).

In summary, an integrated life-cycle approach to the application of computer-aided methods is required. One way of reflecting such an approach is through the concept of MacroCAD illustrated in Figure 5.10. The objective of MacroCAD is to link early life-cycle design decisions with operational outcomes, using simulation methods and mathematical optimization. This approach is feasible because of the large body of knowledge available about operational modeling and simulation. Key is the identification of design dependent system parameters and optimization over design variables. For each set of design dependent system parameters, one can identify an optimal value for the chosen measure of evaluation. By comparing these optimal values for a set of alternatives, a preferred configuration can be selected. The mathematical and algorithmic basis for MacroCAD is developed further in Part Three of this text.

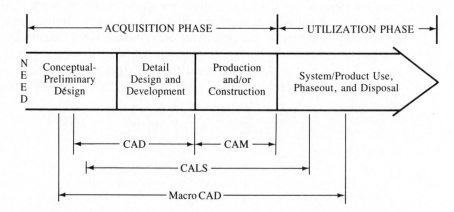

Figure 5.10   The relationship of CAD, CAM, CALS, and MacroCAD.

Design Standards and Documentation

The development and use of design standards and/or design criteria documentation constitutes another form of design aid. Throughout the design process, there are numerous occasions where guidance is required in terms of a given design approach, the selection of an appropriate component part, the reliability and maintainability characteristics of an item, and so on. Design documentation, to be available for reference on a day-to-day basis, may be categorized as follows:

1. *Design standards documentation*—Design standards manuals are developed to cover preferred component parts and supplier data, mathematical and statistical tables, stress-strength tables, engineering drawing practices, and the like.
2. *Design criteria documentation*—Individual criteria documentation is often de-

veloped to cover areas such as reliability, maintainability, human factors, and safety. Documentation here relates to design features to include accessibility, diagnostic and testing provisions, packaging and mounting techniques, interchangeability, controls and panel display methods, human performance capabilities, reliability and safety characteristics, and so on.

Physical Models and Mock-ups

Quite often the design of an item is difficult to visualize in its true perspective when relying on design data alone. As an aid to the designer, three-dimensional scale models or mock-ups are sometimes developed to provide a realistic simulation of a proposed final system or equipment configuration at an early point in the program prior to the development of final design data or actual prototype hardware.

Models or mock-ups can be developed to any desired scale and to varying depths of detail depending on the level of emphasis required. Mock-ups can be developed for large systems as well as small systems and may be constructed of heavy cardboard, wood, metal, or a combination of materials. Mock-ups can be developed on a relatively inexpensive basis and in a short period of time when employing the right materials and personnel services (industrial design, human factors, and/or model shop personnel are usually well oriented to this area and should be utilized to the greatest extent possible). The uses and values of a mock-up are numerous.

1. It provides the design engineer with the opportunity of experimenting with different facility layouts, packaging schemes, panel displays, cable runs, and so on, prior to the preparation of final design data.
2. It provides the reliability-maintainability-human factors engineer with the opportunity to accomplish a more effective review of a proposed design configuration for the incorporation of supportability characteristics. Problem areas readily become evident.
3. It provides the maintainability-human factors engineer with a tool for use in the accomplishment of predictions and detailed task analyses. It is often possible to simulate operator and maintenance tasks to acquire task sequence and time data.
4. It provides the design engineer with an excellent tool for conveying his or her final design approach during a formal design review.
5. It serves as an excellent marketing tool.
6. It can be employed to facilitate the training of system operator and maintenance personnel.
7. It is utilized by production and industrial engineering personnel in developing fabrication and assembly processes and procedures and in the design of factory tooling and associated test fixtures.
8. At a later stage in the system life cycle, it may serve as a tool for the verification of a modification kit design prior to the preparation of formal data and the development of kit hardware.

5.5. SYSTEM PROTOTYPE DEVELOPMENT

Thus far, Part Two has primarily covered system design from the standpoint of a theoretical entity supported by concepts, analyses, drawings, and related documentation. The design function has been able to establish specific system objectives and subsequently assess a given design configuration relative to compliance with these objectives. The assessment up to this point has been analytical in nature, providing a certain level of confidence that all qualitative and quantitative requirements have been met. Although the analytical approach fulfills a definite need throughout the total design process, there is also a need to verify one's concepts and design configuration through the use of actual system components as soon as practicable. In other words, an objective in the detail design phase is to develop hardware, software, and the appropriate elements of logistic support; combine and integrate these elements into the proper system configuration (to the extent possible); and accomplish the necessary test and evaluation to physically demonstrate that system requirements have been met.

Given that design has progressed to the point where definition can be attained through formalized specifications and supporting documentation, the applicable design may be evaluated through the construction and use of physical working models of one type or another. Categories of models are noted below.

1. *Engineering model* represents a working system, or an element of the system, that will exhibit the functional performance characteristics defined in the specification. An engineering model may be developed in either the preliminary system design stage or in the system detail design stage, and is employed primarily to verify the technical feasibility of an item. It does not necessarily represent the system in terms of physical dimensions (i.e., form or fit).

2. *Serivce test model* represents a working system, or an element of the system, that reflects the end product in terms of functional performance and physical dimensions. It is like the operational configuration except that the same component parts may not be incorporated (i.e., substitute components may be used as long as the system functions properly). The service test model may be developed in the detail design stage to verify the functional performance-physical configuration interface.

3. *Prototype model* represents the production configuration of a system in all aspects of form, fit, and function except that it has not been fully qualified in terms of operational and environmental testing. A prototype may evolve through a series of working hardware configurations such as an engineering model and a service test model, or may evolve directly from a mock-up. A prototype of the prime equipment, software, and associated elements of logistic support is produced in the detail design stage for the purpose of final system/equipment test and evaluation prior to entering into the production and/or construction phase. The intent is to verify design adequacy to the maximum extent considered appropriate at this stage in the life cycle.

As part of the overall design process, engineering evaluation functions include the use of engineering models, service test models, and/or prototype models to assist in the verification of technical concepts and various system design approaches. Areas of noncompliance with the specified requirements are identified and correction action is initiated as required.

5.6. FORMAL DESIGN REVIEW

Although the quantity, type, and scheduling of design reviews may vary from program to program, four basic types are common to most programs. They include the conceptual design review discussed in Chapter 3, the system design review discussed in Chapter 4, and the equipment and critical design reviews covered herein. Figure 3.9 illustrates these different reviews in terms of the phases in the system life cycle.

Equipment design reviews are scheduled during the detail design phase when layouts, preliminary mechanical and electrical drawings, functional and logic diagrams, computer data, and component-parts lists are available. In addition, these reviews cover engineering models, service test models, and prototypes. Supporting the design are detail trade-off study reports, reliability analyses and predictions, maintainability analyses and predictions, human factors analyses, and logistic support analyses. The design process at this point has identified specific design constraints, additional or new requirements, and major problem areas. Such reviews are conducted prior to proceeding with finalization of the detail design.

The *critical design review* is scheduled after detail design has been completed but prior to the release of firm design data to production. Such a review is conducted to verify the adequacy and producibility of the design. Design is essentially frozen at this point, and manufacturing methods, schedules, and costs are reevaluated for final approval. The critical design review covers all design effort accomplished subsequent to the completion of the equipment reviews. This includes changes resulting from recommendations for corrective action stemming from the equipment design review and prototype testing. Data requirements include manufacturing drawings and material lists, final reliability and maintainability predictions, engineering test reports, a firm logistic suport analysis, a production/construction plan, a product evaluation plan, and a product use and logistic support plan.

The success of a formal design review is dependent on the depth of planning, organization, and data preparation prior to the review itself. A tremendous amount of coordination is required regarding

1. The items to be reviewed.
2. A selected date for the review.
3. The location or facility where the review is to be conducted.
4. An agenda for the review (including a definition of the basic objectives).
5. A design review board representing the organizational elements and disciplines

affected by the review. Basic design functions, reliability, maintainability, human factors, quality control, manufacturing, and logistic support representation are included. Individual organization responsibilities should be identified. Depending on the type of review, the consumer and/or individual equipment suppliers may be included.

6. Equipment (hardware) requirements for the review. Engineering prototypes and/or mock-ups may be required to facilitate the review process.

7. Design data requirements for the review. This may include all applicable specifications, lists, drawings, predictions and analyses, logistic data, and special reports.

8. Funding requirements. Planning is necessary in identifying sources and a means for providing the funds for conducting the review.

9. Reporting requirements and the mechanism for accomplishing the necessary follow-up actions stemming from design review recommendations. Responsibilities and action-item time limits must be established.

The design review involves a number of different discipline areas and covers a wide variety of design data and in some instances hardware. In order to fulfill its objective expeditiously (i.e., review the design to ensure that all system requirements are met in an optimum manner), the design review must be well organized and firmly controlled by the design review board chairman. Design review meetings should be brief and to the point, and must not be allowed to drift away from the topics on the agenda. Attendance should be limited to those who have a direct interest and can contribute to the subject matter being presented. Specialists who participate should be authorized to speak and make decisions concerning their area of specialty. Finally, the design review must make provisions for the identification, recording, scheduling, and monitoring of corrective actions. Specific responsibility for follow-up action must be designated by the chairman of the design review board.

SELECTED REFERENCES

(1) Beakley, G. C., D. L. Evans, and J. B. Keats, *Engineering—An Introduction to a Creative Profession* (3rd ed.). New York: Macmillan, 1986.

(2) Blanchard, B. S., *Engineering Organization and Management*. Englewood Cliffs, N.J.: Prentice-Hall, 1976.

(3) Blanchard, B. S., *Logistics Engineering and Management* (3rd ed.). Englewood Cliffs, N.J.: Prentice-Hall, 1986.

(4) Chase, W. P., *Management of Systems Engineering*. New York: John Wiley, 1974.

(5) Chestnut, H., *Systems Engineering Methods*. New York: John Wiley, 1967.

(6) Defense Systems Management College (DSMC), *Systems Engineering Management Guide*. Fort Belvoir, Virginia, December 1986.

(7) Dieter, G. E., *Engineering Design: A Materials and Processing Approach*. New York: McGraw-Hill, 1983.

(8) Eisner, H., *Computer-Aided Systems Engineering*. Englewood Cliffs, N.J.: Prentice-Hall, 1988.

(9) Krouse, J. K., *What Every Engineer Should Know About Computer-Aided Design and Computer-Aided Manufacturing*. New York: Marcel Dekker, 1982.

(10) Ostrofsky, B., *Design, Planning, and Development Methodology*. Englewood Cliffs, N.J.: Prentice-Hall, 1977.

(11) Teicholz, E., ed., *CAD/CAM Handbook*. New York: McGraw-Hill, 1985.

(12) Vick, C. R., and C. V. Ramamoorthy, eds., *Handbook of Software Engineering*. New York: Van Nostrand Reinhold, 1984.

(13) Woodson, T. T., *Introduction to Engineering Design*. New York: McGraw-Hill, 1966.

QUESTIONS AND PROBLEMS

1. What are the basic differences among conceptual system design, preliminary system design, and detail design? Are these stages of design applicable to all systems?

2. Describe in brief terms how each of the following affects system design: reliability and maintainability, human factors, safety, supportability, producibility, life-cycle cost. How can these areas be expressed quantitatively (if at all)?

3. Define cost-effectiveness and system-effectiveness. How would you specify and measure each?

4. What impact could the logistic support capability for a given system have on the prime mission equipment operation?

5. What is meant by life-cycle cost? What aspects of cost are considered? How is life-cycle cost used in the design process?

6. Select a system (or element of a system) of your choice and develop a design review checklist that you can use for evaluation purposes.

7. Design is a team effort. True or false? Why?

8. Describe the checks and balances of the design process as you see them.

9. How does system analysis fit into the overall design process? Consider conceptual design, preliminary system design, and detail design.

10. Why is engineering documentation necessary?

11. Why are design standards important?

12. Define computer-aided design (CAD), computer-aided manufacturing (CAM), and computer-aided acquisition and logistic support (CALS). Provide some examples. List some advantages to this approach. List some disadvantages.

13. What are some of the benefits of physical models and/or mock-ups?

14. What benefits are provided by a formal design review?

15. What are the input requirements for the conceptual design review? System design review? Equipment design review? Critical design review?

16. What determines the success of a design review?

17. Refer to the definition of systems engineering in Chapter 2. How does systems engineering apply in the detail design stage? What activities are necessary (if any)?

6

System Test and Evaluation

Evaluation refers to the examination and judgment of a system (or an element of a system) in terms of worth, quality of performance, degree of effectiveness, condition, and the like. Evaluation is a continuous process which begins during conceptual design and extends through the product use and logistic support phase until the system is retired. The purpose is to determine (through a combination of prediction, analysis, and measurement) the *true* characteristics of the system and to ensure that it successfully fulfills its intended mission.

The primary objective of this chapter is to address the measurement and evaluation aspect of the design process as depicted by the last function in Block 3, Figure 2.4. Referring to the figure, one initially defines requirements (Block 1), develops a system (Blocks 2 and 3), and measures the results (Block 3). If the output is in compliance with the initially specified requirements, then the production/construction stage may begin; if not, corrective action is required. System evaluation and its relationship(s) with system requirements definition and with system development is illustrated in Figure 6.1.

6.1. REQUIREMENTS FOR TEST AND EVALUATION

The specific needs for test and evaluation are intially defined during conceptual design when the requirements for the overall system are established. As system-level requirements are identified through feasibility studies, the definition of operational

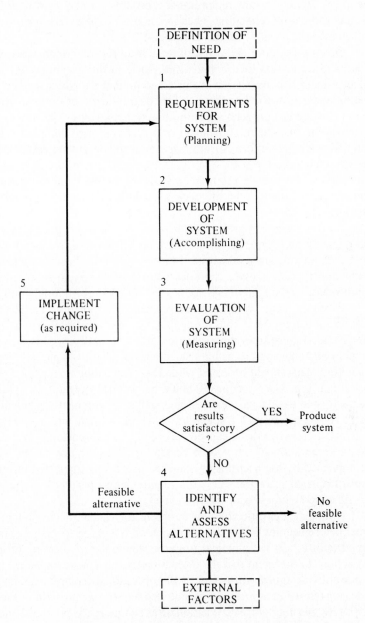

Figure 6.1 System requirements and evaluation.

requirements and the maintenance concept, etc., a method must be established for evaluation and the subsequent determination of whether these requirements have been met. Thus, as a new requirement is established, the question is how will we determine if the requirement has been met and what method of measurement should be implemented to verify that this is the case.

Given this initial definition of system-level requirements, an ongoing iterative process of evaluation then commences. Early in the system design process, analytical techniques may be used to evaluate and predict the anticipated characteristics of a system being brought into being. The definition of operational requirements and maintenance concept, accomplishing functional analyses, performing trade-offs and optimization, and the establishment of a preferred configuration is a continuing design and evaluation effort. Various aspects of the design are reviewed on the basis of the specified requirements, and corrective action is initiated in areas where non-responsiveness to these requirements and/or product improvement are indicated. Changes at this stage in the life cycle are relatively easy to incorporate, and the costs are generally low.

As the design progresses from paper definition to the point where mock-ups and engineering service test models are developed, the evaluation process becomes more meaningful, since actual hardware is available. Further, when prototypes and production models evolve, the effectiveness of the evaluation assumes even more significance, as the system approaches its intended operational configuration.

When viewing overall test requirements, it should be realized that a true test (that which is relevant from the standpoint of assessing total system performance and effectiveness) constitutes the evaluation of a system deployed in an operational environment and subjected to actual use conditions. For example, an aircraft or a power generating plant should be tested while it is performing its intended mission in actual operational use. User personnel should accomplish operator and maintenance functions with the designated field test and support equipment, technical manual procedures, and so on. In such a situation, actual operational and maintenance experience in a realistic environment can be recorded and subsequently evaluated to reflect a true representation of the system design. A demonstration of this type can be best accomplished by the user during standard operations supported through the employment of normal resources (i.e., the requirements specified for the product use and logistic support function of the life cycle).

Although idealistically it is desirable to wait until the system is fully operational before accomplishing an evaluation of system performance, effectiveness, and supportability, it is not practical from the standpoint of allowing for possible corrective action. In the event that the evaluation indicates noncompliance (i.e., the system as presently designed will not meet the operational requirements and fulfill its mission), corrective action should be initiated as early as possible in the system life cycle. Accomplishing corrective action after equipment is produced and fully deployed in the field can result in extensive modification programs, which are costly. Thus, it is feasible to establish an overall test program which allows for the evaluation of hardware, software, and its support elements on an evolutionary basis. As is the case for the earlier design evaluation using analytical techniques, an evaluation of hardware is accomplished, beginning with the development of the first engineering model and extending through the test of equipment deployed in the field. This evaluation process includes various types of demonstrations.

6.2. CATEGORIES OF TEST AND EVALUATION

The evolution of system evaluation is illustrated in Figure 6.2. Various categories of testing are identified by program phase, and the effectiveness of the evaluation effort increases when progressing through Type 1, 2, 3, and 4 testing. These steps are necessary to ensure, on a continuing basis, that system requirements are being achieved. Incorporating changes after the system is in operational use can be extremely costly to both the producer and consumer; thus, it is preferable to eliminate (or avoid) potential problems as early as possible in the life cycle through the best means available.[1]

Type 1 Testing

During the early phases of detail design, breadboards, bench-test models, engineering models, engineering software, and service test models are built with the intent of verifying certain performance and physical design characteristics. These models, representing either an entire system or a designated system component, usually operate functionally (electrically and mechanically) but do not by any means represent production equipment.[2]

Tests may involve equipment operational and logistic support actions which are directly comparable to tasks performed in a true operational situation (e.g., measuring a performance parameter, accomplishing a remove-replace action, accomplishing a servicing requirement). Although these tests are not formal demonstrations in an operational environment, information pertinent to actual system characteristics can be derived and used as an input in the overall system evaluation and assessment. Such testing is usually performed in the producer/supplier's facility by engineering technicians using "jury-rigged" test equipment and engineering notes for test procedures. It is during this initial phase of testing that changes to the design can be incorporated on a minimum-cost basis.

---

[1] The scope and depth of testing will depend on the type of system, whether new design and development are required, and the risks involved. Comprehensive testing may be required where design is new and many unknowns exist, whereas off-the-shelf, proven commodities will not require much testing except to assure the proper integration with other elements of the system. Not too much or too little testing should be accomplished. Thus, good test planning is essential.

[2] Individual components are frequently tested in lieu of testing the entire system. For instance, it may not be feasible to test an entire petroleum distribution system, whereas testing a segment of pipeline may be sufficient to verify design adequacy. The objective at this point is to verify design adequacy to the maximum extent practicable in view of the information obtainable via economical means.

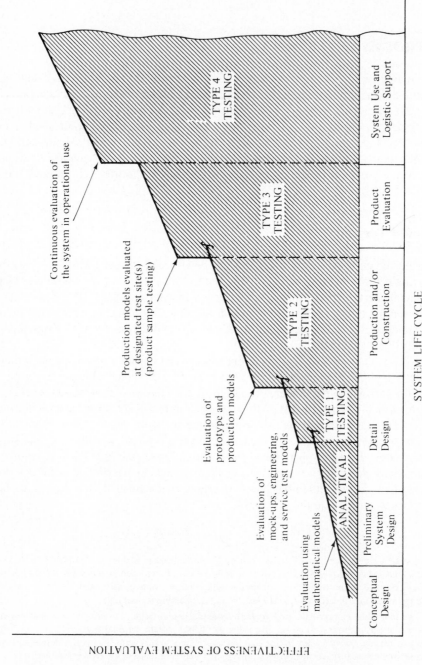

Figure 6.2 Stages of system evaluation during the life cycle.

SYSTEM LIFE CYCLE

EFFECTIVENESS OF SYSTEM EVALUATION

Type 2 Testing

Formal tests and demonstrations are accomplished during the latter part of detail design when preproduction prototype equipment is available. Prototype equipment is similar to production equipment (that which will be deployed for operational use), but it is not necessarily fully qualified at this point in time.[3] A test program may constitute a series of individual tests tailored to the need; such a program might include the following:

1. *Performance tests*—Tests are accomplished to verify individual system performance characteristics. For instance, tests are designed to determine whether the electric motor will provide the necessary output, whether the pipeline will withstand certain fluid pressures, whether the airplane will perform its intended mission successfully, whether the process will provide x widgets per given period of time, and so on. Also, it is necessary to verify form, fit, interchangeability, product safety, and other comparable features.

2. *Evnironmental qualification*—Temperature cycling, shock and vibration, humidity, wind, salt spray, dust and sand, fungus, acoustic noise, pollution emission, explosion proofing, and electromagnetic interference tests are conducted. These factors are oriented to what the equipment will be subjected to during operation, maintenance, and during transportation and handling functions. In addition, the effects of the item being tested on the overall environment will be noted.

3. *Structural tests*—Tests are conducted to determine material characteristics relative to stress, strain, fatigue, bending, torsion, and general decomposition.

4. *Reliability qualification*—Tests are accomplished on one or more equipment items to determine the mean time between failure (MTBF) and the mean time between maintenance (MTBM). Also, special tests are often designed to measure component life, to evaluate degradation, and to determine modes of failure.[4]

5. *Maintainability demonstration*—Tests are conducted on one or more equipment items to determine the values for mean active maintenance time ($\overline{M}$), mean corrective maintenance time ($\overline{M}ct$), mean preventive maintenance time ($\overline{M}pt$), maintenance labor-hours per operation hour (MLH/OH), and so on. In addition, maintenance tasks, task times and sequences, prime equipment-test equipment interfaces, maintenance personnel quantities and skills, maintenance procedures, and maintenance facilities are verified to varying degrees. The elements of logistic support are initially evaluated on an individual basis.[5]

6. *Support equipment compatibility tests*—Tests are often accomplished to verify

---

[3] Qualified equipment refers to a production configuration that has been verified through the successful completion of performance tests, environmental qualification tests (e.g., temperature cycling, shock, vibration), reliability qualification, maintainability demonstration, and compatibility tests. Type 2 testing primarily refers to that activity associated with the qualification of a system for operational use.

[4] Reliability testing is covered further in Chapter 13.

[5] Maintainability demonstration is covered further in Chapter 14.

the compatibility among the prime equipment, test and support equipment, and transportation and handling equipment.

7. *Personnel test and evaluation*—Tests are often accomplished to verify the relationships between people and equipment, the personnel skill levels required, and training needs. Both operator and maintenance tasks are evaluated.

8. *Technical data verification*—The verification of operational and maintenance procedures is accomplished.

9. *Software verification*—The verification of operational and maintenance software is accomplished. This includes hardware-software compatibility, software reliability and maintainability, and packaging and documentation.

The ideal situation is to plan and schedule these individual tests such that they can be accomplished on an integrated basis as one overall test. Data output from one test may be beneficial as an input to another test. The intent is to provide the proper emphasis, consistent with the need, and eliminate redundancy and excessive cost. Proper test planning is essential.

Another aspect of testing in this category involves production sampling tests when multiple quantities of an equipment item are produced. The tests defined above basically qualify the item; that is, the equipment hardware configuration meets the requirements for production and operational use. However, once a system is initially qualified, some assurance must be provided that all subsequent replicas of that item are equally qualified; thus, in a multiple-quantity situation, samples are selected from the production line and tested.[6]

Production sampling tests may cover certain critical performance characteristics, reliability, or any other designated parameter that may significantly vary from one serial-numbered item to the next, or may vary as a result of a production process. Samples may be selected on the basis of a percentage of the total equipment produced or may tie in with x number of pieces of equipment in a given calendar time period. This depends on the peculiarities of the system and the complexities of the production process. From production sampling tests, one can measure system growth (or degradation) throughout the production/construction phase.

Type 2 tests are generally performed in the producer/supplier's facility by personnel at that facility. Test and support equipment, designated for operational use, and preliminary technical manual procedures are employed where possible. User personnel often observe and/or participate in the testing activities. Equipment changes as a result of corrective action are handled through a formalized engineering change procedure.

Type 3 Testing

Formal tests and demonstrations, started after initial system qualification and prior to the completion of production, are accomplished at a designated field test site by user

---

[6] These tests are in addition to the normal performance tests that are accomplished on every piece of equipment after fabrication and assembly and prior to delivery to the consumer.

personnel.[7] Operational test and support equipment, operational spares, and formal operator and maintenance procedures are used. Testing is generally continuous, accomplished over an extended period of time, and covers the evaluation of a number of system elements scheduled through a series of simulated operational exercises.

This is the first time that all elements of the system (i.e., prime equipment and the elements of logistic support) are operated and evaluated on an integrated basis. The compatibility of the prime equipment with logistic support is verified as well as the compatibility of the various elements of logistic support with each other. Turnaround times and logistics transportation times, stock levels, personnel effectiveness factors, and other related operational and logistic parameters are measured. In essence, system performance (based on certain use conditions) and operational readiness characteristics (i.e., operational availability, dependability, system effectiveness, etc.) can be determined to a certain extent.[8]

Type 4 Testing

During the product-use phase, formal tests are sometimes conducted to gain further insight in a specific area. It may be desirable to vary the mission profile or the system utilization rate to determine the impact on total system effectiveness, or it might be feasible to evaluate several alternative support policies to see whether system operational availability can be improved. Even though the system is designed and operational in the field, this is actually the first time that we really know its true capability. Hopefully, the system will accomplish its objective in an efficient manner; however, there is still the possibility that improvements can be realized by varying basic operational and maintenance support policies.

Type 4 testing is accomplished at one or more operational sites (in a realistic environment) by user operator and maintenance personnel, and supported through the normal logistics capability. The elements of logistic support as designated through predictions and analyses, earlier testing, and so on, are evaluated in the context of the total system.

6.3. *FORMAL TEST PLANNING*

Test planning actually begins as part of the advance system planning activity in conceptual design (refer to Chapter 3). If a system requirement is to be specified at that point, there must be a way to evaluate the system later to ensure that the requirement has been met; hence, testing considerations are intuitive at an early point in time.

Throughout the various stages of system development, a number of individual subsystem or equipment tests may be specified. Often there is a tendency to design a test to measure one system characteristic, design another test to measure a different

[7] The test site may constitute a ship at sea, an aircraft or space vehicle in flight, a facility in the arctic or located in the middle of the desert, or a mobile land vehicle traveling between two points.

[8] Type 3 testing does not represent a complete operational situation; however, tests can be designed to provide a close approximation.

parameter, and so on. However, the amount of testing specified may be overwhelming and prove to be quite costly. Test requirements must be considered on an integrated basis. Where possible, individual tests are reviewed in terms of resource requirements and output results and are scheduled in such a manner as to gain the maximum benefit possible. For instance, maintainability data can be obtained from reliability tests that result in a possible reduction in the amount of maintainability testing required. Support equipment compability data and personnel data can be obtained from both reliability and maintainability testing; thus, it might be feasible to schedule reliability qualification testing first and maintainability demonstration second. In some instances, the combining of tests may be feasible as long as the proper characteristics are measured and the data output is compatible with the initial testing objectives.

For each program, an integrated test and evaluation plan is prepared, usually for implementation beginning in the preliminary design phase.[9] Although the specific content may vary somewhat depending on system requirements, the plan will generally include

1. The definition and schedule of all test requirements, including anticipated test output (in terms of what the test is to accomplish) for each individual test and integrated where possible. In determining test requirements, some components of the system may go through each category of the test, while other components may undergo only a very limited amount of testing. This is a function of the degree of design definition and the risks associated with each item in question. An example approach is illustrated in Figure 6.3.
2. The definition of organization, administration, and control responsibilities (organization functions, organizational interfaces, monitoring of test activities, cost control, and reporting).
3. The definition of test conditions and logistic resource requirements (test environment, facilities, test and support equipment, spare and repair parts, test personnel, software, and test procedures).
4. A description of the test preparation phase for each type of testing (selection of specific test method, training of test personnel, acquisition of logistic resource requirements, and preparation of facilities).
5. A description of the formal test phase (test procedures and test data collection, reduction, and analysis methods).
6. A description of conditions and provisions for a retest phase (methods for conducting additional testing as required due to a reject situation).
7. The identification of test documentation (test reporting requirements).

[9] In the defense sector, the Test and Evaluation Master Plan (TEMP) is prepared during conceptual design and usually includes coverage of test objectives, technical and operating characteristics of the system, critical issues and interfaces, developmental test and evaluation requirements, operational test and evaluation requirements, test resource requirements, test procedures, and test reporting.

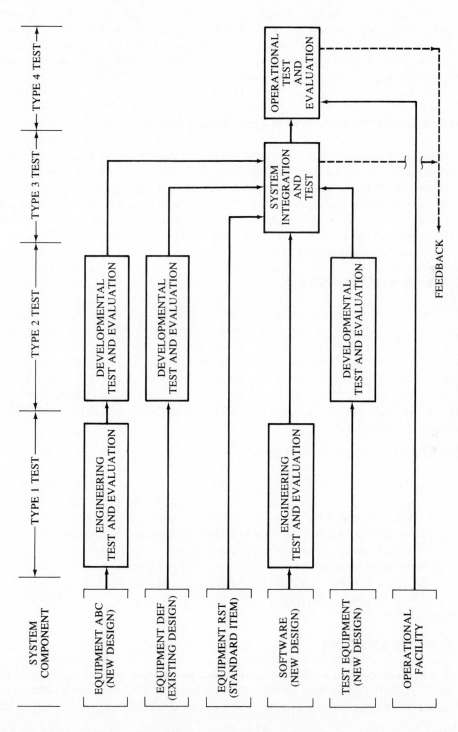

Figure 6.3 Evolution of test requirements.

The basic test plan serves as a valuable reference throughout system design, development, and production. It indicates what is to be accomplished, the requirements for testing, a schedule for the processing of equipment and material for test support, data collection and reporting methods, and so on. It is an integrating device for a number of individual tests in various categories, and a change in any single test requirement will likely affect the total plan.

6.4. PREPARATION FOR SYSTEM TEST AND EVALUATION

After initial planning, and prior to the start of formal evaluation, a period of time is set aside for test preparation. During this period, the proper conditions must be established to ensure effective results. Although there is some variance, depending on the type of evaluation, these conditions or prerequisites include the selection of the item(s) to be tested, establishment of test procedures, test site selection, selection and training of test personnel, preparation of test facilities and resources, the acquisition of support equipment, and test supply support.

Selection of Test Item(s)

The equipment configuration used in the test must be representative of the operational item to the maximum extent possible. For Type 1 tests, engineering models are used that are not often directly comparable with operational equipment; however, most subsequent testing will be accomplished at such a time that equipment (or software) representing the final configuration is available. A prerequisite task involves selecting the test model by serial number, defining the configuration in terms of incorporated versus unincorporated engineering changes (if any), and ensuring that it is available at the time needed.

Test and Evaluation Procedures

Fulfillment of test objectives involves the accomplishment of both operator and maintenance tasks. Completion of these tasks should follow formal approved procedures, which are generally in the form of technical manuals developed during the latter phases of detail equipment design. Following approved procedures is necessary to ensure that the equipment is operated and maintained in a proper manner. Deviation from approved procedures may result in the introduction of personnel-induced failures and will distort maintenance frequencies and task times as recorded in the test data. The identification of the procedures to be used in testing should be included in the evaluation plan.

Test-Site Selection

A single test site may be selected to evaluate the product in a designated environment; several test sites may be selected to evaluate the product under different environmental conditions (e.g., arctic versus tropical areas, mountainous versus flat terrain); or a variety of geographical locations may be selected to evaluate the production terms of different markets and consumer conditions. Any one or combination of these methods may be appropriate, depending on the product type and program requirements.

Test Personnel and Training

Test personnel will include (1) individuals who actually operate and maintain the system and equipment during the test and (2) the supporting engineers, technicians, data recorders, analysts, and test adminstrators, as appropriate. Individuals assigned to the operation and maintenance of the system should possess backgrounds and skill levels similar to consumer personnel who will normally operate and support the system throughout its life cycle. The required proficiency level(s) is attained through a combination of formal and on-the-job training.

Test Facilities and Resources

The necessary facilities, test chambers, capital equipment, environmental controls, special instrumentation, and associated resources (e.g., heat, water, air conditioning, gas, telephone, power, lighting) must be identified and scheduled. In many instances, new design and construction are required, which directly affects the scheduling and duration of the test preparation period. A detailed description of the test facility and the facility layout should be included in test planning documentation and in subsequent test reports.

Test and Support Equipment

Test and support equipment requirements are initially considered in the maintenance concept and later defined through the logistic support analysis. During the latter phases of detail design, the necessary support items are procured and should be available for Types 2, 3, and 4 testing. In the event that the proper type of support equipment is not available and alternative items are required, such items must be identified in the test and evaluation plan. The use of alternative items generally results in distorted test data (e.g., maintenance times), tends to cause personnel-induced failures in the equipment, and often forces a change in test procedures and facility requirements.

Test Supply Support

Supply support constitutes all materials, data, personnel, and related activities associated with the requirements, provisioning, and procurement of spare and repair parts and the sustaining maintenance of inventories for support of the system throughout its life cycle. Specifically, this includes

1. Initial and sustaining requirements for spares, repair parts, and consumables for the prime equipment. Spares are major replacement items and are re-

pairable, while repair parts are nonrepairable, smaller components. Consumables refer to fuel, oil, lubricants, liquid oxygen, nitrogen, and so on.

2. Initial and sustaining requirements for spares, repair parts, material, and consumables for the various elements of logistic support (i.e., test and support equipment, transportation and handling equipment, training equipment, and facilities).

3. Facilities required for the storage of spares, repair parts, and consumables. This involves consideration of space requirements, location, and environmental needs.

4. Personnel requirements for the accomplishment of supply support activities, such as provisioning, cataloging, receipt and issue, inventory management and control, shipment, and disposal of material.

5. Technical data requirements for supply support, which include initial and sustaining provisioning data, catalogs, material stock lists, receipt and issue reports, and material disposition reports.

The types of spare and repair parts needed at each level of maintenance are dependent on the maintenance concept and the logistic support analysis. For Types 3 and 4 testing, spare and repair parts will generally be required for all levels since these tests primarily involve an evaluation of the system as an entity and its total logistic support capability. The complete maintenance cycle, supply support provisions (the type and quantity of spares specified at each level), transportation times, turnaround times, and related factors are evaluated as appropriate. In certain instances, the producer's facility may provide depot-level support. Thus, it is important to establish a realistic supply system (or a close approximation) for the item being tested.

The type and quantity of spare and repair parts and the procedures for spare parts inventory control throughout the test program should be specified in the test and evaluation plan. Usage rates, reorder requirements, procurement lead times, and out-of-stock conditions during the test are recorded and included in the test report.

6.5. TEST PERFORMANCE AND REPORTING

With the necessary prerequisites established, the next step is to commence with the formal test and demonstration of the system. This requires operating and supporting the system and equipment in a prescribed manner as defined in the system test and evaluation plan. Throughout this process, data are collected and analyzed, which leads to assessment of system performance and effectiveness characteristics.

The assessment of performance and effectiveness of a system requires the availability of operational and maintenance histories of the various system elements. Performance and effectiveness parameters are established early in the life cycle with the development of operational requirements and the maintenance concept. These parameters describe the characteristics of the system that are considered paramount in fulfilling the need objectives. With the system in an operational status, the following questions arise:

1. What is the *true* performance and effectiveness of the system?
2. What is the *true* effectiveness of the logistic support capability?
3. Are the initially specified requirements being met?

Providing answers to these questions requires a formalized data-information and feedback capability with the proper output. A data subsystem must be designed, developed, and implemented to achieve a specific set of objectives, and these objectives must relate directly to the preceding questions and/or whatever other questions the engineer or manager needs to answer. The establishment of the data subsystem capability is basically a two-step process: (1) the identification of requirements and the applications for such; and (2) the design, development, and implementation of a capability that will satisfy the identified requirements.

Test Data Requirements

The purpose of a test data and information feedback subsystem is twofold.

1. It provides ongoing data that are analyzed to evaluate and assess the performance, effectiveness, operation, maintenance, logistic support capability, and so on, for the system in the field. The systems engineer or manager needs to know exactly how the system is doing, and needs the answer relatively quickly. Thus, certain types of information must be provided at designated times.
2. It provides historical data (covering existing systems in the field) that are applicable in the design and development of new systems and equipment having a similar function and nature. Engineering potential in the future certainly depends on the ability to capture experiences of the past, and subsequently be able to apply the results in terms of what to do and what not to do in new design situations.

Supporting the above requires a capability that is both responsive to a repetitive need in an expeditious and timely manner (i.e., the manager's need for assessment information), and one that incorporates the provisions for data storage and retrieval. It is necessary to determine the specific elements of data reporting. These factors are combined to identify total volume requirements for the subsystem and the type, quantity, and frequency of data reports.

The elements of data identified are related to the operational and support requirements for the system. An analysis of these elements will provide certain evaluative- and verification-type functions covering the characteristics of the system that are to be assessed. A listing of sample applications appropriate in the assessment of system characteristics is presented in Figure 6.4. When determining the reliability of an item (i.e., the probability that an item will operate satisfactorily in a given environment for a specified period of time), the data required will include the system operating time to failure and the time-to-failure distribution, which can be generated from a history of the particular item or a set of identical and independent items. When verifying spare/repair part demand rates, the data required should include a

1. *General Operational and Support Factors*
   (a) Evaluation of mission requirements (operational scenarios).
   (b) Evaluation of performance factors (capacity, output, range, accuracy, size, weight, volume, mobility, etc.).
   (c) Verification of system utilization (modes of operation and operating hours).
   (d) Verification of cost and system effectiveness, operational availability, dependability, reliability, maintainability, human factors, safety.
   (e) Evaluation of levels and location of maintenance.
   (f) Evaluation of operation and maintenance function/tasks by level and location.
   (g) Verification of repair level policies.
   (h) Verification of frequency distributions for maintenance actions and repair times.
   (i) Evaluation of software and software interfaces with other system elements.

2. *Test and Support Equipment*
   (a) Verification of support equipment type and quantity by maintenance level and location.
   (b) Verification of support equipment availability.
   (c) Verification of support equipment use (frequency of use, location, percent of time used, flexibility of use).
   (d) Evaluation of maintenance requirements for support equipment (scheduled and unscheduled maintenance, downtime, logistic resource requirements).

3. *Supply Support (Spare and Repair Parts)*
   (a) Verification of spare and repair part types and quantities by maintenance level and location.
   (b) Evaluation of supply responsiveness. (Is a spare available when needed?)
   (c) Verification of item replacement rates, condemnation rates, attrition rates.
   (d) Verification of inventory turnaround and supply pipeline times.
   (e) Evaluation of maintenance requirements for shelf items.
   (f) Evaluation of spare and repair part replacement and inventory policies.
   (g) Identification of shortage risks.

4. *Personnel and Training*
   (a) Verification of personnel quantities and skills by maintenance level and location.
   (b) Verification of elapsed times and man-hour expenditures by personnel skill level.
   (c) Evaluation of personnel skill mixes.
   (d) Evaluation of personnel training policies.
   (e) Verification of training equipment and data requirements.

5. *Transportation and Handling*
   (a) Verification of transportation and handling equipment type and quantity by maintenance level and location.
   (b) Verification of availability and utilization of transportation and handling equipment.
   (c) Evaluation of delivery response times.

6. *Facilities*
   (a) Verification of facility adequacy and utilization (operation, maintenance, and training facilities).

(b) Evaluation of logistics resource requirements for support and operation, maintenance, and training facilities.

7. *Technical Data*

    (a) Verification of adequacy of data coverage (level, accuracy, and method of information presentation) in operating and maintenance manuals.

    (b) Verification of adequacy of field data, collection, analysis, and corrective action subsystem.

8. *Operational and Maintenance Software*

    (a) Verification of the compatibility of software with other elements of the system.

    (b) Verification of software reliability and maintainability.

9. *Consumer Response*

    (a) Evaluation of degree of consumer satisfaction.

    (b) Verification that consumer needs are met.

Figure 6.4.   Data information system applications (requirements).

history of all item replacements, system/equipment operating time at replacement, and disposition of the replaced items (whether the item is condemned or repaired and returned to stock). An evaluation of organizational effectiveness will require identification of assigned personnel by quantity and skill level, the tasks accomplished by the organization, and the labor-hours and elapsed time expended in task accomplishment. Assessment objectives such as these (comparable to those listed in the figure) serve as the basis for the idenficiation of the specific data factors needed.

Development of a Data Sub-system

With the overall sub-system objectives defined, the next step is to identify the specific data factors that must be acquired and the method for acquisition. A format for data collection must be developed and should include both success data and maintenance data. *Success data* constitute information covering system operation and utilization on a day-to-day basis, and the information should be comparable to

---

System Operational Information Report

1. Report number, report date, and individual preparing report.
2. System nomenclature, part number, manufacturer, serial number.
3. Description of system operation by date (mission type, profile, and duration).
4. Equipment utilization by date (operating time, cycles of operation, etc.).
5. Description of personnel, transportation and handling equipment, and facilities required for system operation.
6. Recording of maintenance events by date and time (reference maintenance event reports).

Figure 6.5.   System success data.

the factors listed in Figure 6.5. *Maintenance data* cover each event involving scheduled and unscheduled maintenance. The events are recorded and referenced in system operational information reports, and the factors recorded in each instance are illustrated in Figure 6.6.

The format for data collection may vary considerably, as there is no set method
for its accomplishment, and the information desired may be different for each sys-

---

Maintenance Event Report

1. *Administrative data*
   (a) Event report number, report date, and individual preparing report.
   (b) Work order number.
   (c) Work area and time of work (month, day, hour).
   (d) Activity (organization) identification.
2. *System factors*
   (a) Equipment part number and manufacturer.
   (b) Equipment serial number.
   (c) System operating time when event occurred (when discovered).
   (d) Segment of mission when event occurred.
   (e) Description of event (describe symptom of failure for unscheduled actions).
3. *Maintenance factors*
   (a) Maintenance requirement (repair, calibration, servicing, etc.).
   (b) Description of maintenance tasks.
   (c) Maintenance downtime (MDT).
   (d) Active maintenance times
   (e) Maintenance delays (time awaiting spare part, delay for test equipment, work
       stoppage, awaiting personnel assistance, delay for weather, etc.).
4. *Logistics factors*
   (a) Start and stop times for each maintenance technician by skill level.
   (b) Technical manual or maintenance procedure used (procedure number,
       paragraph, date, comments on procedure adequacy).
   (c) Test and support equipment used (item nomenclature, part number,
       manufacturer, serial number, time of item usage, operating time on test
       equipment when used).
   (d) Description of facilities used.
   (e) Description of replacement parts (type and quantity).
       i. Nomenclature, part number, manufacturer, serial number, and operating
          time on replaced item. Describe disposition.
       ii. Nomenclature, part number, manufacturer, serial number, and operating
          time on installed item.
5. *Other information*
   Include any additional data considered appropriate and related to the
   maintenance event.

Figure 6.6.   System maintenance data.

tem. However, most of the factors in Figure 6.5 and 6.6 are common for all systems and must be addressed in the design of a new data subsystem. In any event, the following provisions should apply:

1. The data collection forms should be simple to understand and complete (preferably on single sheets), as the task of recording the data may be accomplished under adverse environmental conditions by a variety of personnel skill levels. If the forms are difficult to understand, they will not be completed properly (if at all), and the needed data will not be available.
2. The factors specified on each form must be clear and concise and not require a lot of interpretation and manipulation to obtain. The right type of data must be collected.
3. The factors specified must have a meaning in terms of application. The usefulness of each factor must be verified.

These considerations are extremely important and cannot be overemphasized. All the analytical methods, prediction techniques, models, and so on, discussed earlier have little meaning without the proper input data. Our ability to evaluate alternatives and predict in the future depends on the availability of good historical data, and the source of such stems from the type of data-information feedback subsystem developed at this stage. This subsystem must not only incorporate the forms for recording the right type of data, but must also consider the personnel factors (skill levels, motivation, etc.) involved in the data recording process. The individual who must complete the appropriate form(s) must understand the system and the purposes for which the data are being collected. If this individual is not properly motivated to do a thorough job in recording events, the resulting data will, of course, be highly suspect.

Once the appropriate data forms are distributed and completed by the responsible line organizations, a means must be provided for the retrieval, formatting, sorting, and processing of the data for reporting purposes. Field data are collected and sent to a designated centralized facility for analysis and processing. The results are disseminated to engineering and management for decision making and entered into a data bank for retention and possible future use.

System Evaluation and Corrective Action

Figure 6.7 illustrates a system evaluation and corrective-action loop. The evaluation aspect responds to the type of subjects listed in Figure 6.4 and can address both the system as an entity or individual segments of the system on an independent basis. Figure 6.8 presents some typical examples of evaluation factors. The evaluation approach and the analytical techniques (i.e., tools) used are basically the same as described earlier for decision making, the only difference being the data input. The evaluation effort can be applied on a continuing basis to provide certain system measures at designated points throughout the life cycle (see [a] and [c] of Figure 6.8), or it may constitute a one-time investigation.

Problem areas are identified at various stages in the evaluation and are re-

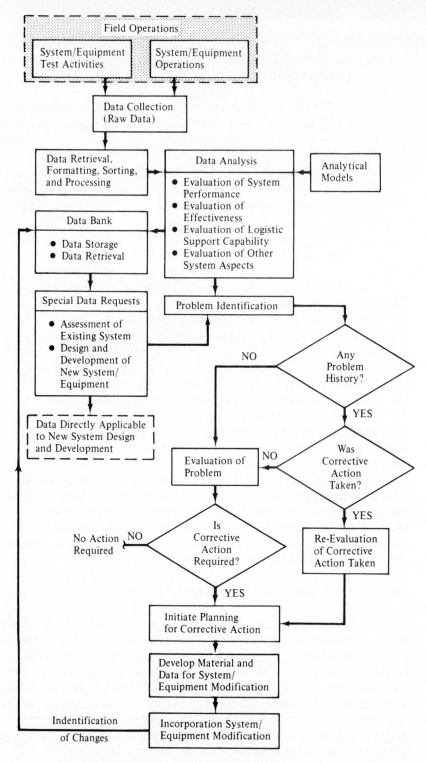

Figure 6.7 System evaluation and corrective action loop.

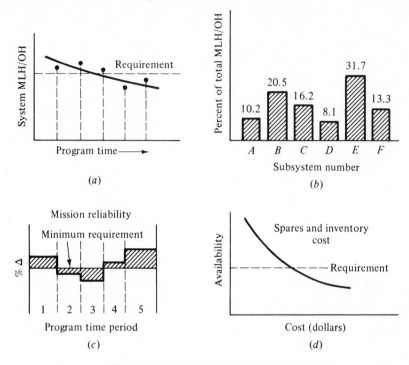

Figure 6.8   Evaluation factors (example).

viewed in terms of the feasibility for corrective action. Referring to Figure 6.8(b), subsystem E is a likely candidate for investigation since the MLH/OH for that item constitutes 31.7% of the total system value. In Figure 6.8(c), one may wish to investigate program time period 3 to determine why the mission reliability was so poor at that time. Corrective action may be accomplished in response to a system/equipment deficiency (i.e., the equipment fails to meet the specified requirements), or may be accomplished to improve system performance, effectiveness, and/or logistic support. If corrective action is to be accomplished, the necessary planning and implementation steps are a prerequisite to ensure the complete compatibility of all elements of the system throughout the change process.

Test Reporting

The final effort in the evaluation process constitutes the preparation of an appropriate test report. A test report should reference the initial system evaluation planning document and should describe all test conditions, incorporated system modifications during the test (if any), test data, and the results of data anlysis. These results may include appropriate recommendations for operation and support of the system as applicable to the utilization phase.

6.6. SYSTEM MODIFICATION

When a change occurs in a procedure, the prime equipment, or an item of logistic support, the change in most instances will affect many different elements of the system. Procedural changes will affect personnel and training requirements and necessitate a change in the technical data (equipment operating and/or maintenance instructions). Hardware changes will affect spare and repair parts, test and support equipment, technical data, and training requirements. Software changes may affect the hardware and the technical data. Each change must be thoroughly evaluated in terms of its impact on other elements of the system prior to a decision on whether or not to incorporate the change. The feasibility of incorporating a change will depend on the extensiveness of the change, its impact on the system's ability to accomplish its mission, the time in the life cycle when the change can be incorporated, and the cost of change implementation. A minimum amount of evaluation and planning are required to make rational decision as to whether the change is feasible.

If a change is to be incorporated, the necessary change control procedures must be implemented. A change control board is established to review and evaluate the potential impact of the change, the effectivity of the change (the serial numbered item on which the change is to be incorporated), retrofit provisions, and so on. The various components of the system (including all the elements of logistic support) must track. Otherwise, the probability of inadequate logistic support and unnecessary waste is high. The concepts and concerns of configuration change control in the production/construction stage are equally as appropriate in the product use and logistic support phase. In fact, a change involving the operational system is more critical, since we are dealing with a later stage in the life cycle. Changes will usually be more costly as the life cycle progresses.[10]

SELECTED REFERENCES

(1) Blanchard, B. S., *Engineering Organization and Management*. Englewood Cliffs, N.J.: Prentice-Hall, 1976.

(2) Blanchard, B. S., *Logistics Engineering and Management* (3rd ed.). Englewood Cliffs, N.J.: Prentice-Hall, 1986.

(3) Buffa, E. S., *Modern Production and Operations Management* (5th ed.). New York: John Wiley, 1987.

(4) Chase, R. B., and N. J. Aquilano, *Production and Operations Management: A Life Cycle Approach*. Homewood, Ill.: Richard D. Irwin, 1981.

(5) Defense Systems Management College (DSMC). *Systems Engineering Management Guide*. Fort Belvoir, Virginia 22060, December 1986.

(6) Duncan, A. J., *Quality Control and Industrial Statistics* (4th ed.). Homewood, IL: Richard D. Irwin, 1974.

[10] Systems engineering must play a major role relative to configuration management and change control if a reasonable baseline is to be maintained throughout system design and development. Otherwise, there is no assurance of an overall unified approach in the system definition process.

(7) Juran, J. M. (ed.), *Quality Control Handbook* (3rd ed.). New York: McGraw-Hill, 1979.

(8) Meister, D., *Behavioral Analysis and Measurement Methods*. New York: John Wiley, 1985.

(9) Ostrofsky, B., *Design, Planning and Development Methodology*. Englewood Cliffs, N.J.: Prentice-Hall, 1977.

(10) Salvendy, G., (ed.), *Handbook of Human Factors*. New York: John Wiley, 1987.

(11) Stevens, R. T., *Operational Test and Evaluation*. New York: John Wiley, 1979.

(12) Vick, C. R., and C. V. Ramamoorthy, *Handbook of Software Engineering*. New York: Van Nostrand Reinhold Co., 1984.

QUESTIONS AND PROBLEMS

1. Describe the system evaluation process. What is the objective?

2. How is system evaluation accomplished in the following phases: Conceptual design? Preliminary system design? Detail design? Production?

3. Define the basic categories of test and their applications. How do they fit into the total system evaluation process?

4. When does the planning for system evaluation commence? What information is included?

5. Select a system of your choice and develop a detailed system test and evaluation plan.

6. How are specific system test requirements determined?

7. Why is it important to establish the proper level of logistic support for system evaluation? What would probably happen in the absence of adequate test procedures? What would happen in the event of inadequate training of test personnel (i.e., the consumer)? What would probably happen in the absence of the proper test and support equipment?

8. A good data subsystem serves what purpose(s)?

9. Why are system success data important? What type of information is required?

10. What data are required to measure the following: cost effectiveness, system effectiveness, operational availability, and life-cycle cost?

11. In the event that system evaluation indicates noncompliance with a specified system requirement, what would you do?

12. If during the evaluation the consumer makes a recommendation for product improvement, what would you do?

13. Why is change control so important? How do system changes affect logistic support?

14. What role does systems engineering fulfill in the system evaluation function?

15. Briefly describe the producer and consumer functions in system evaluation.

16. What benefits are derived from test reporting?

17. How does evaluation fit into the systems engineering process?

PART III TOOLS FOR SYSTEMS ANALYSIS

7

Alternatives and Models in Decision Making

Engineers, whether engaged in research, design, development, construction, production, operations, or many of these activities, are concerned with the efficient use of limited resources. When known opportunities fail to hold sufficient promise for the employment of resources, more promising opportunities are sought. People with vision accept the premise that better opportunities exist than are known to them. This view accompanied by initiative leads to exploratory activities aimed at finding the better opportunities. In such activities, steps are taken into the unknown to find new possibilities which may then be evaluated to determine if they are superior to those that are known.

An understanding of the decision making process usually requires simplification of the complexity facing the decision maker. Conceptual simplifications of reality, or models, are a means to this end. In this chapter a structure for alternatives and a general decision evaluation approach using models is presented to facilitate decision making in design and in operations.

7.1. ALTERNATIVES IN DECISION MAKING

A complete and all-inclusive alternative rarely emerges in its final state. It begins as a hazy but interesting idea. The attention of the individual or group is then directed to analysis and synthesis, and the result is a definite proposal. In its final form, an alternative should consist of a complete description of its objectives and its requirements in terms of benefit and cost.

122

Both different ends and different methods are embraced by the term alternative. All proposed alternatives are not necessarily attainable. Some are proposed for analysis even though there seems to be little likelihood that they will prove feasible. The idea is that it is better to consider many unprofitable alternatives than to overlook one that is profitable. Alternatives that are not considered cannot be adopted, no matter how desirable they may actually be.

Limiting and Strategic Factors

Those factors that stand in the way of attaining objectives are known as *limiting factors*. An important element of the creative engineering process is the identification of the limiting factors restricting accomplishment of a desired objective. Once the limiting factors have been identified, they are examined to locate *strategic factors*, those factors which can be altered to make progress possible.

The identification of strategic factors is important, for it allows the decision maker to concentrate effort on those areas in which success is obtainable. This may require inventive ability, or the ability to put known things together in new combinations, and is distinctly creative in character. The means that will achieve the desired objective may consist of a procedure, a technical process, or a mechanical, organizational, or managerial change. Strategic factors limiting success may be circumvented by operating on engineering, human, and economic factors individually and jointly.

An important element of the process of defining alternatives is the identification of the limiting factors restricting the accomplishment of a desired objective. Once the limiting factors have been identified, they are examined to locate those strategic factors which can be altered in a cost-efficient way so that a selection from among the alternatives may be made.

Comparing Alternatives Equivalently

In order to compare alternatives equivalently, it is important that they be converted to a common measure. This conversion to a common measure permits comparison on the basis of equivalence. Models are essential in the conversion process and permit alternatives to be compared equivalently.

On completion of the conversion step, quantitative and qualitative outputs and inputs for each alternative form the basis for comparison and decision. Quantitative measures should be obtained with the use of suitable models, and decisions between alternatives should be made on the basis of their differences. Thus, all identical factors can be canceled out for the comparison of any two or more alternatives at any step of a decision making process.

After a situation has been carefully analyzed and the possible outcomes have been evaluated as accurately as possible, a decision must be made. Several outcomes are evaluated for each alternative that is examined. We know that all the outcomes cannot occur, and that at any given time any outcome can occur. Thus, decision making or selection from among alternatives is often done under risk or uncertainty.

In addition to the alternatives formally set up for evaluation, another alternative is almost always present—that of making no decision. The decision not to decide may be a result of either active consideration or passive failure to act; it is usually motivated by the thought that there will be opportunities in the future that may prove more profitable than any known at present.

7.2. MODELS IN DECISION MAKING

Models and their manipulation (the process of simulation) are very useful tools in systems analysis. A *model* may be used as a representation of a system to be brought into being, or to analyze a system already in being. Experimental investigation utilizing a model yields design or operational decisions in less time and at less cost than direct manipulation of the system itself. This is particularly true when it is not possible to manipulate reality because the system is not yet in existence, or when manipulation is costly and disruptive as with complex industrial systems.

There is a fundamental difference between models used in science and engineering. Science is concerned with the natural world, whereas engineering is concerned primarily with the man-made world. Science uses models to gain an understanding of the way things are in the natural world. Engineering uses models of the man-made world in an attempt to achieve what ought to be. The validated models of science are used in engineering to establish bounds for engineering creations and to improve the products of such creations.

Classification of Models

When used as a noun the word model implies representation. An aeronautical engineer may construct a wooden model of a possible configuration for a proposed aircraft type. An architect might represent a proposed building with a scale model of the building. An industrial engineer may use templates to represent a proposed layout of equipment in a factory. The word model may also be used an an adjective, carrying with it the implication of ideal. Thus, a man may be referred to as a model husband or a child praised as a model student. Finally, the word model may be used as a verb, as is the case where a person models clothes.

Models are designed to represent a system under study, by an idealized example of reality, in order to explain the essential relationships involved. They can be classified by distinguishing physical, analog, schematic, and mathematical types. Physical models look like what they represent, analog models behave like the original, schematic models graphically describe a situation or process, and mathematical models symbolically represent the principles of a situation being studied. All four model types are used successfully in systems engineering and analysis.

Physical models.   Physical models are geometric equivalents, either as miniatures, enlargements, or duplicates made to the same scale. Globes are one example. They are used to demonstrate the shape and orientation of continents, water bodies, and other geographic features of the earth. A model of the solar system is used to demonstrate the orientation of the sun and planets in space. A model of an atomic structure would be similar in appearance, but at the other extreme in dimensional reproduction. Each of these models represents reality and is used for demonstration.

Some physical models are used in the simulation process. An aeronautical engineer may test a specific tail assembly design with a model airplane in a wind tunnel. A pilot plant might be built by a chemical engineer to test a new chemical process for the purpose of locating operational difficulties before full-scale production. An environmental chamber is often used to create conditions anticipated for a component under test.

The use of templates in plant layout is an example of experimentation with a physical model. Templates are either two- or three-dimensional replicas of machinery and equipment which are moved about on a scale-model area. The relationship of distance is important and the templates are manipulated until a desirable layout is obtained. Such factors as noise generation, vibration, and lighting are also important but are not a part of the experimentation and must be considered separately.

Analog models.   Analog comes from the Greek word analogia, which means proportion. This explains the concept of an *analog model;* the focus is on similarity in relations. Analogs are usually meaningless from the visual standpoint.

Analog models can be physical in nature such as where electric circuits are used to represent mechanical systems, hydraulic systems, or even economic systems. Analog computers use electronic components to model power distribution systems, chemical processes, and the dynamic loading of structures. The analog is represented by physical elements. When a digital computer is used as a model for a system, the analog is more abstract. It is represented by symbols in the computer program and not by the physical structure of the computer components.

The analog may be a partial subsystem or it may be an almost complete representation of the system under study. For example, the tail assembly design being tested in a wind tunnel may be complete in detail but incomplete in the properties being studied. The wind tunnel test may examine only the aerodynamic properties and not the structural, weight, or cost characteristics of the assembly. From this it is evident that only those features of an analog model which serve to describe reality should be considered. These models, like other types, suffer from certain inadequacies.

Schematic models.   A schematic model is developed by reducing a state or event to a chart or diagram. The schematic model may or may not look like the real-world situation it represents. It is usually possible to achieve a much better understanding of the real-world system described by the model through use of an explicit coding process employed in the construction of the model. The execution of a football play may be diagrammed on a blackboard with a simple code. It is the idealized aspect of this schematic model that permits this insight into the football play.

An organizational chart is a common schematic model. It is a representation of the state of formal relationships existing between various members of the organization. A human-machine chart is another example of a schematic model. It is a model of an event, that is, the time-varying interaction of one or more people and one or more machines over a complete work cycle. A flow process chart is a schematic model which describes the order or occurrence of a number of events that constitute an objective, such as the assembly of an automobile from a multitude of component parts.

In each case, the value of the schematic model lies in its ability to describe the essential aspects of the existing situation. It does not include all extraneous actions and relationships but rather concentrates on a single facet. Thus, the schematic model is not in itself a solution but only facilitates a solution. After the model has been carefully analyzed, a proposed solution can be defined, tested, and implemented.

Mathematical models. A mathematical model employs the language of mathematics and, like other models, may be a description and then an explanation of the system it represents. Although its symbols may be more difficult to comprehend than verbal symbols, they do provide a much higher degree of abstraction and precision in their application. Because of the logic it incorporates, a mathematical model may be manipulated in accordance with established mathematical procedures.

Almost all mathematical models are used either to predict or to control. Such laws as Boyle's law, Ohm's law, and Newton's laws of motion are formulated mathematically and may be used to predict certain outcomes when dealing with physical phenomena. Outcomes of alternative courses of action may also be predicted if a measure of evaluation is adopted. For example, a linear programming model may predict the profit associated with various production quantities of a multiproduct process. Mathematical models may be used to control an inventory. In quality control, a mathematical model may be employed to monitor the proportion of defects that will be accepted from a supplier. Such models maintain control over a state of reality.

Mathematical models directed to the study of systems differ from those traditionally used in the physical sciences in two important ways. First, since the system being studied usually involves social and economic factors, these models must often incorporate probabilistic elements to explain their random behavior. Second, mathematical models formulated to explain existing or planned operations incorporate two classes of variables: those under the control of a decision maker and those not directly under control. The objective is to select values for controllable variables so that some measure of effectiveness is optimized. Thus, these models are of great benefit in systems engineering and systems analysis.

Models and Indirect Experimentation

Models and the process of simulation provide a convenient means of obtaining factual information about a system being designed or a system in being. In component design it is customary and feasible to build several prototypes, test them, and then modify the design based on the test results. This is often not possible in systems engineering because of the cost involved and the length of time required over the system life cycle. A major part of the design process requires decisions based on a model of the system rather than decisions derived from the system itself.

Direct and indirect experimentation.   In direct experimentation, the object, state, or event, and/or the environment is subject to manipulation and the results are observed. For example, a couple might rearrange the furniture in their living room by this method. Essentially, they move the furniture and observe the results. This process may then be repeated with a second move and perhaps a third, until all logical alternatives have been exhausted. Eventually, one such move is subjectively judged best; the furniture is returned to this position and the experiment is completed. Direct experimentation such as this may be applied to the rearrangement of equipment in a factory. Such a procedure is time-consuming, disruptive, and costly. Hence, simulation or indirect experimentation is employed, with templates representing the equipment to be moved.

Direct experimentation in aircraft design would involve constructing a full-scale prototype which would be flight-tested under real conditions. Although this is an essential step in the evolution of a new design, it would be very costly as the first step. The usual procedure is evaluating several proposed configurations by building a model of each and then testing in a wind tunnel. This is the process of indirect experimentation, or *simulation*. It is extensively used in situations where direct experimentation is not economically feasible.

In systems analysis indirect experimentation is effected through the formulation and manipulation of decision models. This makes it possible to determine how changes in those aspects of the system under control of the decision maker affect the modeled system. Indirect experimentation enables the systems analyst to evaluate the probable outcome of a given decision without changing the operational system itself. In effect, indirect experimentation in the study of operations provides a means for making quantitative information available to the decision maker without disturbing the operations under his or her control.

Simulation through indirect experimentation.   In most design and operational situations, the objective sought is the optimization of a performance measure economically. Rarely, if ever, can this be done by direct experimentation with a system under development or a system in being.

The primary use of simulation in systems engineering is to explore the effects of alternative system characteristics on system performance without actually producing and testing each candidate system. Most models used will fit the classification given earlier and many will be mathematical. The type used will depend upon the questions to be answered. In some instances, simple schematic diagrams will suffice. In others, mathematical or probabilistic representations will be needed. In many cases, simulation with the aid of an analog or digital computer will be required.

In most systems engineering undertakings, a number of models must be formulated. These models form a hierarchy ranging from considerable aggregation to extreme detail. At the start of a systems project, knowledge of the system is quite sketchy and general. As the design progresses this knowledge becomes more detailed, and consequently, the models used for simulation should be detailed.

There is no available theory by which the best model for a given system simulation can be selected. The choice of an appropriate model is determined as much by the background of the systems analyst as the system itself.

7.3. DECISION EVALUATION THEORY

Decision evaluation is an important part of systems engineering and analysis. Evaluation is needed as a basis for choice among alternatives which arise from design activities as well as for optimizing systems already in operation. In either case, equivalence provides the common evaluation measure upon which choice can be based.[1]

Decision Evaluation Function

An evaluation measure, E, may be derived from a decision evaluation function. This evaluation function is a mathematical model formally linking the evaluation measure with controllable decision variables, X, and system parameters that cannot be controlled by the decision maker, Y. It provides a means for testing decision variables in the presence of system parameters. This test is an indirect experiment performed mathematically which results in an equivalent value for E. The functional relationship, in its unconstrained form, is expressed as

$$E = f(X, Y). \tag{7.1}$$

In constrained form, the evaluation function is subject to a set of constraints expressed as functions of the decision variables and system parameters and a set of constant constraints, C, expressed as

$$g(X, Y) = C,$$

where the equality may also be $\leq$ or $\geq$.

As an example of the evaluation function structure for an unconstrained decision situation, consider the determination of an optimal procurement quantity for inventory operations. Here the evaluation measure is cost, and the objective is to choose a procurement quantity in the face of demand, procurement cost, and holding cost, so that total cost is minimized. The procurement quantity is the variable directly under the control of the decision maker. Demand, procurement cost, and holding cost are not directly under his or her control. The use of a decision model such as this allows the decision maker to arrive at a value for the variable under his or her control that trades off conflicting cost elements.

The evaluation function structure for a constrained decision situation can be illustrated by the determination of an optimal production mix for a number of products. Here the evaluation measure is profit, and the objective is to choose the number of units of each product which should be produced per day to maximize profit. The variables directly under control of the decision maker are the production quanti-

---

[1] A general procedure for the evaluation of alternatives was presented in Figure 4.10.

ties to be specified. Not directly under his or her control are the profits per unit contributed by each unit produced, the amount of scarce machine time consumed by each unit of product in each processing area, and the amount of machine time available in each processing area each day. A decision model such as this is known as linear programming. It allows the decision maker to optimize a linear evaluation function in the face of linear constraints, these constraints arising because the activities compete for scarce resources.

The decision evaluation function may be extended to operational and design decision situations involving alternatives. This extension involves the identification and isolation of decision-dependent system parameters, Y_d, from decision independent system parameters, Y_i. Accordingly, Equation 7.1 can be restated, in unconstrained form as

$$E = f(X, Y_d, Y_i). \qquad (7.2)$$

In constrained form, Equation 7.2 is augmented by a set of constraints expressed as functions of the controllable decision variables, system parameters, and a set of constant constraints, C, expressed as

$$g(X, Y_d, Y_i) = C$$

where the equality may also be $\leq$ or $\geq$.

As an example of the application of this version of the decision evaluation function, consider the establishment of a procurement and inventory system to meet the demand for an item which is available from one of several sources. Decision variables are the procurement level and the procurement quantity. For each source under consideration there exists a set of source dependent parameters. These are the item cost per unit, the procurement cost per procurement, the replenishment rate, and the procurement lead time. Uncontrollable system parameters include the demand rate, the holding cost per unit per period, and the shortage penalty cost. Constraints may exist on warehouse capacity, the procurement quantity, or the maximum investment in inventory. The objective is to determine the procurement level, and procurement quantity, and the procurement source so that total system cost will be minimized.

In design decision making, consider the deployment of a population of repairable equipment units to meet a demand. Three decision variables may be identified, the number of units to deploy, the number of maintenance channels to provide, and the age at which units should be retired. Controllable system parameters include the MTBF, MTTR, energy efficiency, and design life. Uncontrollable system parameters include the cost of energy, the time value of money, and the penalty cost incurred when there are insufficient operational units to meet demand. Constraints may exist on such things as the maximum number of units which may be procured and deployed, the retirement age, or the minimum value of MTBF. Since the controllable parameters are design dependent, the objective is to develop design alternatives in the face of decision variables to identify the alternative which will minimize total system cost.

A summary of the evolution and development of the decision evaluation function is given in Table 7.1. Cited references in the table are given in complete form at

TABLE 7.1   SUMMARY OF THE DECISION EVALUATION FUNCTION

REFERENCES	FUNCTIONAL FORM	APPLICATION
Churchman, Ackoff, & Arnoff (1957)	$E = f(x_i, y_j)$ E = system effectiveness x_i = variables under control y_j = variables not subject to direct control	Operations
Fabrycky, Ghare, & Torgersen (1984)	$E = f(X, Y); g(X, Y) \lesseqgtr B$ E = system effectiveness X = controllable variables Y = uncontrollable variables	Operations
Banks & Fabrycky (1987)	$E = f(X, Y_d, Y_i); g(X, Y_d, Y_i) \lesseqgtr C$ E = system effectiveness X = procurement level and procurement quantity Y_d = source dependent parameters Y_i = source independent parameters	Procurement Operations
MacroCAD Research (see Section 5.4 and Figure 5.9)	$E = f(X, Y_d, Y_i); g(X, Y_d, Y_i) \lesseqgtr C$ E = evaluation measure X = design variables Y_d = design dependent parameters Y_i = design independent parameters	Design Optimization

the end of this chapter. Models in subsequent chapters are developed in accordance with the decision evaluation function as presented in this section and summarized in Table 7.1.

Decision Evaluation Matrix

A particular decision can result in one of several outcomes, depending upon which of several future events takes place. For example, a decision to go sailing can result in a high degree of satisfaction if the day turns out to be sunny or in a low degree of satisfaction if it rains. These levels of satisfaction would be reversed if the decision were made to stay home. Thus, for the two states of nature, sun and rain, there are different payoffs depending upon the alternative chosen.

    A decision evaluation matrix is a formal way of exhibiting the interaction of a finite set of alternatives and a finite set of possible futures (or states of nature). In this usage, alternatives have the meaning presented in Section 7.3; that is, they are courses of action among which a decision maker expects to choose. The states of nature are normally not natural events such as rain, sleet, or snow, but are a wide variety of future outcomes over which the decision maker has no direct control.

    The general decision evaluation matrix is a model depicting the positive and negative results that will occur for each alternative under each possible future. In abstract form, this model is structured as shown in Figure 7.1. Its symbols are defined as follows:

A_i \ P_j, F_j	P_1 F_1	P_2 F_2	...	P_n F_n
A_1	E_{11}	E_{12}	...	E_{1n}
A_2	E_{21}	E_{22}	...	E_{2n}
⋮	⋮	⋮		⋮
A_m	E_{m1}	E_{m2}	...	E_{mn}

Figure 7.1 Decision evaluation matrix.

A_i = an alternative available for selection by the decision maker, where $i = 1, 2, \ldots, m$.
F_j = a future not under control of the decision maker, where $j = 1, 2, \ldots, n$.
P_j = the probability that the jth future will occur, where $j = 1, 2, \ldots, n$.
E_{ij} = evaluation measure (positive or negative) associated with the ith alternative and the jth future determined from Equation 7.2.

Several assumptions underlie the application of this decision evaluation matrix model to decision making under assumed certainty, risk, and uncertainty. Foremost among these is the presumption that all viable alternatives have been considered and all possible futures have been identified. Alternatives not considered cannot be adopted, no matter how desirable they may prove to be. Possible futures not identified can significantly affect the actual outcome relative to the planned outcome.

Evaluation measures in the matrix model are associated with outcomes which may be either objective or subjective. The most common case is one in which the outcome values are objective and therefore subject to quantitative expression in cardinal form. For example, the payoffs may be profits expressed in dollars, yield expressed in pounds, costs (negative payoffs) expressed in dollars, or other desirable or undesirable measures. Subjective outcomes, on the other hand, are those which are valued on an ordinal or ranking scale. Examples are expressions of preference, such as a good corporate image being preferred to a poor image, higher-quality outputs being preferred to those of lower quality, and so forth.

Other assumptions of importance in the evaluation matrix representation of decisions are

1. The occurrence of one future precludes the occurrence of any other future (futures are mutually exclusive).
2. The occurrence of a specific future is not influenced by the alternative selected.
3. The occurrence of a specific future is not known with certainty, even though certainty is often assumed for analysis purposes.

7.4. DECISIONS UNDER ASSUMED CERTAINTY

In dealing with physical aspects of the environment, physical scientists and engineers have a body of systematic knowledge and physical laws upon which to base their reasoning. Such laws as Boyle's law, Ohm's law, and Newton's laws of motion were developed primarily by collecting and comparing many comparable instances and by the use of an inductive process. These laws may then be applied with a high degree of certainty to specific instances. They are supplemented by many models for physical phenomena which enable conclusions to be reached about the physical environment that match the facts with narrow limits. Much is known with certainty about the physical environment.

Much less, particularly of a quantitative nature, is known about the environment within which operational decisions are made. Nonetheless, the primary aim of operations research and management science is to bring the scientific approach to bear to a maximum feasible extent. This is done with the aid of conceptual simplifications and models of reality, the most common being the assumption of a single known future. It is not claimed that knowledge about the future is in hand. Rather, the suppression of risk and uncertainty is one of the ways in which the scientific approach simplifies reality in order to gain insight. Such insight can assist greatly in decision making, provided that its shortcomings are recognized and accommodated.

The evaluation matrix for decision making under assumed certainty is not a matrix at all. It is a vector with as many evaluations as there are alternatives, with the outcomes constituting a single column. This decision vector is a special case of the matrix of Figure 7.1. It appears as in Figure 7.2, with the payoffs represented by E_i, where $i = 1, 2, \ldots, m$. The single future, which is assumed to occur with certainty, actually carries a probability of unity ($P = 1.0$) in the matrix. All other futures are suppressed by carrying probabilities of zero ($P = 0.0$).

When the outcomes, E_i, are stated in monetary terms (cost or profit), the decision rule or principle of choice is quite simple. If the alternatives are equal in all other respects, one would choose the alternative that minimizes cost or maximizes profit. In the case of cost, one would choose

$$\min_i \{E_i\} \qquad \text{for } i = 1, 2, \ldots, m.$$

Figure 7.2 Decision evaluation vector.

For profit, one would choose

$$\max_i \{E_i\} \qquad \text{for } i = 1, 2, \ldots, m.$$

It is often not possible to accept the premise that only the cost or the profit differences are important, with intangibles and irreducibles having little or no effect. Unquantifiable nonmonetary factors may be significant enough to outweigh calculated costs or profit differences among alternatives. In other cases, the outcome is not easily expressed in monetary terms, or even in quantitative terms of some other evaluation measure, such as time, percent of market, and so forth. Valid qualitative comparisions may be made when the quantitative outcomes cannot stand alone and/ or when the outcomes are nonquantitative.

The use of outcome scales often makes possible a somewhat rational choice from among a number of nonquantifiable alternatives, each with an outcome rating determined by expert opinion, estimation, history, or other means. Foremost among these are ordinal comparisons. Where intangibles and irreducibles are significant, ranking each outcome above or below every other outcome leads to a preferred choice. To do this, each outcome can be compared to a common standard, or the outcome can be paired and compared. As an example of the paired approach, suppose that four alternatives are assumed to lead (with certainty) to four outcomes E_1, E_2, E_3, and E_4. Suppose further that the six possible pairs are arranged according to preference as follows.[2]

$$E_1 > E_3 \qquad E_2 > E_3 \qquad E_2 > E_1$$
$$E_2 > E_4 \qquad E_3 > E_4 \qquad E_1 > E_4.$$

In these comparisons, E_2 is preferred three times; E_1 twice; E_3 once; and E_4 not at all. Accordingly, the preference ranking is

$$E_2 > E_1 > E_3 > E_4.$$

7.5. DECISIONS UNDER RISK

There is usually little assurance that predicted futures will coincide with actual futures. The physical and economic elements upon which a course of action depends may vary from their estimated values because of chance causes. Not only are the estimates of future costs problematical but, in addition, the anticipated future worth of most ventures is known only with a degree of assurance. This lack of certainty about the future makes decision making one of the most challenging tasks faced by individuals, industry, and government.

Decision making under risk occurs when the decision maker does not suppress acknowledged ignorance about the future, but makes it explicit through the assignment of probabilities. Such probabilities may be based on experimental evidence, expert opinion, subjective judgment, or a combination of these.

---

[2] The symbol $>$ is used to indicate that the outcome identified first is preferred to its counterpart.

Consider the following example. A computer systems firm has the opportunity to bid on two related contracts being advertised by a municipality. The first pertains to the selection and installation of hardware for a central computing facility together with required software. The second involves the development of a distributed computing network involving the selection and installation of hardware and software. The firm may be awarded either contract C_1 or contract C_2, or both contract C_1 and C_2. Thus, there are three possible futures.

Careful consideration of the possible approaches leads to the identification of five alternatives. The first is for the firm to subcontract the hardware selection and installation, but to develop the software itself. The second is for the firm to subcontract the software development, but to select and install the hardware itself. The third is for the firm to handle both the hardware and software tasks itself. The fourth is for the firm to bid jointly with a partner firm on both the hardware and software projects. The fifth alternative is for the firm to serve only as project manager, subcontracting all hardware and software tasks.

With the possible futures and various alternatives identified, the next step is to determine payoff values. Also to be determined are the probabilities for each of the three futures, where the sum of these probabilities must be unity. Suppose that these determinations lead to the profits and probabilities given in Table 7.2.

Table 7.2 is structured in accordance with the format of the decision evaluation matrix model exhibited in Figure 7.1. It is observed that the firm anticipates a profit of \$100,000 if alternative A_1 is chosen and contract C_1 is secured. If contract C_2 is secured, the profit would also be \$100,000. However, if both contract C_1 and C_2 are secured, the profit anticipated is \$400,000. Similar information is exhibited for the other alternatives, with each row of the matrix representing the outcome expected for each future (column) for a particular alternative.

Before proceeding to the application of criteria for the choice from among alternatives, the decision evaluation matrix should be examined for dominance. Any alternatives that are clearly not preferred, regardless of the future which occurs, may be dropped from consideration. If the outcomes for alternative x are better than the outcomes for alternative y for all possible futures, alternative x is said to *dominate* alternative y, and y can be eliminated as a possible choice.

The computer systems firm, facing the evaluation matrix of Table 7.2, may eliminate A_5 from consideration since it is dominated by all other alternatives. This

TABLE 7.2 DECISION EVALUATION MATRIX (PROFIT IN THOUSANDS OF DOLLARS)

		PROBABILITY:	(0.3)	(0.2)	(0.5)
		FUTURE:	C_1	C_2	$C_1 + C_2$
	A_1		100	100	400
	A_2		−200	150	600
Alternative	A_3		0	200	500
	A_4		100	300	200
	A_5		−400	100	200

means that the possible choice of serving only as project manager is inferior to each and every one of the other alternatives, regardless of the way in which the projects are awarded. Therefore, the matrix can be reduced to that given in Table 7.3. The decision criteria in the sections that follow may be used to assist in the selection from among alternatives A_1 through A_4.

Aspiration Level Criterion

Some form of aspiration level exists in most personal and professional decision making. An aspiration level is some desired level of achievement such as profit, or some undesirable result level to be avoided, such as loss. In decision making under risk, the aspiration level criterion involves selecting some level of achievement that is to be met, followed by a selection of that alternative which maximizes the probability of achieving the stated aspiration level.

The computer systems firm is now at the point of selecting from among alternatives A_1 through A_4, as presented in the reduced matrix of Table 7.3. Under the aspiration level criterion, management must set a minimum aspiration level for profit and possibly a maximum aspiration level for loss. Suppose that the profit level is set to be at least \$400,000 and the loss level is set to be no more than \$100,000. Under these aspiration level choices, alternatives A_1, A_2, and A_3 qualify as to profit potential, but alternative A_2 fails the loss test and must be eliminated. The choice could now be made between A_1 and A_3 by some other criterion, even though both satisfy the aspiration level criterion.

TABLE 7.3 REDUCED DECISION
EVALUATION MATRIX (PROFIT IN
THOUSANDS OF DOLLARS)

PROBABILITY: FUTURE:	(0.3) C_1	(0.2) C_2	(0.5) $C_1 + C_2$
A_1	100	100	400
A_2	-200	150	600
Alternative A_3	0	200	500
A_4	100	300	200

Most Probable Future Criterion

A basic human tendency is to focus on the most probable outcome from among several that could occur. This approach to decision making suggests that all except the most probable future be disregarded. Although somewhat equivalent to decision making under certainty this criterion works well when the most probable future has a significantly high probability so as to partially dominate.

Under the most probable future criterion, the computer systems firm would focus its selection process from among the four alternatives on the profits associated with the future designated $C_1 + C_2$ (both contracts awarded). This is because the probability of this future occuring is 0.5, the most probable possibility. Alternative A_2 is preferred by this approach.

The most probable future criterion could be applied to select between A_1 and A_3, as identified under the aspiration level criterion. If this is done, the firm would choose alternative A_3.

Expected Value Criterion

Many decision makers strive to make choices that will maximize expected profit or minimize expected loss. This is ordinarily justified in repetitive situations where the choice is to be made over and over again with increasing confidence that the calculated expected outcome will be achieved. This criterion is viewed with caution only when the payoff consequences of possible outcomes are disproportionately large, making a result that deviates from the expected outcome a distinct possibility.

The calculation of the expected value requires weighting all payoffs by their probabilities of occurrence. These weighted payoffs are then summed across all futures for each alternative. For the computer systems firm, alternatives A_1 through A_4 yield the following expected profits (in thousands):

$$A_1: \quad \$100(0.3) + \$100(0.2) + \$400(0.5) = \$250$$

$$A_2: \quad \$-200(0.3) + \$150(0.2) + \$600(0.5) = \$270$$

$$A_3: \quad \$0(0.3) + \$200(0.2) + \$500(0.5) = \$290$$

$$A_4: \quad \$100(0.3) + \$300(0.2) + \$200(0.5) = \$190.$$

From this analysis it is clear that alternative A_3 would be selected. Further, if this criterion were to be used to resolve the choice of either A_1 or A_3 under the aspiration level approach, the choice would be alternative A_3.

Comparison of Decisions

It is evident that there is no one best selection when these criteria are utilized for decision making under risk. The decision made is dependent on the decision criterion adopted by the decision maker. For the example of this section, the alternatives selected under each criterion were

Aspiration level criterion: A_1 or A_3
Most probable future criterion: A_2
Expected value criterion: A_3

If the application of the latter two criteria to the resolution of A_1 or A_3 chosen under the aspiration level criterion is accepted as valid, then A_3 is preferred twice and A_2 once. From this it might be appropriate to suggest that A_3 is the best alternative arising from the use of these three criteria.

7.6. DECISIONS UNDER UNCERTAINTY

It may be inappropriate or impossible to assign probabilities to the several futures identified for a given decision situation. Often no meaningful data are available from which probabilities may be developed. In other instances the decision maker may be unwilling to assign a subjective probability, as is often the case when the future could prove to be unpleasant. When probabilities are not available for assignment to future events, the situation is classified as decision making under uncertainty.

As compared with decision making under certainty and under risk, decisions under uncertainty are made in a more abstract environment. In this section several decision criteria will be applied to the example of Section 7.5 to illustrate the formal approaches that are available.

Laplace Criterion

Suppose that the computer systems firm is unwilling to assess the futures in terms of probabilities. Specifically, the firm is unwilling to differentiate between the likelihood at acquiring contract C_1, contract C_2, and contract C_1 and contract C_2. In the absence of these probabilities one might reason that each possible state of nature is as likely to occur as any other. The rationale of this assumption is that there is no stated basis for one state of nature to be more likely than any other. This is called the *Laplace principle* or the *principle of insufficient reason* based on the philosophy that nature is assumed to be indifferent.

Under the Laplace principle, the probability of the occurrence of each future state of nature is assumed to be $1/n$, where n is the number of possible future states. To select the best alternative one would compute the arithmetic average for each. For the decision matrix of Table 7.3, this is accomplished as shown in Table 7.4. Alternative A_3 results in a maximum profit of \$233,000 and would be selected.

TABLE 7.4   COMPUTATION OF AVERAGE PROFIT (THOUSANDS OF DOLLARS)

ALTERNATIVE	AVERAGE PAYOFF
A_1	(\$100 + \$100 + \$400) ÷ 3 = \$200
A_2	(\$−200 + \$150 + \$600) ÷ 3 = \$183
A_3	(\$0 + \$200 + \$500) ÷ 3 = \$233
A_4	(\$100 + \$300 + \$200) ÷ 3 = \$200

Maximin and Maximax Criteria

Two simple decision rules are available for dealing with decisions under uncertainty. The first is the *maximin* rule, based on an extremely pessimistic view of the outcome of nature. The use of this rule would be justified if it is judged that nature will do its worst. The second is the *maximax* rule, based on an extremely optimistic view of the future. Use of this rule is justified if it is judged that nature will do its best.

Because of the pessimism embraced by the maximin rule, its application will lead to the alternative that assures the best of the worst possible outcomes. If E_{ij} is used to represent the payoff for the ith alternative and the jth state of nature, the required computation is

$$\max_i \{\min_j E_{ij}\}.$$

Consider the decision situation described by the decision matrix of Table 7.3. The application of the maximin rule requires that the minimum value in each row be selected. Then the maximum value is identified from these and associated with the alternative that would produce it. This procedure is illustrated in Table 7.5. Selection of either alternative A_1 or A_4 assures the firm of a profit of at least \$100,000 regardless of the future.

TABLE 7.5   PROFIT BY THE MAXIMIN RULE (THOUSANDS OF DOLLARS)

ALTERNATIVE	Min E_{ij} j
A_1	\$ 100
A_2	−200
A_3	0
A_4	100

The optimism of the maximax rule is in sharp contrast to the pessimism of the maximin rule. Its application will choose the alternative that assures the best of the best possible outcomes. As before, if E_{ij} represents the payoff for the ith alternative and the jth state of nature, the required computation is

$$\max_i \{\max_j E_{ij}\}.$$

Consider the decision situation of Table 7.3 again. The application of the maximax rule requires that the maximum value in each row be selected. Then the maximum value is identified from these and associated with the alternative that would produce it. This procedure is illustrated in Table 7.6. Selection of alternative A_2 is indicated. Thus, the decision maker may receive a profit of \$600,000 if the future is benevolent.

TABLE 7.6   PROFIT BY THE MAXIMAX RULE (THOUSANDS OF DOLLARS)

ALTERNATIVE	Max E_{ij} j
A_1	\$400
A_2	600
A_3	500
A_4	300

A decision maker who chooses the maximin rule considers only the worst possible occurrence for each alternative and selects that alternative which promises the best of the worst possible outcomes. In the example where A_1 was chosen, the firm would be assured of a profit of at least \$100,000, but it could not receive a profit any greater than \$400,000. Or, if A_4 were chosen, the firm could not receive a profit any greater than \$300,000. Conversely, the firm that chooses the maximax rule is optimistic and decides solely on the basis of the highest profit offered for each alternative. Accordingly, in the example in which A_2 was chosen, the firm faces the possibility of a loss of \$200,000 while seeking a profit of \$600,000.

Hurwicz Criterion

Because the decision rules presented above are extreme, they are shunned by many decision makers. Most people have a degree of optimism or pessimism somewhere between the extremes. A third approach to decision making under uncertainty involves an index of relative optimism and pessimism. It is called the *Hurwicz rule*.

A compromise between optimism and pessimism is embraced in the Hurwicz rule by allowing the decision maker to select an index of optimism, α, such that $0 \le \alpha \le 1$. When $\alpha = 0$ the decision maker is pessimistic about nature, while an $\alpha = 1$ indiciates optimism about nature. Once α is selected, the Hurwicz rule requires the computation of

$$\max_i \{\alpha[\max_j E_{ij}] + (1 - \alpha)[\min_j E_{ij}]\},$$

where E_{ij} is the payoff for the ith alternative and the jth state of nature.

As an example of the Hurwicz rule, consider the payoff matrix of Table 7.3 with $\alpha = 0.2$. The required computations are shown in Table 7.7 and alternative A_1 would be chosen by the firm.

Additional insight into the Hurwicz rule can be obtained by graphing each alternative for all values of α between zero and one. This makes it possible to identify the value of α for which each alternative would be favored. Such a graph is shown in Figure 7.3. It may be observed that alternative A_1 yields a maximum expected profit for all values of $\alpha \le \frac{1}{2}$. Alternative A_3 exhibits a maximum for $\frac{1}{2} \le \alpha \le \frac{2}{3}$ and alternative A_2 gives a maximum for $\frac{2}{3} \le \alpha \le 1$. There is no value of α for which alternative A_4 would be best except at $\alpha = 0$, where it is as good an alternative as A_1.

TABLE 7.7   PROFIT BY THE HURWICZ RULE WITH $\alpha = 0.2$ (THOUSANDS OF DOLLARS)

ALTERNATIVE	$\alpha[max\ E_{ij}] + (1 - \alpha)[min\ E_{ij}]$
A_1	0.2(\$400) + 0.8(\$100) = \$160
A_2	0.2(\$600) + 0.8(\$-200) = \$-40
A_3	0.2(\$500) + 0.8(0) = \$100
A_4	0.2(\$300) + 0.8(\$100) = \$140

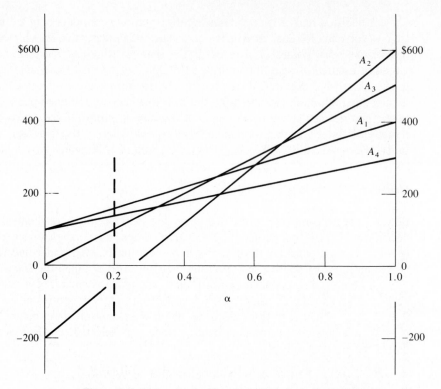

Figure 7.3 Values for the Hurwicz rule for four alternatives.

When $\alpha = 0$, the Hurwicz rule gives the same result as the maximin rule, and when $\alpha = 1$, it is the same as the maximax rule. This may be shown for the case where $\alpha = 0$ as

$$\max_{i} \{0[\max_{j} E_{ij}] + (1 - 0)[\min_{j} E_{ij}]\} = \max_{i} [\min_{j} E_{ij}].$$

For the case where $\alpha = 1$,

$$\max_{i} \{1[\max_{j} E_{ij}] + (1 - 1)[\min_{j} E_{ij}]\} = \max_{i} [\max_{j} E_{ij}].$$

Thus, the maximin rule and the maximax rule are special cases of the Hurwicz rule.

The philosophy behind the Hurwicz rule is that focus on the most extreme outcomes or consequences bounds or brackets the decision. By use of this rule, the decision maker may weight the extremes in such a manner as to reflect their relative importance.

Comparison of Decisions

As was the case for the decision criteria applied for decision making under risk, it is evident that there is no one best criterion for decision making under uncertainty. The decision made is dependent on the decision criterion adopted by the decision maker. For the examples of this section, the alternatives selected were

Laplace criterion: A_3

Maximin criterion: A_1 or A_4

Maximax criterion: A_2

Hurwicz criterion ($\alpha = 0.2$): A_1.

Examination of the selections recommmended by the five decision rules indicates that each has its own merit. Several factors may influence a decision maker's choice of a rule in a given decision situation. The decision maker's attitude toward uncertainty (pessimistic or optimistic) and his or her personal utility function are important influences. Thus, the choice of a particular decision rule for a given decision situation must be based on subjective judgment.

7.7. MODELS IN DESIGN AND OPERATIONS

The delay in the development of models to explain and describe complex systems may be attributed to an early preoccupation with the physical and biological sciences. This is understandable, however, since during much of history the limiting factors in the satisfaction of human wants were predominantly physical and biological. But with the accumulation of scientific knowledge, human beings have been able to design complex systems to meet their needs. Engineers and managers are becoming increasingly aware that experience, intuition, and judgment are insufficient for the effective pursuit of design and operational objectives. Models are useful in design and operations because they take the decision maker part way to the point of decision.

In formulating a mathematical decision model, one attempts to consider all components of the system which are relevant to the system's effectiveness and cost. Because of the impossibility of including all parameters in constructing the evaluation function, it is common practice to consider only those upon which the outcome is believed to depend significantly. This necessary viewpoint sometimes leads to the erroneous conclusion that certain segments of the environment are actually isolated from each other. Although it may be feasible to consider only those relationships that are significantly pertinent, one should remember that all system elements are interdependent.

Models for design decisions and for operational decision making are abstractions of the system under study. Like all abstractions, models involve a number of assumptions: assumptions about the operating characteristics of components, about the behavior of people, and about the nature of the environment. These assumptions must be fully understood and evaluated when models are used for decision making in design and operations.

Manipulation of the model can lead to model modification to reduce the misfit between the model and the real world. This *validation* process has a recurrent pattern analogous to the scientific method. It consists of three steps: (1) postulate a model, (2) test the model prediction or explanation against measurements or observations, and (3) modify the model to reduce the misfit.

The validation process should continue until the model is supported reasonably by evidence from measurement and observation. As a model evolves to a state of validity, it provides postulates of reality which can be depended upon. A model may have achieved a state of validity even though it gives biased results. Here the model is valid because it is consistent; that is, it gives results that do not vary.

A decision model cannot be classified as accurate or inaccurate in any absolute sense. It may be considered to be accurate if it is an idealized substitute of the actual system under study. If manipulation of the model would yield the identical result that manipulation of reality would have yielded, the model would be true. If, however, one knew what the manipulation of reality would have yielded, the process of simulation would be unnecessary. Hence, it is difficult to test a decision model except by an intuitive check for reasonableness.

SELECTED REFERENCES

(1) Ackoff, R. L., *The Art of Problem Solving*. New York: John Wiley, 1978.

(2) Banks, J., and W. J. Fabrycky, *Procurement and Inventory Systems Analysis*. Englewood Cliffs, N.J.: Prentice-Hall, 1987.

(3) Churchman, C. W., R. L. Ackoff, and E. L. Arnoff, *Introduction to Operations Research*. New York: John Wiley, 1957.

(4) Fabrycky, W. J., P. M. Ghare, and P. E. Torgersen, *Applied Operations Research and Management Science*. Englewood Cliffs, N.J.: Prentice-Hall, 1984.

(5) Fabrycky, W. J., and G. J. Thuesen, *Economic Decision Analysis* (2nd ed.). Englewood Cliffs, N.J.: Prentice-Hall, 1980.

(6) Hillier, F. S., and G. J. Lieberman, *Operations Research* (4th ed.). San Francisco: Holden-Day, 1986.

(7) Luce, R. D., and H. Raiffa, *Games and Decisions*. New York: John Wiley, 1957.

(8) Rivett, P., *Principles of Model Building*. New York: John Wiley, 1972.

(9) Rubinstein, M. F., *Tools for Thinking and Problem Solving*. Englewood Cliffs, N.J.: Prentice-Hall, 1986.

QUESTIONS AND PROBLEMS

1. Should decision making be classified as an art or as a science?

2. Discuss the various meanings of the word model.

3. Describe briefly physical models, schematic models, and mathematical models.

4. How do mathematical models directed to decision situations differ from those traditionally used in the physical sciences?

5. Contrast direct and indirect experimentation.

6. Write the general form of the evaluation function and define its symbols.

7. Identify a decision situation and indicate the variables under the control of the decision maker and those not directly under his or her control.

8. Why is it impossible to formulate a model that accurately represents reality?

9. What means may be employed to simplify model construction?

10. Formulate an evaluation matrix for a hypothetical decision situation of your choice.

11. Formulate an evaluation vector for a hypothetical decision situation under assumed certainty.

12. Develop an example to illustrate the application of paired outcomes in decision making among a number of nonquantifiable alternatives.

13. What approaches may be used to assign probabilities to future outcomes?

14. What is the role of dominance in decision making among alternatives?

15. Give an example of an aspiration level in decision making.

16. When would one follow the most probable future criterion in decision making?

17. What drawback exists in using the most probable future criterion?

18. How does the Laplace criterion for decision making under uncertainty actually convert the situation to decision making under risk?

19. Discuss the maximin and the minimax rules as special cases of the Hurwicz rule.

20. Under what conditions may a properly formulated model become useless as an aid in decision making?

21. Explain the nature of the cost components that should be considered in deciding how frequently to review a dynamic environment.

22. What caution must be exercised in the use of models?

23. Discuss several specific reasons why models are of value in decision making.

24. What should be done with those facets of a decision situation that cannot be explained by the model?

25. The cost of developing an internal training program for office automation is unknown but described by the following probability distribution.

COST	PROBABILITY OF OCCURRENCE
$ 80,000	0.20
95,000	0.30
105,000	0.25
115,000	0.20
130,000	0.05

What is the expected cost of the course? What is the most probable cost? What is the maximum cost that will occur with a 95 percent assurance?

26. Net profit has been calcuated for five investment opportunities under three possible futures. Which alternative should be selected under the most probable future criterion; the expected value criterion?

	(0.3) F_1	(0.2) F_2	(0.5) F_3
A_1	$ 100,000	$100,000	$380,000
A_2	−200,000	160,000	590,000
A_3	0	180,000	500,000
A_4	110,000	280,000	200,000
A_5	400,000	90,000	180,000

27. Daily positive and negative payoffs are given for five alternatives and five futures in the matrix below. Which alternative should be chosen to maximize the probability of receiv-

ing a payoff of at least 9? What choice would be made by using the most probable future criterion?

	(0.15) F_1	(0.20) F_2	(0.30) F_3	(0.20) F_4	(0.15) F_5
A_1	12	8	−4	0	9
A_2	10	0	5	10	16
A_3	6	5	10	15	−4
A_4	4	14	20	6	12
A_5	−8	22	12	4	9

28. The following matrix gives the payoffs in utiles (a measure of utility) for three alternatives and three possible states of nature.

		STATE OF NATURE		
		S_1	S_2	S_3
	A_1	50	80	80
Alternative	A_2	60	70	20
	A_3	90	30	60

Which alternative would be chosen under the Laplace principle? The maximin rule? The maximax rule? The Hurwicz rule with $\alpha = 0.75$?

29. The following payoff matrix indicates the costs associated with three decision options and four states of nature.

		STATE OF NATURE			
		S_1	S_2	S_3	S_4
	T_1	20	25	30	35
Option	T_2	40	30	40	20
	T_3	10	60	30	25

Select the decision option that should be selected for the maximin rule; the maximax rule; the Laplace rule; the minimax regret rule; and the Hurwicz rule with $\alpha = 0.2$. How do the rules applied to the cost matrix differ from those that are applied to a payoff matrix of profits?

30. The following matrix gives the expected profit in thousands of dollars for five marketing strategies and five potential levels of sales.

		LEVEL OF SALES				
		L_1	L_2	L_3	L_4	L_5
	M_1	10	20	30	40	50
	M_2	20	25	25	30	35
Strategy	M_3	50	40	5	15	20
	M_4	40	35	30	25	25
	M_5	10	20	25	30	20

Which marketing strategy would be chosen under the maximin rule? The maximax rule? The Hurwicz rule with $\alpha = 0.4$?

31. Graph the Hurwicz rule for all values of α using the payoff matrix of Problem 30.

32. The following decision evaluation matrix gives the expected savings in maintenance costs (in thousands of dollars) for three policies of preventive maintenance and three levels of operation of equipment. Given the probabilities of each level of operation, $P_1 = 0.3$, $P_2 = 0.25$, and $P_3 = 0.45$, determine the best policy based on the most probable future criterion.

POLICY	LEVEL OF OPERATION		
	L_1	L_2	L_3
M_1	10	20	30
M_2	22	26	26
M_3	40	30	15

Also, determine the best policy under uncertainty, using the Laplace rule, the Maximax rule, and the Hurwicz rule with $\alpha = 0.2$.

8
Models for Economic Evaluation

Economic considerations are very important in systems engineering, the process of bringing systems into being. For systems already in existence, economic considerations often provide a basis for the analysis of system operation and retirement. There are numerous examples of structures, processes, and systems that exhibit excellent physical design but have little economic merit. The essential prerequisite of successful engineering application is economic feasibility.

Many design and operational alternatives may be described in terms of their receipts and disbursements over time. When this is the case, reducing these monetary values to a common base is essential in decision making. This chapter presents interest factor applications and related techniques of economic analysis useful in comparing alternatives on an equivalent economic basis.

8.1. INTEREST AND INTEREST FORMULAS

The time value of money in the form of an interest rate is an important element in most decision situations involving money flow over time. Because money can earn at a certain interest rate, it is recognized that a dollar in hand at present is worth more than a dollar to be received at some future date. A lender may consider interest received as a gain or profit, whereas a borrower usually considers interest to be a charge or cost.

The interest rate is the ratio of the borrowed money to the fee charged for its use over a period of time, usually one year. This ratio is expressed as a percentage.

For example, if $100 is paid for the use of $1,000 for one year, the interest rate is 10%. In compound interest, the interest earned at the end of the interest period is either paid at that time or earns interest upon itself. This compounding assumption is usually used in the economic evaluation of alternatives.

A schematic model for money flow over time is shown in Figure 8.1. This money flow model is the basis for the derivation of interest factors. It may be applied to any phase of the system life cycle for the purpose of life-cycle cost and/or income analysis. Let

i = nominal annual rate of interest
n = number of interest periods, usually annual
P = principal amount at a time assumed to be the present
A = single amount in a series of n equal amounts at the end of each interest period
F = amount, n interest periods hence, equal to the compound amount of P, or the sum of the compound amounts of A, at the interest rate i.

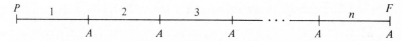

Figure 8.1   The money flow model.

Single Payment Compound Amount Formula

When interest is permitted to compound, the interest earned during each interest period is added to the principal amount at the beginning of the next interest period. Using the terms defined, the relationship among F, P, n, and i can be developed as shown in Table 8.1. The resulting factor, $(1 + i)^n$, is known as the *single payment compound amount factor* [1] and is designated as

$$\begin{pmatrix} F/P, i, n \\ \end{pmatrix}.$$

This factor may be used to express the equivalence between a present amount, P, and a future amount, F, at an interest rate i for n years. The formula is

$$F = P(1 + i)^n.$$

TABLE 8.1   SINGLE-PAYMENT COMPOUND-AMOUNT FORMULA

YEAR	AMOUNT AT BEGINNING OF YEAR	INTEREST EARNED DURING YEAR	COMPOUND AMOUNT AT END OF YEAR
1	P	Pi	$P + Pi = P(1 + i)$
2	$P(1 + i)$	$P(1 + i)i$	$P(1 + i) + P(1 + i)i = P(1 + i)^2$
3	$P(1 + i)^2$	$P(1 + i)^2 i$	$P(1 + i)^2 + P(1 + i)^2 i = P(1 + i)^3$
n	$P(1 + i)^{n-1}$	$P(1 + i)^{n-1} i$	$P(1 + i)^{n-1} + P(1 + i)^{n-1} i = P(1 + i)^n = F$

[1] Values for this factor and those which follow are tabulated in Appendix D.

or

$$F = P(\overset{F/P,i,n}{\quad}). \tag{8.1}$$

The compound amount of \$1,000 in six years at 7% interest compounded annually may be found from Equation 8.1 as

$$F = \$1,000 (1 + 0.07)^n$$

$$= \$1,000(1.501) = \$1,501.00$$

or by use of the factor designation and its tabular value from Appendix D, Table D.2,

$$F = \$1,000(\overset{F/P,7,6}{1.501}) = \$1,501.00.$$

Single-Payment Present-Amount Formula

The single-payment compound-amount formula may be solved for P and expressed as

$$P = F\left[\frac{1}{(1 + i)^n}\right].$$

The resulting factor, $1/(1 + i)^n$, is known as the *single-payment present-amount factor* and is designated

$$(\overset{P/F,i,n}{\quad}).$$

This factor may be used to express the equivalence between a future amount, F, and a present amount, P, at an interest rate i for n years. The formula is

$$P = F(\overset{P/F,i,n}{\quad}). \tag{8.2}$$

As an example, assume that it is desired to accumulate \$1,000 in 4 years. If the interest rate is 8%, the amount of money that must be deposited may be found from Equation 8.2 as

$$P = \$1,000(\overset{P/F,8,4}{0.7350}) = \$735.00.$$

Equal-Payment Series Compound-Amount Formula

In some situations, a series of receipts or disbursements occuring uniformly at the end of each year may be encountered. The sum of the compound amounts of this series may be determined by reference to Figure 8.1.

The A dollars deposited at the end of the nth year will earn no interest and will contribute only A dollars to F. The A dollars deposited at the end of period $n - 1$ will earn interest in the amount of Ai, and $A(1 + i)$ will be contributed to the sum. The amount at the end of period $n - 2$ will contribute $A(1 + i)^2$. The sum of this series will be

$$F = A(1) + A(1 + i) + A(1 + i)^2 + \cdots + A(1 + i)^{n-2} + A(1 + i)^{n-1}.$$

Multiplying this series by $(1 + i)$ gives

$$F(1 + i) = A[(1 + i) + (1 + i)^2 + (1 + i)^3 + \cdots + (1 + i)^{n-1} + (1 + i)^n].$$

Subtracting the first expression from the second gives

$$F(1 + i) - F = A[(1 + i)^n - 1]$$

$$Fi = A[(1 + i)^n - 1]$$

$$F = A\left[\frac{(1 + i)^n - 1}{i}\right].$$

The resulting factor, $[(1 + i)^n - 1]/i$, is known as the *equal-payment series compound-amount factor* and is designated

$$\left(\overset{F/A,i,n}{}\right).$$

This factor may be used to express the equivalence between an equal-payment series, A, and a future amount F, at an interest rate i for n years. The formula is

$$F = A\left(\overset{F/A,i,n}{}\right). \tag{8.3}$$

Consider an example where \$500 amounts are invested at the end of each year for four years at 9% interest. The total that will be accumulated at the end of the four-year period is

$$F = \$500(\overset{F/A,9,4}{4.5730}) = \$2{,}286.50.$$

Equal-Payment Series Sinking-Fund Formula

The equal-payment series compound-amount formula may be solved for A and expressed as

$$A = F\left[\frac{i}{(1 + i)^n - 1}\right].$$

The resulting factor, $i/[(1 + i)^n - 1]$, is known as the *equal-payment series sinking-fund factor* and is designated

$$\left(\overset{A/F,i,n}{}\right).$$

This factor may be used to express the equivalence between a future amount, F, and an equal-payment series, A, at an interest rate i for n years. The formula is

$$A = F\left(\overset{A/F,i,n}{}\right). \tag{8.4}$$

As an example, suppose that it is desired to deposit a series of uniform, year-end amounts over ten years in order to provide a total of $5,000. The amount that should be deposited each year at an interest rate of 8% is

$$A = \$5,000(\overset{A/F,8,10}{0.0690}) = \$345.00.$$

Equal-Payment Series Capital-Recovery Formula

The substitution of $P(1 + i)^n$ for F in the equal-payment series sinking-fund formula results in

$$A = P(1 + i)^n\left[\frac{i}{(1 + i)^n - 1}\right]$$

$$= P\left[\frac{i(1 + i)^n}{(1 + i)^n - 1}\right].$$

The resulting factor, $i(1 + i)/[(1 + i)^n - 1]$, is known as the *equal-payment series capital-recovery factor* and is designated

$$(\overset{P/A,i,n}{\quad}).$$

This factor may be used to express the equivalence between future equal-payment series, A, and a present amount, P, at an interest rate i for n years. The formula is

$$A = P(\overset{A/P,i,n}{\quad}). \tag{8.5}$$

As an example of the application of this formula, assume that a deposit of $100,000 at 8% for eight years is made in a savings account. This will enable a uniform amount to be withdrawn every year, for the next eight years. The magnitude of this amount is

$$A = \$100,000(\overset{A/P,8,8}{0.1740}) = \$17,400.00.$$

Equal-Payment Series Present-Amount Formula

The equal-payment series capital-recovery formula can be solved for P and expressed as

$$P = A\left[\frac{(1 + i)^n - 1}{i(1 + i)^n}\right].$$

The resulting factor $[(1 + i)^n - 1]/(1 + i)^n$, is known as the *equal-payment series present-amount factor* and is designated

$$(\overset{P/A,i,n}{\quad}).$$

This factor may be used to express the equivalence between future equal-payment series, A, and a present amount, P, at an interest rate i for n years. The formula is

$$P = (\overset{P/A,\,i,\,n}{}).\tag{8.6}$$

As an example of the application of this formula, assume that an expenditure now will save \$4,000 per year in operating costs over the next ten years. If an interest rate of 15% is used, the present equivalent amount of these savings is

$$P = \$4,000(\overset{P/A,\,15,\,10}{5.0188}) = \$20,075.20.$$

Summary of Interest Formulas

The interest factors derived in the previous paragraphs express relationships among P, A, F, i, and n. In any given application, an equivalence between F and P, P and F, F and A, A and F, P and A, or A and P is expressed for an interest rate i and a number of years n. Table 8.2 may be used to select the interest formula needed in a given situation. The factor designations summarized in the last column make it possible to set up a problem symbolically before determining the values of the factors involved.

Interest tables based on the designations summarized in Table 8.2 make it unnecessary to remember the name of each formula. Such tables are given in Appendix D for a range of values for i and n. Each column in the table is headed by a designation that identifies the function of the tabular value as established by the money-flow diagram. Each table is based on \$1 for a specific value of i, a range of n, and all six factors.

TABLE 8.2   SUMMARY OF INTEREST FORMULAS

FORMULA NAME	FUNCTION	FORMULA	DESIGNATION
Single-payment compound amount	Given P Find F	$F = P(1 + i)^n$	$F = P(^{F/P,\,i,\,n})$
Single-payment present amount	Given F Find P	$P = F\left[\dfrac{1}{(1 + i)^n}\right]$	$P = F(^{P/F,\,i,\,n})$
Equal-payment series compound amount	Given A Find F	$F = A\left[\dfrac{(1 + i)^n - 1}{i}\right]$	$F = A(^{F/A,\,i,\,n})$
Equal-payment series sinking fund	Given F Find A	$A = F\left[\dfrac{i}{(1 + i)^n - 1}\right]$	$A = F(^{A/F,\,i,\,n})$
Equal-payment series present amount	Given A Find P	$P = A\left[\dfrac{(1 + i)^n - 1}{i(1 + i)^n}\right]$	$P = A(^{P/A,\,i,\,n})$
Equal-payment series capital recovery	Given P Find A	$A = P\left[\dfrac{i(1 + i)^n}{(1 + i)^n - 1}\right]$	$A = P(^{A/P,\,i,\,n})$

8.2. DETERMINING ECONOMIC EQUIVALENCE

If two or more situations are to be compared, their characteristics must be placed on an equivalent basis. Two things are said to be equivalent when they have the same effect. For instance, the torques produced by applying forces of 100 lb and 200 lb, 2 ft and 1 ft, respectively, from the fulcrum of the lever are equivalent, since each produces a torque of 200 ft-lb.

Two monetary amounts are equivalent when they have the same value in exchange. Three factors are involved in the equivalence of sums of money: (1) the amount of the sums, (2) the time of occurrence of the sums, and (3) the interest rate. Symbolically, the economic equivalence function may be stated as[2]

$$E = f(F_t, i, n) \tag{8.7}$$

where

E = the present equivalent, annual equivalent, or future equivalent amount
F_t = a positive or negative sum of money at the end of the year t
i = the annual rate of interest
n = the number of years.

Interest Formula Equivalence Calculations

Equivalence calculations based on simple applications of the interest formulas derived in Section 8.1 are illustrated in this section. More comprehensive applications are presented later. In each case, the determination of economic equivalence is in accordance with Equation 8.7.

At an interest rate of 10% with $n = 8$ years, a P of $1 is equivalent to an F of $2.144. This may be stated as

$$F = \$1(\overset{F/P,10,8}{2.1440}) = \$2.1440.$$

A practical application of this statement of equivalence is that $1 spent today must result in a revenue receipt, or avoid a cost of $2.144 eight years hence, if the interest rate is 10%.

A reciprocal situation is where $i = 12\%$, $n = 10$ years, and $F = \$1$. The P that is equivalent to $1 is

$$P = \$1(\overset{P/F,12,10}{0.3220}) = \$0.3220.$$

A practical application of this statement of equivalence is that no more than $0.322 can be spent today to secure a revenue receipt or to avoid a cost of $1 ten years hence if the interest rate is 12%.

---

[2] This equivalence function is related to the decision evaluation function given by Equation 7.1 in Section 7.3. It is directly applicable to alternatives that are described by money flows over time according to the convention established in Figure 8.1.

If $i = 8\%$ and $n = 20$ years, the F that is equivalent to an A of $1 is

$$F = \overset{F/A, 8, 20}{\$1(45.7620)} = \$45.7620.$$

A practical application of this equivalence statement is that $1 spent each year for 20 years must result in a revenue receipt, or it must avoid a cost of $45.762, 20 years hence if the interest rate is 8%.

A reciprocal situation is where $i = 12\%$, $n = 6$ years, and $F = \$1$. The A that is equivalent to $1 is

$$A = \overset{A/F, 12, 6}{\$1(0.1232)} = \$0.1232.$$

A practical application of this equivalence statement is that $0.1232 must be received each year for six years to be equivalent to the receipt of $1 six years hence.

If $i = 9\%$, $n = 10$ years, and A is $1, the P that is equivalent to $1 is

$$P = \overset{P/A, 9, 10}{\$1(6.4177)} = \$6.4177.$$

A practical application of this statement of equivalence is that an investment of $6.4177 today must yield an annual benefit of $1 each year for ten years if the interest rate is 9%.

A reciprocal situation is where $i = 14\%$, $n = 7$ years, and $P = \$1$. The A that is equivalent to $1 is

$$A = \overset{A/P, 14, 7}{\$1(0.2332)} = \$0.2332.$$

A practical application of this statement of equivalence is that $1 can be spent today to capture an annual saving of $0.2332 per year over seven years if the interest rate is 14%.

Equivalence Function Diagrams

A useful technique in the determination of equivalence is to plot present value as a function of the interest rate. For example, what value of i will make a P of $1,500 equivalent to an F of $5,000 if $n = 9$ years? Symbolically,

$$\$5,000 = 1,500(\overset{F/P, i, 9}{\quad}).$$

The solution is illustrated graphically in Figure 8.2, showing that i is slightly less than 13%.

Present value may also be plotted as a function of n as a useful technique for determining equivalence. For example, what value of n will make a P of $4,000 equivalent to an F of $8,000 if $i = 8\%$? Symbolically,

$$\$8,000 = \$4,000(\overset{F/P, 8, n}{\quad}).$$

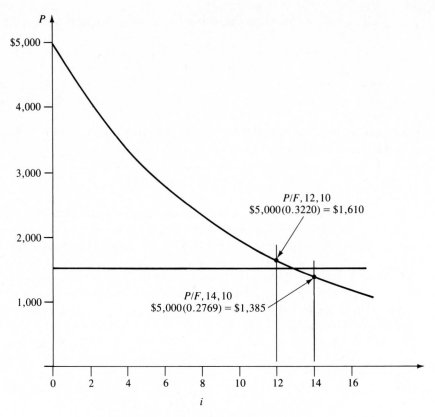

Figure 8.2   Equivalence function diagram for *i*.

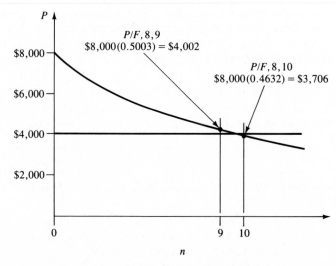

Figure 8.3   Equivalence function diagram for *n*.

The solution is illustrated graphically in Figure 8.3, showing that n is between nine and ten years.

Equivalence functions can be plotted for A or F as a function of i or n if desired. In these cases, equivalence would be found between the two quantities A or F as a function of i or n.

8.3. EVALUATING A SINGLE ALTERNATIVE UNDER CERTAINTY

In most cases, a decision is reached by selecting one option from among two or more available alternatives. Sometimes, however, the decision is limited to acceptance or rejection of a single alternative. In such a case the decision will be based on the relative merit of the alternative and other opportunities believed to exist, even though none of the latter have been crystallized into definite proposals.

When only one specified possibility exists, it should be evaluated within a framework that will permit its desirability to be compared to other oportunities that may exist but are unspecified. The most common bases for comparison are the present-equivalent evaluation, the annual-equivalent evaluation, and future-equivalent evaluation, the rate-of-return evaluation, and the service-life evaluation.

The following example illustrates the several bases of evaluation of a single alternative under certainty about future outcomes. Assume that a waste heat recirculation system is contemplated for a small building. The anticipated costs and savings of this project are given in Table 8.3. Consider an interest rate of 12% as the cost of money and take January 1, 19×0, to be the present.

For convenience, costs and savings that may occur during the year are assumed to occur at the end of that year or the start of the next year, considered to be the same point in time. There is some small error in the practice of considering money flows to be year-end amounts. This error is insignificant, however, in comparison to the usual errors in estimates, except under very high interest rates. The costs and savings can be represented by the money flow diagram shown in Figure 8.4.

TABLE 8.3 DISBURSEMENTS AND SAVINGS FOR A SINGLE ALTERNATIVE

ITEM	DATE	DISBURSEMENTS	SAVINGS
Initial cost	1-1-19×0	$28,000	—
Saving, first year	1-1-19×1	—	$9,500
Saving, second year	1-1-19×2	—	9,500
Overhaul cost	1-1-19×2	2,500	—
Saving, third year	1-1-19×3	—	9,500
Saving, fourth year	1-1-19×4	—	9,500
Salvage value	1-1-19×4	—	8,000

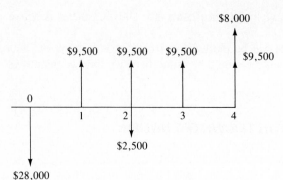

Figure 8.4 Money flow diagram from Table 8.3.

Present Equivalent Evaluation

The present equivalent evaluation is based on finding a present equivalent amount, *PE*, that represents the difference between present equivalent savings and present equivalent costs for an alternative at a given interest rate. Thus, the present equivalent amount at an interest rate i over n years is found by applying Equation 8.2.

$$PE(i) = F_0(\overset{P/F,i,0}{\quad}) + F_1(\overset{P/F,i,1}{\quad}) + F_2(\overset{P/F,i,2}{\quad}) + \cdots + F_n(\overset{P/F,i,n}{\quad})$$

$$= \sum_{t=0}^{n} F_t(\overset{P/F,i,t}{\quad}). \tag{8.8}$$

But since $(\overset{P/F,i,t}{\quad}) = (1 + i)^{-t}$,

$$PE(i) = \sum_{t=0}^{n} F_t(1 + i)^{-t} \quad (-1 < i < \infty). \tag{8.9}$$

For the waste heat recirculation system example, the present equivalent amount at 12% using Equation 8.7 is

$$PE(12) = -\$28,000(\overset{P/F,12,0}{1.0000}) + \$9,500(\overset{P/F,12,1}{0.8929}) + \$7,000(\overset{P/F,12,2}{0.7972})$$

$$+ \$9,500(\overset{P/F,12,3}{0.7118}) + \$17,500(\overset{P/F,12,4}{0.6355})$$

$$= \$3,946.$$

Or by using Equation 8.8

$$PE(12) = -\$28,000(1.12)^0 + \$9,500(1.12)^{-1} + \$7,000(1.12)^{-2}$$

$$+ \$9,500(1.12)^{-3} + \$17,500(1.12)^{-4}$$

$$= \$3,946.$$

Since the present equivalent amount is greater than zero, this is a desirable undertaking at 12%.

Annual Equivalent Evaluation

The annual equivalent evaluation is similar to the present-equivalent evaluation except that the difference between savings and costs is now expressed as an annual equivalent amount, AE, at a given interest rate as follows:

$$AE(i) = PE(i)(\overset{A/P, i, n}{}).\tag{8.10}$$

But since

$$PE(i) = \sum_{t=0}^{n} F_t(1 + i)^{-t}$$

and

$$(\overset{A/P, i, n}{}) = \left[\frac{i(1 + i)^n}{(1 + i)^n - 1}\right]$$

$$AE(i) = \left[\sum_{t=0}^{n} F_t(1 + i)^{-t}\right]\left[\frac{i(1 + i)^n}{(1 + i)^n - 1}\right].\tag{8.11}$$

For the waste-heat recirculation example, the annual equivalent evaluation using Equation 8.10 gives

$$AE(12) = \$3,946(\overset{A/P, 12, 4}{0.3292}) = \$1,299.$$

Or, by using Equation 8.11

$$AE(12) = [-\$28,000(1.12)^0 + \$9,500(1.12)^{-1} +$$

$$+ \$7,000(1.12)^{-2} + \$9,500(1.12)^{-3} +$$

$$+ \$17,500(1.12)^{-4}]\left[\frac{0.12(1.12)^4}{(1.12)^4 - 1}\right]$$

$$= \$1,299.$$

These results mean that if \$28,000 is invested on January 1, 19×0, a 12% return will be received plus an equivalent of \$1,299 on January 1, 19×1, 19×2, 19×3, and 19×4.

Future Equivalent Evaluation

The future equivalent basis for comparison is based on finding an equivalent amount, FE, that represents the difference between the future equivalent savings and future equivalent costs for an alternative at a given interest rate. Thus, the future equivalent amount at an interest rate i over n years is formed by applying Equation 8.1.

$$FE(i) = F_0 \left(\overset{F/P,i,n}{} \right) + F_1 \left(\overset{F/P,i,n-1}{} \right) + \cdots + F_{n-1} \left(\overset{F/P,i,1}{} \right) + F_n \left(\overset{F/P,i,0}{} \right)$$

$$= \sum_{t=0}^{n} F_t \left(\overset{F/P,i,n-t}{} \right). \tag{8.12}$$

But since

$$\left(\overset{F/P,i,n-t}{} \right) = (1 + i)^{n-t}$$

$$FE(i) = \sum_{t=0}^{n} F_t (1 + i)^{n-t} \tag{8.13}$$

or,

$$FE(i) = PE(i) \left(\overset{F/P,i,n}{} \right). \tag{8.14}$$

For the waste heat recirculation example, the future equivalent evaluation using Equation 8.12 gives

$$FE(12) = -\$28,000 \left(\overset{F/P,12,4}{1.5740} \right) + \$9,500 \left(\overset{F/P,12,3}{1.4050} \right) + \$7,000 \left(\overset{F/P,12,2}{1.2542} \right)$$

$$+ \$9,500 \left(\overset{P/F,12,1}{1.1208} \right) + \$17,500 \left(\overset{P/F,12,0}{1.0000} \right)$$

$$= \$6,211.$$

Or, by the use of Equation 8.14,

$$FE(12) = \$3,946 \left(\overset{F/P,12,4}{1.5740} \right) = \$6,211.$$

Since the future equivalent amount is greater than zero, this is a desirable venture at 12%.

Rate-of-Return Evaluation

The rate-of-return evaluation is probably the best method for comparing a specific proposal with other opportunities believed to exist but not delineated. The rate-of-return is a universal measure of economic success since the return from different classes of opportunities are usually well established and generally known. This permits comparison of an alternative against accepted norms. This characteristic makes the rate-of-return comparison well adapted to the situation where the choice is to accept or reject a single alternative.

Rate-of-return is a widely accepted index of profitability. It is defined as the interest rate that causes the equivalent receipts of a money flow to be equal to the equivalent disbursements of that money flow. The interest rate that reduces the $PE(i)$, $AE(i)$, or $FE(i)$ of a series of receipts and disbursements to zero is another way of defining the rate-of-return. Mathematically, the rate-of-return for an investment proposal is the interest rate i^* that satisfies the equation

$$0 = PW(i^*) = \sum_{t=0}^{n} F_t(1 + i^*)^{-t} \qquad (8.15)$$

where the proposal has a life of n years.

As an example of the rate-of-return evaluation, the energy saving system can be evaluated on the basis of the rate of return that would be secured from the invested funds. In effect, a rate of interest will be specified that makes the receipts and disbursements equivalent. This can be done either by equating present equivalent, annual equivalent, or future equivalent amounts.

Equating the present-equivalent amount of receipts and disbursements at $i = 15\%$ gives

$$[\$9,500(\overset{P/A,15,4}{2.8850}) + \$8,000(\overset{P/F,15,4}{0.5718})] = [28,000 + \$2,500(\overset{P/F,15,2}{0.7562})]$$

$$[\$27,408 + \$4,574] = [\$28,000 + \$1,890]$$

$$\$31,982 = \$29,800$$

At $i = 20\%$

$$[\$9,500(\overset{P/A,20,4}{2.5887}) + \$8,000(\overset{P/F,20,4}{0.4823})] = [\$28,000 + \$2,500(\overset{P/F,20,2}{0.6945})]$$

$$[\$24,592 + \$3,858] = [\$28,000 + \$1,736]$$

$$\$28,450 = \$29,736$$

Interpolating,

$$i = 15\% + (5)\left[\frac{[31,982 - 29,800] - 0}{[31,982 - 29,800] - [28,450 - 29,736]}\right]$$

$$i = 18.15\%.$$

From these results, a present-equivalent graph can be developed as shown in Figure 8.5. From the graph it is evident that the rate of return on the venture is just over 18%. This means that the investment of $28,000 in the system should yield a 18.15% rate of return over the four year period.

Payout Evaluation

Often a proposed system can be evaluated in terms of how long it will take the system to pay for itself from benefits, revenues, or savings. Systems that tend to pay for themselves quickly are desirable because there is less uncertainty with estimates of short duration.

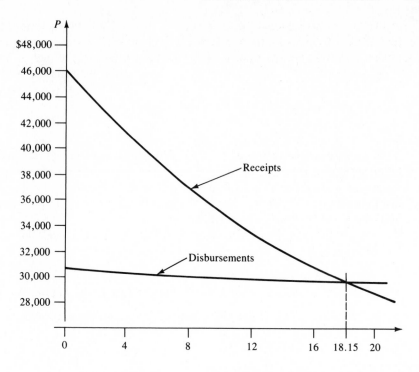

Figure 8.5   Solution for the rate of return.

The payout period is the amount of time required for the difference in the present value of receipts (savings) to equal the present value of the disbursements (costs); the annual equivalent receipts and disbursements may also be equated. For the present equivalent approach

$$0 \le \sum_{t=0}^{n^*} F_t(1 + i)^{-t}. \qquad (8.16)$$

The smallest value of n^* that satisfied the above expression is the payout duration.

For a duration of three years the equation for present equivalent of savings and disbursements is

$$[\$9{,}500(\overset{P/A,\,12,\,3}{2.4018}) + \$8{,}000(\overset{P/F,\,12,\,3}{0.7118})] = [\$28{,}000 + \$2{,}500(\overset{P/F,\,12,\,2}{0.7972})]$$

$$[\$22{,}817 + \$5{,}694] = [\$28{,}000 + \$1{,}993]$$

$$\$28{,}511 = \$29{,}993.$$

For a duration of four years, the equation for present amount of savings and disbursements is

$$\overset{P/A,\,12,\,4}{[\$9,\!500(\;3.0374\;)} + \overset{P/F,\,12,\,4}{\$8,\!000(\;0.6355\;)]} = [\$28,\!000 + \overset{P/F,\,12,\,2}{\$2,\!500(\;0.7972\;)]}$$

$$[\$28,\!855 + \$5,\!084] = [\$28,\!000 + \$1,\!993]$$

$$\$33,\!939 = \$29,\!993.$$

Interpolating, we have

$$n = 2 + \frac{[22,\!433 - 26,\!007] - 0}{[22,\!433 - 26,\!007] - [28,\!511 - 26,\!220]}$$

$$n = 3.3 \text{ years.}$$

From these results, a present-equivalent graph can be developed, as shown in Figure 8.6. From this graph it is evident that the payout period on the energy-saving system is just over three years.

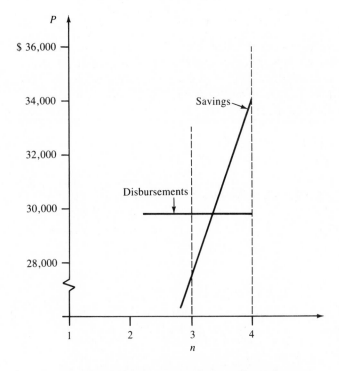

Figure 8.6 Solution for the payout period.

Annual Equivalent Cost of an Asset

A very useful application of the annual equivalent evaluation approach pertains to the cost of an asset. The cost of any asset is made up of two components, the cost of depreciation and the cost of interest on the undepreciated balance. It can be shown that the annual equivalent cost of an asset is independent of the depreciation function chosen to represent the value of the asset over time.

As an example of the cost of depreciation plus the cost of interest on the unde-preciated balance, consider the following example based on straight line deprecia-tion. An asset has a first cost of $5,000, a salvage value of $1,000, a service life of five years, and the interest rate is 10%. The annual costs are shown in Figure 8.7.

The present equivalent cost of depreciation plus interest on the the undepreci-ated balance from Equation 8.8 is

$$= \$1,300(\overset{P/F, 10, 1}{0.9091}) + \$1,220(\overset{P/F, 10, 2}{0.8265}) + \$1,140(\overset{P/F, 10, 3}{0.7513})$$

$$+ \$1,060(\overset{P/F, 10, 4}{0.6830}) + \$980(\overset{P/F, 10, 5}{0.6209})$$

$$= \$4,379$$

and the annual equivalent cost of the asset from Equation 8.10 is

$$\$4,379(\overset{A/P, 10, 5}{0.2638}) = \$1,155.$$

Regardless of the depreciation function that describes the reduction in value of a physical asset over time, the annual equivalent cost of an asset may be expressed as the annual equivalent first cost minus the annual equivalent salvage value. This an-nual equivalent cost is the amount an asset must earn each year if the invested capital is to be recovered along with a return on the investment. The annual equivalent cost is derived as follows.

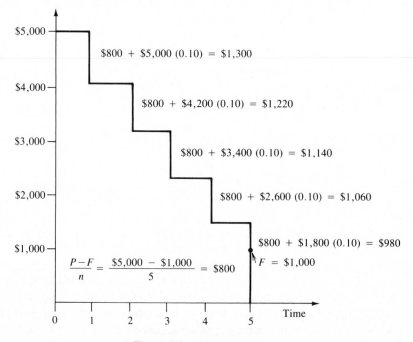

Figure 8.7 Value-time function.

$$A = P(\overset{A/P,i,n}{\quad}) - F(\overset{A/F,i,n}{\quad})$$

$$(\overset{A/F,i,n}{\quad}) = (\overset{A/P,i,n}{\quad}) - i.$$

But since

$$A = P(\overset{A/P,i,n}{\quad}) - F[(\overset{A/P,i,n}{\quad}) - i]$$

$$A = (P - F)(\overset{A/P,i,n}{\quad}) + F(i). \tag{8.17}$$

This means that the annual equivalent cost of any asset may be found from knowledge of its first cost, P, its anticipated service life, n, its estimated salvage value, F, and the interest rate, i. For example, an asset with a first cost of $5,000, a service life of five years, and a salvage value of $1,000 will lead to an annual equivalent cost of

$$(\$5,000 - \$1,000)(\overset{A/P,10,5}{0.2638}) + \$1,000(0.10) = \$1,155$$

if the interest rate is 10%.

The annual equivalent cost formula is very useful in a "should cost" study. For example, suppose that a contractor is to use equipment, tooling, and so on, for a certain defense contract which will cost $3,250,000 and which will depreciate to $750,000 at the end of the four year production contract. If the money costs the contractor 10%, the annual equivalent cost of the contractor's equipment should be

$$(\$3,250,000 - \$750,000)(\overset{A/P,10,4}{0.3155}) + \$750,000(0.10) = \$863,750.$$

8.4. EVALUATING MULTIPLE ALTERNATIVES UNDER CERTAINTY

When a number of mutually exclusive alternatives provide service of equal value, it is desirable to compare them directly to each other. Where service of unequal value is provided by multiple alternatives, each alternative must be evaluated as a single alternative and is accepted or rejected on the basis of one or more of the comparisons suggested in the preceding section. In many cases, however, the available alternatives do provide outputs that are identical or equal in value. Under this condition, the objective is to select the alternative that provides the desired service at least cost.

Assume that a defense contractor is considering the purchase of new test equipment. Semiautomatic test equipment will cost $110,000 and can be expected to last six years, with a salvage value of $10,000. Operating costs will be $28,000 per year. Fully automatic equipment will cost $160,000, should last six years and have a salvage value of $14,000. The operating costs will be $12,000 per year. The services provided by the machines will be identical. With a desired interest rate of 14%, the alternative that meets the criterion of least cost should be selected.

Present Equivalent Evaluation

Under this method, the two alternatives may be compared on the basis of equivalent cost at a time taken to be the present. The present equivalent cost of the semi-automatic test equipment is

$$\$110{,}000 + \$28{,}000(\overset{P/A,\,14,\,6}{3.8887}) - \$10{,}000(\overset{P/F,\,14,\,6}{0.4556}) = \$214{,}328.$$

The present equivalent cost of the fully automatic equipment is

$$\$160{,}000 + \$12{,}000(\overset{P/A,\,14,\,6}{3.8887}) - \$14{,}000(\overset{P/F,\,14,\,6}{0.4556}) = \$200{,}286.$$

This comparison shows the present equivalent cost of the fully automatic test equipment to be less than the present equivalent cost of the semiautomatic test equipment by \$14,042 (\$214,328 − \$200,286).

Annual Equivalent Evaluation

The annual equivalent costs are taken as an equal-cost series over the life of the assets. The annual equivalent cost of the semiautomatic test equipment is

$$\$110{,}000(\overset{A/P,\,14,\,6}{0.2572}) + \$28{,}000 - \$10{,}000(\overset{A/F,\,14,\,6}{0.1175}) = \$55{,}118$$

and the annual equivalent cost of the fully automatic test equipment is

$$\$160{,}000(\overset{A/P,\,14,\,6}{0.2572}) + \$12{,}000 - \$14{,}000(\overset{A/F,\,14,\,6}{0.1175}) = \$51{,}507.$$

The annual equivalent difference of \$55,118 minus \$51,507 or \$3,611 is the annual equivalent cost superiority of the fully automatic test equipment. As a verification, the annual equivalent amount of the present equivalent difference is

$$\$14{,}041(\overset{A/P,\,14,\,6}{0.2572}) = \$3{,}611.$$

Rate-of-Return Evaluation

The previous cost comparisons indicated that the fully automatic test equipment was more desirable at an interest rate of 14%. At some higher interest rate, the two alternatives will be identical in equivalent cost, and beyond that interest rate, the semiautomatic test equipment will be less expensive because of its lower initial cost.

The interest rate at which the costs of the two alternatives are identical can be determined by setting the present equivalent amounts for the alternatives equal to each other and solving for the interest rate, i. Thus,

$$\$110{,}000 + \$28{,}000(\overset{P/A,i,6}{\quad}) - \$10{,}000(\overset{P/F,i,6}{\quad})$$

$$= \$160{,}000 + \$12{,}000(\overset{P/A,i,6}{\quad}) - \$14{,}000(\overset{P/F,i,6}{\quad})$$

$$\$16{,}000(\overset{P/A,i,6}{\quad}) + \$4{,}000(\overset{P/F,i,6}{\quad}) = \$50{,}000.$$

For $i = 20\%$

$$\$16{,}000(\,\overset{P/A,20,6}{3.3255}\,) + \$4{,}000(\,\overset{P/F,20,6}{0.3349}\,) = \$54{,}548$$

For $i = 25\%$

$$\$16{,}000(\,\overset{P/A,25,6}{2.9514}\,) + \$4{,}000(\,\overset{P/F,25,6}{0.2622}\,) = \$48{,}271$$

Then, by interpolation,

$$i = 20 + (5)\frac{\$[54{,}548 - \$50{,}00]}{\$[54{,}548 - \$48{,}271]}$$

$$i = 23.6\%.$$

When funds earn less than 23.6%, the fully automatic test equipment will be most desirable. When funds earn more than 23.6%, the semiautomatic equipment would be preferred.

Payout Evaluation

The service life of six years for each of the two test equipment alternatives is only the result of estimates and may be in error. If the services are needed for shorter or longer periods of time and if the assets are capable of providing the service for a longer period of time, the advantage may pass from one alternative to the other. Just as there is an interest rate at which two alternatives may be equal, there may be a service life at which the equivalent costs may be identical. This service life may be obtained by setting the alternatives equal to each other and solving for the life, n. Thus, for an interest rate of 14%,

$$\$110{,}000 + \$28{,}000(\overset{P/A,14,n}{\quad}) - \$10{,}000(\overset{P/F,14,n}{\quad})$$

$$= \$160{,}000 + \$12{,}000(\overset{P/A,14,n}{\quad}) - \$14{,}000(\overset{P/F,14,n}{\quad})$$

$$\$16{,}000(\overset{P/A,14,n}{\quad}) + \$4{,}000(\overset{P/F,14,n}{\quad}) = \$50{,}000.$$

For $n = $ four years

$$\$16{,}000(\,\overset{P/A,14,4}{2.9137}\,) + \$4{,}000(\,\overset{P/F,14,4}{0.5921}\,) = \$48{,}988.$$

For n = five years

$$\overset{P/A,\,14,5}{\$16,000(\ 3.4331\ )} + \overset{P/F,\,14,5}{\$4,000(\ 0.5194\ )} = \$57,007.$$

Then, by interpolation,

$$n = 4 + (1)\frac{[\$50,000 - \$48,988]}{[\$57,007 - \$48,988]}$$

$$n = 4.12 \text{ years.}$$

Thus if the desired service were to be used for less than 4.12 years, the semiautomatic equipment would be the most desirable.

Life-Cycle Economic Evaluation

An example of the evaluation of multiple alternatives under certainty where the equivalence measure is not determined at a traditional point in time is considered next. Suppose that there are two design concepts being considered for the acquisition of a complex system for which the life-cycle costs (in millions of dollars) are estimated to be as shown in Figure 8.8.

Although other choices are available the design concepts may be compared by finding their equivalent costs at the end of the production phase (the point in time when the system goes into operation). For Design A this equivalent cost at 10% is

$$\overset{F/P,\,10,1}{\$10 + \$5(\ 1.100\ )} + \overset{F/P,\,10,2}{\$1(\ 1.210\ )} + \overset{F/P,\,10,3}{\$0.8(\ 1.331\ )} + \overset{F/P,\,10,4}{\$0.6(\ 1.464\ )}$$

$$+ \overset{P/A,\,10,7}{\$1(\ 4.8684\ )} = \$23.52 \text{ million.}$$

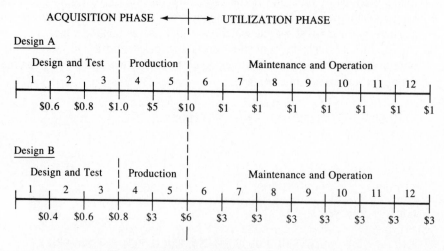

Figure 8.8 Two design alternatives for a complex system.

And for Design B, the equivalent life-cycle cost is

$$\$6 + \$3(\overset{F/P,\,10,\,1}{1.100}) + \$0.8(\overset{F/P,\,10,\,2}{1.210}) + \$0.6(\overset{F/P,\,10,\,3}{1.331}) + \$0.4(\overset{F/P,\,10,\,4}{1.464})$$

$$+ \$3(\overset{P/A,\,10,\,7}{4.8684}) = \$26.26 \text{ million.}$$

From the above analysis it is evident that Design A would be preferred since $23.52 is less than $26.26. These are decision values and are not to be considered the costs of Design A and Design B.

8.5. EVALUATING MULTIPLE ALTERNATIVES WITH MULTIPLE FUTURES

An engineering firm is engaged in the design of a unique machine tool. Since the new machine tool is technologically advanced, the firm is considering setting up a new production facility separate from its other operations.

Since the firm's own finances are not adequate to set up this new facility, there is a need to borrow capital from financial sources. The firm has identified three sources, namely, A, B, and C, which would lend money for this venture.

Each of these financial sources will lend money based on the anticipated sales of the product and will use a varying interest rate based on the anticipated demand for the product. If the product demand is anticipated to be low over the next few years, the financial backers will charge a high interest rate. This is because the backers feel that the firm may not borrow money in the future if the product fails due to low demand.

On the other hand, if the financial sources feel that the demand will be high, they will be willing to lend the money at a lower interest rate. This is because of the possibility of future business with the firm owing to its present product being popular.

The engineering firm has fairly accurate knowledge of the receipts and disbursements associated with each demand level and the financial sources under consideration. The firm has knowledge about the initial outlay and anticipated receipts over three years for each demand level as is summarized in Table 8.4. The interest rates quoted by each financial source are summarized in Table 8.5.

The engineering firm wishes to select a financing source in the face of three demand futures, not knowing which demand future will occur. Accordingly, a deci-

TABLE 8.4 ANTICIPATED DISBURSEMENTS AND RECEIPTS FOR MACHINE TOOL PRODUCTION

DEMAND LEVEL	INITIAL OUTLAY	ANNUAL RECEIPTS
Low	$ 500	$400
Medium	1,300	700
High	2,000	900

TABLE 8.5   ANTICIPATED INTEREST RATES FOR PRODUCTION FINANCING

FINANCING SOURCE	DEMAND LEVEL	INTEREST RATE
A	Low	15
	Medium	13
	High	7
B	Low	14
	Medium	12
	High	8
C	Low	15
	Medium	11
	High	6

sion evaluation matrix is developed as a first step. Using the present equivalent approach, nine evaluation values are derived and displayed in Table 8.6, one for each financing source for each demand level. For example, if financing comes from Source A, and the demand level is low, the present equivalent payoff is

$$PE(15) = \$-500 + \$400(\overset{P/A,\,15,\,3}{2.2832}) = \$413.$$

If the firm applied probabilities to each of the demand futures, the financing decision becomes one under risk. If the probability of a low demand is 0.3, the probability of a medium demand is 0.2, and the probability of a high demand is 0.5, the expected present equivalent amount for each financing source is calculated as follows.

Source A: $413(0.3) + $353(0.2) + $362(0.5) = $376

Source B: $429(0.3) + $382(0.2) + $320(0.5) = $365

Source C: $343(0.3) + $411(0.2) + $406(0.5) = $388

Accordingly, financing Source C is selected by this decision rule.

In the event the firm has no basis for assigning probabilities to demand futures, the financing decision must be made under uncertainty. Under the Laplace criterion, each future is assumed to be equally likely and the present equivalent payoffs are calculated as follows.

TABLE 8.6   PRESENT EQUIVALENT PAYOFFS FOR THREE FINANCING SOURCES

Source	DEMAND LEVEL		
	Low	Medium	High
A	$413	$353	$362
B	429	382	320
C	343	411	406

$$\text{Source A: } \$(413 + 353 + 362) \div 3 = \$376$$

$$\text{Source B: } \$(429 + 382 + 320) \div 3 = \$377$$

$$\text{Source C: } \$(343 + 411 + 406) \div 3 = \$386$$

Financing Source C would be selected by this decision rule.

By the Maximin rule, the firm calculates the minimum present equivalent amount which could occur for each financing source and then selects the source which provides a maximum. The minimums for each source are

Source A: $353

Source B: $320

Source C: $343

Financing Source A would be selected by this decision rule as the one which will maximize the minimum present equivalent amount.

By the Maximax rule, the firm calculates the maximum present equivalent amount which could occur for each financing source and then selects the source which provides a maximum. The maximums for each source are

Source A: $413

Source B: $429

Source C: $411

Financing Source B would be selected by this decision rule as the one which will maximize the present equivalent amount.

A summary of the sources selected by each decision rule is given in Table 8.7.

TABLE 8.7 SUMMARY OF FINANCING SOURCE SELECTIONS

	FINANCING SOURCE		
Decision rule	A	B	C
Expected Value			X
Laplace			X
Maximin	X		
Maximax		X	

8.6. *BREAK-EVEN ECONOMIC EVALUATIONS*

Break-even analysis may be graphical or mathematical in nature. This economic evaluation technique is useful in relating fixed and variable costs to the number of hours of operation, the number of units produced, or other measures of operational activity. In each case, the break-even point is of primary interest in that it identifies the range of the decision variable within which the most desirable economic outcome may occur.

When the cost of two or more alternatives is a function of the same variable, it is usually useful to find the value of the variable for which the alternatives incur equal cost. Several examples will be presented in this section.

Make-or-Buy Evaluation

Often a manufacturing firm has the choice of making or buying a certain component for use in the product being produced. When this is the case, the firm faces a make-or-buy decision.

Suppose, for example, that a firm finds that it can buy from a vendor the electric power supply for the system it produces for $8 per unit. Alternatively, suppose that it can manufacture an equivalent unit for a variable cost of $4 per unit. It is estimated that the additional fixed cost in the plant would be $12,000 per year if the unit is manufactured. The number of units per year for which the cost of the two alternatives breaks even would help in making the decision.

First, the total annual cost is formulated as a function of the number of units for the make alternative. It is

$$TC_M = \$12,000 + \$4N$$

and the total annual cost for the buy alternative is

$$TC_B = \$8N.$$

Break-even occurs when $TC_M = TC_B$ or

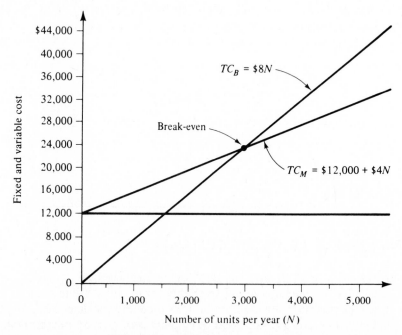

Figure 8.9 Break-even for make or buy.

$$\$12,000 + \$4N = \$8N$$

$$\$4N = \$12,000$$

$$N = 3,000 \text{ units.}$$

These cost functions and the break-even point are shown in Figure 8.9. For requirements in excess of 3,000 units per year, the make alternative would be more economical. If the rate of use is likely to be less than 3,000 units per year, the buy alternative should be chosen.

If the production requirement changes during the course of the production program, the break-even choice in Figure 8.9 can be used to guide the decision of whether to make or to buy the power unit. Small deviations below and above 3,000 units per year make little difference. However, the difference can be significant when the production requirement is well above or below 3,000 units per year.

Lease-or-Buy Evaluation

As another example of break-even analysis, consider the decision to lease or buy a piece of equipment. Assume that a small electronic computer is needed for data processing in an engineering office. Suppose that the computer can be leased for $50 per day, which includes the cost of maintenance. Alternatively, the computer can be purchased for $25,000.

The computer is estimated to have a useful life of 15 years with a salvage value of $4,000 at the end of that time. It is estimated that annual maintenance costs will be $2,800. If the interest rate is 9% and it costs $50 per day to operate the computer, how many days of use per year are required for the two alternatives to break even?

First, the annual cost if the computer is leased is

$$TC_L = (\$50 + \$50)N$$

$$= \$100N$$

and the annual equivalent total cost if the computer is bought is

$$TC_B = (\$25,000 - \$4,000)(\overset{A/P,9,15}{0.1241}) + \$4,000(0.09) + \$2,800 + \$50N$$

$$= \$2,606 + \$360 + \$2,800 + \$50N$$

$$= \$5,766 + \$50N.$$

The first three terms represent the fixed cost and the last term is the variable cost. Break-even occurs when $TC_L = TC_B$ or

$$\$100N = \$5,766 + \$50N$$

$$\$50\,N = \$5,766$$

$$N = 115 \text{ days.}$$

A graphical representation of this decision situation is shown in Figure 8.10.

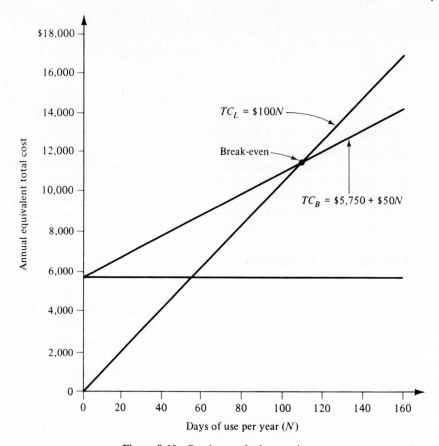

Figure 8.10   Break-even for lease or buy.

For all levels of use exceeding 115 days per year, it would be more economical to purchase the computer. If the level of use is anticipated to be below 115 days per year, the computer should be leased.

Equipment Selection Evaluation

Suppose that a fully automatic controller for a machine center can be fabricated for $140,000 and that it will have an estimated salvage value of $20,000 at the end of four years. Maintenance cost will be $12,000 per year, and the cost of operation will be $85 per hour.

As an alternative, a semiautomatic controller can be fabricated for $55,000. This device will have no salvage value at the end of a four-year service life. The cost of operation and maintenance is estimated to be $140 per hour.

With an interest rate of 10%, the annual equivalent total cost for the automatic attachment as a function of the number of hours of use per year is

$$TC_A = (\$140{,}000 - \$20{,}000)(\overset{A/P,\,10,\,4}{0.3155}) + \$20{,}000(0.10) + \$12{,}000 + \$85N$$

$$= \$51{,}800 + \$85N.$$

and the annual equivalent total cost for the semiautomatic attachmenmt as a function of the number of hours of use per year is

$$TC_S = \$55{,}000(\overset{A/P,\,10,\,4}{0.3155}) + \$140N$$

$$= \$17{,}400 + \$140N.$$

Break-even occurs when $TC_A = TC_S$, or

$$\$51{,}800 + \$85N = \$17{,}400 + \$140N$$

$$\$55N = \$34{,}400$$

$$N = 625 \text{ hrs.}$$

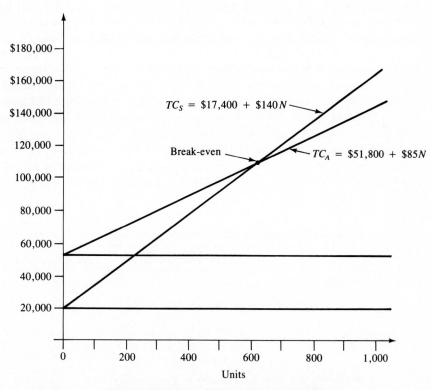

Figure 8.11   Break-even for equipment selection.

Figure 8.11 shows the two cost functions and the break-even point. For rates of use exceeding 625 hours per year, the automatic controller would be more economical. However, if it is anticipated that the rate of use will be less than 625 hours per year, the semiautomatic controller should be fabricated.

Profitability Evaluation

There are two aspects of production operations. One consists of assembling the production system of facilities, material, and people, and the other consists of the sale of the goods produced. The economic success of an enterprise depends upon its ability to carry on these activities to the end that there may be a net difference between receipts and the cost of production.

Linear break-even analysis is useful in evaluating the effect on profit of proposals for new operations not yet implemented and for which no data exist. Consider a proposed activity consisting of the manufacturing and marketing of a certain plastic item for which the sale price per unit is estimated to be $30,000. The machine required in the operation will cost $140,000 and will have an estimated life of eight years. It is estimated that the cost of production, including power, labor, space, and selling expense, will be $21 per unit sold. Material will cost $9.50 per unit. An interest rate of 8% is considered necessary to justify the required investment. The estimated costs associated with this activity are as given in Table 8.8.

The difficulty of making a clear-cut separation between fixed and variable costs becomes apparent when attention is focused on the item for repair and maintenance. In practice it is very difficult to distinguish between repairs that are a result of deterioration that takes place with the passage of time and those that result from the wear and tear of use. However, in theory, the separation can be made as shown in this example and is in accord with fact, with the exception, perhaps, of the assumption that repairs from wear and tear will be in direct proportion to the number of units manufactured. To be in accord with actualities, depreciation also undoubtedly should have been separated so that a part would appear as variable cost.

TABLE 8.8 FIXED AND VARIABLE MANUFACTURING COSTS

	FIXED COSTS	VARIABLE COSTS
Capital recovery with return		
$140,000 ($\overset{A/P,12,8}{0.1877}$)	$26,278	
Insurance and taxes	2,000	
Repairs and maintenance	1,722	$0.50/unit
Materials		9.50/unit
Labor, electricity, space		11.00/unit
Total	$30,000	$21.00/unit

In this example, $F = \$30,000$, $TC = \$30,000 + \$21N$, and $I = \$30N$. If F, TC, I are plotted as N varies from 0 to 6,000 units, the results will be as shown in Figure 8.12.

The cost of producing the plastic item will vary with the number made per year. The production cost per unit is given by $F/N + V$. If production cost per unit, variable cost per unit, and income per unit are plotted as N varies from 0 to 6,000, the results will be as shown in Figure 8.13. It will be noted that the fixed cost per unit may be infinite. Thus, in determining unit costs, fixed cost has little meaning unless the number of units to which it applies are known.

Income for most enterprises is directly proportional to the number of units sold. However, the income per unit may easily be exceeded by the sum of the fixed and the variable cost per unit for low production volumes. This is shown by comparing the total cost per unit curve and the income per unit curve shown in Figure 8.13.

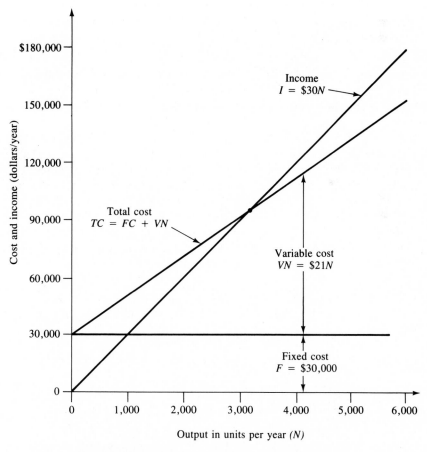

Figure 8.12   Fixed cost, variable cost, and income per year.

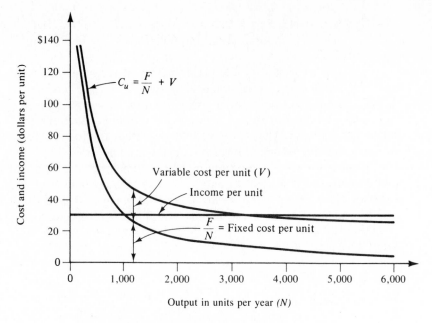

Figure 8.13　Fixed cost, variable cost, and income per unit.

8.7. BREAK-EVEN ANALYSIS UNDER RISK

In Section 8.6, the make-or-buy analysis was aided by Figure 8.9 showing the break-even point upon which a make-or-buy decision is based. The analysis was based on certainty about the demand for the electric power supply. Ordinarily, the decision maker does not know with certainty what the demand level will be.

Consider demand ranging from 1,500 units to 4,500 units with probabilities as given in Table 8.9. Entries for the cost to make and the cost to buy are calculated from the equations in Section 8.6 as was exhibited in Figure 8.9.

The expected demand may be calculated from the probabilities associated with each demand level as given in Table 8.7.

$$1,500(0.05) + 2,000(0.10) + 2,500(0.15) + 3,000(0.20) + 3,500(0.25)$$
$$+ 4,000(0.15) + 4,500(0.10) = 3,175.$$

From the equations for this decision situation, the expected total cost to make is

$$TC_M = \$12,000 + \$4(3,175)$$
$$= \$24,700$$

and the expected total cost to buy is

$$TC_B = \$8(3,175)$$
$$= \$25,400.$$

TABLE 8.9 DEMAND RANGE FOR MAKE OR BUY

	DEMAND IN UNITS						
	1500	2000	2500	3000	3500	4000	4500
Probability	0.05	0.10	0.15	0.20	0.25	0.15	0.10
Cost to make	$18,000	20,000	22,000	24,000	26,000	28,000	30,000
Cost to buy	$12,000	16,000	20,000	24,000	28,000	32,000	36,000

Accordingly, the power supply should be made. Alternately, this decision might be based on the most probable future criterion. From Table 8.9 it is evident that a demand level of 3,500 is most probable. Therefore, the cost to make of $26,000 is compared with cost to buy of $28,000 and the decision would be to make the item. This is the same decision as for the expected demand approach.

SELECTED REFERENCES

(1) Collier, C. A., and W. B. Ledbetter, *Enginering Economics and Cost Analysis*. New York: Harper and Row, 1988.

(2) English, J. M., *Project Evaluation*. New York: MacMillan, 1984.

(3) Jelin, F. C., and J. H. Black, *Cost and Optimization Engineering* (2nd ed.). New York: McGraw-Hill, 1983.

(4) Kurtz, M., *Handbook of Engineering Economics*. New York: McGraw-Hill, 1984.

(5) Sprague, J. C., and J. D. Whittaker, *Economic Analysis for Engineers and Managers*. Englewood Cliffs, N.J.: Prentice-Hall, 1986.

(6) Thuesen, G. J., and W. J. Fabrycky, *Engineering Economy* (7th ed.). Englewood Cliffs, N.J.: Prentice-Hall, 1989.

(7) White, J. A., M. H. Agee, and K. E. Case, *Principles of Engineering Economic Analysis* (3rd ed.). New York: John Wiley, 1989.

QUESTIONS AND PROBLEMS

1. What amount will be accumulated by each of the following investments?
   (a) $8,000 at 12% compounded annually over ten years.
   (b) $52,500 at 8% compounded annually over five years.

2. How much money must be invested to accumulate $1,000 in eight years at 9% compounded annually?

3. What is the present equivalent amount of a year-end series of receipts of $600 over five years at 8% compounded annually?

4. What interest rate compounded annually is involved if $4,000 results in $10,000 in six years?

5. How many years will it take for $4,000 to grow to $7,000 at an interest rate of 10% compounded annually?

6. What interest rate is necessary for a sum of money to double itself in eight years?

7. An asset was purchased for $5,200 with the anticipation that it would serve for 12 years and be worth $600 as scrap. After 5 years of operation the asset was sold for $1,800. The interest rate is 14%.
   (a) What was the anticipated annual equivalent cost of capital recovery plus return?
   (b) What was the actual annual equivalent cost of capital recovery plus return?

8. A cement mixer purchased for $3,300 has an estimated salvage value of $500 and an expected life of three years. An average of 25 cubic yards of concrete per month will be produced by the mixer.
   (a) Calculate the annual equivalent cost of capital recovery plus return with an interest rate of 8%.
   (b) Calculate the annual equivalent cost of capital recovery plus return per cubic yard of concrete with an interest rate of 12%.

9. The table below shows the receipts and disbursements for a given venture. Determine the desirability of the venture for a 14% interest rate based on the present equivalent comparison and the annual equivalent comparison.

END OF THE YEAR	RECEIPTS	DISBURSEMENTS
0	$ 0	$25,000
1	600	0
2	500	400
3	500	0
4	1,200	100

10. A new automatic controller can be installed for $30,000 and will have a $3,000 salvage value after ten years. This controller is expected to decrease operating cost by $4,000 per year.
   (a) What rate of return is expected if the controller is used for ten years?
   (b) For what life will the controller give a return of 15%?

11. Transco plans on purchasing a bus for $75,000 which will have a capacity of 80 passengers. As an alternative, a larger bus can be purchased for $95,000 which will have a capacity of 100 passengers. The salvage value of either bus is estimated to be $8,000 after a ten year life. If an annual net profit of $200 can be realized per passenger, which alternative should be recommended using an interest rate of 15%?

12. An office building and its equipment are insured to $710,000. The present annual insurance premium is $0.85 per $100 of coverage. A sprinkler system with an estimated life of 20 years and no salvage value can be installed for $18,000. Annual maintenance and operating cost is estimated to be $360. The premium will be reduced to $0.40 per $100 coverage if the sprinkler system is installed.
   (a) Find the rate of return if the sprinkler system is installed.
   (b) With interest at 12%, find the payout period for the sprinkler system.

13. The design of a system is to be pursued from one of two available alternatives. Each alternative has a life-cycle cost associated with an expected future. The costs for the corresponding futures are given in the table below (in millions of dollars). If the probabilities of occurrence of the futures are 30%, 50%, and 20%, respectively, which alternative is most desirable from an expected cost viewpoint using an interest rate of 10%.

DESIGN 1						YEARS					
Future	1	2	3	4	5	6	7	8	9	10	11	12
Optimistic	0.4	0.6	5.0	7.0	0.8	0.8	0.8	0.8	0.8	0.8	0.8	0.8
Expected	0.6	0.8	1.0	5.0	10.0	1.0	1.0	1.0	1.0	1.0	1.0	1.0
Pessimistic	0.8	0.9	1.0	7.0	10.0	1.2	1.2	1.2	1.2	1.2	1.2	1.2

DESIGN 2						YEARS					
Future	1	2	3	4	5	6	7	8	9	10	11	12
Optimistic	0.4	0.4	0.4	1.0	3.0	2.5	2.5	2.5	2.5	2.5	2.5	2.5
Expected	0.6	0.8	1.0	3.0	6.0	3.0	3.0	3.0	3.0	3.0	3.0	3.0
Pessimistic	0.6	0.8	1.0	5.0	6.0	3.1	3.1	3.1	3.1	3.1	3.1	3.1

14. Prepare a decision evaluation matrix for the design alternatives in Problem 13 and then choose the alternative which is best under the following decision rules: Laplace, Maximax, Maximin, and Hurwicz with $a = 0.6$. Assume that the choice is under uncertainty.

15. A campus building can be air conditioned by piping chilled water from a central refrigeration plant. Two competing proposals are being considered for the piping system, as outlined in the table. On the basis of a ten-year life, find the number of hours of operation per year for which the cost of the two systems will be equal if the interest rate is 9%.

	6-IN. SYSTEM	8-IN. SYSTEM
Motor size in horsepower	30	15
Installed cost of pump and pipe	$3,200	$4,400
Installed cost of motor	450	300
Salvage value of system	500	600
Energy cost per hour of operation	$ 0.32	$ 0.18

16. Fence posts for a large cattle ranch are currently purchased for $2.19 each. It is estimated that equivalent posts can be cut from timber on the ranch for a variable cost of $0.77 each, which is made up of the value of the timber plus labor cost. Annual fixed cost for required equipment is estimated to be $800. Two thousand posts will be required each year. What will be the annual saving if posts are cut?

17. A manufacturer can buy a required component from a supplier for $96 per unit delivered. Alternatively, the firm can manufacture the component for a variable cost of $46 per unit. It is estimated that the additional fixed cost would be $8,000 per year if the part is manufactured. Find the number of units per year for which the cost of the two alternatives will break even.

18. A marketing company can lease a fleet of automobiles for its sales personnel for $35 per day plus $0.09 per mile for each vehicle. As an alternative, the company can pay each salesperson $0.30 per mile to use his or her own automobile. If these are the only costs to the company, how many miles per day must a salesperson drive for the two alternatives to break even?

19. An electronics manufacturer is considering the purchase of one of two types of laser trimming machines. The sales forecast indicated that at least 8,000 units will be sold per year. Machine A will increase the annual fixed cost of the plant by $20,000 and will reduce variable cost by $5.60 per unit. Machine B will increase the annual fixed cost by

$5,000 and will reduce variable cost by $3.60 per unit. If variable costs are now $20 per unit produced, which machine should be purchased?

20. Machine A costs $20,000, has zero salvage value at any time, and has an associated labor cost of $1.14 for each piece produced on it. Machine B costs $36,000, has zero salvage value at any time, and has an associated labor cost of $0.85. Neither machine can be used except to produce the product described. If the interest rate is 10% and the annual rate of production is 20,000 units, how many years will it take for the cost of the two machines to break even?

21. An electronics manufacturer is considering two methods for producing a circuit board. The board can be hand-wired at an estimated cost of $0.98 per unit and an annual fixed equipment cost of $1,000. A printed equivalent can be produced using equipment costing $18,000 with a service life of nine years and salvage value of $1,000. It is estimated that the labor cost will be $0.32 per unit and that the processing equipment will cost $450 per year to maintain. If the interest rate is 13%, how many circuit boards must be produced each year for the two methods to break even?

22. It is estimated that the annual sales of a new product will be 2,000 the first year and increased by 1,000 per year until 5,000 units are sold during the fourth year. Proposal A is to purchase equipment costing $12,000 with an estimated salvage value of $2,000 at the end of four years. Proposal B is to purchase equipment costing $28,000 with an estimated salvage value of $5,000 at the end of four years. The variable cost per unit under proposal A is estimated to be $0.80, but is estimated to be only $0.25 under proposal B. If the interest rate is 9%, which proposal should be accepted for a four-year production period?

23. The fixed cost of a machine (capital recovery, interest, maintenance, space charges, supervision, insurance, and taxes) is F dollars per year. The variable cost of operating the machine (power, supplies, and other items, but excluding direct labor) is V dollars per hour of operation. If N is the number of hours the machine is operated per year, TC the annual total cost of operating the machine, TC_h the hourly cost of operating the machine, t the time in hours to process 1 unit of product, and M the machine cost of processing 1 unit of product per year, write expressions for (a) TC, (b) TC_h and (c) M.

24. In problem 23, $F = 600 per year, $t = 0.2$ hr, $V = 0.50 per hr, and N varies from 0 to 10,000 in increments of 1,000.
    (a) Plot values of M as a function of N.
    (b) Write an expression for the total cost of direct labor and machine cost per unit, TC_u, using the symbols in Problem 23 and letting W equal the hourly cost of direct labor.

25. A certain firm has the capacity to produce 800,000 units per year. At present it is operating at 75% of capacity. The income per unit is $0.10 regardless of output. Annual fixed costs are $28,000 and the variable cost is $0.06 per unit. Find the annual profit or loss at this capacity and the capacity for which the firm will break even.

26. An arc welding machine that is used for a certain joining process costs $10,000. The machine has a life of five years and a salvage value of $1,000. Maintenance, taxes, insurance, and other fixed costs amount to $500 per year. The cost of power and supplies is $3.20 per hour of operation and the operator receives $15.80 per hour. If the cycle time per unit of product is 60 min and the interest rate is 8%, calculate the cost per unit if (a) 200, (b) 600, (c) 1,200, (d) 2,500 units of product are made per year.

27. A certain firm has the capacity to produce 650,000 units of product per year. At present, it is operating at 65% of capacity. The firm's annual income is $416,000. Annual fixed costs are $192,000 and the variable costs are $0.38 per unit of product.

(a) What is the firm's annual profit or loss?

(b) At what volume of sales does the firm break even?

(c) What will be the profit or loss at 70, 80, and 90% of capacity on the basis of constant income per unit and constant variable cost per unit?

28. A chemical company owns two plants, A and B, that produce an identical product. The capacity of plant A is 60,000 gallons while that of B is 80,000 gallons. The annual fixed cost of plant A is $260,000 per year and the variable cost is $3.20 per gallon. The corresponding values for plant B are $280,000 and $3.90 per gallon. At present, plant A is being operated at 35% of capacity and plant B is being operated at 40% of capacity.

(a) What would be the total cost of production of plant A and plant B?

(b) What is the total cost and the average cost of the total output of both plants?

(c) What would be the total cost to the company and cost per gallon if all production were transferred to plant A?

(d) What would be the total cost to the company and cost per gallon if all production were transferred to plant B?

9
Optimization in Design and Operations

The design of complex systems that are optimum in some sense is a most important challenge facing the systems engineer. This challenge is paralleled by the frequent task of optimizing the operation of systems already in being. In the former case, optimum values of design variables are sought. In the latter, optimum values for policy variables are desired. This distinction is important only as a means for classifying the engineering design and analysis effort in relation to the system life cycle. The modeling approaches are essentially the same.

In this chapter the general approach to the formulation and manipulation of mathematical models is presented. The specific mathematical methods used vary in degree of complexity and depend upon the system under study. Examples will be used to illustrate design situations for static systems and for dynamic systems. Both unconstrained and constrained examples will be presented. Classical optimization techniques will be illustrated first, followed by some nonclassical approaches, including techniques of linear and dynamic programming.

9.1. CLASSICAL OPTIMIZATION THEORY

Optimization is the process of seeking the best. In systems engineering and analysis, this process is applied to each alternative and then across alternatives in accordance with the decision evaluation theory presented in Section 7.3. In so doing, the general decision evaluation function given by Equation 7.1 is used along with the decision evaluation matrix of Figure 7.1. Before presenting specific applications of optimization in design and operations, this section provides a review of calculus-based methods for optimizing functions with one or more decision variables.

The Slope of a Function

The slope of a function $y = f(x)$ is defined as the rate of change of the dependent variable, y, divided by the rate of change of the independent variable, x. If a positive change in x results in a positive change in y, the slope is positive. Conversely, a positive change in x resulting in a negative change in y indicates a negative slope.

If $y = f(x)$ defines a straight line, the difference between any two points x_1 and x_2 represents the change in x and the difference between any two points y_1 and y_2 represents the change in y. Thus, the rate of change of y with respect to x is $(y_1 - y_2)/(x_1 - x_2)$; the slope of the straight line. This slope is constant for all points on the straight line; $\Delta y / \Delta x = $ constant.

For a nonlinear function, the rate of change of y with respect to changes in x is not constant, but changes with changes in x. The slope must be evaluated at each point on the curve. This can be done by assuming an arbitrary point on the function p, for which the x and y values are x_0 and $f(x_0)$, as shown in Figure 9.1. The rate of change of y with respect to x at point p is equal to the slope of a line tangent to the function at that point. It is observed that the rate of change of y with respect to x differs from that at p at other points on the curve.

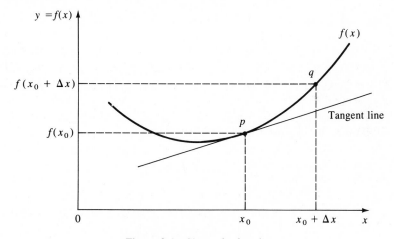

Figure 9.1 Slope of a function.

Differential Calculus Fundamentals

Differential calculus is a mathematical tool for finding successively better and better approximations for the slope of the tangent line shown in Figure 9.1. Consider another point on $f(x)$ designated q, situated at an x distance from point p equal to Δx and situated at a y distance from point p equal to Δy. Then the slope of a line segment through points p and q would have a slope $\Delta x / \Delta y$. But this is the average slope of $f(x)$ between the designated points. In classical optimization, it is the instantaneous rate of change at a given point that is sought.

The instantaneous rate of change in $f(x)$ at $x = x_0$ can be found by letting $\Delta x \to 0$. Referring to Figure 9.1, this can be stated as

$$\underset{\Delta x \to 0}{\text{limit}} \frac{f(x_0 + \Delta x) - f(x_0)}{\Delta x}. \qquad (9.1)$$

It is noted that as Δx becomes smaller and smaller, the line segment passing through points p and q approaches the tangent line to point p. Thus, the slope of the function at x_0 is given by Equation 9.1 when Δx is infinitesimally small.

Equation 9.1 is an expression for slope of general applicability. It is called a *derivative,* and its application is a process known as *differentiation.* For the function $y = f(x)$, the symbolism dy/dx or $f'(x)$ is most often used to denote the derivative.

As an example of the process of differentiation utilizing Equation 9.1, consider the function $y = f(x) = 8x - x^2$. Substituting into Equation 9.1 gives

$$\underset{\Delta x \to 0}{\text{limit}} = \frac{[8(x + \Delta x) - (x + \Delta x)^2] - [8x - x^2]}{\Delta x}$$

$$= \frac{8x + 8\Delta x - x^2 - 2x\,\Delta x - \Delta x^2 - 8x + x^2}{\Delta x}$$

$$= \frac{8\Delta x - 2x\,\Delta x - \Delta x^2}{\Delta x}$$

$$= 8 - 2x - \Delta x.$$

As $\Delta x \to 0$ the derivative, $dy/dx = 8 - 2x$, the slope at any point on the function. For example, at $x = 4$, $dy/dx = 0$, indicating that the slope of the function is zero at $x = 4$.

An alternative example of the process of differentiation, not tied to Equation 9.1, might be useful to consider. Begin with a function of some form such as $y = f(x) = 3x^2 + x + 2$. Let x increase by an amount Δx. Then y will increase by an amount Δy as shown in Figure 9.2, giving

$$y + \Delta y = 3(x + \Delta x)^2 + (x + \Delta x) + 2$$

$$= 3[x^2 + 2x(\Delta x) + (\Delta x)^2] + (x + \Delta x) + 2$$

$$\Delta y = 3[x^2 + 2x(\Delta x) + (\Delta x)^2] + (x + \Delta x) + 2 - y$$

but $y = 3x^2 + x + 2$, giving

$$\Delta y = 3x^2 + 6x(\Delta x) + 3(\Delta x)^2 + x + \Delta x + 2 - 3x^2 - x - 2$$

$$= 6x(\Delta x) + 3(\Delta x)^2 + \Delta x.$$

The average rate of change of Δy with respect to Δx is

$$\frac{\Delta y}{\Delta x} = \frac{6x(\Delta x)}{\Delta x} + \frac{3(\Delta x)^2}{\Delta x} + \frac{\Delta x}{\Delta x}$$

$$= 6x + 3(\Delta x) + 1.$$

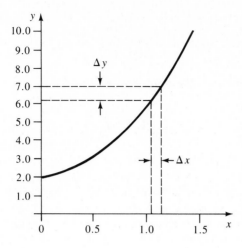

Figure 9.2 Function $3x^2 + x + 2$.

But since

$$\underset{\Delta x \to 0}{\text{limit}} = \frac{dy}{dx} = f'(x),$$

the instantaneous rate of change is $6x + 1$.

Instead of proceeding as above for each case encountered, certain rules of differentiation have been developed for a range of common functional forms. Some of these are

1. Derivative of a constant: If $f(x) = k$, a constant, then $dy/dx = 0$.
2. Derivative of a variable: If $f(x) = x$, a variable, then $dy/dx = 1$.
3. Derivative of a straight line: If $f(x) = ax + b$, a straight line, then $dy/dx = a$.
4. Derivative of a variable raised to a power: If $f(x) = x^n$, a variable raised to a power, then $dy/dx = nx^{n-1}$.
5. Derivative of a constant times a function: If $f(x) = k[g(x)]$, a constant times a function, then

$$\frac{dy}{dx} = k\left[\frac{dg(x)}{dx}\right].$$

6. Derivative of the sum or difference of two functions: If $f(x) = g(x) \pm h(x)$, a sum or difference of two functions, then

$$\frac{dy}{dx} = \frac{dg(x)}{dx} \pm \frac{dh(x)}{dx}.$$

7. Derivative of the product of two functions: If $f(x) = g(x)h(x)$, a product of two functions then

$$\frac{dy}{dx} = g(x)\frac{dh(x)}{dx} + h(x)\frac{dg(x)}{dx}.$$

8. Derivative of an exponential function: If $f(x) = e^x$, an exponent of e, then

$$\frac{dy}{dx} = e^x.$$

9. Derivative of the natural logarithm: If $f(x) = \ln x$, natural log of x, then

$$\frac{dy}{dx} = \frac{1}{x}.$$

The methods presented above may be used to find *higher-order* derivatives. When a function is differentiated, another function results which may also be differentiated if desired. For example, the derivative of $8x - x^2$ was $8 - 2x$. This was the slope of the function. If the slope is differentiated, the higher-order derivative is found to be -2. This is the rate of change in dy/dx and is designated d^2y/dx^2, of $f''(x)$. The process of differentiation can be extended to even higher orders as long as there remains some function to differentiate.

Partial Differentiation

Partial differentiation is the process of finding the rate of change whenever the dependent variable is a function of more than one independent variable. In the process of partial differentiation, all variables except the dependent variable, y, and the chosen independent variable are treated as constants and differentiated as such. The symbolism $\partial y/\partial x_1$ is used to represent the partial derivative of y with respect to x_1.

As an example, consider the following function of two independent variables, x_1 and x_2:

$$4x_1^2 + 6x_1x_2 - 3x_2.$$

The partial derivative of y with respect to x_1 is

$$\frac{\partial y}{\partial x_1} = 8x_1 + 6x_2 - 0.$$

The partial derivative of y with respect to x_2 is

$$\frac{\partial y}{\partial x_2} = 0 + 6x_1 - 3.$$

Partial differentiation permits a view of the slope along one dimension at a time of an *n*-dimensional function. By combining this information from such a function, the slope at any point on the function can be found. Of particular interest is the point on the function where the slope along all dimensions is zero. For the function above, this point is found by setting the partial derivatives equal to zero and solving for x_1 and x_2 as follows:

$$8x_1 + 6x_2 = 0$$

$$6x_1 - 3 = 0$$

from which

$$x_1 = 1/2 \text{ and } x_2 = -2/3.$$

The meaning of this point is explained in the paragraphs which follow.

Unconstrained Optimization

Unconstrained optimization simply means that no constraints are placed on the function under consideration. Under this assumption a *necessary* condition for $x*$ to be an optimum point on $f(x)$ is that the first derivative be equal to zero. This is stated as

$$\left.\frac{df(x)}{dx}\right|_{x=x*} = 0. \tag{9.2}$$

The first derivative being zero at the stationary point $x*$ is not *sufficient*, since it is possible for the derivative to be zero at a point of inflection on $f(x)$. It is sufficient for $x*$ to be an optimum point if the second derivative at $x*$ is positive or negative; $x*$ being a minimum if the second derivative is positive and a maximum if the second derivative is negative.

If the second derivative is also zero, a higher-order derivative is sought until the first nonzero one is found at the nth derivative as

$$\left.\frac{d^n f(x)}{dx^n}\right|_{x=x*} = 0. \tag{9.3}$$

If n is odd, then $x*$ is a point of inflection. If n is even and if

$$\left.\frac{d^n f(x)}{dx^n}\right|_{x=x*} < 0 \tag{9.4}$$

then $x*$ is a local maximum. But if

$$\left.\frac{d^n f(x)}{dx^n}\right|_{x=x*} > 0 \tag{9.5}$$

then $x*$ is a local minimum.

As an example, suppose that the minimum value of x is sought for the function $f(x) = 2x^3 - 3x^2 + 1$ exhibited in Figure 9.3. The necessary condition for a minimum to exist is

$$\frac{df(x)}{dx} = 6x^2 - 6x = 0.$$

Both $x = 0$ and $x = 1$ satisfy the condition above. Accordingly, the second derivative is taken as

$$\frac{d^2 f(x)}{dx^2} = 12x - 6$$

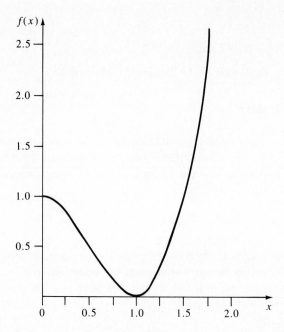

Figure 9.3   Function $2x^3 - 3x^2 + 1$.

and evaluated at $x = 0$ and $x = 1$. At $x = 0$ the second derivative is -6, indicating that this is a maximum point. At $x = 1$ the second derivative is 6, indicating that this is the minimum sought. These points are shown on Figure 9.2.

For functions of more than one independent variable, the vector $x^* = [x_1, x_2, \ldots, x_n]$ will have a stationary point on a function $f(x)$ if all elements of the vector of first partial derivatives evaluated at x^* are zero. Symbolically,

$$F(x) = \left[\frac{\partial F(x)}{\partial x_1}, \frac{\partial F(x)}{\partial x_2}, \ldots, \frac{\partial F(x)}{\partial x_n} \right] = 0. \tag{9.6}$$

Consider a function in two variables, x_1 and x_2, given as

$$f(x_1, x_2) = x_1^2 - 8x_2 + 2x_2^2 - 6x_1 + 30.$$

The necessary condition requires that

$$\frac{\partial f(x_1, x_2)}{\partial x_1} = 2x_1 - 6 = 0$$

and

$$\frac{\partial f(x_1, x_2)}{\partial x_2} = 4x_2 - 8 = 0.$$

The optimal value x_1^* is found to be 3 and the optimal value x_2^* is found to be 2. These may be substituted into $f(x_1, x_2)$ to give the optimal value.

$$f(x_1, x_2) = (3)^2 - 8(2) + 2(2)^2 - 6(3) + 30 = 13.$$

To determine whether this value for $f(x_1, x_2)$ is a maximum or a minimum, the second partials are required. These are

$$\frac{\partial(2x_1 - 6)}{\partial x_1} = 2 \quad \text{and} \quad \frac{\partial(4x_2 - 8)}{\partial x_2} = 4.$$

Each is greater than zero, indicating that a minimum may have been found. To verify this, the following relationship must be satisfied:

$$\left(\frac{\partial^2 y}{\partial x_1^2}\right)\left(\frac{\partial^2 y}{\partial x_2^2}\right) - \left(\frac{\partial^2 y}{\partial x_1 \, \partial x_2}\right)^2 > 0, \tag{9.7}$$

where

$$\frac{\partial^2 y}{\partial x_1 \, \partial y_2} = \frac{\partial(\partial y / \partial x_1)}{\partial x_2}.$$

Since

$$\frac{\partial y}{\partial x_1} = 2x_1 - 6$$

$$\frac{\partial^2 y}{\partial x_1 \, \partial x_2} = \frac{\partial(2x_1 - 6)}{\partial x_2} = 0.$$

Substituting back into Equation 9.7 gives $(2)(4) - 0 = 8$, verifying that the value found for $f(x_1, x_2)$ is a minimum.

9.2. UNCONSTRAINED CLASSICAL OPTIMIZATION

There are a number of situations in systems engineering to which classical optimization theory can be applied. Many design and operational decisions are characterized by two or more factors that are affected differently by common design or policy variables. For example, certain cost components may vary directly with an increase in the value of a variable while others may vary inversely. When the total cost of an alternative is a function of increasing and decreasing cost components, a value may exist for the common variable which will result in a minimum cost for the alternative.

This section presents selected examples from classical situations which will illustrate the application of unconstrained optimization theory. These examples are from bridge design for single and multiple alternatives, electrical conductor design for single and multiple alternatives, inventory operations for a single replenishment source, and then multisource inventory operations.

Optimum Bridge Design

Before engaging in detail design, it is important to optimally allocate the anticipated capital investment to superstructure and to piers. This can be accomplished by recognizing that there exists an inverse relationship between the cost of superstructure and the number of piers. As the number of piers increases, the cost of superstructure decreases. Conversely, the cost of superstructure increases as the number of piers decreases. Pier cost for the bridge is directly related to the number specified. This is a classical design situation involving increasing and decreasing cost components, the sum of which will be a minimum for a certain number of piers (or the span between piers).

A general mathematical model may be derived for the evaluation of investment in superstructure and piers. The decision evaluation function of Equation 7.1 applies as

$$E = f(X, Y)$$

where

E = evaluation measure (total system cost)
X = design variable of the span between piers
Y = system parameters of the bridge length, the superstructure weight, the erected cost of superstructure per pound, and the cost of piers per pier.

Equation 7.2 is the design evaluation function applicable to the evaluation of alternative design configurations, where design dependent parameters are the key to selecting the best alternative. Both versions of the function are utilized in their unconstrained forms in this section.

Bridge design evaluation model. A general evaluation model for bridge design may be derived and then applied to specific situations. Let

L = bridge length in feet
W = superstructure weight in pounds per foot
S = span between piers in feet
C_s = erected cost of superstructure in dollars per pound
C_p = installed cost of piers in dollars per pier.

Assume that the weight of the superstructure is linear over a certain span range, $W = AS + B$, with the parameters A and B having been established by a sound statistical procedure. Accordingly, the superstructure cost, SC, will be

$$SC = (AS + B)(L)(C_s). \qquad (9.8)$$

The total cost of piers, PC, will be

$$PC = \left(\frac{L}{S} + 1\right)(C_p). \qquad (9.9)$$

where two abutments are included as though they were piers.

The total cost of the bridge is expressed as

$$TC = SC + PC$$

$$= (AS + B)(L)(C_s) + \left(\frac{L}{S} + 1\right)(C_p)$$

$$= ASLC_s + BLC_s + \frac{LC_p}{S} + C_p. \qquad (9.10)$$

To find the optimum span between piers, differentiate Equation 9.10 with respect to S and equate the result to zero as follows:

$$\frac{dTC}{dS} = ALC_s - \frac{LC_p}{S^2} = 0$$

$$S* = \sqrt{\frac{C_p}{AC_s}}.$$

(9.11)

The minimum total cost for the bridge is found by substituting Equation 9.11 for S in Equation 9.10 to obtain

$$TC* = 2\sqrt{AC_p L^2 C_s} + BLC_s + C_p.$$

(9.12)

Equation 9.11 can be used to find the optimal pier spacing for a given bridge design. However, to evaluate alternative bridge designs, Equation 9.12 would be used first, followed by a single application of Equation 9.11 to the best alternative. This procedure is illustrated by examples in the sections which follow.

Single design alternative.   Assume that a bridge to serve a 1,000 foot crossing is to be fabricated from steel with a certain girder design configuration. For this design alternative, the weight of the superstructure in pounds per foot is estimated to be linear (over a limited span range) and is expressed as

$$W = 16S + 600.$$

Also, assume that the superstructure is expected to cost \$0.65 per foot erected. Piers are anticipated to cost \$80,000 each in place and this amount will also be used as the estimated cost of each abutment.

From Equation 9.11, the optimum span between piers is found to be

$$S* = \sqrt{\frac{\$80,000}{16 \times \$0.65}} = 87.7 \text{ ft.}$$

This result is a theoretical spacing and must be adjusted to obtain an integer number of piers. The required adjustment gives 12 piers (11 spans) for a total cost from Equation 9.12 of \$2,295,454. The span will be 90.9 feet, slightly greater than the theoretical minimum. A check can be made by considering 13 piers (12 spans) with each span being 83.3 feet. In this case, the total cost is higher (\$2,296,666) so the 12 pier design would be adopted.

Total cost as a function of the pier spacing and the number of piers is summarized in Table 9.1. Note that a minimum occurs when 12 piers are specified; actually, 10 piers and two abutments. The optimal pier spacing is 90.9 feet.

Multiple design alternatives.   To introduce optimal design for multiple alternatives, assume that there is another superstructure configuration under consideration for the bridge design described in the previous section. The weight in pounds per foot for the alternative configuration is estimated from parametric methods to be

$$W = 22S.$$

TABLE 9.1   TC AS A FUNCTION OF THE SPAN BETWEEN PIERS

SPAN (FEET)	NUMBER OF PIERS	PIER COST ($)	SUPERSTRUCTURE COST ($)	TOTAL COST ($)
142.8	8	$640,000	$1,875,714	$2,515,714
111.1	10	800,000	1,545,555	2,345,555
100.0	11	880,000	1,430,000	2,310,000
90.9	12	960,000	1,335,454	2,295,455
83.3	13	1,040,000	1,256,666	2,296,667
76.9	14	1,120,000	1,190,000	2,310,000
71.4	15	1,200,000	1,132,857	2,332,857
66.6	16	1,280,000	1,083,333	2,363.333
58.8	18	1,440,000	1,001,765	2,441,765

Assume that all other factors are the same as for the previous design.

A choice between the design alternatives is made by finding the optimum total cost for each alternative using Equation 9.12. This gives $2,294,280 for Design A and $2,219,158 for Design B. Thus, Design B would be chosen.

For Design B, the optimum span between piers is found from Equation 9.11 to be

$$S^* = \sqrt{\frac{\$80,000}{22 + \$0.65}} = 74.8 \text{ ft.}$$

And the lowest cost integer number of piers is found to be 13 (14 spans) with a span of 76.9 feet.

Table 9.2 summarizes this design decision by giving the total cost as a function of the number of piers for each alternative. The theoretical total cost function for each alternative is exhibited in Figure 9.4 along with an indication of the number of piers.

TABLE 9.2   TC FOR TWO BRIDGE DESIGN ALTERNATIVES

NO. OF PIERS	PIER COST ($)	SUPERSTRUCTURE COST ($)		TOTAL COST ($)	
		DESIGN A	DESIGN B	DESIGN A	DESIGN B
8	640,000	1,875,714	2,042,857	2,515,714	2,682,857
10	800,000	1,545,555	1,588,889	2,345,555	2,388,889
11	880,000	1,430,000	1,430,000	2,310,000	2,310,000
12	960,000	1,335,454	1,300,000	2,295,454	2,260,000
13	1,040,000	1,256,666	1,191,666	2,296,667	2,231,666
14	1,120,000	1,190,000	1,100,000	2,310,000	2,220,000
15	1,200,000	1,132,857	1,021,438	2,332,857	2,221,438
16	1,280,000	1,083,333	953,333	2,363,333	2,233,333
18	1,440,000	1,001,765	841,176	2,441,765	2,281,176

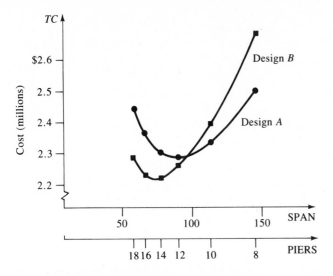

Figure 9.4 TC functions for bridge design alternatives.

Optimum Conductor Design

Another classical optimization situation is the design of an electrical conductor, a static system with a flow component. The design variable of interest is the cross-sectional area, with increasing and decreasing cost components related thereto. As the area increases, so does the installed cost of the conductor. However, the power loss cost is inversely proportional to the area. The sum of these costs will be a minimum at the optimum value for the cross-sectional area.

An evaluation model may be derived for determining total cost as a function of the cross-sectional area of a conductor. The decision evaluation function of Equation 7.1 applies as

$$E = f(X, Y)$$

where

E = evaluation measure (total system cost)
X = design variable of the cross-sectional area
Y = system parameters of the conductor length, the electrical load, the resistivity of the conductor material, the cost of energy, the cost of the conductor, and the cost of money.

If alternative designs (material types) are under consideration, Equation 7.2 is the applicable design evaluation function. Design dependent parameters arise from the material choice and are key to the selection from among design alternatives. Both versions of the function are utilized in their unconstrained forms in this section.

Conductor design evaluation model. A general evaluation model for electrical conductor design may be derived and then applied to specific situations.

Let

A = conductor cross-sectional area (in^2)
L = length of the conductor (feet)
I = transmission load through the conductor (amperes)
H = number of hours conductor is utilized per year
i = interest rate (%)
n = conductor useful life
C_e = cost of electricity ($/kwh)
C_i = fixed installation cost ($)
C_m = unit cost of the conductor ($/lb)
R_m = resistance of the conductor (ohms)
D_m = density of the conductor material (lbs/ft^3)
W_m = weight of the conductor material (lbs)
F_m = salvage value of the conductor material ($/lb).

In the above notation, the subscript m will designate the material selected for a given design alternative.

The equivalent cost, which is annual equivalent total cost (AETC), is composed of the cost due to power loss and the capital investment cost as

Annual Equivalent Total Cost (AETC) = Power Loss Cost + Investment Cost

$$(9.13)$$

The power loss in dollars per year is stated as

$$\text{Power Loss Cost} = C_e I^2 R_m \frac{H}{1000A}. \qquad (9.14)$$

The annual equivalent capital investment cost (capital recovery) in dollars is stated as

$$\text{Investment Cost} = C_i(\overset{A/P,i,n}{\quad}) + (C_m - F_m)(\overset{A/P,i,n}{\quad})W_m + F_m W_m i \qquad (9.15)$$

where

$$W_m = \frac{LAD_m}{144}. \qquad (9.16)$$

Therefore, the annual equivalent total cost is

$$\text{AETC} = C_i(\overset{A/P,\,i,\,n}{\quad}) + (C_m - F_m)(\overset{A/P,\,i,\,n}{\quad})\frac{LAD_m}{144} + F_m\frac{LAD_m i}{144} + C_e I^2 R_m \frac{H}{1000A}. $$

$$(9.17)$$

In order to optimize the above equation, the minimum cross-sectional area must be found. Therefore, by taking the derivative of Equation 9.17 with respect to A and setting it equal to zero, one obtains

$$\frac{d(\text{AETC})}{dA} = (C_m - F_m)(\overset{A/P,i,n}{\quad})\frac{LD_m}{144} + F_m i\frac{LD_m}{144} - C_e I^2 R_m\frac{H}{1000A^2} = 0 \qquad (9.18)$$

from which

$$C_e I^2 R_m\frac{H}{1000A^2} = (C_m - F_m)(\overset{A/P,i,n}{\quad})\frac{LD_m}{144} + iF_m\frac{LD_m}{144} \qquad (9.19)$$

$$A^2 = \frac{C_e I^2 R_m\dfrac{H}{1000}}{(C_m - F_m)(\overset{A/P,i,n}{\quad})\dfrac{LD_m}{144} + iF_m\dfrac{LD_m}{144}} \qquad (9.20)$$

$$A^* = \sqrt{\frac{C_e I^2 R_m\dfrac{H}{1000}}{(C_m - F_m)(\overset{A/P,i,n}{\quad})L\dfrac{D_m}{144} + iF_m L\dfrac{D_m}{144}}}. \qquad (9.21)$$

By substituting Equation 9.21 into Equation 9.17, the minimum AETC (AETC*) is found as

$$\text{AETC*} = C_i(\overset{A/P,i,n}{\quad})$$

$$+ (C_m - F_m)(\overset{A/P,i,n}{\quad})\frac{LD_m}{144}\sqrt{\frac{C_e I^2 R_m\dfrac{H}{1000}}{(C_m - F_m)(\overset{A/P,i,n}{\quad})L\dfrac{D_m}{144} + F_m i L\dfrac{D_m}{144}}}$$

$$+ F_m i\frac{LD_m}{144}\sqrt{\frac{C_e I^2 R_m\dfrac{H}{1000}}{(C_m - F_m)(\overset{A/P,i,n}{\quad})L\dfrac{D_m}{144} + F_m i L\dfrac{D_m}{144}}}$$

$$+ C_e I^2 R_m\frac{H}{1000}\sqrt{\frac{(C_m - F_m)(\overset{A/P,i,n}{\quad})L\dfrac{D_m}{144} + F_m i L\dfrac{D_m}{144}}{C_e I^2 R_m\dfrac{H}{1000}}}. \qquad (9.22)$$

To insure that the above equation is truly a minimum, the second derivative test is applied. Taking the second derivative of Equation 9.17 yields the following.

$$\frac{d^2(\text{AETC})}{dA^2} = 2C_e I^2 R_m\frac{H}{1000A^3}. \qquad (9.23)$$

Since the second derivative is positive, A is a minimum and Equation 9.22 is indeed optimal.

Single design alternative. Assume that a copper conductor is being designed to transmit 2,000 amperes continuously throughout the year for ten years over a distance of 120 feet. The resistivity of a copper conductor is 0.000982 ohms, and copper has a density of 555 pounds per cubic foot. Other parameters are

$$C_i = \$300$$

$$C_c = \$1.30/\text{lb}$$

$$F_c = \$0.78/\text{lb}$$

$$C_e = \$0.052/\text{kwh}$$

$$i = 15\%$$

Using Equations 9.14 through 9.17, Table 9.3 is obtained. The major costs are presented graphically in Figure 9.5.

Next, using Equations 9.21 and 9.22, the optimal cross-sectional area is found to be 4.19 in^2 at an AETC* of $914.39 as shown in Figure 9.5. At this point, Design Alternative 1 is considered to be the current best or baseline design.

Multiple design alternatives. Conductors can be fabricated from materials other than copper. Assume that aluminum is being considered as an alternative to the copper choice in the previous section. The resistivity of an aluminum conductor is 0.001498 ohms, and aluminum has a density of 168 pounds per cubic foot. Other parameters are

$$C_i = \$300$$

$$C_a = \$1.10/\text{lb}$$

$$F_a = \$0.66/\text{lb}$$

$$C_e = \$0.052/\text{kwh}$$

$$i = 15\%$$

Using Equations 9.14 through 9.17, Table 9.4 is obtained. The optimal cross-sectional area is found from Euqation 9.21 to be 10.22 in^2 at an AETC* of $593.98. At this point the designer is ready to make a decision. The annual equivalent cost curves for both the copper design alternative and the aluminum design alternative

TABLE 9.3 AETC FOR DESIGN ALTERNATIVE 1 AS A FUNCTION OF CROSS-SECTIONAL AREA

Cost	CROSS-SECTIONAL AREA (in²)					
	2	4	6	8	10	12
Investment Cost	$ 263.88	467.97	672.05	876.14	1,080.23	1,284.32
Power Loss Cost	$ 894.64	447.32	298.21	223.66	178.93	149.11
Annual Equiv. Cost	$1,158.52	915.29	970.27	1,099.80	1,259.13	1,433.43

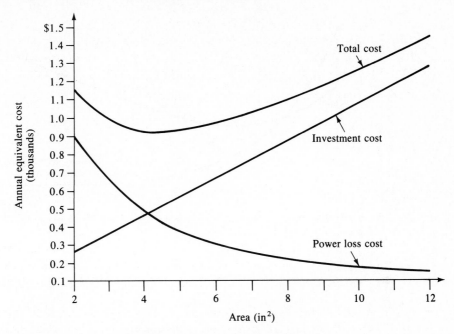

Figure 9.5   Cost curves for Design Alternative 1.

are presented graphically in Figure 9.6. Clearly the aluminum design, with an equivalent annual cost of $593.98, provides the lower cost alternative and is selected. Therefore, Design Alternative 2 becomes the current best.

Design decision reversal.   The designer would like to know how much the price of copper would have to drop in order for copper to become the preferred conductor material. Holding the price of aluminum at $1.10 per pound and varying the cost of copper, the annual equivalent total costs (AETC) in Table 9.5 are obtained. In each case the salvage value is assumed to be in the same proportion to the price of copper as the original estimate; that is, 0.60 ($0.78/$1.3). The optimal cross-sectional area and the corresponding annual equivalent total costs (AETC*) are provided in Table 9.5.

TABLE 9.4   AETC FOR DESIGN ALTERNATIVE 2 AS A FUNCTION OF CROSS-SECTIONAL AREA

Cost	CROSS-SECTIONAL AREA (in²)					
	2	4	6	8	10	12
Investment Cost	$ 112.06	164.34	216.61	268.89	321.16	373.43
Power Loss Cost	$1,364.74	682.37	454.91	341.18	272.95	227.46
Annual Equiv. Cost	$1,476.80	846.71	671.52	610.07	594.11	600.89

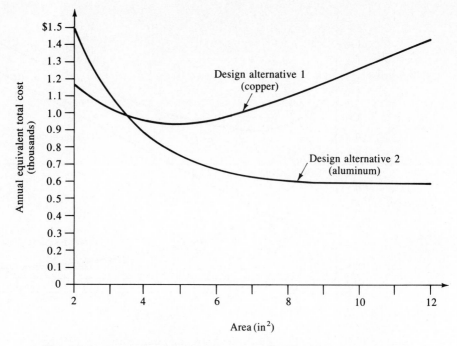

Figure 9.6   Total cost curves for Design Alternative 1 and Design Alternative 2.

The results of Table 9.5 are presented graphically in Figure 9.7 which shows the points of design decision reversal as the price of copper changes. Figure 9.7 demonstrates that, as the price of copper decreases, the annual equivalent total cost curves shift downward indicating lower optimal annual costs (AETC*). In addition, Figure 9.7 shows that for copper to become the preferred material selection, its price must be approximately \$0.50/lb or lower. That is, at a price less than \$0.50/lb the AETC* of copper becomes lower than that of aluminum. The point at which the copper conductor becomes the lower cost alternative is defined as a decision switch point. If

$$C_c > \$0.50/\,\text{lb} \qquad \text{select aluminum}$$

$$C_c < \$0.50/\,\text{lb} \qquad \text{select copper}$$

$$C_c = \$0.50/\,\text{lb} \qquad \text{use either.}$$

TABLE 9.5   OPTIMAL AETC FOR GIVEN PRICES OF COPPER

PRICE OF COPPER (\$/lb)	OPTIMUM AREA (in²)	AETC* (\$/yr)
1.30	4.19	914.39
1.00	4.77	809.32
0.70	5.71	686.90
0.50	6.75	589.79
0.40	7.55	533.84

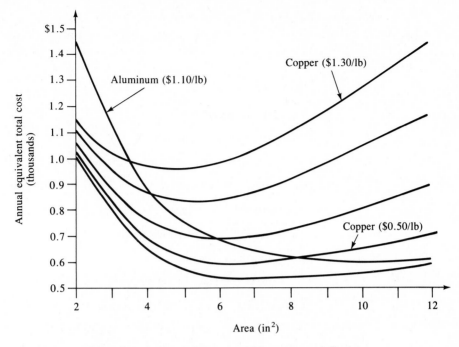

Figure 9.7   Design decision reversal for copper and aluminum.

Optimizing Inventory Operations

The bridge design and conductor design examples illustrated the application of classical optimization theory to systems having structure without activity. In this section classical optimization theory is applied to a dynamic system with single-item, multisource inventory operations. Here the objective is to determine the optimum procurement and inventory policy so that the system will operate at minimum cost over time.

When to procure and how much to procure are policy variables of importance in the single-source system. The decision evaluation function of Equation 7.1 applies as

$$E = f(X, Y)$$

where

E = evaluation measure (total system cost)
X = policy variables of when to procure and how much to procure
Y = system parameters of demand, procurement lead time, replenishment rate, item cost, procurement cost, holding cost, and shortage cost.

Equation 7.2 is the decision evaluation function applicable to the multisource system, where source dependent parameters are the key to determining the best procurement source. Both versions of the function are used in their unconstrained forms in this section.

  Inventory decision evaluation model. A mathematical model can be formulated for the general deterministic case (demand and lead time constant) from the inventory flow geometry shown in Figure 9.8. The following symbolism will be adopted:

$$TC = \text{total system cost per period}$$

$$L = \text{procurement level}$$

$$Q = \text{procurement quantity}$$

$$D = \text{demand rate in units per period}$$

$$T = \text{lead time in periods}$$

$$N = \text{number of periods per cycle}$$

$$R = \text{replenishment rate in units per period}$$

$$C_i = \text{item cost per unit}$$

$$C_p = \text{procurement cost per procurement}$$

$$C_h = \text{holding cost per unit per period}$$

$$C_s = \text{shortage cost per unit short per period}$$

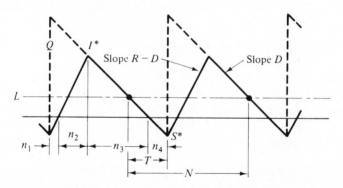

Figure 9.8   Deterministic inventory system geometry.

It is assumed that the replenishment rate is greater than the demand rate and that unsatisfied demand is not lost. From Figure 9.8, the number of periods per cycle can be expressed as

$$N = \frac{Q}{D}. \tag{9.24}$$

Also, the following relationships are evident:

$$(n_1 + n_2)(R - D) = (n_3 + n_4)D \tag{9.25}$$

$$n_1 + n_2 = \frac{Q}{R} \tag{9.26}$$

and

$$n_3 + n_4 = \frac{I' + DT - L}{D}. \tag{9.27}$$

From Equations 9.25, 9.26, and 9.27,

$$I' = Q\left(1 - \frac{D}{R}\right) + L - DT. \tag{9.28}$$

The total number of unit periods of stock on hand during the inventory cycle, I, is

$$I = \frac{I'}{2}(n_2 + n_3)$$

$$= \frac{I'^2}{2(R - D)} + \frac{I'^2}{2D}. \tag{9.29}$$

Substituting Equation 9.28 for I' gives

$$I = \frac{[Q(1 - D/R) + L - DT]^2}{2}\left(\frac{1}{R - D} + \frac{1}{D}\right). \tag{9.30}$$

The total number of unit periods of shortage during the cycle, S, is

$$S = \frac{S'}{2}(n_1 + n_4)$$

$$= \frac{S'^2}{2(R - D)} + \frac{S'^2}{2D}.$$

But, since $S' = DT - L$,

$$S = \frac{(DT - L)^2}{2}\left(\frac{1}{R - D} + \frac{1}{D}\right). \tag{9.31}$$

The total system cost per period will be a summation of the item cost for the period, the procurement cost for the period, the holding cost for the period, and the shortage cost for the period, or

$$TC = IC + PC + HC + SC. \tag{9.32}$$

Item cost for the period will be the product of the item cost per unit and the demand rate in units per period, or

$$IC = C_i D. \tag{9.33}$$

The procurement cost for the period will be the procurement cost per procurement divided by the number of periods per inventory cycle, or

$$PC = \frac{C_p}{N}$$

$$= \frac{C_p D}{Q}. \tag{9.34}$$

Holding cost for the period will be the product of the holding cost per unit per period and the average number of units on hand during the period, or

$$HC = \frac{C_h I}{N}$$

$$= \frac{C_h D}{Q}\left\{ \frac{[Q(1 - D/R) + L - DT]^2}{2}\left(\frac{1}{R - D} + \frac{1}{D}\right)\right\}.$$

Note the existence of the following relationship:

$$\frac{D}{Q}\left(\frac{1}{R - D} + \frac{1}{D}\right) = \frac{1}{Q(1 - D/R)}. \tag{9.35}$$

Using Equation 9.34, the holding-cost component becomes

$$HC = \frac{C_h}{2Q(1 - D/R)}\left[Q\left(1 - \frac{D}{R}\right) + L - DT\right]^2. \tag{9.36}$$

Shortage cost for the period will be the product of the shortage cost per unit short per period and the average number of units short during the period, or

$$SC = \frac{C_s S}{N}$$

$$= \frac{C_s D}{Q}\left[\frac{(DT - L)^2}{2}\left(\frac{1}{R - D} + \frac{1}{D}\right)\right]. \tag{9.37}$$

Equation 9.37 may be written as

$$SC = \frac{C_s(DT - L)^2}{2Q(1 - D/R)}. \tag{9.38}$$

The total system cost per period will be a summation of the four cost components developed above and may be expressed as

$$TC = C_i D + \frac{C_p D}{Q} + \frac{C_h}{2Q(1 - D/R)}\left[Q\left(1 - \frac{D}{R}\right) + L - DT\right]^2$$

$$+ \frac{C_s(DT - L)^2}{2Q(1 - D/R)}. \tag{9.39}$$

The minimum-cost procurement level and procurement quantity may be found by setting the partial derivatives equal to zero and solving the resulting equations. Modifying Equation 9.39 gives

$$TC = C_i D + \frac{C_p D}{Q} + \frac{C_h Q(1 - D/R)}{2} - C_h(DT - L)$$

$$+ \frac{C_h(DT - L)^2}{2Q(1 - D/R)} + \frac{C_s(DT - L)^2}{2Q(1 - D/R)}. \tag{9.40}$$

By taking the partial derivative of Equation 9.40 with respect to Q and then with respect to $DT - L$, and setting both equal to zero, one obtains

$$\frac{\partial TC}{\partial Q} = -\frac{C_p D}{Q^2} + \frac{C_h(1 - D/R)}{2}$$
$$-\frac{C_h(DT - L)^2}{2Q^2(1 - D/R)} - \frac{C_s(DT - L)^2}{2Q^2(1 - D/R)} = 0 \qquad (9.41)$$

$$\frac{\partial TC}{\partial(DT - L)} = -C_h + \frac{C_h(DT - L)}{Q(1 - D/R)} + \frac{C_s(DT - L)}{Q(1 - D/R)} = 0. \qquad (9.42)$$

Equation 9.42 may be expressed as

$$\frac{DT - L}{Q} = \frac{C_h(1 - D/R)}{C_h + C_s} \qquad (9.43)$$

Substituting Equation 9.43 into Equation 9.41 gives

$$-\frac{C_p D}{Q^2} + \frac{C_h(1 - D/R)}{2} - \frac{C_h^2(1 - D/R)}{2(C_h + C_s)^2} - \frac{C_s C_h^2(1 - D/R)}{2(C_h + C_s)^2} = 0.$$

$$\frac{C_p D}{Q^2} = \frac{C_h C_s(1 - D/R)}{2(C_h + C_s)}$$

Solving for Q gives

$$Q* = \sqrt{\frac{1}{1 - D/R}} \sqrt{\frac{2C_p D}{C_h} + \frac{2C_p D}{C_s}}. \qquad (9.44)$$

Substituting Equation 9.44 into Equation 9.43 gives

$$L* = DT - \frac{C_h(1 - D/R)}{C_h + C_s} \sqrt{\frac{1}{1 - D/R}} \sqrt{\frac{2C_p D}{C_h} + \frac{2C_p D}{C_s}}$$

$$L* = DT - \sqrt{1 - \frac{D}{R}} \sqrt{\frac{2C_p D}{C_s(1 + C_s/C_h)}}. \qquad (9.45)$$

Equations 9.44 and 9.45 may now be substituted back into Equation 9.40 to give an expression for the minimum total system cost.[1] After several steps, the result is

$$TC* = C_i D + \sqrt{1 - \frac{D}{R}} \sqrt{\frac{2C_p C_h C_s D}{C_h + C_s}}. \qquad (9.46)$$

---

[1] The policy variables, $Q*$ and $L*$, supposedly occur at the minimum of the total cost function. Actually, there are three possibilities that may occur when the first partials of a function are set equal to zero and solved: a maximum, a minimum, or a point of inflection may be found. Although the nature of the total cost function derived here is such that a minimum is always found, a formal test could be presented to show that a minimum does exist using Equation 9.7.

Single source alternative. As an example of an inventory system with only one procurement source alternative, consider a situation in which an item having the following system parameters is to be manufactured:

$$D = 10 \text{ units per period}$$

$$R = 20 \text{ units per period}$$

$$T = 8 \text{ periods}$$

$$C_i = \$4.82 \text{ per unit}$$

$$C_p = \$100.00 \text{ per procurement}$$

$$C_h = \$0.20 \text{ per unit per period}$$

$$C_s = \$0.10 \text{ per unit per period}$$

Total system cost may be tabulated as a function of Q and L by use of Equation 9.39. The resulting values in the region of the minimum-cost point are given in Table 9.6. The minimum-cost Q and L may be found by inspection. As in the previous examples, the surface generated by the TC values is seen to be rather flat. Also, the selection of integral values of Q and L does not affect the total cost appreciably.

TABLE 9.6 TOTAL SYSTEM COST AS A FUNCTION OF L AND Q

Q / L	242	243	244	245	246	247	248
−6	56.399	56.394	56.390	56.386	56.383	56.380	56.377
−5	56.387	56.383	56.380	56.377	56.374	56.372	56.371
−4	56.378	56.375	56.372	56.370	56.368	56.367	56.366
−3	56.371	56.369	56.367	56.366	56.365	56.364	56.364
−2	56.366	56.365	56.364	56.363	56.363	56.364	56.364
−1	56.364	56.364	56.364	56.364	56.365	56.366	56.367
0	56.364	56.365	56.366	56.367	56.368	56.370	56.373
1	56.367	56.369	56.370	56.372	56.374	56.377	56.380
2	56.373	56.375	56.377	56.380	56.383	56.386	56.390
3	56.381	56.383	56.386	56.390	56.394	56.398	56.403
4	56.391	56.394	56.398	56.403	56.407	56.442	56.418

The minimum-cost procurement quantity and minimum-cost procurement level may be found directly by substituting into Equations 9.44 and 9.45 as follows.

$$Q^* = \sqrt{\frac{1}{1 - 10/20}} \sqrt{\frac{2(\$100.00)(10)}{\$0.20} + \frac{2(\$100.00)(10)}{\$0.10}} = 244.95$$

$$L^* = 10(8) - \sqrt{1 - \frac{10}{20}} \sqrt{\frac{2(\$100.00)(10)}{\$0.10(1 + \$0.10/\$0.20)}} = -1.65$$

The resulting total system cost at the minimum-cost procurement quantity and procurement level may be found from Equation 9.46 as

$$TC^* = \$4.82(10) + \sqrt{1 - \frac{10}{20}} \sqrt{\frac{2(\$100.00)(\$0.20)(\$0.10)(10)}{\$0.20 + \$0.10}} = \$56.363.$$

Accordingly, the item should be manufactured in lots of 245 units. Procurement action should be initiated when the stock on hand is negative; that is, when the backlog is two units. The total system cost will be $56.36 per period under these optimum operating conditions.

Multiple source alternatives.   Consider the following situation. A single-item inventory is maintained to meet demand. When the number of units on hand falls to a predetermined level, action is initiated to procure a replenishment quantity from one of several possible sources as is illustrated in Figure 9.9. The object is to determine the procurement level, the procurement quantity, and the procurement source in the light of system and cost parameters, so that the sum of all costs associated with the system will be minimized.

The procurement lead time, rate of replenishment, item cost, and procurement cost are source dependent. All other system and cost parameters are source independent. Explicit recognition of source-dependent parameters permits the use of Equation 9.46 to find the minimum cost procurement source. The procedure is to compare TC^* for all sources and to choose that source that results in a minimum TC^*. Equations 9.45 and 9.44 may then be applied to find the minimum cost procurement level and minimum cost procurement quantity.

An example of the procedure is the situation in which a system manager is experiencing a demand of four units per period for an item that may be either manufactured or purchased from one of two vendors. Holding cost per unit per period is $0.24 and shortage cost per unit short per period is $0.17. Values for source-dependent parameters are given in Table 9.7.

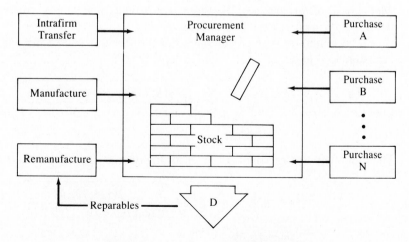

Figure 9.9   Multiple replenishment sources.

TABLE 9.7   SOURCE-DEPENDENT PARAMETERS FOR MULTI-SOURCE SYSTEM

PARAMETER	MANUFACTURE	VENDOR A	VENDOR B
R	12.00	∞	∞
T	6.00	3.00	4.00
C_i	$19.85	$17.94	$18.33
C_p	$17.32	$18.70	$17.50

The procurement source resulting in a minimum total system cost can be found from Equation 9.46. For the manufacturing alternative, it is

$$TC^* = \$19.85(4) + \sqrt{1 - \frac{4}{12}}\sqrt{\frac{2(\$17.32)(\$0.24)(\$0.17)(4)}{\$0.24 + \$0.17}}$$

$$= \$82.43.$$

For the alternative designated vendor A, it is

$$TC^* = \$17.94(4) + \sqrt{1 - \frac{4}{\infty}}\sqrt{\frac{2(\$18.70)(\$0.24)(\$0.17)(4)}{\$0.24 + \$0.17}}$$

$$= \$75.62.$$

For the alternative designated vendor B, it is

$$TC^* = \$18.33(4) + \sqrt{1 - \frac{4}{\infty}}\sqrt{\frac{2(\$17.50)(\$0.24)(\$0.17)(4)}{\$0.24 + \$0.17}}$$

$$= \$77.05.$$

On the basis of this analysis, the alternative designated vendor A would be chosen as the minimum cost procurement source.

The minimum cost procurement quantity for this source may be found from Equation 9.44 to be

$$Q^* = \sqrt{1 - \frac{4}{\infty}}\sqrt{\frac{2(\$18.70)(4)}{\$0.24} + \frac{2(\$18.70)(4)}{\$0.17}}$$

$$= 38.78.$$

The minimum cost procurement level for this source, from Equation 9.45 is

$$L^* = 4(3) - \sqrt{1 - \frac{4}{\infty}}\sqrt{\frac{2(\$18.70)(4)}{\$0.17(1 + \$0.17/\$0.24)}}$$

$$= -10.69.$$

Thus, the system manager would initiate procurement action when the stock level falls to -10.69 units, for a procurement quantity of 38.78 units, from the procurement source designated vendor A.

9.3. *CONSTRAINED CLASSICAL OPTIMIZATION*

In design and operations alike, physical and economic limitations often exist which act to limit system optimization globally. These limitations arise for a variety of reasons and generally cannot be removed by the decision maker. Accordingly, there may be no choice except to find the best or optimum solution subject to the constraints.

In this section, the example applications already presented will be revisited to illustrate three physical constraint situations: span constrained bridge design, dimension constrained conductor design, and space constrained inventory system operations. Constrained optimization by linear and dynamic programming for both physical and economic constraints is treated in Sections 9.4 and 9.5.

Span Constrained Bridge Design

Table 9.2 gave total costs as a function of the number of piers for two bridge design alternatives. Suppose that there exists a constraint on the pier spacing, expressed as a minimum spacing of 110 feet to permit the safe passage of barge traffic. With this requirement, it is evident from Table 9.2 and Figure 9.10 that Design A now replaces Design B as the minimum cost choice.

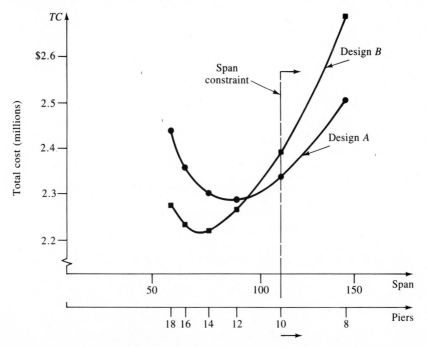

Figure 9.10   Span constrained bridge design.

This particular constraint leads to a design decision reversal and to a penalty in total system cost. Although a reversal will not always occur, a penalty in total cost usually will. In this example, the penalty is found from Table 9.2 to be $2,345,000 minus $2,220,000 or $125,555.

Dimension Constrained Conductor Design

Figure 9.6 illustrated the total cost as a function of the cross-sectional area for two conductor alternatives. The best choice was found to be aluminum with a cross-sectional area of 10.22 square inches.

Suppose that cross sections greater than 6 square inches are not feasible due to installation considerations. Under this constraint, the aluminum choice would still be best (no decision reversal), but a penalty in total cost will be incurred of $671.52 minus $593.98 or $77.54. This situation is illustrated in Figure 9.11.

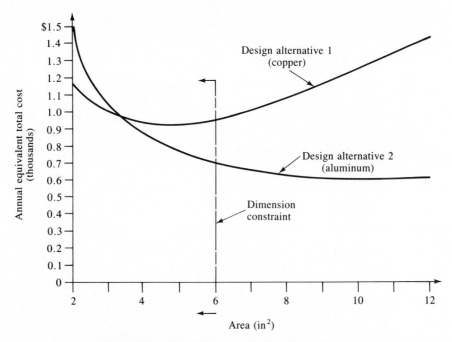

Figure 9.11   Dimension constrained conductor design.

Space Constrained Inventory Operations

Warehouse space is often a scarce resource. It may be expressed in cubic units designated W. Each unit in stock consumes a certain amount of space designated w. The units on hand require a total amount of space which must not exceed the amount available. In general, this may be formally stated as $I'w \leq W$, where I' is the maximum accumulation of stock.

The space constrained inventory model.   Consider the inventory operations of Figure 9.9. The total cost for this model was given by Equation 9.39. For this model, the maximum inventory, I', was given by Equation 9.28. If $I'w \leq W$, then $I' \leq W/w$, or

$$I' = Q\left(1 - \frac{D}{R}\right) + L - DT \leq \frac{W}{w}.$$

If the warehouse restriction is active, then $I' = W/w$, or

$$I' = Q\left(1 - \frac{D}{R}\right) + L - DT - \frac{W}{w} = 0. \tag{9.47}$$

For an active warehouse restriction, the minimum total cost will occur along the intersection of the surface defined by Equation 9.39 and the plane defined by Equation 9.47. The new *TC* equation is determined by substituting Equation 9.47 into Equation 9.39 giving

$$TC = C_iD + \frac{C_pD}{Q} + \frac{C_h(W/w)^2}{2Q(1 - D/R)} + \frac{C_s[Q(1 - D/R) - W/w]^2}{2Q(1 - D/R)}. \tag{9.48}$$

After some manipulation,

$$TC = a_1 + \frac{a_2}{Q} = a_3Q \tag{9.49}$$

where

$$a_1 = C_iD - C_s\left(\frac{W}{w}\right)$$

$$a_2 = C_pD + \frac{(C_h + C_s)(W/w)^2}{2(1 - D/R)}$$

and

$$a_3 = \frac{C_s(1 - D/R)}{2}.$$

The optimal value of Q can be obtained by differentiating Equation 9.49 with respect to Q, setting the result equal to zero, and solving for Q^* as

$$\frac{\partial TC}{\partial Q} = -\frac{a_2}{Q^2} + a_3 = 0$$

and

$$Q^* = \sqrt{\frac{a_2}{a_3}}$$

which can be written in terms of the original parameters as

$$Q^* = \sqrt{\frac{2C_pD}{C_s(1 - D/R)} + \frac{(C_h + C_s)(W/w)^2}{C_s(1 - D/R)^2}} \tag{9.50}$$

The optimal L^* can be determined from Equation 9.47 with $Q = Q^*$ as

$$L^* = DT + \frac{W}{w} - Q^*\left(1 - \frac{D}{R}\right). \tag{9.51}$$

Finally, the minimum total cost is given by substituting the optimal values of Q into Equation 9.48, or using Equation 9.49 with

$$TC^* = a_1 + \frac{a_2}{Q^*} + a_3 Q^*. \tag{9.52}$$

Space constrained single-source alternative. As an example of the application of Equations 9.47 through 9.52, suppose that a procurement manager will purchase an item having the following parameters:

$$D = 6 \text{ units per period}$$

$$R = \infty \text{ units per period}$$

$$T = 7 \text{ periods}$$

$$C_i = \$34.75 \text{ per unit}$$

$$C_p = \$23.16 \text{ per procurement}$$

$$C_h = \$0.30 \text{ per unit per period}$$

$$C_s = \$0.30 \text{ per unit per period}$$

$$W = 100 \text{ cubic units of space}$$

$$w = 24 \text{ cubic units per item}$$

With no restrictions, the optimal $Q^* = 43.04$ using Equation 9.44, $L^* = 20.48$ using Equation 9.45, and $TC^* = \$214.96$ using Equation 9.46. If $R = \infty$, Equation 9.47 can be written as

$$I'^* = Q^* + L^* - DT. \tag{9.53}$$

When the optimal values are used, Equation 9.53 yields

$$I'^* = 43.04 + 20.48 - 6(7) = 21.52.$$

Since $W/w = 100/24 < 21.52 = I'^*$, the warehouse restriction is active.

Next, we determine the optimal policy values for the constrained warehouse where $I'^* = W/w = 100/24 = 4.17$. From Equation 9.50 with $1 - D/R \rightarrow$ unity,

$$Q^* = \sqrt{\frac{2(\$23.16)(6)}{\$0.30} + \frac{(\$0.30 + \$0.30)(4.17)^2}{\$0.30}} = 30.57.$$

Next, using Equation 9.51 with $1 - D/R \rightarrow$ unity, we determine L^* as

$$L^* = 6(7) + 4.17 - 30.57 = 15.60.$$

Finally, we determine the optimal total cost, using Equation 9.48 with $1 - D/R \rightarrow$ unity, as

$$TC^* = \$34.75(6) + \frac{\$23.16(6)}{30.57} + \frac{\$0.30(4.17)^2}{2(30.57)}$$

$$+ \frac{\$0.30[6(7) - 15.17]^2}{2(30.57)} = \$216.55.$$

The difference in TC at the optimum is approximately \$1.60 per period. The total cost as a function of the procurement quantity both without and with the warehouse restriction are shown in Figure 9.12.

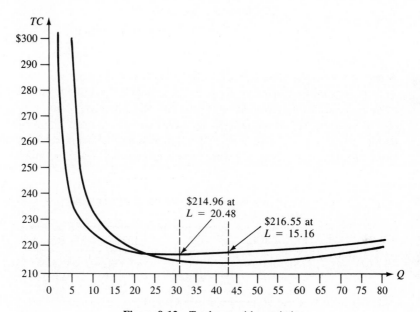

Figure 9.12   Total cost with restriction.

Space constrained multiple-source alternatives.   The explicit recognition of source dependent parameters permits the use of models for the constrained system to be used for the single-item, multi-source system. The procedure is to evaluate each source and to find that source which will result in a minimum total cost subject to the restriction.

As an example, suppose that the system is constrained by a total warehouse space of 1000 cubic units. Also suppose that the item requires 120 cubic units of space. We determine if the restriction is active for vendor A using Equation 9.53 as

$$I'^* = 38.7806 - 10.6917 - 4(3) = 16.0889.$$

Since $W/w = 1000/120 < 16.0889 = I'^*$, the warehouse restriction is active.

Next, determine the optimal policy values for the constrained situation with

$I'* = 1000/120 = 8.3333$. From Equation 9.50 with $1 - D/R \rightarrow$ unity, $Q*$ is found as

$$Q* = \sqrt{\frac{2(\$18.70)4}{\$0.17} + \frac{(\$0.24 + \$0.17)8.3333}{\$0.17}} = 32.3648$$

Using Equation 9.51 with $1 - D/R \rightarrow$ unity, $L*$ is found to be

$$L* = 4(3) + 8.3333 - 32.3648 = -12.0315.$$

As the last step, determine the optimal total cost using Equation 9.48 with $1 - D/R \rightarrow$ unity, as

$$TC* = \$17.94(4) + \frac{\$18.70(4)}{32.3648} + \frac{\$0.24(8.3333)^2}{2(32.3648)}$$

$$+ \frac{\$0.17[(32.3648) - 8.3333]^2}{2(32.3648)} = \$75.85.$$

A penalty of $75.85 less $75.62 = $0.23 occurs due to the warehouse constraint.

It is possible for the restricted source choice to differ from the source choice when no restriction exists. Accordingly, the other source possibilities in the example should be evaluated in the face of the warehouse restriction. The results of this evaluation are given in Table 9.8. Inspection reveals that vendor A remains the best choice for the case where the warehouse capacity is 1000 cubic units.

TABLE 9.8   OPTIMAL POLICIES FOR RESTRICTED MULTI-SOURCE SYSTEM

SOURCE	$L*$	$Q*$	$TC*$
Manufacture	5.6710	40.0000	$82.5158
Vendor A	−12.0315	32.3648	75.8454
Vendor B	−7.1470	31.4820	77.2550

9.4. *CONSTRAINED OPTIMIZATION BY LINEAR PROGRAMMING*[2]

The general linear programming model may be stated mathematically as that of optimizing the evaluation function,

$$E = \sum_{j=1}^{n} e_j x_j \qquad (9.54)$$

subject to the constraints

---

[2] The mathematical formulation of the general linear programming model, together with the simplex optimization algorithm, was developed by George Dantzig in 1947.

$$\sum_{j=1}^{n} a_{ij}x_j = c_i \qquad i = 1, 2, \ldots, m$$

$$x_j \geq 0 \qquad j = 1, 2, \ldots, n.$$

Optimization requires either maximization or minimization, depending upon the measure of evaluation involved. The decision maker has control of the vector of variables, x_j. Not directly under his or her control is the vector of evaluation coefficients, e_j, the matrix of constraints, a_{ij}, and the vector of constants, c_i. This structure is in accordance with the constrained version of the decision evaluation function in Equation 7.1.

Graphical Optimization Methods

The mathematical statement of the general linear programming model may be explained in graphical terms. There exist n variables that define an n-dimensional space. Each constraint corresponds to a hyperplane in this space. These constraints surround the region of feasible solution by hypersurfaces so that the region is the interior of a convex polyhedron. Since the evaluation function is linear in the n variables, the requirement that this function have some constant value gives a hyperplane that may or may not cut through the polyhedron. If it does, one or more feasible solutions exist. By changing the value of this constant, a family of hyperplanes parallel to each other is generated. The distance from the origin to a member of this family is proportional to the value of the evaluation function.

Two limiting hyperplanes may be identified: one corresponds to the largest value of the evaluation function for which the hyperplane just touches the polyhedron; the other corresponds to the smallest value. In most cases, the limiting hyperplanes just touch a single vertex of the limiting polyhedron. This outermost limiting point is the solution that optimizes the evaluation function.

Although this is a graphical description, a graphical solution is not convenient when more than three variables are involved. The following examples illustrate the general linear programming model and its graphical optimization method for operations having two and three activities.

Graphical maximization for two activities.   As an example of the case where two activities compete for scarce resources, consider the following production system. Two products are to be manufactured. A single unit of product A requires 2.4 min of punch press time and 5.0 min of assembly time. The profit for product A is \$0.60 per unit. A single unit of product B requires 3.0 min of punch press time and 2.5 min of welding time. The profit for product B is \$0.70 per unit. The capacity of the punch press department available for these products is 1,200 min per week. The welding department has idle capacity of 600 min per week and the assembly department can supply 1,500 min of capacity per week. The manufacturing and marketing data for this production system are summarized in Table 9.9.

In this example, two products compete for the available production time. The objective is to determine the quantity of product A and the quantity of product B to

TABLE 9.9 MANUFACTURING AND MARKETING DATA FOR
TWO PRODUCTS

DEPARTMENT	PRODUCT A	PRODUCT B	CAPACITY
Punch Press	2.4	3.0	1,200
Welding	0.0	2.5	600
Assembly	5.0	0.0	1,500
Profit	$0.60	$0.70	

produce so that total profit will be maximized. This will require maximizing

$$TP = \$0.60A + \$0.70B$$

subject to

$$2.4A + 3.0B \leq 1,200$$

$$0.0A + 2.5B \leq 600$$

$$5.0A + 0.0B \leq 1,500$$

$$A \geq 0 \quad \text{and} \quad B \geq 0.$$

The graphical equivalent of the algebraic statement of this two-product system is shown in Figure 9.13. The set of linear restrictions define a region of feasible solutions. This region lies below $2.4A + 3.0B = 1,200$ and is restricted further by the requirements that $B \geq 240$, $A \geq 300$, and that both A and B be nonnegative. Thus, the scarce resources determine which combinations of the activities are feasible and which are not feasible.

The production quantity combinations of A and B that fall within the region of feasible solutions constitute feasible production programs. That combination or combinations of A and B which maximize profit are sought. The relationship between A and B is $1.167A = B$. This relationship is based on the relative profit of each product. The total profit realized will depend upon the production quantity combination chosen. Thus, there is a family of isoprofit lines, one of which will have at least one point in the region of the feasible production quantity combinations and be a maximum distance from the origin. The member that satisfies this condition intersects the region of feasible solutions at the extreme point $A = 300$, $B = 160$. This is shown as a dashed line in Figure 9.13 and represents a total profit of $\$0.60(300) + \$0.70(160) = \$292$. No other production quantity combination would result in a higher profit.

Alternative production programs with the same profit might exist in some cases. This would occur when the isoprofit line lies parallel to one of the limiting restrictions. For example, if the relative profits of product A and product B were $A = 1.25B$, the isoprofit line in Figure 9.13 would coincide with the restriction $2.4A + 3.0B = 1,200$. In this case, the isoprofit line would touch the region of feasible solutions along a line instead of at a point. All production quantity combinations along the line would maximize profit.

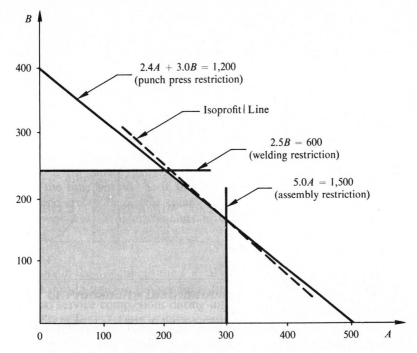

Figure 9.13   Graphical optimization for a two product production system.

Graphical maximization for three activities.   When three activities compete for scarce resources, a three-dimensional space is involved. Each constraint is a plane in this space, and all constraints taken together identify a volume of feasible solutions. The evaluation function is also a plane, its distance from the origin being proportional to its value. The optimum value for the evaluation function occurs when this plane is located so that it is at the extreme point of the volume of feasible solutions.

As an example, suppose that the production operations for the previous system are to be expanded to include a third product, designated C. A single unit of product C will require 2.0 min of punch press time, 1.5 min of welding time, and 2.5 min of assembly time. The profit associated with product C is \$0.50 per unit. Manufacturing and marketing data for this revised production situation are summarized in Table 9.10.

TABLE 9.10   MANUFACTURING AND MARKETING DATA FOR THREE PRODUCTS

DEPARTMENT	PRODUCT A	PRODUCT B	PRODUCT C	CAPACITY
Punch Press	2.4	3.0	2.0	1,200
Welding	0.0	2.5	1.5	600
Assembly	5.0	0.0	2.5	1,500
Profit	\$0.60	\$0.70	\$0.50	

In this example, three products compete for the available production time. The objective is to determine the quantity of product A, the quantity of product B, and the quantity of product C to produce so that total profit will be maximized. This will require maximizing

$$TP = \$0.60A + \$0.70B + \$0.50C$$

subject to

$$2.4A + 3.0B + 2.0C \leq 1,200$$

$$0.0A + 2.5B + 1.5C \leq 600$$

$$5.0A + 0.0B + 2.5C \leq 1,500$$

$$A \geq 0, \quad B \geq 0, \quad \text{and} \quad C \geq 0.$$

The graphical equivalent of the algebraic statement of this three-product production situation is shown in Figure 9.14. The set of restricting planes defines a volume of feasible solutions. This region lies below $2.4A + 3.0B + 2.0C = 1,200$ and is restricted further by the requirement that $2.5B + 1.5C \leq 600$, $5.0A + 1.5C \leq 1,500$, and that A, B, and C be nonnegative. Thus, the scarce resources determine which combinations of the activities are feasible and which are not feasible.

The production quantity combinations of A, B, and C that fall within the volume of feasible solutions constitute feasible production programs. That combination or combinations of A, B, and C which maximizes total profit is sought. The expression $0.60A + 0.70B + 0.50C$ gives the relationship among A, B, and C based on

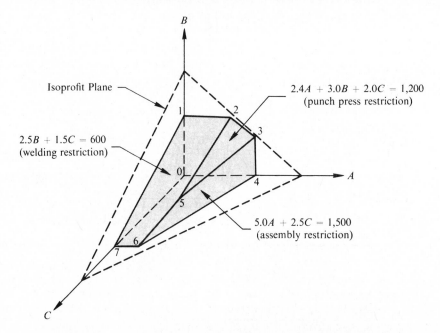

Figure 9.14   Graphical optimization for a three product production system.

the relative profit of each product. The total profit realized will depend upon the production quantity combination chosen. Thus, there exists a family of isoprofit planes, one for each value of total profit. One of these planes will have at least one point in the volume of feasible solutions and will be a maximum distance from the origin. The plane that maximizes profit will intersect the volume at an extreme point. This calls for the computation of total profit at each extreme point as given in Table 9.11. The coordinates of each extreme point were found from the restricting planes, and the associated profit was calculated from the total profit equation.

TABLE 9.11 TOTAL PROFIT
COMPUTATIONS AT EXTREME POINTS
OF FIGURE 9.14

	COORDINATE			
Point	A	B	C	Profit
0	0	0	0	$ 0
1	0	240	0	$168.00
2	200	240	0	$288.00
3	300	160	0	$292.00
4	300	0	0	$180.00
5	180	96	240	$295.20
6	100	0	400	$260.00
7	0	0	400	$200.00

Inspection of the total profit values in Table 9.11 indicates that profit is maximized at point 5, which has the coordinates 180, 96, and 240. This means that if 180 units of product A, 96 units of product B, and 240 units of product C are produced, profit will be maximized. No other production quantity combination will result in a higher profit. Also, no alternative optimum solutions exist, since the total profit plane intersects the volume of feasible solutions at only a single point.

Addition of a third product to the two-product production system increased total profit from $292.00 per week to $295.20 per week. This increase in profit results from reallocations of the available production time and from greater utilization of the idle capacity of the welding department. The number of units of product A was reduced from 200 to 180 and the number of units of B was reduced from 240 to 96. This made it possible to add 240 units of product C to the production program, with the resulting increase in profit.

The Simplex Optimization Algorithm

Many problems to which the general linear programming model may be applied are n-dimensional in that n activities compete for scarce resources. In these cases graphical optimization methods cannot be applied and a numerical optimization technique known as the simplex algorithm can be used.

The *simplex optimization algorithm* is an iterative technique that begins with a feasible solution, tests for optimality, and proceeds toward an improved solution. It can be shown that the algorithm will finally lead to an optimal solution if such a solution exists. In this section, the simplex method will be applied to the three-product production problem presented above. Reference to the graphical solution will explain certain facets of the computational procedure.

The simplex matrix. The three-product production problem required the maximization of a total profit equation subject to certain constraints. These constraints must be converted to equalities of the form specified by the general linear programming model. This requires the addition of three "slack" variables to remove the inequalities. Thus, the constraints become

$$2.4A + 3.0B + 2.0C + S_1 = 1,200$$

$$0.0A + 2.5B + 1.5C + S_2 = 600$$

$$5.0A + 0.0B + 2.5C + S_3 = 1,500$$

The amount of departmental time not used in the production program is represented by a slack variable. Thus, each slack variable takes on whatever value is necessary for the equality to exist. If nothing is produced, the slack variables assume values equal to the total production time available in each department. This gives an initial feasible solution expressed as $S_1 = 1,200$, $S_2 = 600$, and $S_3 = 1,500$. Each slack variable is one of the x_j variables in the model. Since, however, no profit can be derived from idle capacity, total profit may be expressed as

$$TP = \$0.60A + \$0.70B + \$0.50C + \$0S_1 + \$0S_2 + \$0S_3.$$

The initial matrix required by the simplex algorithm may now be set up as shown in Table 9.12. The first column is designated e_i and gives the coefficients applicable to the initial feasible solution. These are all zero, since the initial solution involves the allocation of all production time to the slack variables. The second column is designated *Sol* and gives the variables in the initial solution. These are the slack variables that were introduced. The third column is designated b and gives the number of minutes of production time associated with the solution variables

TABLE 9.12 INITIAL MATRIX FOR A THREE-PRODUCT PRODUCTION PROBLEM

e_j			0	0	0	0.60	0.70	0.50	
e_i	Sol	b	S_1	S_2	S_3	A	B	C	θ
0	S_1	1,200	1	0	0	2.4	3.0	2.0	400
0	S_2	600	0	1	0	0	2.5	1.5	240
0	S_3	1,500	0	0	1	5.0	0	2.5	∞
E_j		0	0	0	0	0	0	0	
$E_j - e_j$			0	0	0	-0.60	-0.70	-0.50	

r

k

of the previous column. These reflect the total production capacity in the initial solution. Each of the next three columns is headed by slack variables with elements of zero or unity, depending upon which equation is served by which slack variable. The e_j heading for these columns carries an entry of zero, corresponding to a zero profit. The last three columns are headed by the activity variables, with elements entered from the restricting equations. The e_j heading for these columns is the profit associated with each activity variable. The last column, designated θ, is utilized during the computational process.

Testing for optimality.   After an initial solution has been obtained, it must be tested to see if a program with a higher profit can be found. The optimality test is accomplished with the aid of the last two rows in Table 9.12. The required steps are

1. Enter values in the row designated E_j from the expression $E_j = \Sigma\ e_i a_{ij}$, where a_{ij} are the matrix elements in the ith row and the jth column.
2. Calculate $E_j - e_j$, for all positions in the row designated $E_j - e_j$.
3. If $E_j - e_j$ is negative for at least one j, a better program is possible.

Application of the optimality test to the initial feasible solution is shown in the last two rows of Table 9.12. The first element in the E_j row is calculated as $0(1,200) + 0(600) + 0(1,500) = 0$. The second is $0(1) + 0(0) = 0$. All values in this row will be zero, since all e_i values are zero in the initial feasible solution. The first element in the $E_j - e_j$ row is $0 - 0 = 0$, the second is $0 - 0 = 0$, the third is $0 - 0 = 0$, the fourth is $0 - 0.60 = -0.60$, and so forth. Since $E_j - e_j$ is negative for at least one j, this initial solution is not optimal.

Iteration toward an optimal program.   If the optimality test indicates that an optimal program has not been found, the following iterative procedure may be employed.

1. Find the minimum value of $E_j - e_j$ and designate this column k. The variable at the head of this column will be the incoming variable.
2. Calculate entries for the column designated θ from $\theta_i = b_i/a_{ik}$.
3. Find the minimum positive value of θ_i and designate this row r. The variable to the left of this row will be the outgoing variable.
4. Set up a new matrix with the incoming variable substituted for the outgoing variable. Calculate new elements, a'_{ij}, as $a'_{rj} = a_{rj}/a_{rk}$ for $i = r$ and $a'_{ij} = a_{ij} - a_{ik}a'_{rj}$ for $i \neq r$.
5. Perform the optimality test.

Apply steps 1, 2, and 3 of the foregoing procedure to the initial matrix. In Table 9.12 step 1 designates B as the incoming variable. Values for θ_i are calculated from step 2. Step 3 designates S_2 as the outgoing variable. The affected column and row are marked with a k and an r, respectively, in Table 9.12.

Steps 4 and 5 require a new matrix, as shown in Table 9.13. The incoming variable B, together with its associated profit, replaces the outgoing variable S_2 with its profit. All other elements in this row are calculated from the first formula of step 4. Elements in the remaining two rows are calculated from the second formula of step 4. The optimality test indicates that an optimal solution has not yet been reached. Note that after this iteration, the profit at point 1 of Figure 9.14 appears. Comparison of the results in Table 9.12 and Table 9.13 with the total profit computations in Table 9.11 indicates that the isoprofit plane, which began at the origin, has now moved away from this initial position to point 1. The gain from this iteration was $168 − $0 = $168.

TABLE 9.13   FIRST ITERATION FOR A THREE-PRODUCT PRODUCTION PROBLEM

e_j			0	0	0	0.60	0.70	0.50	
e_j	Sol	b	S_1	S_2	S_3	A	B	C	θ
0	S_1	480	1	−1.20	0	2.40	0	0.20	200
0.70	B	240	0	0.40	0	0	1	0.60	∞
0	S_3	1,500	0	0	1	5.00	0	2.50	300
E_j		168	0	0.28	0	0	0.70	0.42	
$E_j − e_j$			0	0.28	0	−0.60	0	−0.08	

Since the first iteration did not yield an optimal solution, it is necessary to repeat steps 1 through 5. Steps 1, 2, and 3 are applied to Table 9.13, designating A as the incoming variable and S_1 as the outgoing variable. The incoming variable, together with its associated profit, replaces the outgoing variable as shown in Table 9.14. All other elements in this new matrix are calculated from the formulas in step 4. Applications of the optimality test indicate that the solution indicated is still not optimal. Table 9.11 and Figure 9.14 show that the isoprofit plane is now at point 2. The gain from this iteration was $288 − $168 = $120.

Table 9.14 did not yield an optimal solution, requiring the reapplication of

TABLE 9.14   SECOND ITERATION FOR A THREE-PRODUCT PRODUCTION PROBLEM

e_j			0	0	0	0.60	0.70	0.50	
e_j	Sol	b	S_1	S_2	S_3	A	B	C	θ
0.60	A	200	0.416	−0.50	0	1	0	0.084	2,381
0.70	B	240	0	0.40	0	0	1	0.60	400
0	S_3	500	−2.08	2.50	1	0	0	2.08	240
E_j		288	0.25	−0.02	0	0.60	0.70	0.47	
$E_j − e_j$			0.25	−0.02	0	0	0	−0.03	

steps 1 through 5. Steps 1, 2, and 3 designate C as the incoming variable and S_3 as the outgoing variable. This incoming variable, together with its associated profit, replaces the outgoing variable as shown in Table 9.15. All other elements in the matrix are calculated from the formulas of step 4. Application of the optimality test indicates that the solution exhibited by Table 9.15 is optimal. Table 9.11 and Figure 9.14 indicate that the isoprofit plane is now at point 5. The gain from this iteration was \$295.20 − \$288 = \$7.20.

TABLE 9.15   THIRD ITERATION FOR A THREE-PRODUCT PRODUCTION PROBLEM

e_j			0	0	0	0.60	0.70	0.50	
e_i	Sol	b	S_1	S_2	S_3	A	B	C	θ
0.60	A	180	0.50	−0.60	−0.04	1	0	0	
0.70	B	96	0.60	−0.32	−0.29	0	1	0	
0.50	C	240	−1	1.20	0.48	0	0	1	
E_j		295.20	0.22	0.016	0.016	0.60	0.70	0.50	
$E_j - e_j$			0.22	0.016	0.016	0	0	0	

Minimizing by the simplex algorithm.   The computational algorithm just presented may be used without modification for problems requiring minimization if the signs of the cost coefficients are changed from positive to negative. The principle that maximizing the negative of a function is the same as minimizing the function then applies. If these coefficients are entered in the simplex matrix with their negative signs, the value of the solution will decrease as the computations proceed.

9.5. CONSTRAINED OPTIMIZATION BY DYNAMIC PROGRAMMING[3]

The dynamic programming model is based on the principle of optimality, which states the following: *An optimal policy has the property that, whatever the initial state and initial decisions are, the remaining decisions must constitute an optimal policy with regard to the state resulting from the first decision.* This principle implies a sequential decision process in which a problem involving several variables is broken down in a sequence of simpler problems each involving a single variable, and prior decisions are not affected by subsequent decisions. Then, each component problem can be solved by the best available procedure, usually enumeration when the evaluation functions are not simple functions. Thus, by using the principle of optimality, a sequence of univariate enumerations is substituted for multivariate enumeration. This almost always results in a significant increase in computational efficiency.

---

[3] Dynamic programming owes much of its development to the pioneering work of Richard Bellman and his associates at the Rand Corporation during the 1950s.

Consider a problem involving three decision variables, each capable of assuming 10 different values. A multivariate enumeration involves $10 \times 10 \times 10$, or 1,000 evaluations. The sequence of three univariate enumerations involves only $10 + 10 + 10$, or 30 evaluations. Best results are achieved when the effectiveness functions are additive, that is, when it can be assumed that the returns from different activities can be measured in a common unit. In addition, the return from any activity must be independent of the level of other activities, and the total return must be the sum of returns from individual activities. The discussion in this section and examples in subsequent sections are based on the assumption of separate evaluation functions, the sequential decision process, and the application of the principle of optimality.

The Dynamic Programming Model

A general dynamic programming model may be developed quite easily from the principle of optimality for single-dimensional processes. The programming situation involves a certain quantity of an economic resource, such as machines, space, money, or personnel, which can be allocated to a number of different activities. A conflict arises from the numerous ways in which the allocation can be made. A certain return is derived from the allocation of all or part of the resource to a given activity. The magnitude of the return depends jointly upon the specific activity and the quantity of the resource allocated. The objective is to allocate the available resource among the various activities so as to maximize the total return.

Let the number of activities be designated by N, enumerated in fixed order, $1, 2, \ldots, N$. Associated with each activity is a return function which gives the dependence of the return from the activity upon the quantity of the resource allocated. The quantity of the resource allocated to the ith activity may be designated x_i, and $g_i(x_i)$ denotes the return function. Since it was assumed that the activities are independent and that the returns from all activities are additive, the total return may be expressed as

$$R(x_1, x_2, \ldots, x_N) = g_1(x_1) + g_2(x_2) + \cdots + g_N(x)_N. \tag{9.55}$$

The limited quantity of the resource, Q, leads to the constraint

$$Q = x_1 + x_2 + \cdots + x_N \qquad x_i \geq 0.$$

The objective is to maximize the total return over all x_i, subject to the foregoing constraint.

The problem of maximizing the total return from the allocation process is viewed as one member of a sequence of allocation processes. The quantity of the resource, Q, and the number of activities, N, will not be assumed to be fixed, but may take on any value subject to the restriction that N be an integer. It is artificially required that the allocations be made one at a time; that is, a quantity of the resource is allocated to the Nth activity, then to the $(N - 1)$th activity, and so on.

Since the maximization of R depends upon Q and N, the dependence must be explicitly stated by a sequence of functions

$$f_N(Q) = \max_{x_i} \{R(x_1, x_2, \ldots, x_N)\} \tag{9.56}$$

The function $f_N(Q)$ represents the maximum return from the allocation of the resource Q to the N activities. If it is assumed that $g_i(0) = 0$ for each activity, it is evident that

$$f_N(0) = 0 \qquad N = 1, 2, \ldots, \text{ and so forth}$$

and that

$$f_1(Q) = g_1(Q). \tag{9.57}$$

These two statements express the return to be expected from the Nth activity if no resources are allocated and the return to be expected from the first activity if all the resources are allocated to it.

A functional relation connecting $f_N(Q)$ and $f_{N-1}(Q)$ for arbitrary values of N and Q may be developed by the following procedure. If x_N, $0 \le x_N \le Q$, is the allocation of the resource to the Nth activity, then, regardless of the value of x_N, a quantity of resource $Q - x_N$ will remain. The return from the $N - 1$ activities may be expressed as $f_{N-1}(Q - x_N)$. Therefore, the total return from the N activity process may be expressed as

$$g_N(x_N) + f_{N-1}(Q - x_N).$$

An optimal choice of x_N would be that choice which maximizes the foregoing function. Thus, the fundamental dynamic programming model may be written as

$$f_N(Q) = \max_{0 \le x_N \le Q} \{g_N(x_N) + f_{N-1}(Q - x_N)\} \qquad N = 2, 3, \ldots \tag{9.58}$$

where $f_1(Q)$ is determined from Equation 9.57. This model is based on the concept stated as the principle of optimality.

Dynamic Programming Applications

Equation 9.58 provides a method for obtaining the sequence $f_N(Q)$ once $f_1(Q)$ is known. Since $f_1(Q)$ determines $f_2(Q)$, it is also true that $f_2(Q)$ leads to an evaluation of $f_3(Q)$. This recursive relationship progresses in this manner until $f_{N-1}(Q)$ determines $f_N(Q)$, at which time the process stops.

As a general application of this procedure, consider the network of Figure 9.15. The maximum path through the network is desired. There are three possible starting points (A, B, and C) and three possible ending points (a, b, and c). The problem might be solved by identifying all possible paths and calculating the value of each. It is evident, however, that this method might result in the omission of a path. In addition, this process requires much calculation.

By utilizing the concept of dynamic programming, the problem is viewed as a stagewise process. The maximum path from stage 1 to stage 2 is entered for each terminal point of stage 2. This occurs because whatever the starting point chosen, the path to the terminal point must be a maximum for an optimal two-stage policy to exist. Thus, the optimal two-stage policy is easily identified with little calculation.

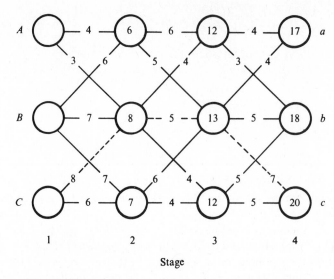

Figure 9.15 A network illustration of dynamic programming.

The optimal three-stage policy is found by again calculating the maximum of each terminal point from a knowledge of the results of the optimal two-stage policies. Continuing to the fourth stage with the same philosophy results in an optimal four-stage policy. This maximum path will have a value of 20 and is indicated by the dashed line.

Consider the resource allocation situation next. The entire set of values $f_N(Q)$ in the range 0 to Q may be found by assuming a finite grid of points. Each element of the sequence $f_N(Q)$ may be evaluated and tabulated at each of these grid points. Maximization of Equation 9.58 may be performed when $N = 1$, since $f_1(Q) = g_1(Q)$. A set of values $f_1(Q)$ may be computed and tabulated. Equation 9.58 may then be used to compute $f_2(Q)$ for $N = 2$; that is,

$$f_2(Q) = \max_{0 \le x_2 \le Q} \{g_2(x_2) + f_1(Q - x_2)\}.$$

The maximization process begins by evaluating $g_2(0) + f_1(Q)$ and $g_2(\Delta) + f_1(Q - \Delta)$. The larger of these values is retained and compared with the values $g_2(2\Delta) + f_1(Q - 2\Delta)$. As before, the larger of these values is retained and compared with $g_2(3\Delta) + f_1(Q - 3\Delta)$. This process is continued until all values of Δ are considered. The result is a table of values for $f_2(Q)$ at the Q values of 0, Δ, 2Δ, . . . , as given by Table 9.16. For any given value of Q, subject to the grid established, the table gives the corresponding allocation, x_2, for the second activity, $N - 2$, since only a two-stage process was considered. Once the allocation x_2 to be the second activity is determined, the allocation of the remaining resources, $Q - x_2$, to the first activity may be determined. The computational procedure can be extended for N stages, giving an expanded table and a means for determining all allocations.

TABLE 9.16 TABULAR REPRESENTATION OF
COMPUTATIONAL SCHEME

Q	x_1	$f_1(Q)$	x_2	$f_2(Q)$	$\ldots$
0					
Δ					
2Δ					
.					
.					
.					

Capital investment application.   As a simple application of the concept developed to this point consider the following situation. A conflict arises from the numerous ways in which a fixed amount of capital can be allocated to a number of activities. A certain return is derived as a result of allocating all or part of the capital to a given activity. The total return depends upon the manner in which the allocation is made. Therefore, the objective is to find that allocation which maximizes the total return.

Suppose that 8 units of capital are available and that three different activities are under consideration. The return functions for each venture are given in Table 9.17, with each return being measured in a common unit. The return from each activity is independent of the allocations to the other activities. In addition, the total return is the sum of the individual returns. Each activity exhibits a return function that increases for a portion of its range and then levels off. This is due to the law of diminishing returns and is typical of many activities.

A finite grid of points is established from this problem by the discrete nature of the return functions. Direct enumeration of all ways in which the 8 units of capital can be allocated to the three activities would be possible. The dynamic programming model given by Equation 9.58 may, however, be used with a saving in computational effort.

TABLE 9.17 RETURN
FUNCTIONS FOR CAPITAL
INVESTMENT

Q	$g_1(Q)$	$g_2(Q)$	$g_3(Q)$
0	0	0	0
1	5	5	4
2	15	15	26
3	40	40	40
4	80	60	45
5	90	70	50
6	95	73	51
7	98	74	52
8	100	75	53

The return expected from the first activity if all the available capital is allo-cated to it is determined from Equation 9.57 as

$$f_1(0) = g_1(0) = \quad 0$$
$$f_1(1) = g_1(1) = \quad 5$$
$$f_1(2) = g_1(2) = \quad 15$$
$$f_1(3) = g_1(3) = \quad 40$$
$$f_1(4) = g_1(4) = \quad 80$$
$$f_1(5) = g_1(5) = \quad 90$$
$$f_1(6) = g_1(6) = \quad 95$$
$$f_1(7) = g_1(7) = \quad 98$$
$$f_1(8) = g_1(8) = 100$$

This completes the computation of $f_1(Q)$. Each value is entered in the first stage of Table 9.18.

TABLE 9.18   TABULAR SOLUTION FOR
A CAPITAL INVESTMENT PROBLEM

Q	x_1	$f_1(Q)$	x_2	$f_2(Q)$	x_3	$f_3(Q)$
0	0	0	0	0	0	0
1	1	5	1	5	0	5
2	2	15	2	15	2	26
3	3	40	3	40	3	40
4	4*	80	0	80	0	80
5	5	90	0	90	0	90
6	6	95	2	95	2	106
7	7	98	3	120	3	120
8	8	100	4*	140	0*	140

From the results of $f_1(Q), f_2(Q)$ may be computed by use of Equation 9.58. It is evident that when $Q = 0, f_2(Q) = 0$. When $Q = 1$,

$$f_2(1) = \max_{0 \le x_2 \le 1} g_2(x_2) + f_1(1 - x_2).$$

For values of x_2 ranging from 0 to 1, this gives

$$f_2(1) = \max \begin{Bmatrix} g_2(0) + f_1(1) = 5 \\ g_2(1) + f_1(0) = 5 \end{Bmatrix}.$$

When $Q = 2$,

$$f_2(2) = \max_{0 \le x_2 \le 2} \{g_2(x_2) + f_1(2 - x_2)\}.$$

For values of x_2 ranging from 0 to 2, this gives

$$f_2(2) = \max \begin{cases} g_2(0) + f_1(2) = 15 \\ g_2(1) + f_1(1) = 10 \\ g_2(2) + f_1(0) = 15 \end{cases}.$$

When $Q = 3$,

$$f_2(3) = \max_{0 \le x_2 \le 3} \{g_2(x_2) + f_1(3 - x_2)\}.$$

For values of x_2 ranging from 0 to 3, this gives

$$f_2(3) = \max \begin{cases} g_2(0) + f_1(3) = 40 \\ g_2(1) + f_1(2) = 20 \\ g_2(2) + f_1(1) = 20 \\ g_2(3) + f_1(0) = 40 \end{cases}$$

This process is continued until $f_2(8)$ is evaluated. The maximum value of $f_2(Q)$ is identified for each value of Q and entered in the second stage of Table 9.18 together with its associated value of x_2. An arbitrary choice may be made when a tie is involved.

The third stage is considered next. Using the results of $f_2(Q)$, $f_3(Q)$ may be computed by use of Equation 9.58. As before, when $Q = 0$, $f_3(Q) = 0$. When $Q = 1$,

$$f_3(1) = \max_{0 \le x_3 \le 1} \{g_3(x_3) + f_2(1 - x_3)\}.$$

For values of x_3 ranging from 0 to 1, this gives

$$f_3(1) = \max \begin{cases} g_3(0) + f_2(1) = 5 \\ g_3(1) + f_2(0) = 4 \end{cases}.$$

When $Q = 2$,

$$f_3(2) = \max_{0 \le x_3 \le 2} \{g_3(x_3) + f_2(2 - x_3)\}.$$

For values of x_3 ranging from 0 to 2, this gives

$$f_3(2) = \max \begin{cases} g_3(0) + f_2(2) = 15 \\ g_3(1) + f_2(1) = 9 \\ g_3(2) + f_2(0) = 26 \end{cases}.$$

When $Q = 3$,

$$f_3(3) = \max_{0 \le x_3 \le 3} \{g_3(x_3) + f_2(3 - x_3)\}.$$

For values of x_3 ranging from 0 to 3, this gives

$$f_3(3) = \max \begin{cases} g_3(0) + f_2(3) = 40 \\ g_3(1) + f_2(2) = 19 \\ g_3(2) + f_2(1) = 31 \\ g_3(3) + f_2(0) = 40 \end{cases}$$

Again this process is continued until $f_3(8)$ is evaluated. The maximum value of $f_3(Q)$ is identified for each value of Q and entered in the third stage of Table 9.18 together with its associated value of x_3.

Table 9.18 may now be used to find the maximum return allocation of capital. This maximum is found in the third stage of the table to be 140 units. The allocation of capital associated with this return may be found by noting that $x_3 = 0$ at $f_3(Q) = 140$. Therefore, 8 units of capital remain for the two-stage process giving a value for x_2 of 4 units. This leaves 4 units for the first stage: $x_1 = 4$. Each allocation is indicated with an asterisk in Table 9.18.

Note that Table 9.18 can be used to find the maximum return investment policy for investments ranging from 1 to 8 units of capital. It may also be used to find the optimal policy for a reduced number of activities. For example, if 6 units of capital are to be invested in activities 1 and 2, the solution will be found by noting that $x_2 = 2$ at $f_2(Q) = 95$. This means that 4 units will remain for the first activity; $x_1 = 4$. Thus, the maximization of R depends on Q and N, as was expressed in Equation 9.56.

Shipping application. A given number of items, each with a different weight and value, are to make up a shipment the total weight of which must not exceed a certain maximum. The objective is to select the number of each item to include in the shipment so that its value will be a maximum. Problems such as this arise in shipping operations where the value of the shipment may be measured in terms of its worth to the receiver or where its value is a function of the shipping revenue. The latter case will be considered in the example of this section.

Suppose that four items with different weights and values are to form a shipment with a total weight of not more than 9 tons. Table 9.19 gives the weight and net profit to be derived from each item. In this problem, the weight of the shipment is a restriction and constitutes a resource to be distributed or allocated to the four items. Thus, the concept outlined previously is applicable.

The first step in the solution is to determine the return functions for each individual item. This is accomplished by considering different weights to be allocated to each item and by identifying the whole number of items that can be accommodated within these weights. By reference to Table 9.19, the return functions of Table 9.20 are developed. Each function gives the return to be expected from allocating designated amounts of the scarce resource (shipping weight) to each activity (item).

TABLE 9.19 WEIGHTS AND VALUES OF ITEMS

ITEM	WEIGHT (TONS)	NET PROFIT
1	2	$ 50
2	4	120
3	5	170
4	3	80

TABLE 9.20 RETURN FUNCTIONS
FOR SHIPPING PROBLEMS

Q	$g_1(Q)$	$g_2(Q)$	$g_3(Q)$	$g_4(Q)$
0	0	0	0	0
1	0	0	0	0
2	50	0	0	0
3	50	0	0	80
4	100	120	0	80
5	100	120	170	80
6	150	120	170	160
7	150	120	170	160
8	200	240	170	160
9	200	240	170	240

First, the return to be expected from the first item if all the available weight is allocated to it must be determined from Equation 9.57 as

$$f_1(0) = g_1(0) = \quad 0$$

$$f_1(1) = g_1(1) = \quad 0$$

$$f_1(2) = g_1(2) = \quad 50$$

$$f_1(3) = g_1(3) = \quad 50$$

$$f_1(4) = g_1(4) = 100$$

$$f_1(5) = g_1(5) = 100$$

$$f_1(6) = g_1(6) = 150$$

$$f_1(7) = g_1(7) = 150$$

$$f_1(8) = g_1(8) = 200$$

$$f_1(9) = g_1(9) = 200$$

This completes the computation of $f_1(Q)$. Each value is entered in the first stage of Table 9.21.

From the results of $f_1(Q)$, $f_2(Q)$ may be computed by the use of Equation 9.58. When $Q = 0$, $f_2(0) = 0$. When $Q = 1$,

$$f_2(1) = \max_{0 \le x_2 \le 1} \{g_2(x_2) + f_1(1 - x_2)\}.$$

For values of x_2 ranging from 0 to 1, this gives

$$f_2(1) = \max \begin{cases} g_2(0) + f_1(1) = 0 \\ g_2(1) + f_1(0) = 0 \end{cases}.$$

TABLE 9.21 TABULAR SOLUTION FOR A SHIPPING PROBLEM

Q	x_1	$f_1(Q)$	x_2	$f_2(Q)$	x_3	$f_3(Q)$	x_4	$f_4(Q)$
0	0*	0	0	0	0	0	0	0
1	1	0	0	0	0	0	0	0
2	2	50	0	50	0	50	0	50
3	3	50	1	50	1	50	3	80
4	4	100	4*	120	0	120	0	120
5	5	100	5	120	5	170	0	170
6	6	150	4	170	6	170	1	170
7	7	150	5	170	5	220	0	220
8	8	200	8	240	0	240	3	250
9	9	200	9	240	5*	290	0*	290

When $Q = 2$,

$$f_2(2) = \max_{0 \le x_2 \le 2} \{g_2(x_2) + f_1(2 - x_2)\}.$$

For values of x_2 ranging from 0 to 2, this gives

$$f_2(2) = \max \left\{ \begin{array}{l} g_2(0) + f_1(2) = 50 \\ g_2(1) + f_1(1) = 0 \\ g_2(2) + f_1(0) = 0 \end{array} \right\}.$$

When $Q = 3$,

$$f_2(3) = \max_{0 \le x_2 \le 3} \{g_2(x_2) + f_1(3 - x_2)\}.$$

For values of x_2 ranging from 0 to 3, this gives

$$f_2(3) = \max \left\{ \begin{array}{l} g_2(0) + f_1(3) = 50 \\ g_2(1) + f_1(2) = 50 \\ g_2(2) + f_1(1) = 0 \\ g_2(3) + f_1(0) = 0 \end{array} \right\}.$$

This process is continued until $f_2(9)$ is evaluated. The maximum values of $f_2(Q)$ are identified for each value of Q and entered in the second stage of Table 9.21 together with their associated values of x_2.

The third stage is considered next. Using the results of $f_2(Q)$, $f_3(Q)$ may be computed by use of Equation 9.58. As before, when $Q = 0$, $f_3(Q) = 0$. When $Q = 1$,

$$f_3(1) = \max_{0 \le x_3 \le 1} \{g_3(x_3) + f_2(1 - x_3)\}.$$

For values of x_3 ranging from 0 to 1, this gives

$$f_3(1) = \max \left\{ \begin{array}{l} g_3(0) + f_2(1) = 0 \\ g_3(1) + f_2(0) = 0 \end{array} \right\}.$$

When $Q = 2$,

$$f_3(2) = \max_{0 \leq x_3 \leq 2} \{g_3(x_3) + f_2(2 - x_3)\}.$$

For values of x_3 ranging from 0 to 2, this gives

$$f_3(2) = \max \begin{Bmatrix} g_3(0) + f_2(2) = 50 \\ g_3(1) + f_2(1) = 0 \\ g_3(2) + f_2(0) = 0 \end{Bmatrix}.$$

When $Q = 3$,

$$f_3(3) = \max_{0 \leq x_3 \leq 3} \{g_3(x_3) + f_2(3 - x_3)\}.$$

For values of x_3 ranging from 0 to 3, this gives

$$f_3(3) = \max \begin{Bmatrix} g_3(0) + f_2(3) = 50 \\ g_3(1) + f_2(2) = 50 \\ g_3(2) + f_2(1) = 0 \\ g_3(3) + f_2(0) = 0 \end{Bmatrix}$$

Again, this process is continued until $f_3(9)$ is evaluated. The maximum values of $f_3(Q)$ are identified for each value of Q and entered in the third stage of Table 5.17 together with their associated values of x_3.

By continuing this pattern for the fourth stage, Table 9.21 is completed. The maximum profit is found in stage 4 to be \$290. The number of tons to be allocated to item 4 is found by noting that $x_4 = 0$ at $f_4(Q) = \$290$. Since 9 tons is still available, x_3 will be 5. This leaves 4 tons for item 2 and none for item 1. Thus, the shipment resulting in a maximum profit will contain one each of items 2 and 3.

SELECTED REFERENCES

(1) Banks, J., and W. J. Fabrycky, *Procurement and Inventory Systems Analysis,* Englewood Cliffs, N.J.: Prentice-Hall, 1987.

(2) Beightler, C. S., D. T. Phillips, and D. J. Wilde, *Foundations of Optimization* (2nd ed.). Englewood Cliffs, N.J.: Prentice-Hall, 1979.

(3) Chestnut, H., *Systems Engineering Tools*. New York: John Wiley, 1965.

(4) Chestnut, H., *Systems Engineering Methods*. New York: John Wiley, 1967.

(5) DeNeufville, R. and J. H. Stafford, *Systems Analysis for Engineers and Managers*. New York: McGraw-Hill, 1971.

(6) Fabrycky, W. J., P. M. Ghare, and P. E. Torgersen, *Applied Operations Research and Management Science,* Englewood Cliffs, N.J.: Prentice-Hall, 1984.

(7) Jelen, F. C. and J. H. Black, *Cost and Optimization Engineering* (2nd ed.). New York: McGraw-Hill, 1983.

(8) Meredith, D. D., and others, *Design and Planning of Engineering Systems*. Englewood Cliffs, N.J.: Prentice-Hall, 1973.

(9) Rubinstein, M. F., *Patterns of Problem Solving*. Englewood Cliffs, N.J.: Prentice-Hall, 1975.

(10) Thuesen, G. J. and W. J. Fabrycky, *Engineering Economy* (7th ed.). Englewood Cliffs, N.J.: Prentice-Hall, 1989.

QUESTIONS AND PROBLEMS

1. The cost per unit produced at a certain facility is represented by the function

$$UC = 2x^2 - 10x + 50$$

   where x is in thousands of units produced. For what value of x would unit cost be minimized (other than zero)? What is the minimum cost at this volume? Show that the value found is truly a minimum.

2. Advertising expenditures have been found to relate to profit approximately in accordance with the function

$$P = x^3 - 100x^2 + 3,125x$$

   where x is the expenditure in thousands of dollars. What advertising expenditure would produce the maximum profit? What profit is expected at this expenditure? Show that the derived result is truly a maximum.

3. The production cost for a certain firm is $\$10,900 + 65x + 1,500x^{1/2}$, where x is the number of units produced per year. The selling cost is $\$500y^{1/2}$, where y is the number of units sold during the year. If the selling price is $180 per unit, find the level of production that will maximize profit.

4. Specify the dimensions of the sides of a rectangle of perimeter p so that the area it encloses will be maximum.

5. Ethyl acetate is made from acetic acid and ethyl alcohol. Let x = pounds of acetic acid input, y = pounds of ethyl alcohol input, and z = pounds of ethyl acetate output. The relationship of output to input is

$$\frac{z^2}{(1.47x - z)(1.91y - z)} = 3.9$$

   (a) Determine the output of ethyl acetate per pound of acetic acid, where the ratio of acetic acid of ethyl alcohol is 2.0, 1.0, and 0.67, and graph the result.
   (b) Graph the cost of material per pound of ethyl acetate for each of the ratios given and determine the ratio for which the material cost per pound of ethyl acetate is a minimum if acetic acid costs $0.80 per pound and ethyl alcohol costs $0.92 per pound.

6. It has been found that the heat loss through the ceiling of a building is 0.13 Btu per hour per square foot of area per degree Fahrenheit. If the 2,200-ft^2 ceiling is insulated, the heat loss in Btu per hour per degree temperature difference per square foot of area is taken as

$$\frac{1}{(1/0.13) + (t/0.27)}$$

   where t is the thickness in inches. The in-place cost of insulation 2, 4, and 6 in. thick is $0.18, $0.30, and $0.44 per square foot, respectively. The building is heated to 75°F

3,000 hrs per year by a gas furnace with an efficiency of 50%. The mean outside temperature is 45°F and the natural gas used in the furnace costs $4.40 per 1,000 ft^3 and has a heating value of 2,000 Btu per 1,000 ft^3. What thickness of insulation, if any, should be used if the interest rate is 10% and the resale value of the building 6 years hence is enhanced $850 if insulation is added, regardless of the thickness?

7. An overpass is being considered for a certain crossing. The superstructure design under consideration will be made of steel and will have a weight per foot depending upon the span between piers in accordance with $W = 32(S) + 1,850$. Piers will be made of concrete and will cost $185,000 each. The superstructure will be erected at a cost of $0.70 per pound. If the number of piers required is to be one less than the number of spans, find the number of piers that will result in a minimum total cost for piers and superstructure if $L = 1,250$ ft.

8. Two girder designs are under consideration for a bridge for a 1,200-foot crossing. The first is expected to result in a superstructure weight per foot of $22(S) + 800$, where S is the span between piers. The second should result in superstructure weight per foot of $20(S) + 1,000$. Piers and two required abutments are estimated to cost $220,000 each. The superstructure will be erected at a cost of $0.55 per pound. Choose the girder design that will result in a minimum cost and specify the optimum number of piers.

9. What is the cost advantage of choosing the best girder design for the bridge described in Problem 8? If the number of piers is determined from the best girder design alternative, but the other design alternative is adopted, what cost penalty is incurred?

10. An hourly electric load of 1,600 A (amperes) is to be transmitted from a generator to a transformer in a certain power plant. A copper conductor 150 ft long can be installed for $380 + $1.15 per pound, will have an estimated life of 20 years, and can be salvaged for $0.96 per pound. Power loss from the conductor will be a function of the cross-sectional area and may be expressed as $25,875 \div A$ kilowatt-hours per year. Energy lost is valued at $0.06 per kilowatt-hour, taxes, insurance, and maintenance are negligible, and the interest rate is 8%. Copper weighs 555 lb per ft^3.
   (a) Plot the total annual cost of capital recovery with a return and power loss cost for conductors for cross sections of 1, 2, 3, 4, and 5 in.^{2}.
   (b) Find the minimum-cost cross section mathematically and check the result against the minimum point found in part (a).

11. The daily electrical load to be transmitted by a conductor in a laboratory is 1,900 amperes per day for 365 days per year. Two conductor materials are under consideration, copper and aluminum. The information listed in the table is available for the competing materials.

	COPPER	ALUMINUM
Length	120 ft	120 ft
Installed cost	$410 + $0.88	$410 + $0.48
Estimated life	10 yr	10 yr
Salvage value	$0.72	$0.40
Electrical resistance of conductor		
120 ft by 1 in.^{2} cross section	0.000982 Ω	0.001498 Ω
Density	555 lb/ft^3	168 lb/ft^3

The energy loss in kilowatt-hours in a conductor due to resistance is equal to I^2R times the number of hours divided by 1,000, where I is the current flow in amperes and R is the resistance in the conductor in ohms. The electrical resistance is inversely proportional to the area of the cross section. Lost energy is valued at $0.048 per kilowatt-hour.

(a) Plot the total annual cost of capital recovery and return plus power loss cost for each material for cross sections of 3, 4, 5, 6, 7, and 8 in^2 if the interest rate is 12%.

(b) Recommend the minimum-cost conductor material and specify the cross sectional area.

12. Plot an inventory flow diagram similar to Figure 9.8 if $R = \infty$. Derive optimum values for Q, L, and TC under this assumption and verify the result by substituting into Equations 9.44, 9.45, and 9.46.

13. Plot an inventory flow diagram similar to Figure 9.8 if $R = \infty$ and $C_s = \infty$. Derive optimum values of Q, L, and TC under these assumptions and verify the result by substituting into Equations 9.44, 9.45, and 9.46.

14. An engine manufacturer requires 82 pistons per day in its assembly operations. No shortages are to be allowed. The machine shop can produce 500 pistons per day. The cost associated with initiating manufacturing action is $400, and the holding cost is $0.45 per piston per day. The manufacturing cost is $105 per piston.

(a) Find the minimum-cost production quantity.

(b) Find the minimum-cost procurement level if production lead time is 8 days.

(c) Calculate the total system cost per day.

15. A subcontractor has been found who can supply pistons to the manufacturer described in Problem 14. Procurement cost will be $90 per purchase order. The cost per unit is $108.

(a) Calculate the minimum total system cost per day for purchasing from the subcontractor.

(b) What is the economic advantage of adopting the minimum-cost source?

16. The demand for a certain item is 12 units per period. No shortages are to be allowed. Holding cost is $0.02 per unit per period. Demand can be met either by purchasing or manufacturing, with each source described by the data given in the table.

	PURCHASE	MANUFACTURE
Procurement lead time	18 periods	13 periods
Item cost	$11.00	$9.60
Procurement cost	$20.00	$90.00
Replenishment rate	∞	25 units/period

(a) Find the minimum-cost procurement source and calculate its economic advantage over its alternative source.

(b) Find the minimum-cost procurement quantity.

(c) Find the minimum-cost procurement level.

17. An electronic equipment manufacturing firm has a demand of 250 units per period. It costs $400 to initiate manufacturing action to produce at the rate of 600 units per period. The unit production cost is $90. The holding cost is $0.15 per unit, and the shortage cost is $3.25 per unit short per period for unsatisfied demand. Determine (a) the minimum-cost manufacturing quantity, (b) the minimum-cost procurement level if production lead time is 12 periods, and (c) the total system cost per period.

18. Derive expressions for the minimum-cost procurement quantity and minimum-cost procurement level when C_h is assumed to be infinite. Name a real-world situation where such a model would apply.

19. If the span between piers must be at least 200 feet in Problem 7, what cost penalty is incurred for this constraint?

20. Suppose that no more than 6 piers can be utilized in the bridge design of Problem 7. What is the cost penalty incurred for this constraint if these piers cost $210,000 each?

21. The cross sectional area of the conductor in Problem 10 must not exceed 3 square inches. What is the cost penalty incurred for this constraint?

22. An item with a demand of 300 units per period is to be purchased and no shortages are allowed. The item costs $1.30, holding cost is $0.02 per unit per period, and it costs $28.00 to process a purchase order. Each item consumes 2 cubic feet of warehouse space. The warehouse contains 1,500 cubic feet of space. Find the minimum cost procurement quantity and procurement level under this restriction if the lead time is 2 periods. What is the cost penalty due to the warehouse restriction?

23. The demand for a certain item is 20 units per period. Unsatisfied demand causes a shortage cost of $0.60 per unit per period. The cost of initiating manufacturing action is $48.00 and the holding cost is $0.04 per unit per period. Production cost is $7.90 per unit and the item may be produced at the rate of 60 units per period. Each item consumes 3 cubic units of warehouse space. The warehouse space reserved for this item is limited to 300 cubic units. Find the minimum cost procurement quantity and procurement level under this restriction if the lead time is 3 periods. What is the cost penalty per period due to the warehouse restriction?

24. Solve graphically for the values of x and y that maximize the function

$$Z = 2.2x + 3.8y$$

subject to the constraints

$$2.4x + 3.2y \leq 140$$
$$0.0x + 2.6y \leq 80$$
$$4.1x + 0.0y \leq 120$$
$$x \geq 0 \quad \text{and} \quad y \geq 0$$

25. A small machine shop has capability in turning, milling, drilling, and welding. The machine capacity is 16 hr per day in turning, 8 hr per day in milling, 16 hr per day in drilling, and 8 hr per day in welding. Two products, designated A and B, are under consideration. Each will yield a net profit of $0.35 per unit and will require the amount of machine time shown in the table. Solve graphically for number of units of each product that should be scheduled to maximize profit.

	PRODUCT A	PRODUCT B
Turning	0.046	0.124
Milling	0.112	0.048
Drilling	0.040	0.000
Welding	0.000	0.120

26. Solve graphically for the values of x, y, and z that maximize the function

$$P = 7.8x + 9.4y + 2.6z$$

subject to the constraints

$$4.2x + 11.7y + 3.5z \leq 1{,}800$$

$$0.8x + 4.3y + 1.9z \leq 2{,}700$$

$$12.7x + 3.8y + 2.5z \leq 950$$

$$x \geq 0, \quad y \geq 0, \quad \text{and} \quad z \geq 0$$

27. Use the simplex method to find the values for a, b, c, d, and e that maximize total profit expressed as

$$TP = 0.80a + 0.50b + 0.70c + 0.30d + 0.20e$$

subject to the constraints

$$0.50a + 0.40b + 0.20c + 0.20d + 0.10e = 0.8$$

$$0.60a + 0.10b + 0.30c + 0.20d + 0.30e = 1.7$$

$$0.20a + 0.00b + 0.60c + 0.20d + 0.10e \geq 1.0$$

$$a > 0, \quad b \geq 0, \quad c \geq 0, \quad d \geq 0, \quad \text{and} \quad e \geq 0$$

28. A soap manufacturing company produces three grades of detergents: soft, medium, and super. Each grade requires additives A, B, and C which are available in the amount of 3,200, 2,800, and 1,200 lbs per day, respectively. One pound of soft detergent requires 0.15 lbs of additive A, 0.55 lbs of additive B, and 0.30 lbs of additive C. One pound of medium detegent requires 0.25 lbs of additive A, 0.65 lbs of additive B, and 0.10 lbs of additive C. One pound of super detergent requires 0.30 lbs of additive A, 0.66 lbs of additive B, and 0.04 lbs of additive C. A profit of \$0.015, \$0.027, and \$0.038 per pound can be realized for soft, medium, and super, respectively. How many pounds of each grade of detergent should be produced each week to maximize profit?

29. A manufacturer has three engine lathes and two profile mills available for machining four products. The unit machining times in hours per unit, the capacities of each machine in hours per week, and the profit per week are summarized in the table. Each product needs processing on only one lathe and/or one mill. The unit times given apply if the designated choice is made. Determine the maximum profit production program.

MACHINE	PRODUCT A	PRODUCT B	PRODUCT C	PRODUCT D	CAPACITY
Lathe 1	0.6	0.2	0.5	0.4	30
Lathe 2	0.5	—	—	0.6	50
Lathe 3	0.4	0.3	—	—	24
Mill 1	—	0.5	0.8	0.7	36
Mill 2	0.2	0.6	—	—	40
Profit	\$0.15	\$0.25	\$0.40	\$0.30	

30. Determine the shortest path through the given network from 1 to 8 using the concept of dynamic programming.

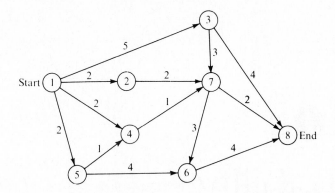

31. Use the concept of dynamic programming to find the minimum path through the given network. What is the minimum path to stage 3?

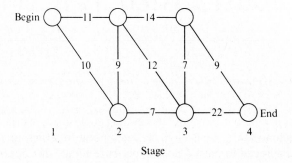

Stage

32. Eight units of capital can be invested in three activities with the return from each activity given in the accompanying table. Determine the capital allocation to each activity that will maximize the total return. What will be the total return if the available capital is reduced by 2 units?

Q	$g_1(Q)$	$g_2(Q)$	$g_2(Q)$
0	0	0	0
1	2	2	2
2	3	4	4
3	4	5	6
4	6	7	6
5	8	7	8
6	9	8	9
7	10	9	9
8	10	11	12

10

Probability
and Statistical Methods

Variation is inherent in nature and in the people-made world. This variation presents a challenge to the systems engineer attempting to design to a need or demand. Most systems deployed to meet a need experience inputs that occur randomly in time. For example, requests for information, units arriving for service, equipment failures, and the time to service a specified unit are random. In these situations it is necessary to describe system operational characteristics in terms of probability and statistical methods.

Some design models will give satisfactory results if variation is not incorporated. However, such models usually deal with physical phenomena. Models formulated for operational man-made systems must incorporate probabilistic elements to be useful in the design process. This chapter presents concepts of probability, probability distribution theory, descriptive statistics, inferential statistics, and an introduction to Monte Carlo analysis. Each of these topics will be applied in subsequent chapters to illustrate their role in systems engineering and analysis.

10.1. PROBABILITY CONCEPTS AND THEORY

If one tosses a coin, the outcome will not be known with certainty until either a head or a tail is observed. Prior to the toss one can assign a probability to the outcome from knowledge of the physical characteristics of the coin. One may know that the diameter of an acorn ranges between 0.80 and 3.20 cm, but the diameter of a specific acorn to be selected from an oak tree will not be known until the acorn is measured. Experiments such as tossing a coin and selecting an acorn provide outcomes called *random events*. Most events in the decision environment are random and probability theory provides a means for quantifying these events.

The Universe and the Sample

The terms *universe* and *population* are used interchangeably. A *universe* consists of all possible objects, stages, and events within an arbitrarily defined boundary. A universe may be finite or it may be infinite. If it is finite, the universe may be very large or very small. If the universe is large, it may sometimes be assumed to be infinite for computational purposes. A universe need not always be large; it may be defined as a dozen events or as only one object. The relative usefulness of the universe as an entity will be paramount in its definition.

A *sample* is a part or portion of a universe. It may range in size from 1 to one less than the size of the universe. A sample is drawn from the population, and observations are made. This is done either because the universe is infinite in size or scope or because the population is large and/or inaccessible as a whole. The sample is used because it is smaller, more accessible, and more economical, and because it suggests certain characteristics of the population.

It is usually assumed that the sample is typical of the population in regard to the characteristics under consideration. The sample is then assessed, and inferences are made in regard to the population as a whole. To the extent that the sample is representative of the population, these inferences may be correct. The problem of selecting a representative sample from a population is an area in statistics to which an entire chapter might be devoted.

Subsequent discussion assumes that the sample is a *random sample,* that is, one in which each object or state or event that consititutes the population has an equally likely chance or probability of being selected and represented in the sample. It is rather simple to state this definition; it may be much more difficult to implement it in practice.

The Probability of an Event

A measure of the relative certainty of an event, before the occurrence of the event, is its probability. The usual representation of a probability is a number $P(A)$ assigned to the outcome A. This number has the following property: $0 \le P(A) \le 1$, with $P(A) = 0$ if the event is certain not to occur and $P(A) = 1$ if the event is certain to occur.

Since probability is only a measure of the certainty (or uncertainty) associated with an event, its definition is rather tenuous. The concept of relative frequency is sometimes employed to establish the number $P(A)$. Sometimes probabilities are established a priori. Other times they are simply a subjective estimate. Consider the example of tossing a fair coin. In a lengthy series of tosses, the coin may have come up heads as often as tails. Then the limiting value of the relative frequency of a head will be 0.5 and will be stated as $P(H) = 0.5$.

Two definitions pertaining to events are needed in the development of probability theorems.

1. Events A and B are said to be *mutually exclusive* if both cannot occur at the same time.
2. Event A is said to be *independent* of event B if the probability of the occurrence of A is the same regardless of whether or not B has occurred.

The Addition Theorem

The probability of the occurrence of either one or another of a series of mutually exclusive events is the sum of probabilities of their separate occurrences. If a fair coin is tossed, and success is defined as the occurrence of either a head or a tail, then the probability of a head or a tail is

$$P(H + T) = P(H) + P(T)$$
$$= 0.5 + 0.5 = 1.0. \tag{10.1}$$

The key to use of the addition theorem is the proper definition of mutually exclusive events. Such events must be distinct one from another. If one event occurs, it must be impossible for the second to occur at the same time. For example, assume that the probability of having a flat tire during a given time period on each of four tires on an automobile is 0.3. Then the probability of having a flat tire on any of the four tires during this time period is not given by the addition of these four probabilities. If $P(T_1) = P(T_2) = P(T_3) = P(T_4) = 0.3$ are the respective probabilities of failure for each of the four tires, then

$$P(T_1 + T_2 + T_3 + T_4) \neq P(T_1) + P(T_2) + P(T_3) + P(T_4)$$
$$\neq 0.3 + 0.3 + 0.3 + 0.3 = 1.2.$$

This cannot be true because the failure of tires is not mutually exclusive. During the time period established, two or more tires may fail, whereas in the example of coin tossing, it is not possible to obtain a head and a tail on the same toss.

The Multiplication Theorem

The probability of occurrence of independent events is the product of the probabilities of their separate events. Implicit in this theorem is the successful occurrence of two events simultaneously or in succession. Thus, the probability of the occurrence of two heads in two tosses of a coin is

$$P(H \cdot H) = P(H)P(H)$$
$$= (0.5)(0.5) = 0.25. \tag{10.2}$$

   The tire-failure problem can now be resolved by considering the probabilities of each tire not failing. The probability of each tire not failing is given by $P(\overline{T}_i) = 0.7$. The probability of no tire failing is then given by

$$P[(\overline{T}_1)(\overline{T}_2)(\overline{T}_3)(\overline{T}_4)] = P(\overline{T}_1)P(\overline{T}_2)P(\overline{T}_3)P(\overline{T}_4)$$

$$= (0.7)(0.7)(0.7)(0.7) = 0.2401.$$

And the probability of a tire failing, or of one or more tires failing, is

$$P(T_1 + T_2 + T_3 + T_4) = 1 - 0.2401 = 0.7599.$$

This approach is valid, since the probability of one tire not failing is independent of the success or failure of the other three tires.

The Conditional Theorem

The probability of the occurrence of two dependent events is the probability of the first event times the probability of the second event, given that the first has occurred. This may be expressed as

$$P(W_1 \cdot W_2) = P(W_1)P(W_2 \mid W_1). \tag{10.3}$$

This theorem is similar to the multiplication theorem, except that consideration is given to the lack of independence between events.

   As an example, consider the probability of selecting two successive white balls from an urn containing three white and two black balls. This problem reduces to a calculation of the product of the probability of selecting a white ball times the probability of selecting a second white ball, given that the first attempt has been successful.

$$P(W_1 \cdot W_2) = \left(\frac{3}{5}\right)\left(\frac{2}{4}\right) = \frac{3}{10}.$$

The conditional theorem makes allowances for a change in probabilities between two successive events. This theorem will be helpful in constructing finite discrete probability distributions.

10.2. PROBABILITY DISTRIBUTION MODELS

The pattern of the distribution of probabilities over all possible outcomes is called a probability distribution. *Probability distribution models* provide a means for assigning the likelihood of occurrences of all possible values. Variables described in terms of a probability distribution are conveniently called *random variables*. The specific value of a random variable is determined by the distribution.

   A probability distribution is completely defined when the probability associated with every possible outcome is defined. In most instances the outcomes themselves are represented by numbers or different values of a variable, such as the diameter of an acorn. When the pattern of the probability distribution is expressed as a function of this variable, the resulting function is called a *probability distribution function*.

An example empirical probability distribution function may be developed as follows. A maintenance mechanic attends four machines and his services are needed only when a machine fails. He would like to estimate how many machines will fail each shift. From previous experience, and using the relative frequency concept of probability, the mechanic knows that 40% of the time only one machine will fail at least once during the shift. Further, 30% of the time two machines will fail, three machines will fail 20% of the time, and all four will fail 10% of the time.

The probability distribution of the number of failed machines may be expressed as $P(1) = 0.4$, $P(2) = 0.3$, $P(3) = 0.2$, and $P(4) = 0.1$. This probability distribution is exhibited in Figure 10.1.

The probability distribution function for this case may be defined as

$$p(x) = \frac{5 - x}{10} \qquad \text{if } x = 1, 2, 3, 4$$

$$P(x) = 0 \qquad \text{otherwise.}$$

Although the function $P(x) = (5 - x)/10$ uniquely represents the probability distribution pattern for the number of failed machines, the function itself belongs to a wider class of functions of the type $P(x) = (a - x)/b$. All functions of this type indicate similar patterns: yet each pair of numbers (a, b) uniquely defines a specific probability distribution. These numbers (a, b) are called *parameters*.

In the sense that they serve to define the probability distribution function, it is possible to look upon parameters as properties of the distribution function. The choice of representation of parameters is not unique, and the most desirable representation would reflect a measure of the properties of the universe under study. Two most commonly sought measures are the *mean*, an indication of central tendency, and the *variance*, a measure of dispersion.

The probability distribution just presented is discrete in that it assigns probabilities to an event that can only take on integer values. Continuous probability distributions are used to define the probability of the occurrence of an event that may take on values over a continuum. Under certain conditions, it may be desirable to use a

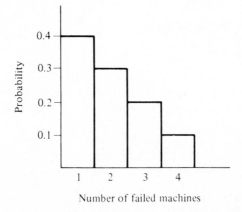

Figure 10.1   A probability distribution of the number of failed machines.

continuous probability distribution to approximate a discrete probability distribution. By so doing, tedious summations may be replaced by integrals. In other instances, it may be desirable to make a continuous distribution discrete as when calculations are to be performed on a digital computer. Several discrete and continuous probability distribution models are presented below.

The Binomial Distribution

The binomial distribution is a basic discrete sampling distribution. It is applicable where the probability is sought of exactly x occurrences in n trials of an event that has a constant probability of occurrence p. The requirement of a constant probability of occurrence is satisfied when the population being sampled is infinite in size, or where replacement of the sampled unit takes place.

The probability of exactly x occurrences in n trials of an event that has a constant probability of occurrence p is given as

$$P(x) = \frac{n!}{x!(n - x)!}p^x q^{n-x} \qquad 0 \leq x \leq n \qquad (10.4)$$

where $q = 1 - p$. The mean and variance of this distribution are given by np and npq, respectively.

As an example of the application of the binomial distribution, assume that a fair coin is to be tossed five times. The probability of obtaining exactly two heads is

$$P(2) = \frac{5!}{2!(5 - 2)!}(0.5)^2(1 - 0.5)^3$$

$$= 10(0.03125) = 0.3125.$$

A probability distribution may be constructed by solving for the probability of exactly zero, one, two, three, four, and five heads in five tosses. If $p = 0.5$, as in this example, the resulting distribution is symmetrical. If $p < 0.5$, the distribution is skewed to the right; if $p > 0.5$, the distribution is skewed to the left.

The Poisson Distribution

The Poisson is a discrete distribution useful in its own right and as an approximation to the binomial. It is applicable when the opportunity for the occurrence of an event is large but when the actual occurrence is unlikely. The probability of exactly x occurrences of an event of probability p in a sample n is

$$P(x) = \frac{(\mu)^x e^{-\mu}}{x!} \qquad 0 \leq x \leq \infty. \qquad (10.5)$$

The mean and variance of this distribution are equal and given by μ, where $\mu = np$. Cumulative Poisson probabilities are tabulated for values of μ ranging up to 24 in Appendix E, Table E.1.

As an example of the application of the Poisson distribution, assume that a sample of 100 items is selected from a population of items which are 1% defective.

The probability of obtaining exactly three defectives in the sample is found from Equation 10.5 or Table E.1 as

$$P(3) = \frac{(1)^3(2.72)^{-1}}{3!} = 0.061.$$

The Poisson distribution may be used as an approximation to the binomial distribution. Such an approximation is good when n is relatively large, p is relatively small, and in general, $pn < 5$. These conditions were satisfied in the previous example.

The Uniform Distribution

The uniform or rectangular probablity distribution may be either discrete or continuous. The continuous form of this simple distribution is

$$f(x) = \frac{1}{a} \qquad 0 \le x \le a. \tag{10.6}$$

The discrete form divides the interval 0 to a into $n + 1$ cells over the range 0 to n, with $1/(n + 1)$ as the unit probabilities. The mean and variance of the rectangular probability distribution are given as $a/2$ and $a^2/12$ for the continuous case, and as $n/2$ and $n^2/12 + n/6$ for the discrete case.

The general form of the rectangular probability distribution is shown in Figure 10.2. The probability that a value of x will fall between the limits 0 and a is equal to unity. One may determine the probability associated with a specific value of x, or a range of x, by integration for the continuous case. The probability associated with a specific value of x for the discrete distributions of the previous section was found from the functions given. Determination of the probability associated with a range of x required a summation of individual probabilities. This is a fundamental difference in dealing with discrete and continuous probability distributions.

Values drawn at random from the rectangular distribution with x allowed to take on values ranging from 0 through 9 are given in Appendix E, Table E.2. These random rectangular variates may be used to randomize a sample or to develop values drawn at random from other probability distributions as will be illustrated in the last section of this chapter.

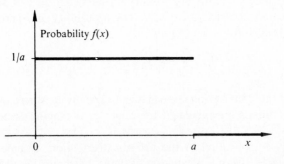

Figure 10.2 The general form of the rectangular distribution.

The Exponential Distribution

The exponential probability distribution is given by

$$f(x) = \frac{1}{a}e^{-x/a} \qquad 0 \leq x \leq \infty. \tag{10.7}$$

The mean and variance of this distribution are given by a and a^2, respectively. Its form is illustrated in Figure 10.3.

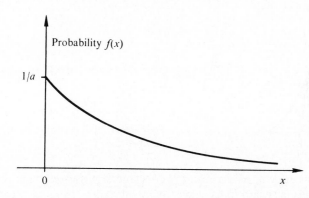

Figure 10.3 The general form of the exponential distribution.

As an example of the application of the exponential probability distribution, consider the selection of a light bulb from a population of light bulbs whose life is known to be exponentially distributed with a mean $\mu = 1,000$ hr. The probability of the life of this sample bulb not exceeding 1,000 hr would be expressed as $P(x \leq 1,000)$. This would be the proportional area under the exponential function over the range $x = 0$ to $x = 1,000$, or

$$P(x \leq 1,000) = \int_0^{1,000} f(x)dx$$

$$= \int_0^{1,000} \frac{1}{1,000} e^{-x/1,000} \, dx$$

$$= -e^{-x/1,000} \Big|_0^{1,000}$$

$$= 1 - e^{-1} = 0.632.$$

Note that 0.632 is that proportion of the area of an exponential distribution to the left of the mean. This illustrates that the probability of the occurrence of an event exceeding the mean value is only $1 - 0.632 = 0.368$.

The Normal Distribution

This normal or Gaussian probability distribution is one of the most important of all distributions. It is defined by

$$f(x) = \frac{1}{\sigma\sqrt{2\pi}} e^{[-(x-\mu)^2/2\sigma^2]} \qquad -\infty \le x \le +\infty. \qquad (10.8)$$

The mean and variance of μ and σ^2, respectively. Variation is inherent in nature, and much of this variation appears to follow the normal distribution, the form of which is given in Figure 10.4.

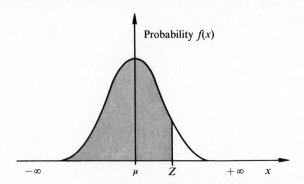

Figure 10.4   The normal probability distribution.

The normal distribution is symmetrical about the mean and possesses some interesting and useful properties in regard to its shape. Where distances from the mean are expressed in terms of standard deviations, σ, the relative areas defined between two such distances will be constant from one distribution to another. In effect, all normal distributions, when defined in terms of a common value of μ and σ, will be identical in form, and corresponding probabilities may be tabulated. Normally, cumulative probabilities are given from $-\infty$ to any value expressed as standard deviation units as in Appendix E, Table E.3. This table gives the probability from $-\infty$ to Z, where Z is a standard normal variate defined as

$$Z = \frac{x - \mu}{\sigma}. \qquad (10.9)$$

This is shown as the shaded area in Figure 10.4.

The area from $-\infty$ to -1σ is indicated as the shaded area in Figure 10.5. From Table B.3, the probability of x falling in this range in 0.1587. Likewise, the area from $-\infty$ to $+2\sigma$ is 0.9773. If the probability of a value falling in the interval -1σ to $+2\sigma$ is required, the following computations are made.

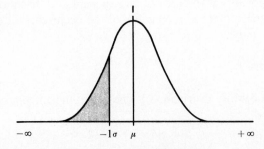

Figure 10.5   The area from $-\infty$ to -1σ under the normal distribution.

$$P(\text{area} - \infty \text{ to} + 2\sigma) = 0.9773$$
$$-P(\text{area} - \infty \text{ to} - 1\sigma) = 0.1587$$
$$\overline{P(\text{area} - 1\sigma \text{ to} + 2\sigma) = 0.8186}$$

This situation is shown in Figure 10.6. The probabilities associated with any normal probability distribution can be calculated by the use of Table E.3.

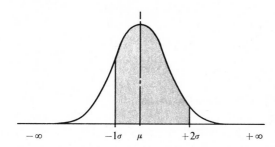

$-\infty \qquad\qquad -1\sigma \quad \mu \qquad +2\sigma \qquad\qquad +\infty$

Figure 10.6 The area from -1σ to $+2\sigma$ under the normal distribution.

Summary of Probability Distributions

The mean, variance, and distribution function for each probability distribution presented previously is summarized in Table 10.1. In most cases the Poisson distribution may be used as an approximation to the binomial when n is relatively large, p is relatively small, and $np < 5$. The normal distribution may also be used as an approximation to the binomial. This approximation will usually be satisfactory if p is close to 0.5 or if n is large. If $n \geq 50$, p may be as small as 0.20 as or large as 0.80 before the approximation ceases to be reasonably good. The ease with which the Poisson and the normal distribution may be used justifies their application in many instances.

TABLE 10.1 SUMMARY OF PROBABILITY DISTRIBUTIONS

PROBABILITY DISTRIBUTION	FUNCTION	MEAN	VARIANCE
Binomial	$P(x) = \dfrac{n!}{x!(n-x)!}p^x q^{n-x}$	np	npq
Poisson	$P(x) = \dfrac{\mu^x e^{-\mu}}{x!}$	μ	μ
Uniform (discrete)	$P(x) = \dfrac{1}{a+1}$	$\dfrac{a}{2}$	$\dfrac{a^2}{12} + \dfrac{a}{6}$
Uniform (continuous)	$f(x) = \dfrac{1}{a}$	$\dfrac{a}{2}$	$\dfrac{a^2}{12}$
Exponential	$f(x) = \dfrac{1}{a}e^{-x/a}$	a	a^2
Normal	$f(x) = \dfrac{1}{\sigma\sqrt{2\pi}}e^{-(x-\mu)^2/2\sigma^2}$	μ	σ^2

Derivations leading to the parameters of probability distributions follow a common pattern. The expectation or expected value of a function $h(x)$ of a random variable x is defined as

$$E[h(x)] = \sum_{R} h(x)P(x) \tag{10.10}$$

if x is discrete, and

$$E[h(x)] = \int_{R} h(x)f(x)dx \tag{10.11}$$

if x is continuous. In each, $\sum_R$ or $\int_R$ denotes the summation or integration over the entire range of possible values of the outcome x.

The mean of the random variable x is defined as $\mu = E(x)$, and the variance of the random variable x is defined as $\sigma^2 = E[(x - \mu)^2]$. For computational ease it is sometimes desirable to express the variance in an alternative manner as

$$\sigma^2 = E(x^2) - [E(x)]^2. \tag{10.12}$$

The derivations for the mean and the variance can be illustrated using the Poisson distribution as an example. If a random variable x exhibits a Poisson distribution function,

$$P(x) = \frac{\mu^x e^{-\mu}}{x!}$$

where $x > 0$ and is an integer. Then

$$\text{mean of } x = E(x)$$

$$= \sum_{R} xP(x) \qquad \text{since } x \text{ is discrete}$$

$$= \sum_{0}^{\infty} x\frac{\mu^x e^{-\mu}}{x!}$$

$$= e^{-\mu} \sum_{1}^{\infty} \frac{\mu^x}{(x-1)!}.$$

Letting $k = x - 1$,

$$\text{mean} = e^{-\mu} \sum_{0}^{\infty} \mu\frac{\mu^k}{k!} = \mu.$$

Since

$$\sum_{0}^{\infty} \frac{\mu^k}{k!} = e^{\mu}$$

then

$$\text{variance of } x = E(x^2) - [E(x)]^2$$

$$= \sum_0^\infty x^2 \frac{\mu^x e^{-\mu}}{x!} - \mu^2$$

$$= e^{-\mu} \sum_0^\infty \left[\frac{\mu^x}{(x-2)!} + \frac{\mu^x}{(x-1)!} \right] - \mu^2$$

$$= e^{-\mu}(\mu^2 e^\mu + \mu e^\mu) - \mu^2$$

$$= \mu^2 + \mu - \mu^2 = \mu.$$

Thus, both the mean and the variance of the Poisson distribution have the value μ as given in Table 10.1.

10.3. DESCRIPTIVE STATISTICS

It is unusual for an entire population to be observed and described by a probability distribution. Normally, a sample from the population is observed with individual sample values being either discrete or continuous. Information about the population must be obtained from sample information. To do this one must note the characteristics of the sample and use these to estimate the corresponding values of the population. *Descriptive statistics* is the body of analysis techniques directed to the description of sample data.

Frequency Distributions

In raw form, a mass of data communicates very little information. Therefore, it is often desirable to present a frequency distribution that describes these data in a more compact form. A frequency distribution also gives a broad estimate of the probability distribution pattern for the population underlying the sample. Also, it is common practice to calculate a measure of central tendency and a measure of dispersion for the data.

Both discrete and continuous data can be expressed in compact form by grouping into frequency distributions. Table 10.2 gives the number of arrivals observed

TABLE 10.2 NUMBER OF ARRIVALS PER HOUR FOR A 30-HOUR PERIOD

HOUR	ARRIVALS	HOUR	ARRIVALS	HOUR	ARRIVALS
1	2	11	1	21	1
2	0	12	4	22	1
3	1	13	2	23	2
4	1	14	2	24	0
5	1	15	1	25	0
6	3	16	1	26	1
7	0	17	4	27	0
8	0	18	3	28	2
9	2	19	1	29	3
10	1	20	0	30	1

each hour for a 30-hr period. These discrete data are exhibited in frequency distribution form in Table 10.3 and graphically by Figure 10.7.

An example involving continuous data is given by the outside diameter of a machined part. Table 10.4 gives the diameter in inches for 50 similarly produced items. These continuous data are exhibited in frequency distribution form in Table 10.5 and graphically by Figure 10.8.

These frequency distributions indicate that the data tend to cluster near the middle, and that the frequency of occurrence of higher and lower values decreases.

TABLE 10.3 DISTRIBUTION OF THE NUMBER OF ARRIVALS PER HOUR

ARRIVALS	NUMBER OF HOURS	FRACTION	RELATIVE FREQUENCY
0	7	7/30	0.2333
1	12	12/30	0.4000
2	6	6/30	0.2000
3	3	3/30	0.1000
4	2	2/30	0.0667
Total	30	30/30	1.0000

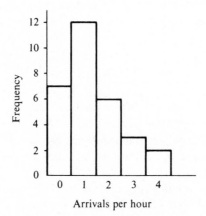

Figure 10.7 Frequency distribution of the number of arrivals per hour.

TABLE 10.4 DIAMETER OF A MACHINED PART (INCHES)

1.0003	0.9982	1.0007	1.0013	0.9968
1.0017	0.9970	1.0003	1.0006	0.9970
0.9987	1.0025	1.0021	0.9976	0.9993
1.0010	0.9990	1.0035	0.9997	1.0001
1.0015	0.9996	0.9998	1.0019	1.0007
1.0004	0.9989	0.9975	1.0023	1.0000
1.0001	1.0006	0.9991	0.9993	1.0028
0.9991	0.9997	1.0012	0.9979	0.9998
0.9987	0.9987	1.0015	1.0026	1.0010
1.0048	1.0041	0.9975	0.9963	0.9996

TABLE 10.5 DISTRIBUTION OF THE DIAMETER OF A MACHINED PART (INCHES)

CLASS INTERVAL	NUMBER OF MEASUREMENTS	FRACTION	RELATIVE FREQUENCY
0.9955–0.9964	1	1/50	0.02
0.9965–0.9974	3	3/50	0.06
0.9975–0.9984	5	5/50	0.10
0.9985–0.9994	9	9/50	0.18
0.9995–1.0004	12	12/50	0.24
1.0005–1.0014	8	8/50	0.16
1.0015–1.0024	6	6/50	0.12
1.0025–1.0034	3	3/50	0.06
1.0035–1.0044	2	2/50	0.04
1.0045–1.0054	1	1/50	0.02
Total	50	50/50	1.00

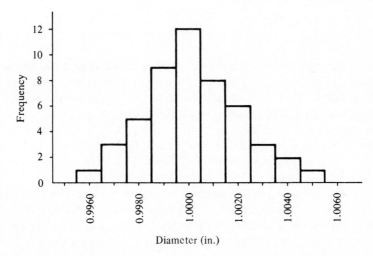

Figure 10.8   Frequency distribution of the outside diameter of a machined part.

Tabular and graphical presentations such as these convey more meaning than the data in raw form. If the ordinate is changed from absolute to relative frequency, the area under the distribution will be equal to unity. This transformation is shown in the last column of Tables 10.4 and 10.5.

Measures of Central Tendency

A number of measures may be used to describe the central tendency of a mass of data. Of these, the *mean* is the most commonly used and may be expressed as

$$\bar{x} = \frac{\sum\limits_{i=1}^{n} x_i}{n} = \frac{x_1 + x_2 + x_3 + \cdots + x_n}{n}. \tag{10.13}$$

The mean of the data given in Table 10.4 may be calculated from Equation 10.13 as

$$\bar{x} = \frac{1.0003 + 1.0017 + \cdots + 0.9996}{50} = 1.0001.$$

Where data have already been grouped into class intervals, as in Table 10.5, the following equation may be used to yield an approximate value of the mean:

$$\bar{x} = \frac{\sum\limits_{i=1}^{k} f_i x_i}{\sum\limits_{i=1}^{k} f_i} = \frac{f_1 x_1 + f_2 x_2 + \cdots + f_{k-1} x_{k-1} + f_k x_k}{f_1 + f_2 + \cdots + f_{k-1} + f_k}. \tag{10.14}$$

Equation 10.14 requires that the midpoint of each class interval, x_i, be multiplied by the number of units in that class interval, f_i, and summed over k, the number of cells. For the data of Table 10.5

$$\bar{x} = \frac{(1)(0.996) + (3)(0.997) + \cdots + (1)(1.005)}{50} = 1.0002.$$

The *median* is a measure of central tendency defined as the value lying in the middle of an ordered set of data. Its computation requires that data be ranked from the smallest value to the largest, or from the largest to the smallest. The median is that value lying in the middle, if the data consist of an odd number of values. Should the data consist of an even number of values, there is no middle value, and the median is the mean of the two central values. In the data from Table 10.4, the twenty-fifth value in ascending order is 1.0001. The twenty-sixth value is also 1.0001; hence, the median is (1.0001 + 1.0001)/2, or 1.0001.

The *mode* is another measure of central tendency. It is defined as that value which occurs most frequently. In continuous data, there may be no value occuring more than once. The mode is then specified as the midpoint of the class interval of greatest frequency. Referring to Figure 10.8, the modal value is 1.0000.

Measures of Dispersion

In addition to information concerning the central tendency of data, it is desirable to describe the extent to which the data cluster about their central value. For this purpose, two measures of variation or dispersion are customarily employed. Of these, the *range* is the least complex and is obtained by calculating the arithmetic difference between the largest and the smallest value. The range is not a very stable measure of variation, since it depends upon only two values. Its advantage lies in the ease with which it may be calculated. The range is equal to 1.0048 − 0.9963, or 0.0085, for the data of Table 10.4.

The *variance* is much more stable measure of dispersion. It may be computed as

$$s^2 = \frac{\sum\limits_{i=1}^{n} (x_1 - \bar{x})^2}{n - 1} = \frac{(x_1 - \bar{x})^2 + (x_2 - \bar{x})^2 + \cdots + (x_n - \bar{x})^2}{n - 1}. \tag{10.15}$$

This definitional formula for the variance requires a lengthy series of calculations. Fortunately, the numerator can be modified by algebraic manipulation, and the variance expressed as

$$s^2 = \frac{\sum\limits_{i=1}^{n} x_i^2 - \left[\left(\sum\limits_{i=1}^{n} x_i\right)^2 \Big/ n\right]}{n - 1}. \tag{10.16}$$

Although Equation 10.16 looks more complex than Equation 10.15, it permits more rapid computation, since the necessity for successive subtraction is eliminated.

The variance of the data given in Table 10.4 is found from Equation 10.15 as

$$s^2 = \frac{(1.0003 - 1.0001)^2 + (1.0017 - 1.0001)^2 + \cdots + (0.9996 - 1.0001)^2}{49}$$

$$= 0.00000365$$

and use of Equation 10.16 gives

$$s^2 = \frac{[(1.0003)^2 + \cdots + (0.9996)^2] - [(1.0003 + \cdots + 0.9996)^2/50]}{49}$$

$$= 0.00000365.$$

Where data have been converted to a frequency distribution, Equation 10.16 becomes

$$s^2 = \frac{\sum\limits_{i=1}^{k} f_i(x_i)^2 - \left[\left(\sum\limits_{i=1}^{k} f_i x_i\right)^2 \Big/ n\right]}{n - 1}. \tag{10.17}$$

10.4. INFERENTIAL STATISTICS

Inferential statistics differs from descriptive statistics in that it is concerned with sampling, and with the drawing of an inference about a population from the sample. This inference usually takes one of two forms. It may deal with an inference about the population parameters or the estimation of parameter values, or it may deal with the testing of one or more hypotheses about the population. If the sample is truly random so that each unit in the population is equally likely to be included in the sample, then the larger the sample, the more certain will be the inference made. However, since the cost of sampling is proportional to the sample size, one must balance the cost against the degree of confidence required.

Parameter Estimation

A *universe parameter* is usually designated by a Greek letter. Thus, the mean of a probability distribution is expressed as μ, and the standard deviation as σ. When the mean and standard deviation of the sample are found, the notations $\bar{x}$ and s are used.

This convention is used to differentiate between the population and a sample from the population.

    With the assumption of random sampling, attributes of the distribution pattern of the sample can be used to estimate the corresponding attributes of the distribution pattern of the universe. These estimates or *estimators*, being functions of observed sample values which are random variables, are themselves random variables. A desirable estimator is one whose expected value coincides with the population parameter it estimates (unbiased) and has a small dispersion (minimum variance). As an example, the sample mean $\bar{x}$ and the sample variance s^2 are usually the unbiased, minimum variance estimates of the population mean μ and variance σ^2.

    Assume that the data given in Table 10.4 represent a sample of machined diameters. Also assume that these data are from a population of units whose diameters are normally distributed. The mean of this sample has been calculated by Equation 10.14 and can be assumed to be a good estimate of the population mean. The sample variance was calculated by Equation 10.15 and is also assumed to be an estimate of the variance of the population. These are summarized as

$$\mu \simeq \bar{x} = 1.0001$$

$$\sigma^2 \simeq s^2 = 0.00000365$$

$$\sigma \simeq s = 0.00191.$$

It is now assumed that their properties define a specific population. The probability of a unit being smaller than 0.9990 in. in diameter can be found. Alternatively, the probability of a unit being within a set of specification limits can be calculated. In both cases, it is necessary to convert these values to the standard normal distribution. In the first case, x_i is 0.9990 and this variation from the mean of the reference distribution is

$$Z = \frac{0.9990 - 1.0001}{0.00191} = -0.576.$$

A unit exactly 0.9990 in. in diameter corresponds to a value lying at -0.57σ on the standard normal distribution. The probability of a unit being smaller than 0.9990 in. is the area under the normal distribution from $-\infty$ to -0.57σ, or 0.283.

    The probability of a unit falling within the specification limits of 1,000 $\pm$ 0.0040 in. is approached the same way. These two limits must first be defined as values of the standard normal distribution. For $x_1 = 0.9960$,

$$Z_1 = \frac{0.9960 - 1.0001}{0.00191} = -2.146$$

for $x_2 = 1.0040$,

$$Z_2 = \frac{1.0040 - 1.0001}{0.00191} = +2.042.$$

The probability of a unit falling within the specification limits is now calculated as

$$P(\text{area} - \infty \quad \text{to} + 2.042\sigma) = 0.9794$$
$$\underline{-P(\text{area} - \infty \quad \text{to} - 2.146\sigma) = 0.0160}$$
$$P(\text{area} - 2.14\sigma \text{ to} + 2.042\sigma) = 0.9634$$

The Distribution of Sample Means

The relationship between the variance of the universe and the variance of sample means from the universe is given by

$$\sigma_{\bar{x}}^2 = \frac{\sigma^2}{n} \tag{10.18}$$

where

$$\sigma_{\bar{x}}^2 = \text{variance of sample means}$$

$$\sigma^2 = \text{variance of the universe}$$

$$n = \text{sample size}$$

This relationship can be viewed as follows in Figure 10.9. Consider the following example. A sample of $n = 4$ is taken from a population of known mean, $\mu = 100$, and variance, $\sigma^2 = 64$. The probability that the mean of that sample will exceed a value $\bar{x} = 105$ may be found by first computing

$$\sigma_{\bar{x}}^2 = \frac{64}{4} = 16.$$

The probability that a sample mean will exceed a value of $\bar{x} = 105$ is P(the area from Z to $+\infty$), where

$$Z = \frac{\bar{x} - \mu}{\sigma_{\bar{x}}} = \frac{105 - 100}{4} = +1.25$$

so P(the area from $+1.25\sigma$ to $+\infty$) = 0.1056.

In the preceding example, it was implicitly assumed that the variation in sample means followed a normal distribution. It might have been inferred that the universe from which these samples were obtained also follows a normal distribution.

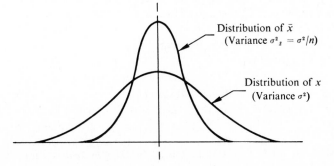

Distribution of $\bar{x}$
(Variance $\sigma_{\bar{x}}^2 = \sigma^2/n$)

Distribution of x
(Variance σ^2)

Figure 10.9 Comparison of the probability distributions of x and $\bar{x}$.

Although many real-world variables are normally distributed, this latter assumption cannot be universally applied. In the preceding example, however, a relationship known as the *central limit theorem* permits one to accept the first assumption without need to accept the second. The central limit theorem states: If x has a distribution for which the moment-generating function exists, then the variable $\bar{x}$ has a distribution that approaches normality as the size of the sample tends toward infinity.[1] The sample size required for any desired degree of covergence is a function of the shape of the parent distribution. Fairly good results have, however, been demonstrated with a sample of $n = 4$ for both the rectangular and triangular distributions.

The central limit theorem greatly enhances the value of the normal distribution in statistical analysis. As an example, a control chart can be constructed for the means of samples and limits placed on this chart which were developed from a normal probability density function. A normal distribution is often used in the analysis of errors because it can be assumed that each individual error is the sum of many small independent errors.

Statistical Errors and Hypothesis Testing

In logic and mathematics, a proof can be demonstrated with certainty after some procedures of proof have been defined. Thus, it is possible to state in absolute terms that the sum of the angles of a triangle is 180 degrees. Such a proof is not possible in inferential statistics. A *hypothesis* must first be stated, and if it is rejected as stated. it is rejected with a *level of significance* or *degree of confidence* that represents the chance of its being correct or in error. The level of significance is defined as a value of α, such as 0.05, and this value represents the probability that an error has been made in rejecting the hypothesis. This is referred to as *Type I error* of probability α.

In addition to the possibility of making a Type I error, it is possible to accept a hypothesis when it false. This is called a *Type II* or *β error*. The probability of a β error is more difficult to determine. Whereas the value of α is specified in advance, the value of β is a function of a value of α which was specified, the size of the sample which was collected, and the magnitude of the error that was not detected.

Hypothesis testing is a formal method by which statistical inferences are made. These inferences are first stated as hypotheses which are then tested through the collection of data and are finally accepted or rejected, based on the results of those data. A hypothesis is usually specified as a null hypothesis, H_0, concerning the population parameter. It can then be rejected or not be rejected with the consequences given in Table 10.6. Note that it is not possible to make both an α and a β error. If the decision is made to reject the null hypothesis, the probability of being in error is α, and the probability of being correct is $1 - \alpha$. If the null hypothesis is not rejected, the probability of being in error is β. This cannot be quantitatively defined. This probability of a Type II error, in addition to being determined by n and α, is a function of the magnitude of the differences of that which was formulated as a hypothesis of no differences. The probability of being correct, $1 - \beta$, also cannot be

---

[1] For a proof of this theorem, see A. W. Drake, *Fundamentals of Applied Probability Theory* (New York: McGraw-Hill, 1967).

TABLE 10.6 OUTCOME POSSIBILITIES FOR HYPOTHESIS TESTING

		REALITY	
		H_0 TRUE	H_0 FALSE
DECISION	Reject H_0	Type I error (probability $= \alpha$)	Correct decision
	Do not reject H_0	Correct decision	Type II error (probability $= \beta$)

calculated. This probability can, however, be defined, with α and n, as a function over a range of these differences. Such a curve is referred to as a *power function*.

The essential steps in hypothesis testing will be demonstrated with a coin-tossing example. Suppose that a coin is suspected of being biased. An experiment to resolve this suspicion will require the following steps.

1. *Formulate the null hypothesis*—A statement that there is difference in the probability of a head and the probability of a tail, $P(H) + P(T) = 0.5$. In almost all cases the null hypothesis must be rejected to verify a conjecture.
2. *Specify the level of significance*—Choose $\alpha = 0.01$ and $n = 12$, for example. This is an arbitrary choice and should be an attempt to balance the costs of sampling and the costs associated with an incorrect conclusion.
3. *Construct the statistical test*—The probability of obtaining x heads or tails in n tosses is given by the binomial distribution. Assuming 4,096 trials of 12 tosses each, the expected frequencies for 0 through 12 heads or tails are given in Table 10.7.

TABLE 10.7 SAMPLING DISTRIBUTION OF THE NUMBER OF HEADS IN 4,096 TRIALS

NUMBER OF HEADS OR TAILS	EXPECTED FREQUENCY	PROBABILITY
12	1	0.00024
11	12	0.00293
10	66	0.01611
9	220	0.05371
8	495	0.12085
7	792	0.19336
6	924	0.22558
5	792	0.19336
4	495	0.12085
3	220	0.05371
2	66	0.01611
1	12	0.00293
0	1	0.00024

4. *Determine the rejection region*—If the coin is fair, 6 is the most likely number of heads that will occur in 12 tosses. It is less likely but still reasonable to expect 5 or 7 heads or tails. It is, however, very unlikely for 0 or 12 heads or tails to occur unless the coin is biased. The probability of obtaining 12 heads or tails from a fair coin is $1/4{,}096 = 0.00024$. This is less than half the significance level, since $\alpha/2 = 0.005$. The probability of obtaining 11 or 12 heads or tails is $1/4{,}096 + 12/4{,}096 = 0.00317$, which is still less than the significance level. The probability of obtaining 10, 11, or 12 heads or tails is $1/4{,}096 + 12/4{,}096 + 66/4{,}096 = 0.01928$. This now exceeds the value of $\alpha/2$. Thus, the rejection region will be defined as 11 or 12 heads or tails, which results in an $\alpha/2$ of 0.0032. This adjustment in the originally established level of confidence is necessary since a discrete distribution is involved.

5. *Make the decision*—The final step is testing the hypothesis requires that the experiment be performed. In this case, the coin would be tossed 12 times and the results observed. If either 11 or 12 heads or tails are obtained, the null hypothesis is rejected. This would verify the conjecture that the coin is biased subject to a Type I error.

The preceding example illustrates a *two-tail test*, since it was not known whether the coin was biased in favor of heads or tails. A *one-tail* test is used when it is known that rejection might occur in one direction. A one-tail test is more powerful; it is more likely to detect a bias if one exists because of a larger rejection region composed of the sum of the probabilities otherwise allocated to each of the two tails.

10.5. REGRESSION AND CORRELATION

Some variables appear to be related to each other. As a variable changes and assumes different values, a second variable may also change. These two variables may be linked in a direct causal relationship, they may both be related to a third variable, or they may appear to have no known and intuitively logical relationship. Whatever the apparent existence or lack of causality, if the relationship between the paired variables follows a consistent linear pattern, this relationship can be described through linear regression. A measure of the degree of this relationship can then be obtained through correlation analysis.

The Scatter Diagram

Consider the actual and observed rating data in Table 10.8. Each pair of values records a person's success in a work-measurement rating exercise. The individual rates film, sequences of industrial work operations and attempts to judge the speed at which a worker is accomplishing a job. If the worker is performing at a referenced normal pace, he or she should be judged at 100%. If the worker moves faster or slower, the rating should be judged proportionally higher or lower. It is assumed that actual values are given for each film sequence and that the data of Table 10.8

TABLE 10.8 ACTUAL AND OBSERVED DATA IN A WORK-MEASUREMENT RATING EXERCISE

ACTUAL	OBSERVED	ACTUAL	OBSERVED	ACTUAL	OBSERVED	ACTUAL	OBSERVED	ACTUAL	OBSERVED
50	65	130	100	180	130	180	150
160	140	140	130	110	100	150	130
110	120	85	95	115	110	115	120
125	110	90	85	110	130	160	130
100	90	185	150	165	160	150	150
180	180	60	65	60	85	175	155
175	135	170	145	100	100	70	90
150	130	80	85	40	60	190	165
70	75	70	55	50	85	50	50
170	150	80	70	130	130	145	115

represent these actual values plus the estimated or observed values. A perfect rating for each film sequence would require that the observed value be identical to the actual value. The relative success of the individual attempting the rating can be seen from the difference in actual and observed values.

Data from two variables that are related may be exhibited on a scatter diagram. One of these two variables is usually classified as the *dependent variable* and is recorded as the ordinate. The other variable is called the *independent variable* and it is recorded as the abscissa. In this framework, y is usually predicted from x, although no causal relationship is necessarily implied. Data may be recorded by plotting point values or by dividing the ordinate and abscisssa into intervals and then counting and recording the number of observations falling into each cell. Figure 10.10 gives an example of plotted point values from Table 10.8.

Some assumptions are necessary for a linear regression equation and correlation coefficient to have meaning. The data are assumed to be linearly related. For a given value of an independent variable, x, the distribution of dependent values, y, is assumed to be normal. Also, the distribution of x values for a given y value is assumed normal. Actually, a two-variable normal distribution is involved. It is also assumed that the variance of y for all values of x is constant, and conversely, the variance of x for all values of y is constant. This property of the distribution is referred to as *homoscedasticity*. Finally, it is assumed that the failure of the means of y values for any given values of x to fall on a straight line is due to chance fluctuations in those means. This assumption also holds true for the mean values of x for given values of y.

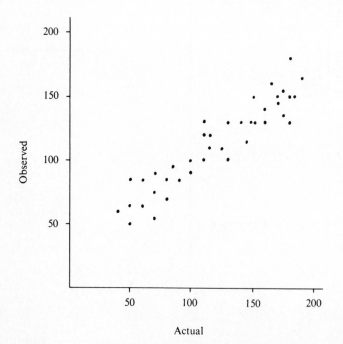

Figure 10.10 Example of plotted point values from Table 6.8.

Linear Regression

In *linear regression,* we seek a straight line that passes through the data of Figure 10.10 and that minimizes the sum of the squares of the errors of estimate. The equation of a straight line for this application may be expressed as

$$y' = b + a(x - \bar{x})$$

where

$$b = y \text{ intercept on the } \bar{x} \text{ axes}$$

$$b - a\bar{x} = y' \text{ intercept on the } y \text{ axes}$$

$$a = \text{slope of the line}$$

$$\bar{x} = \text{mean value of the } x \text{ values}$$

$$y = \text{value of the variable to be estimated}$$

$$y' = \text{corresponding linear value}$$

$$y - y' = \text{error of estimate}$$

The straight line, together with the associated notation, is shown in Figure 10.11.

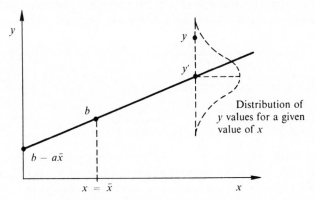

Figure 10.11   The straight line $y' = b + a(x - \bar{x})$.

Specific values for a and b in Equation 10.18 that minimize $\Sigma\,(y - y')^2$ are

$$b = \frac{\Sigma\,y}{n} = \bar{y} \qquad \text{and} \qquad a = \frac{\Sigma\,y(x - \bar{x})}{\Sigma\,(x - \bar{x})^2} = \frac{\Sigma\,xy - n\bar{x}\bar{y}}{\Sigma\,x^2 - n\bar{x}^2}.$$

Substituting the foregoing expressions into Equation 10.18 gives

$$y' - \bar{y} = a(x - \bar{x}). \tag{10.19}$$

The use of Equation 10.19 can be illustrated with the data of Table 10.8. The slope is

$$a = \frac{601{,}300 - 40(120.63)(113.0)}{664{,}425 - 40(14{,}550.39)} = 0.68$$

and the value for b is

$$b = \frac{4,520}{40} = 113.$$

Therefore,

$$y' - 113 = 0.68(x - 120.63)$$
$$y' = 0.68x + 30.97.$$

In this example, the objective of the rating exercise is for the observed values to equal the actual values. Thus, $y' = x$ is the desired line of best fit. The individual under consideration has a positive intercept and a slope less than 1. She tends to rate slower workers faster than they are actually working and to penalize the faster workers by not giving them due credit. She is most accurate in the middle ranges of an actual rate of 100%. This equation should show her that, in future rating efforts, she should be more severe in her estimates of slower workers and more generous in her estimates of faster workers. This will tend to lower the intercept to zero and increase the slope to 1.

Statistical Correlation

The *correlation coefficient*, r, expresses the degree of relationship between two variables and may be defined as

$$r = \frac{\Sigma (x - \bar{x})(y - \bar{y})}{n\sigma_x\sigma_y}. \tag{10.20}$$

For ease of computation, this may be converted to

$$r = \frac{n\Sigma xy - \Sigma x \Sigma y}{\sqrt{n\Sigma x^2 - (\Sigma x)^2} \sqrt{n\Sigma y^2 - (\Sigma y)^2}}. \tag{10.21}$$

Correlation coefficients vary from -1.00 to $+1.00$, with these two extremes representing a perfect relationship. If $r = 0$, no relationship whatsoever exists between the paired variables. In effect, r may be considered to vary from $r = 0$ to $r = +1.00$, from no predictive relationship to the case where every paired value falls on a straight line. Negative values of r indicate the same degree of relationship as corresponding positive values, except that a negative correlation coefficient indicates one value is decreasing while the other is increasing. Any linear regression equation with a negative slope will also possess a negative correlation coefficient.

Using Equation 10.20, a correlation may be obtained for the data of Table 10.8.

$$r = \frac{40(601,300) - (4,825)(4,520)}{\sqrt{40(664,425) - (4,825)^2} \sqrt{40(554,850) - (4,520)^2}}$$

$$= \frac{2,243,000}{2,410,984} = +0.93.$$

This is a fairly high degree of relationship and one that should be achieved in this type of exercise. Should another individual obtain a higher or lower coefficient, this would reflect a higher or lower level of consistency or precision in rating. Note, however, that this is a rating about the equation developed in linear regression, rather than the ideal of $y = x$.

10.6. MONTE CARLO ANALYSIS

The decision environment is made up of many random variables. Thus, models used to explain operational systems must often incorporate probabilistic elements. In some cases, formal mathematical solutions are difficult or impossible to obtain from these models. Under such conditions it may be necessary to use a method known as *Monte Carlo analysis*. When applied to an operational system. Monte Carlo analysis provides a powerful means of simulation.

As an introduction to the idea of Monte Carlo analysis, consider its application to the determination of the area of a circle with a diameter of 1 in. Proceed as follows:

1. Enclose the circle of a 1 in. square as shown in Figure 10.12.
2. Divide two adjoining sides of the square into tenths, or hundredths, or thousandths, and so on, depending upon the accuracy desired.
3. Secure a sequence of pairs of random rectangular variates (random numbers).

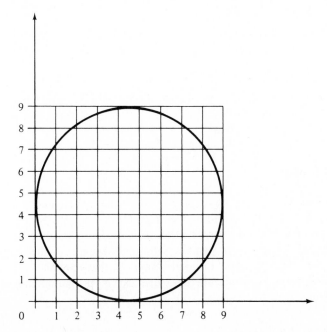

Figure 10.12 Area of circle by Monte Carlo analysis.

4. Use each pair of rectangular variates to determine a point within the square and possibly within the circle. This process is illustrated in Table 10.9 for 100 trials.

TABLE 10.9 DETERMINING THE AREA
OF A CIRCLE

TRIAL	RANDOM NUMBER	IN	OUT
1	73	x	
2	26	x	
3	19		x
4	84	x	
5	81	x	
6	47	x	
7	18		x
8	44	x	
.	.	.	.
.	.	.	.
.	.	.	.
100	35	x	
Total		79	21

5. Compute a ratio of the number of times a point falls within the circle to the total number of trials. The value of this ratio is an approximate area for the circle expressed as a fraction of the 1 in^2 represented by the square. It is 79/100, or 0.79, in this example.

The example just presented has a well-known mathematical solution as follows:

$$A = \pi r^2 = 3.1416(0.50)^2 = 0.7854.$$

Most models used to explain operational systems incorporate probabilistic elements, and for this reason Monte Carlo analysis is often utilized. A step-by-step procedure used in Monte Carlo analysis is presented next.

Formalize the System Logic

The system under study is usually assumed to operate in accordance with a certain logical pattern. Before beginning the actual Monte Carlo process, it is necessary to formalize the operational procedure by the construction of a model. This may require the development of a step-by-step flow diagram outlining the logic. If the actual simulation process is to be performed on a digital computer, it is mandatory to prepare an accurate logic diagram. From this, the computer can be programmed to pattern the process under study.

Consider the determination of the distribution of the random variable *LTD* (lead time demand) which is expressed as

$$LTD = \sum^{L} D$$

where D is a random variable representing demand in units per day and L is a random variable expressing lead time in days. The mathematical expression shown above establishes the computational procedure required.

Determine the Distributions

Each random variable in the model refers to an event in the system being studied. Therefore, an important step in Monte Carlo analysis is determining the behavior of these random variables. This involves the development of empirical frequency distributions to describe the relevant variables by the collection of historical data. Once this is done, the frequency distribution for each variable may be studied statistically to ascertain whether it conforms to a known theoretical distribution.

For the example under consideration, assume that data for random variable L have been collected and studied. Suppose that it is concluded that this random variable conforms to the exponential distribution with a mean of 4. This is a distribution model from Equation 10.7 expressed as

$$f(L) = \frac{1}{4}e^{-L/4}.$$

Likewise, assume that data for random variable D have been collected. Suppose that the resulting frequency distribution is as shown in Figure 10.13.

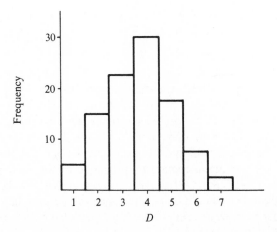

Figure 10.13   Distribution of the random variable D.

Develop the Cumulative Distributions

The probability distributions describing the random variables will be theoretical or empirical or both. Those expressed theoretically may be manipulated mathematically in order to obtain the required cumulative probability distributions. The cumulative distribution for random variable L may be expressed as follows.

$$
\begin{aligned}
F(L) &= \int_0^x f(L)\, dx \\
&= \int_0^x \frac{1}{4} e^{-L/4}\, dx \\
&= -e^{-L/4} \Big|_0^L \\
&= 1 - e^{-L/4}.
\end{aligned}
\tag{10.22}
$$

This cumulative exponential distribution is shown in Figure 10.14.

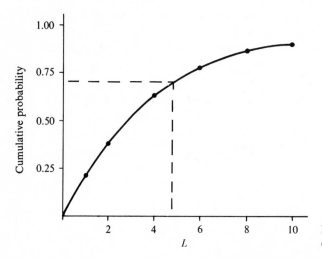

Figure 10.14 Cumulative exponential distribution of mean $L = 4$.

   Those distributions expressed empirically cannot be converted to cumulative distributions by mathematical means. This is the case for random variable D shown in Figure 10.13. The Monte Carlo analysis, however, requires a cumulative distribution for each random variable. Therefore, graphical means must be used. Figure 10.15 exhibits the cumulative distribution corresponding to the distribution of Figure 10.13. It is developed by summing the probabilities from left to right and plotting the results. This is the same as the mathematical process used for the exponential distribution.

   The cumulative distribution of Figure 10.14 and Figure 10.15 may be used to transform random rectangular variates to values drawn at random from the basic distributions. By this means, specific values are determined for random variable L and

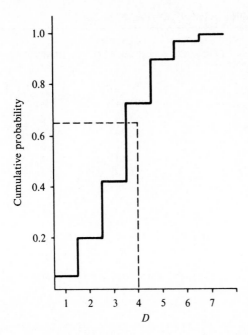

Figure 10.15 Cumulative distribution of the random variable D.

random variable D. Thus, a random rectangular variate with a value of 68 gives a value for the random variable L of 4.57, as shown in Figure 10.14. Similarly, a random rectangular variate with a value of 654 gives a value for the random variable D of 4, as shown in Figure 10.15. Use of Figure 10.14 may be bypassed, since its mathematical equivalent is available. Equation 10.22 may be used to transform the random rectangular variate 681 to a random exponential variable directly. The result is

$$0.681 = 1 - e^{-L/4}$$

$$L = 4.572.$$

Perform the Monte Carlo Analysis

The example under consideration requires that L random demand variates be added together to form one random LTD value. This is recognized to be the lead-time demand resulting from demand distributed as illustrated in Figure 10.13 and lead time distributed exponentially as illustrated in Figure 10.14.

To perform the Monte Carlo analysis, L could be arbitrarily rounded upward to the nearest integer. Thus, for the value of L generated in Figure 10.14, set $L = 5$ and draw five values at random from the demand distribution, D. Suppose that these are 3, 1, 5, 2, and 4. The first trial then yields a value of LTD of 15. This process is continued numerous times to develop the distribution of LTD as shown in Table 10.10. The distribution of LTD would be very useful in setting the order level in an inventory system.

TABLE 10.10   DETERMINING THE LEAD-TIME DEMAND

TRIAL	RANDOM NUMBER	L ROUNDED	RANDOM NUMBER	DEMAND	LTD
1	68	5	28	3	
			03	1	
			81	5	
			18	2	
			54	4	15
2	31	2	92	6	
			63	4	10
3	61	4	05	1	
			90	5	
			33	3	
			82	5	14
4	78	7	62	4	
			04	1	
			89	5	
			37	3	
			17	2	
			19	2	
			94	6	23
.	.	.	.	.	.
.	.	.	.	.	.
.	.	.	.	.	.

SELECTED REFERENCES

(1) Ackoff, R. L., *The Art of Problem Solving*. New York: John Wiley, 1978.

(2) Bowker, A. H., and G. J. Lieberman, *Engineering Statistics*. Englewood Cliffs, N.J.: Prentice-Hall, 1972.

(3) Drake, A. W., *Fundamentals of Applied Probability Theory*. New York: McGraw-Hill, 1967.

(4) Fabrycky, W. J., Ghare, P. M., and P. E. Torgersen, *Applied Operations Research and Management Science*. Englewood Cliffs, N.J.: Prentice-Hall, 1984.

(5) Gibra, I. N., *Probability and Statistical Inferences for Scientists and Engineers*. Englewood Cliffs, N.J.: Prentice-Hall, 1973.

(6) Grant, E. L., and R. S. Leavenworth, *Statistical Quality Control*. New York: McGraw-Hill, 1986.

(7) Hogg, R. V., and J. Ledolter, *Engineering Statistics*. New York: Macmillian, 1987.

QUESTIONS AND PROBLEMS

1. Discuss the difference between a universe and a sample.

2. Give several ways of assigning a probability to an event.

3. Under what conditions are events mutually exclusive; independent?

4. If one card is drawn at random from an ordinary deck, what is the probability that it is (a) an ace, (b) the king of diamonds, and (c) not a diamond?

5. If a die is rolled, what is the probability of a 5 or a 6 appearing?

6. If two dice are rolled, what is the probability that the first shows at least 3 and the second shows not more than 4?

7. If two coins are tossed, what is the probability of two heads; two tails; one head and one tail?

8. A hand of 13 cards is dealt from an ordinary deck. Compute the probability that there are at least two aces in the hand.

9. In a certain maintenance facility, only 80% of a certain electronic test device are usable and only 20% are being used. What is the probability that a test device selected at random is being used.?

10. If a die is rolled 10 times, what is the probability of obtaining no sixes; all sixes?

11. A bowl contains four black chips and seven blue chips. A second bowl contains eight black chips and three blue chips. One of the two bowls is selected at random and one chip is drawn from that bowl.
(a) Calculate the probability that the chip is blue.
(b) Calculate the conditional probability that this blue chip is drawn from the first bowl.

12. An amplifier consists of three transistors, five capacitors, and seven resistors of which 6, 3, and 2%, respectively, are initially defective. What is the probability of (a) exactly one defective component and (b) at least one defective component?

13. Solve Problem 12 using the Poisson distribution to obtain an approximate answer.

14. The data in the table represent the service time in hours at a system support facility. Plot a frequency distribution of these data.

12.1	11.4	14.0	12.7	10.5	12.0	11.0	12.5
12.5	10.0	11.3	10.7	10.5	11.2	12.3	12.0
10.9	10.7	12.5	13.7	12.2	12.2	10.9	10.5
12.0	12.4	12.0	9.9	12.4	13.1	12.9	11.4
11.8	12.6	11.8	14.3	10.7	11.5	11.7	13.9
12.7	11.8	10.7	10.6	11.9	10.7	14.4	12.3
10.5	12.3	12.5	12.7	12.7	11.3	12.6	10.8
12.0	12.9	11.6	13.8	11.7	12.7	11.3	12.6
11.5	13.3	14.7	10.7	12.3	13.3	10.7	11.7
12.4	12.0	10.7	11.4	11.9	14.9	11.9	12.7
13.9	13.1	12.6	12.3	12.9	10.4	12.7	13.3
11.4	12.7	11.9	12.9	13.7	11.3	11.4	11.0
12.4	13.4	12.9	13.7				

15. Calculate the mean, the median, and the mode of the data in Problem 14.

16. Calculate the range and the variance of the data in Problem 14.

17. If the data in Problem 14 come from an underlying distribution that is normal, what would the probability be of the next value exceeding 12 hr?

18. If a random sample of size 5 were drawn from the population described in Problem 14, what would be the variance of the sample means?

19. Discuss the difference between descriptive statistics and inferential statistics.

20. Give an example of a possible null hypothesis and describe the Type I (α) error and the Type II (β) error.

21. Test the hypothesis that a coin is not biased by tossing it 20 times. Use $\alpha = 0.10$ and a two-tail test.

22. The following proficiency ratings were given to a group of operators on system I and system II. Show how the ratings on system I may be used to predict the ratings on system II and give the precision of the prediction.

I	II	I	II	I	II
23	21	31	24	25	22
21	22	27	25	24	23
30	24	21	17	33	31
15	17	23	27	33	24
20	28	25	26	21	23
23	25	28	30	21	19
32	23	24	15	29	30
17	22	27	15	33	28
26	27	27	20	24	28
30	24	31	22	27	24

23. Convert six random rectangular variates to values drawn at random from the continuous distribution $f(x) = 0.6x, \ 0 < x < 3$.

24. Develop a random sequence of failed machines by drawing 10 values at random from the distribution of Figure 10.1.

25. Generate six random variates from a Poisson distribution with a mean of 4.

26. If A is distributed exponentially with a mean of 2 and B is distributed as in Problem 23, find the distribution of C, where $C = A + B$.

11
Queuing Theory and Analysis

The *queuing* or *waiting-line system* under study within this chapter may be described as follows. A facility or group of facilities is maintained to meet the demand for service created by a population of individuals or units. These individuals form a queue or waiting line and receive service in accordance with a predetermined waiting-line discipline. In most cases, the serviced units rejoin the population and again become candidates for service. In other cases, the individuals form a waiting line at the next stage in the system.

Systems having these characteristics are common in many operations in which people, materials, equipment, or vehicles form waiting lines. The public forms waiting lines at cafeterias, doctor's offices, and theaters. In production, the flow of items in process produces a waiting line at each machine center. In maintenance, equipment to be repaired waits for service at maintenance facilities. In transportation, waiting lines form at toll gates, traffic signals, docks, and loading ramps. In each case, the objective is to determine the capacity of the service facility in the light of the relevant costs and the characteristics of the arrival pattern so that the sum of all costs associated with the queuing system will be minimized.

11.1. THE QUEUING SYSTEM

A multiple-channel queuing system is illustrated schematically in Figure 11.1. It exists because the population shown requires service. In satisfying the demand upon the system, the decision maker must establish the level of service capacity to provide. This will involve increasing or decreasing the service capacity by altering the service rate at existing channels or by adding or deleting channels. The following paragraphs describe the components of the waiting-line system and indicate their importance in the decision-making process.

The Arrival Mechanism

The demand for service is the primary stimulus on the waiting-line system and the justification for its existence. As previously indicated, the waiting-line system may exist to meet the service demand created by people, materials, equipment, or vehicles. The characteristics of the arrival pattern depend upon the nature of the population giving rise to the demand for service.

Waiting-line systems come into being because there exists a population of individuals or units requiring service from time to time. Usually, the arrival population is best thought of as a group of items, some of which depart and join the waiting line. For example, if the population is composed of all airborne aircraft, then flight schedules and random occurrences will determine the number of aircraft that will join the landing pattern of a given airport during a given time interval. If the population is composed of telephone subscribers, then the time of day and the day of the week, as well as many other factors, will determine the number of calls placed upon an exchange. If the population consists of production machines, the deterioriation, wear-out, and use rates will determine the departure mechanism causing the machines to join a waiting line of machines requiring repair service.

The arrival population, although always finite in size, may be considered infinite under certain conditions. If the departure rate is small relative to the size of the population, the number of units that potentially may require service will not be seriously depleted. Under this condition, the population may be considered infinite. Models used to explain the behavior of such systems are much easier to formulate than are models for the finite population. Examples of populations that may usually be treated as infinite are automobiles that may require passage over a bridge, customers who may potentially patronize a theater, telephone subscribers who may place a call, and production orders that may require processing at a specific machine center.

In some cases the proportion of the population requiring service may be fairly large when compared with the population itself. In these cases, the population is seriously depleted by the departure of individuals to the extent that the departure rate will not remain stable. Since models used to explain waiting-line systems depend upon the stability of the arrival rate, finite cases must be given special treatment. Examples of waiting-line operations that might be classified as finite are equipment items that may require repair, semiautomatic production facilities that require operator attention, and company cafeterias that serve a captive population. Waiting-line systems for the case where the infinite population assumption does not hold will be treated in Section 11.6.

The Waiting Line

In any queuing system, a departure mechanism exists which governs the rate at which individuals leave the population and join the queue. This departure mechanism is responsible for the formation of the waiting line and the need to provide service. Formation of the queue is a discrete process. Individuals or items joining the waiting line do so as integer values. The number of units in the waiting line at any point in time is an integer value. Rarely, if ever, is the queuing process continuous.

Individuals or items becoming a part of the waiting line take a position in the queue in accordance with a certain waiting-line discipline. The most common discipline is that of first come, first served. Other priority rules that may exist are the random selection process; the relative urgency rule; first come, last served; and disciplines involving a combination of these. In addition, individuals may remain in the queue for a period of time and then rejoin the population. This behavior is called *reneging*.

When a unit joins the waiting line, or is being serviced, a waiting cost is incurred. Waiting cost per unit per period will depend upon the units in question. If expensive equipment waits for operator attention, the loss of profit may be sizable. Vehicles waiting in queue at a toll gate incur a waiting cost due to interruption of trip progress. Customers waiting at a checkout counter become irritated and the proprietor suffers a loss of good will.

Increasing the level of service capacity will cause a decrease in both the length of the waiting line and the time for each service. As a result, the waiting time will be decreased. Since waiting cost to the system is a product of the number of units waiting and the time duration involved, this action will decrease this cost component. But since increasing the service facility capacity increases the service cost, it is appropriate to seek a reduction in waiting cost for the system only up to the point where the saving justifies the added facility cost.

The Service Mechanism

The rate at which units requiring service are serviced is assumed to be a variable directly under the control of the decision maker. This parameter can be assigned a specific value to create a minimum-cost waiting-line system.

Service is the process of providing the activities required by the units in the waiting line. It may consist of collecting a toll, filling an order, providing a necessary repair, or completing a manufacturing operation. In each case, the act of providing the service causes a unit decrease in the waiting line. The service mechanism, like the arrival mechanism, is discrete, since items are processed on a unit basis.

The service facility may consist of a single channel, or it may consist of several channels in parallel, as in Figure 11.1. If it consists of only a single channel, all arrivals must eventually pass through it. If several channels are provided, items may

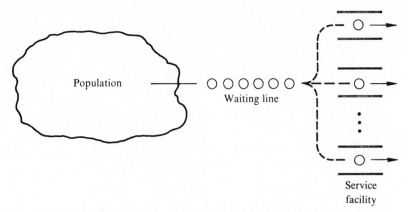

Figure 11.1   A multiple-channel queuing system.

move from the waiting line into the first channel that becomes empty. The rate at which individuals are processed depends upon the service capacity provided at the individual channels and upon the number of channels in the system.

The service may be provided by human beings only, by human beings aided by tools and equipment, or by equipment alone. For example, collecting a fee is essentially a clerk's task which requires no tools or equipment. Repairing a vehicle, on the other hand, requires a mechanic aided by tools and equipment. Processing a phone call dialed by the subscriber seldom requires human intervention and is usually paced by the automatic equipment. These examples indicate that service facilities can vary widely with respect to the person-machine mix utilized to provide the required service.

Each channel of the service facility represents a capital investment plus operating and maintenance costs. In addition, wages for personnel may be involved, together with associated overhead rates. The capability of the channel to process units requiring service is a function of the resources expended at the channel. For example, the channel may consist of a single repairperson with modest tools, or it may be a crew of workers with complex tools and equipment. The cost of providing such a facility will depend on the characteristics of the personnel and equipment employed.

Since increasing the service capacity will result in a reduction in the waiting line, it is appropriate to adjust service capacity so that the sum of waiting cost and service cost is a minimum. The general structure of decision models directed to this objective will be presented next.

The Decision Model

The primary objective of the queuing system is to meet the demand for service at minimum cost. This requires the establishment of an appropriate level of service capacity by constructing and manipulating a mathematical model of the form presented in Equation 7.1,

$$E = f(X, Y)$$

where

E = measure of evaluation (minimize total system cost)
X = policy variable concerning the level of service capacity to provide
Y = system parameters of the arrival pattern, the waiting cost, and the service facility cost

The following sections are devoted to developing decision models with the preceding characteristics. Let

TC = total system cost per period

A = number of periods between arrivals

S = number of periods to complete one unit of service

C_w = cost of waiting per unit per period

C_f = service facility cost for servicing one unit

Additional notation will be adopted and defined as required for deriving specific decision models.

11.2. MONTE CARLO ANALYSIS OF QUEUING

Decision models for probabilistic waiting-line systems are usually based upon certain assumptions regarding the mathematical forms of the arrival and service-time distributions. Monte Carlo analysis, however, does not require that these distributions obey certain theoretical forms. Waiting-line data are produced as the system is simulated over time. Conclusions can be reached from the output statistics, whatever the form of the distributions assumed. In addition, the detailed numerical description that results from Monte Carlo analysis assists greatly in understanding the probabilistic queuing process. This section illustrates the application of Monte Carlo analysis to an infinite population, single-channel waiting-line system.

Arrival- and Service-Time Distributions

The probabilistic waiting-line system usually involves both an arrival time and a service-time distribution. Monte Carlo analysis requires that the form and parameters of these distributions be specified. The cumulative distributions may then be developed and used as a means for generating arrival- and service-time data.

For the example under consideration, assume that the time between arrivals, A_x, has an empirical distribution with a mean of 6.325 periods ($A_m = 6.325$). Service time, S_x, will be assumed to have a normal distribution with a mean of 5.000 periods ($S_m = 5.000$) and a standard deviation of 1 period. These distributions are exhibited in Figure 11.2. The probabilities associated with each value of A_x and S_x are indicated. By summing these individual probabilities from left to right, the cu-

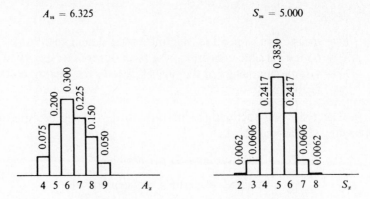

Figure 11.2 Arrival and service time distributions.

mulative distributions of Figure 11.3 result. These distributions may be used with a table of random rectangular variates to generate arrival- and service-time random variables.

The Monte Carlo Analysis

The queuing process under study is assumed to begin when the first arrival occurs. A unit will move immediately into the service facility if it is empty. If the service facility is not empty, the unit will wait in the queue. Units in the waiting line will enter the service facility on a first-come, first-served basis. The objective of the Monte Carlo analysis is to simulate this process over time. The number of unit periods of waiting in the queue and in service may be observed for each of several service rates. The service rate resulting in a minimum-cost system may then be adopted.

The waiting-line process resulting from the arrival and service time distributions of Figure 11.2 is shown in Figure 11.4. The illustration reads from left to right, with the second line being a continuation of the first, and so forth. The interval between vertical lines represents two periods, the heavy dots represent arrivals, the slanting path a unit in service, and the arrows a service completion. When a unit

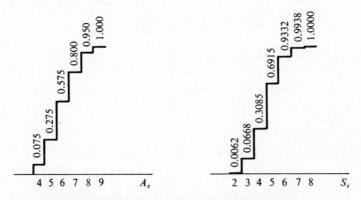

Figure 11.3 Cumulative arrival and service time distributions.

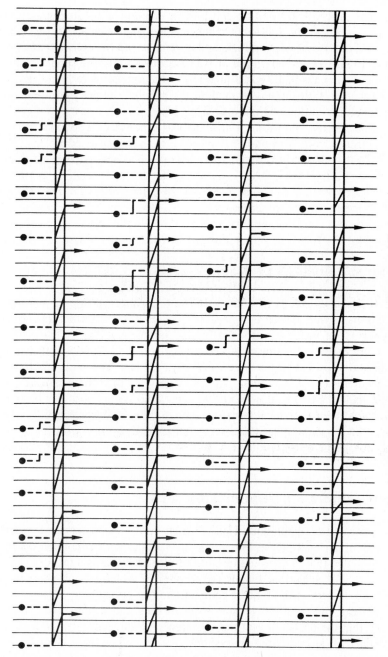

Figure 11.4 Single channel queuing analysis by Monte Carlo.

cannot move directly into the service channel it waits in the queue, which is represented by a horizontal path.

The probabilistic waiting-line process of Figure 11.4 involves 400 periods and was developed in the following manner. First, the sequence of arrivals was established by the use of random rectangular variates and the cumulative arrival distribution of Figure 11.3. Next, each arrival was moved into the service channel if it was available. This availability is a function of the arrival pattern and the service durations selected with the aid of random rectangular variates and the cumulative service-time distribution of Figure 11.3.

Specifically, the Monte Carlo analysis proceeded as follows. Random rectangular variates from Appendix E, Table E.2 were chosen as 5668, 3513, 2060, 7804, 0815, 2639, 9845, 6549, 6353, 7941, and so on. These correspond to arrival intervals of 6, 6, 5, 7, 5, 5, 9, 7, 7, 7, and so forth. Next, random rectangular variates were chosen as 323, 249, 404, 275, 879, 404, 740, 779, 441, 384, and so on. These correspond to service durations of 5, 4, 5, 4, 6, 5, 6, 6, 5, 5, and so forth. These service times determine the time an arrival enters the service channel and the time it is discharged. By proceeding in this manner, the results of Figure 11.4 are obtained.

Economic Analysis of Results

The 400 periods simulated produced a waiting pattern involving 337 unit periods of waiting in service and 23 unit periods of waiting in the queue. The total number of unit periods of waiting for the 400-period sample was 360.

Suppose that waiting cost per unit per period is $9.60 and that it costs $16.10 per period to provide the service capability indicated by the service-time distribution of Figure 11.4. The total system cost for the study period is therefore $9.60(360) + $16.10(400) = $9,896. This total system cost may be compared with the total system cost for alternative service policies by performing a Monte Carlo simulation for the alternative policies. Although this process is time-consuming, it is applicable to many situations that cannot be treated by mathematical means.

11.3. SINGLE-CHANNEL QUEUING MODELS

Assume that the population of units which may demand service is infinite with the number of arrivals per period a random variable with a Poisson distribution. It is also assumed that the time required to service each unit is a random variable with an exponential distribution. Events are recognized to have occurred at the time of arrival of a unit or at the time of completion of a service.

Under the assumption of Poisson arrivals and exponential service, it can be shown that the probability of the occurrence of an event (arrival or service completion) during a specific time interval does not depend on the time of the occurrence of the immediately preceding event of the same kind. The expected number of arrivals per period may be expressed as $1/A_m$, defined as λ, and the expected number of service completions per period may be expressed as $1/S_m$, defined as μ.

Probability of n Units in the System

Under the foregoing assumptions, the probability that an arrival occurs between the time t and time $t + \Delta t$ is $\lambda \Delta t$. Similarly, the probability that a service completion occurs between time t and time $t + \Delta t$, given that a unit is being serviced at time t, is $\mu \Delta t$. Let

$$n = \text{number of units in the system at time } t,$$
$$\text{including the unit being served, if any}$$
$$P_n(t) = \text{probability of } n \text{ units in the system at time } t.$$

Since the time interval Δt is small, it can be assumed that the probability of more than one arrival or service completion during the interval is negligible. Consider the event that there are n units in the system at time $t + \Delta t$ with $n \geq 1$ expressed as

Event $\{n$ units in the system at time $t + \Delta t\}$
$\quad$ = Event $\{n$ units in the system at time t, no arrivals during interval Δt, and no service completions during interval $\Delta t\}$, or
$\quad$ Event $\{n + 1$ units in the system at time t, no arrivals during interval Δt, and one service completion during interval $\Delta t\}$, or
$\quad$ Event $\{n - 1$ units in the system at time t, one arrival during interval Δt, and no service completion during interval $\Delta t\}$

The probability of the event n units in the system at time $t + \Delta t$ can be written as the sum of the probabilities of these three mutually exclusive events as

$$P_n(t + \Delta t) = \{P_n(t)[1 - \lambda \Delta t][1 - \mu \Delta t]\}$$
$$+ \{P_{n+1}(t)[1 - \lambda \Delta t]\mu \Delta t\} + \{P_{n-1}(t)\lambda \Delta t[1 - \mu \Delta t]\}$$
$$= P_n(t) - (\lambda + \mu)P_n(t)\Delta t + \lambda\mu P_n(t)(\Delta t)^2 + \mu P_{n+1}(t)\Delta t \qquad (11.1)$$
$$- \lambda\mu P_{n+1}(t)(\Delta t)^2 + \lambda P_{n-1}(t)\Delta t - \lambda\mu P_{n-1}(t)(\Delta t)^2.$$

Terms involving $(\Delta t)^2$ can be neglected. Subtracting $P_n(t)$ from both sides and dividing by Δt yields

$$\frac{P_n(t + \Delta t) - P_n(t)}{\Delta t} = -(\lambda + \mu)P_n(t) + \mu P_{n+1}(t) + \lambda P_{n-1}(t).$$

In the limit,

$$\lim_{\Delta t \to 0} \frac{P_n(t + \Delta t) - P_n(t)}{\Delta t} = \frac{d}{dt}P_n(t) =$$
$$- (\lambda + \mu)P_n(t) + \mu P_{n+1}(t) + \lambda P_{n-1}(t). \qquad (11.2)$$

For the special case $n = 0$,

Event {0 units in the system at time $t + \Delta t$}
         = Event {0 units in the system at time t and no arrivals during the interval Δt}, or
         Event {1 unit in the system at time t, no arrivals during the interval Δt, and one service completion during the interval Δt}

The probability of no units in the system at time $t + \Delta t$ can be written as the sum of the probabilities of these two mutually exclusive events as

$$
\begin{aligned}
P_0(t + \Delta t) &= \{P_0(t)[1 - \lambda\,\Delta t]\} + \{P_1(t)[1 - \lambda\,\Delta t]\mu\,\Delta t\} \\
&= P_0(t) - \lambda P_0(t)\Delta t + \mu P_1(t)\Delta t - \lambda\mu P_1(t)(\Delta t)^2 .
\end{aligned}
\tag{11.3}
$$

Again neglecting terms involving $(\Delta t)^2$, subtracting $P_0(t)$ from both sides, and dividing by Δt, we obtain

$$
\frac{P_0(t + \Delta t) - P_0(t)}{\Delta t} = -\lambda P_0(t) + \mu P_1(t).
$$

In the limit,

$$
\frac{d}{dt}P_n(t) = \lim_{\Delta t \to 0} \frac{P_0(t + \Delta t) - P_0(t)}{\Delta t} = -\lambda P_0(t) + \mu P_1(t).
\tag{11.4}
$$

Equations 11.2 and 11.4 are called the *governing equations* of a Poisson arrival and exponential service single-channel queue. The differential equations constitute an infinite system for which the general solution is rather difficult to obtain. Figure 11.5 shows an example of the nature of the solution for a particular case.

As long as the probabilities $P_n(t)$ are changing with time, the queue is considered to be in a transient state. From Figure 11.5 it should be noted that this change in $P_n(t)$ becomes smaller and smaller as the time increases. Eventually, there will be little change in $P_n(t)$ and the queue will have reached a steady state. In the steady state the rate of change $dP_n(t)/dt$ can be considered to be zero and the probabilities considered to be independent of time. The steady-state governing equations can be written as

$$
(\lambda + \mu)P_n = \mu P_{n+1} + \lambda P_{n-1}
\tag{11.5}
$$

and

$$
\lambda P_0 = \mu P_1 .
\tag{11.6}
$$

Equations 11.5 and 11.6 constitute an infinite system of algebraic equations that can be solved by substituting $P_{n+1} = P_n \cdot \rho$ into Equation 11.5, giving

$$
(\lambda + \mu)\rho P_{n-1} = (\rho^2 \mu + \lambda)P_{n-1}
$$

or

$$
(\lambda + \mu)\rho = \rho^2 \mu + \lambda
$$

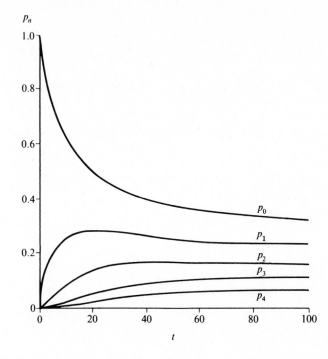

Figure 11.5   Transient solution to governing equations.

or

$$\rho = \frac{\lambda}{\mu}.$$

A substitution in Equation 11.6 gives the same result. Using this substitution, the general solution can be written as

$$P_1 = \rho P_0$$

and

$$P_n = \rho P_{n-1}$$
$$= \rho^n P_0. \tag{11.7}$$

However, since $\Sigma_{n=0}^{\infty} P_n = 1$,

$$1 = \sum_{n=0}^{\infty} P_0 \rho^n = P_0 \sum_{n=0}^{\infty} \rho^n$$
$$= P_0 \left(\frac{1}{1-\rho} \right). \tag{11.8}$$

Hence,

$$P_0 = 1 - \frac{\lambda}{\mu}$$

and

$$P_n = \left(1 - \frac{\lambda}{\mu}\right)\left(\frac{\lambda}{\mu}\right)^n. \tag{11.9}$$

The requirement for the convergence of the sum $\sum_{n=0}^{\infty} (\lambda/\mu)^n$ is that λ/μ be less than 1. This implies that the arrival rate λ must be less than the service rate μ for the queue to reach steady state.

As an example of the significance of Equation 11.9 in waiting-line operations, suppose that a queue is experiencing Poisson arrivals with a mean rate of $1/10$ unit per period and that the service duration is distributed exponentially with a mean of 4 periods. The service rate is, therefore, $1/4$, or 0.25 unit per period. Probabilities associated with each value of n may be calculated as follows:

$$P_0 = (0.6)(0.4)^0 = 0.600$$

$$P_1 = (0.6)(0.4)^1 = 0.240$$

$$P_2 = (0.6)(0.4)^2 = 0.096$$

$$P_3 = (0.6)(0.4)^3 = 0.039$$

$$P_4 = (0.6)(0.4)^4 = 0.015$$

$$P_5 = (0.6)(0.4)^5 = 0.006$$

$$P_6 = (0.6)(0.4)^6 = 0.003$$

$$P_7 = (0.6)(0.4)^7 = 0.001$$

Figure 11.6 exhibits the probability distribution of n units in the system. Certain important characteristics of the waiting-line system can be extracted from this distribution. For example, the probability of 1 or more units in the system is 0.4, the probability of no units in the system is 0.6, the probability of more than 4 units in the system is 0.01, and so forth. Such information as this is useful when there is a restriction on the number of units in the system. By altering the arrival population or the service rate or both, the probability of the number of units in the system exceeding a specified value may be controlled.

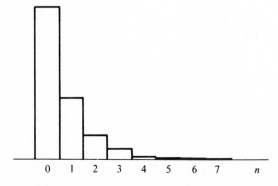

0   1   2   3   4   5   6   7 n

Figure 11.6 Probability distribution of n units in the system.

Mean Number of Units in the System

The mean number of units in the system may be expressed as

$$n_m = \sum_{n=0}^{\infty} nP_n = \sum_{n=0}^{\infty} n(1 - \rho)\rho^n$$

$$= (1 - \rho) \sum_{n=0}^{\infty} n\rho^n.$$

Let $g = \sum_{n=0}^{\infty} n\rho^n$, then

$$\rho g = \sum_{n=0}^{\infty} n\rho^{n+1} = \sum_{n=1}^{\infty} (n - 1)\rho^n.$$

Subtracting ρg from g,

$$(1 - \rho)g = \sum_{n=0}^{\infty} n\rho^n - \sum_{n=1}^{\infty} (n - 1)\rho^n$$

$$= \sum_{n=1}^{\infty} n\rho^n - \sum_{n=1}^{\infty} n\rho^n + \sum_{n=1}^{\infty} \rho^n$$

$$= \sum_{n=1}^{\infty} \rho^n = \rho \sum_{n=0}^{\infty} \rho^n = \frac{\rho}{1 - \rho}.$$

Hence,

$$n_m = (1 - \rho)g = \frac{\rho}{1 - \rho}.$$

Substituting $\rho = \lambda/\mu$ yields

$$n_m = \frac{\lambda}{\mu - \lambda}. \tag{11.10}$$

For the example given previously, the mean number of units in the system is

$$n_m = \frac{0.10}{0.25 - 0.10} = 0.667.$$

Average Length of the Queue

The average length of the queue m_m can be expressed as the average number of units in the system less the average number of units being serviced

$$m_m = \frac{\lambda}{\mu - \lambda} - \frac{\lambda}{\mu}$$

$$= \frac{\lambda^2}{\mu(\mu - \lambda)}. \tag{11.11}$$

For the previous example, the average length of queue is

$$m_m = \frac{(0.10)^2}{0.25(0.25 - 0.10)} = 0.267.$$

that is, the average length of a nonempty waiting line. The probability that the queue is nonempty is given by

$$P(m > 0) = 1 - P_0 - P_1$$
$$= 1 - (1 - \rho) - (1 - \rho)\rho = \rho^2. \tag{11.12}$$

And the average length of the nonempty queue is

$$(m \,|\, m > 0)_m = \frac{m_m}{P(m > 0)}$$

$$= \frac{\lambda^2/\mu(\mu - \lambda)}{\rho^2} \tag{11.13}$$

$$= \frac{\mu}{\mu - \lambda}.$$

For the previous numerical example, the probability that the queue is nonempty is

$$P(m > 0) = \rho^2 = \frac{(0.10)^2}{(0.25)^2} = 0.16$$

and the average length of the nonempty queue is

$$\frac{0.25}{0.25 - 0.10} = 1.667.$$

Distribution of Waiting Time

In a probabilistic queuing system, waiting time spent by a unit before it goes into service is a random variable that depends upon the status of the system at the time of arrival and also on the times required to service the units already waiting for service. In the case of a single-channel system, an arriving unit can go immediately into service only if there are no other units in the system. In all other cases the arriving unit will have to wait.

Two different events can be identified under the Poisson arrival and exponential service assumption for the waiting time, w:

1. Event $\{w = 0\}$ is identical to Event $\{0$ units in the system$\}$.
2. Event $\{$waiting time is in the interval w and $w + \Delta w\}$ is a composite event of there being n units in the system at the time of the arrival, $n - 1$ services being completed during the time w, and the last service being completed within the interval w and $w + \Delta w$.

The probability of the first event occuring is given by

$$P(w = 0) = P_0$$
$$= 1 - \frac{\lambda}{\mu}. \tag{11.14}$$

The second event is illustrated in Figure 11.7. There will be one such event for every n. Furthermore,

$$P(w \leq \text{waiting time} \leq w + \Delta w) = f(w)\Delta w$$

$$= \sum_{n=1}^{\infty} \{P_n \cdot P[(n - 1)\text{services in time } w]\}$$

$$\cdot P(\text{one service completion in time } \Delta w)$$

$$= \sum_{n=1}^{\infty} \left(1 - \frac{\lambda}{\mu}\right)\left(\frac{\lambda}{\mu}\right)^n \left[\frac{(\mu w)^{n-1} e^{-\mu w}}{(n - 1)!}\right] \mu \Delta w$$

$$= \sum_{n=1}^{\infty} \left(\frac{\lambda}{\mu}\right)^n \left[\frac{(\mu w)^{n-1}}{(n - 1)!}\right] e^{-\mu w} \left(1 - \frac{\lambda}{\mu}\right)\mu \Delta w.$$

Let $k = n - 1$. Then

$$f(w)\Delta w = \sum_{k=0}^{\infty} \left(\frac{\lambda}{\mu}\right)^k \frac{\mu^k w^k}{k!} \left(\frac{\lambda}{\mu}\right) e^{-w} \left(1 - \frac{\lambda}{\mu}\right)\mu \Delta w. \tag{11.15}$$

However,

$$\sum_{k=0}^{\infty} \left(\frac{\lambda}{\mu}\right)^k \frac{\mu^k w^k}{k!} = \sum_{k=0}^{\infty} \left(\frac{\lambda w}{k!}\right)^k = e^{\lambda w}. \tag{11.16}$$

Substituting Equation 11.16 into Equation 11.15 gives

$$f(w)\Delta w = e^{\lambda w}\left(\frac{\lambda}{\mu} e^{-\mu w}\right)\mu \Delta w \left(1 - \frac{\lambda}{\mu}\right)$$

$$= \lambda\left(1 - \frac{\lambda}{\mu}\right)e^{-(\mu - \lambda)w}\Delta w$$

or

$$f(w) = \lambda\left(1 - \frac{\lambda}{\mu}\right)e^{-(\mu - \lambda)w}. \tag{11.17}$$

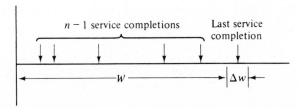

Figure 11.7   Waiting time when there are n units in the system.

Equations 11.14 and 11.17 describe the waiting-time distribution for a single-channel queue with Poisson arrivals and exponential service. Note that this distribution is partly discrete and partly continuous.

The mean time an arrival spends waiting for service can be obtained from the waiting-time distribution as

$$w_m = 0 \cdot P(w = 0) + \int_{w>0} w\lambda \left(1 - \frac{\lambda}{\mu}\right) e^{-(\mu-\lambda)w} dw.$$

After several steps,

$$w_m = \frac{\lambda}{\mu(\mu - \lambda)}. \qquad (11.18)$$

For the numerical example presented previously,

$$w_m = \frac{0.10}{0.25(0.25 - 0.10)}$$

$$= 2.667 \text{ periods.}$$

The average time that a unit spends in the waiting-line system is composed of the average waiting time and the average time required for service, or

$$t_m = \frac{\lambda}{\mu(\mu - \lambda)} + \frac{1}{\mu}$$

$$= \frac{\lambda + \mu - \lambda}{\mu(\mu - \lambda)} = \frac{1}{\mu - \lambda}. \qquad (11.19)$$

For the example above, the average time spent in the system will be

$$t_m = \frac{1}{0.25 - 0.10} = 6.667 \text{ periods.}$$

Minimum-Cost Service Rate

The expected total system cost per period is the sum of the expected waiting cost per period and the expected facility cost per period, that is,

$$TC_m = WC_m + FC_m.$$

The expected waiting cost per period is obtained as the product of cost of waiting per period, the expected number of units arriving per period, and the average time each unit spends in the system, or

$$WC_m = C_w(\lambda)\frac{1}{\mu - \lambda}$$

$$= \frac{C_w \lambda}{\mu - \lambda}. \qquad (11.20)$$

Alternatively, the expected waiting cost per period can be obtained as the product of cost of waiting per period and the mean number of units in the system during the period, or

$$WC_m = C_w(n_m)$$

$$= \frac{C_w\lambda}{\mu - \lambda}.$$

(11.21)

The expected service cost per period is the product of the cost of servicing one unit and the service rate in units per period, or

$$FC_m = C_f(\mu).$$

(11.22)

The expected total system cost per period is the sum of these cost components and may be expressed as

$$TC_m = \frac{C_w\lambda}{\mu - \lambda} + C_f(\mu).$$

(11.23)

A minimum-cost service rate may be found by differentiating with respect to μ, setting the result equal to zero, and solving for μ as follows:

$$\frac{dTC_m}{d\mu} = -C_w\lambda(\mu - \lambda)^{-2} + C_f = 0$$

$$(\mu - \lambda)^2 C_f = \lambda C_w$$

$$\mu = \lambda + \sqrt{\frac{\lambda C_w}{C_f}}.$$

(11.24)

As an application of the preceding model, consider the following Poisson arrival- and exponential service-time situation. The mean time between arrivals is eight periods, the cost of waiting is \$0.10 per unit per period, and the facility cost for serving 1 unit is \$0.165. The expected waiting cost per period, the expected facility cost per period, and the expected total system cost per period are exhibited as a function of μ in Table 11.1. The expected waiting cost per period is infinite when

TABLE 11.1   COST COMPONENTS FOR EXPONENTIAL SERVICE DURATION

μ	WC_m	FC_m	TC_m
0.125	\$ ∞	\$0.0206	\$ ∞
0.150	0.5000	0.0248	0.5248
0.200	0.1667	0.0330	0.1997
0.250	0.1000	0.0413	0.1413
0.300	0.0714	0.0495	0.1209
0.400	0.0455	0.0660	0.1115
0.500	0.0333	0.0825	0.1158
0.600	0.0263	0.0990	0.1253
0.800	0.0185	0.1320	0.1505
1.000	0.0143	0.1650	0.1793

$\mu = \lambda$ and decreases as μ increases. The expected facility cost per period increases with increasing values of μ. The minimum expected total system cost occurs when μ is 0.4 unit per period.

The minimum-cost service rate may be found directly by substituting into Equation 11.24 as follows:

$$\mu = 0.125 + \sqrt{\frac{(0.125)(\$0.10)}{0.165}}$$

$$= 0.125 + 0.275 = 0.400 \text{ unit per period.}$$

11.4. MULTIPLE-CHANNEL QUEUING MODELS

In the previous sections the arriving units were assumed to have been serviced through a single service facility. In many practical situations, however, there are several alternative service facilities. One example is the toll plaza on the turnpike, where several toll booths may serve the arriving traffic.

In the multiple-channel case the service facility will have c service channels, each capable of serving one unit at a time. An arriving unit will go to the first available service channel that is not busy. If all channels are busy, additional arrivals will form a single queue. As soon as any busy channel completes service and becomes available, it accepts the first unit in the queue for service. The steady-state probabilities in such a system are defined as

$P_{m,n}(t) = $ probability that there are n units waiting in queue
and m channels are busy at time t.

It must be noted that m can only be an integer between 0 and c, and that n is zero unless $m = c$.

The determination of the steady-state probabilities follows the logic used in the single-channel case, although the solution becomes quite involved. The results are given below for the Poisson arrival and exponential service situation. Define

$$\rho = \frac{\lambda}{c\mu}$$

then

$$P_{c,n} = P_{0,0}\left(\frac{\lambda}{\mu}\right)^c \frac{1}{c!}\rho^n$$

$$P_{m,0} = P_{0,0}\left(\frac{\lambda}{\mu}\right)^m \frac{1}{m!} \qquad\qquad (11.25)$$

$$P_{0,0} = \frac{1}{(\lambda/\mu)^c(1/c!)[1/(1-\rho)] + \sum_{r=0}^{r=c-1}(\lambda/\mu)^r(1/r!)}$$

As an example, consider a three-channel system with Poisson arrivals at a mean rate of 0.50 units per period and exponential service at each channel with a mean service rate of 0.25 units per period. Under these conditions ρ is $[0.50/(3 \times 0.25)] = 2/3$ and

$$P_{0,0} = \cfrac{1}{(0.50/0.25)^3(1/3!)[1/(1 - 2/3)] + \sum_{0}^{2} (0.50/0.25)^r(1/r!)}$$

$$= \frac{1}{4 + 1 + 2 + 2} = \frac{1}{9}.$$

Average Length of the Queue

The average queue length is obtained from the expression

$$m_m = P_{0,0} \frac{(\lambda/\mu)^{c+1}}{(c - 1)!(c - \lambda/\mu)^2}. \tag{11.26}$$

For the example considered,

$$m_m = \frac{1}{9}\left[\frac{(0.50/0.25)^4}{2!(3 - 0.50/0.25)^2}\right]$$

$$= \frac{1}{9}\left[\frac{(2)^4}{2}\right] = \frac{8}{9} = 0.89 \text{ units}.$$

Mean Number of Units in the System

The mean number of units in the system is

$$n_m = m_m + \frac{\lambda}{\mu}. \tag{11.27}$$

For the example considered,

$$n_m = 0.89 + \frac{0.50}{0.25} = 2.89 \text{ units}.$$

Mean Waiting Time

The mean waiting time can be obtained from the expression for m_m as

$$w_m = \frac{m_m}{\lambda}. \tag{11.28}$$

In the illustrative example, $w_m = 0.89/0.50$, or 1.78 periods.

Average Delay or Holding Time

The average delay is obtained as the sum of waiting and service times as

$$d_m = w_m + \frac{1}{\mu}. \tag{11.29}$$

In the example, average delay is $1.78 + 4$, or 5.78 periods.

Probability that an Arriving Unit Must Wait

The probability of a delay is the same as the probability that all channels are occupied

$$Pr(w > 0) = \sum_0^\infty P_{c,n}$$

$$= P_{0,0}\left(\frac{\lambda}{\mu}\right)^c \frac{1}{c!(1 - \rho)}. \tag{11.30}$$

In the example, the probability that an arriving unit has to wait is

$$P(w > 0) = \frac{1}{9}\left(\frac{0.50}{0.25}\right)^3 \frac{1}{3!(1 - 2/3)}$$

$$= \frac{1}{9}(8)\frac{1}{2} = \frac{4}{9} = 0.444.$$

11.5. QUEUING WITH NONEXPONENTIAL SERVICE

The assumption that the number of arrivals per period obeys a Poisson distribution has a sound practical basis. Although if cannot be said that the Poisson distribution always adequately describes the distribution of the number of arrivals per period, much evidence exists to indicate that this is often the case. Intuitive considerations add support to this assumption, since arrival rates are usually independent of time, queue length, or any other property of the waiting-line system. Evidence in support of the exponential distribution of service durations is not as strong. Often this distribution is assumed for mathematical convenience, as in the previous sections. When the service-time distribution is nonexponential, the development of decision models is quite difficult. Therefore, this section will present models with nonexponential service without proof.

Poisson Arrivals with Constant Service Times

When service is provided automatically by mechanical means, or when the service operation is mechanically paced, the service duration might be a constant. Under these conditions, the service-time distribution has a variance of zero. The mean number of units in the system is given by

$$n_m = \frac{(\lambda/\mu)^2}{2[1 - (\lambda/\mu)]} + \frac{\lambda}{\mu} \tag{11.31}$$

and the mean waiting time is

$$w_m = \frac{\lambda/\mu}{2\mu[1 - \lambda/\mu)]} + \frac{1}{\mu}. \tag{11.32}$$

The expected total system cost per period is the sum of the expected waiting cost per period and the expected facility cost per period; that is,

$$TC_m = WC_m + FC_m.$$

The expected waiting cost per period is the product of the cost of waiting per unit per period and the mean number of units in the system during the period, or

$$WC_m = C_w(n_m)$$

$$= C_w\left\{\frac{(\lambda/\mu)^2}{2[1 - (\lambda/\mu)]} + \frac{\lambda}{\mu}\right\}.$$

The expected facility cost per period is the product of the cost of servicing one unit and the service rate in units per period, or

$$FC_m = C_f(\mu).$$

The expected total system cost per period is the sum of these cost components and may be expressed as

$$TC_m = C_w\left\{\frac{(\lambda/\mu)^2}{2[1 - (\lambda/\mu)]} + \frac{\lambda}{\mu}\right\} + C_f(\mu). \tag{11.33}$$

As an application of the foregoing model, consider the example of the previous section. Instead of the parameter μ being the expected value from an exponential distribution, however, assume that it is a constant. The expected waiting cost per period, the expected service cost per period, and the expected total system cost per period is exhibited as a function of μ in Table 11.2. Although the expected waiting-

TABLE 11.2 COST COMPONENTS FOR CONSTANT
SERVICE DURATION

μ	WC_m	FC_m	TC_m
0.1250	$\ \infty$	$0.0206	$\ \infty$
0.1500	0.2913	0.0248	0.3161
0.2000	0.1145	0.0330	0.1475
0.2500	0.0750	0.0413	0.1163
0.3000	0.0566	0.0495	0.1061
0.4000	0.0383	0.0660	0.1043
0.5000	0.0292	0.0825	0.1117
0.6000	0.0236	0.0990	0.1226
0.8000	0.0170	0.1320	0.1490
1.0000	0.0134	0.1650	0.1785

cost function differs from the previous example, the minimum-cost service interval is still 0.4 units per period.

The example may be more easily compared by graphing the expected total cost functions as shown in Figure 11.8. The upper curve is the expected total system cost function when μ is an expected value from an exponential distribution. The lower curve is the expected total system cost when μ is a constant. No significant difference in the minimum-cost policy is evident for the example considered.

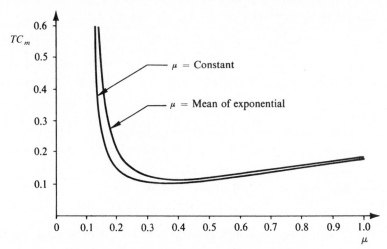

Figure 11.8 TC_m as a function of the service rate.

Poisson Arrivals with Any Service-Time Distribution

For further generality, it is desirable to have expressions for pertinent system characteristics regardless of the form of the service-time distribution. If σ^2 is the variance of the service-time distribution, the mean number of units in the system is given by

$$n_m = \frac{(\lambda/\mu)^2 + \lambda^2\sigma^2}{2[1 - (\lambda/\mu)]} + \frac{\lambda}{\mu} \tag{11.34}$$

and the mean waiting time is

$$w_m = \frac{(\lambda/\mu^2) + \lambda\sigma^2}{2[1 - (\lambda/\mu)]} + \frac{1}{\mu}. \tag{11.35}$$

Equation 11.34 reduces to Equation 11.31 and Equation 11.35 reduces to Equation 11.32 when $\sigma^2 = 0$. In addition, since the variance of an exponential distribution is $(1/\mu)^2$, Equation 11.34 reduces to Equation 11.10 and Equation 11.35 reduces to Equation 11.23 when this substitution is made.

The expected total system cost per period is the sum of the expected waiting cost per period and the expected facility cost per period; that is,

$$TC_m = WC_m + FC_m.$$

The expected waiting cost per period is the product of the cost of waiting per unit per period and the mean number of units in the system during the period. The expected facility cost per period may be taken as the product of the cost of serving one unit and the service rate in units per period. Therefore, the expected total system cost per period is

$$TC_m = C_w \left\{ \frac{(\lambda/\mu)^2 + \lambda^2 \sigma^2}{2[1 - (\lambda/\mu)]} + \frac{\lambda}{\mu} \right\} + C_f(\mu). \tag{11.36}$$

As an example of the application of this model, consider the following situation. The number of arrivals per hour has a Poisson distribution with a mean of 0.2 unit. The cost of waiting per unit per hour is \$2.10 and the cost of servicing one unit is \$4.05. The decision maker may choose one of two service policies. The first will result in a service rate of 0.4 units per hour with a service time variance of 3 hrs. The second will result in a service rate of 0.5 units per hour with a service time variance of 4 hours. The first policy will result in an expected total system cost of

$$TC_m = \$2.10 \left\{ \frac{(0.2/0.4)^2 + (0.2)^2(3)}{2[1 - (0.2/0.4)]} + \frac{0.2}{0.4} \right\} + \$4.05(0.4).$$

$$= \$1.83 + \$1.62 = \$3.45.$$

The second policy will result in an expected total system cost of

$$TC_m = \$2.10 \left\{ \frac{(0.2/0.5)^2 + (0.2)^2(4)}{2[1 - (0.2/0.5)]} + \frac{0.2}{0.5} \right\} + \$4.05(0.5).$$

$$= \$1.40 + \$2.03 = \$3.43.$$

From these results, it is evident that it makes little difference which policy is adopted.

11.6. *FINITE POPULATION QUEUING MODELS*

Finite waiting-line models must be applied to those waiting-line systems where the population is small relative to the arrival rate. In these systems, units leaving the population significantly affect the characteristics of the population and the arrival probabilities. It is assumed that both the time between calls for service for a unit of the population and the service times are distributed exponentially.

Finite Queuing Theory

Units leave the population when they fail, and the number of these failed units can be determined by an approach similar to that used in Section 11.3. Let

N = number of units in the population

M = number of service channels in the repair facility

λ = failure rate of an item, 1/MTBF

μ = repair rate of a repair channel, $1/\text{MTTR}$

n = number of failed items

P_n = steady-state probability of n failed items

P_0 = probability that no items failed

$M\mu$ = maximum possible repair rate

λ_n = failure rate when n items already failed

μ_n = repair rate when n items already failed

The failure rate of an item is expressed as $\lambda = 1/\text{MTBF}$ and the failure rate of the entire population when n items already failed can be expressed as $\lambda_n = (N - n)\lambda$, where $N - n$ is the number of operational units, each of which fails at a rate of λ. Similarly, the repair rate of a repair channel is expressed as $\mu = 1/\text{MTTR}$, and the repair rate of the entire repair facility when n items have already failed can be expressed as

$$\mu_n = \begin{cases} n\mu & \text{if } n \in 1, 2, \ldots, M - 1 \\ M\mu & \text{if } n \in M, M + 1, \ldots, N \end{cases} \tag{11.37}$$

An analysis using birth-death processes is employed to determine the probability distribution, P_n, for the number of failed items. In the birth-death process, the state of the system is the number of failed items (state $= 0, 1, 2, \ldots, N$). The rates of change between the states are the breakdown rate, λ_n, and the repair rate, μ_n. This gives

λ_n: $\qquad N\lambda \qquad (N - 1)\lambda \qquad (N - M + 2)\lambda \quad (N - M + 1)\lambda \qquad \lambda$

State: $\quad 0 \quad\rightleftharpoons\quad 1 \quad\rightleftharpoons\quad 2 \cdots M - 2 \;\; M - 1 \qquad M \quad \cdots N - 1 \rightleftharpoons N$

μ_n: $\qquad \mu \qquad 2\mu \qquad\qquad (M - 1)\mu \qquad M\mu \qquad\qquad M\mu$

If we assume steady-state operation of the system, this yields

$$N\lambda P_0 = \mu P_1$$

$$N\lambda P_0 + 2\mu P_2 = [\mu + (N - 1)\lambda]P_1$$

$$(N - 1)\lambda P_1 + 3\mu P_3 = [2\mu + (N - 2)\lambda]P_2$$

$$\cdot$$
$$\cdot$$
$$\cdot$$

$$(N - M + 2)\lambda P_{M-2} + M\mu P_M = [(M - 1)\mu + (N - M + 1)\lambda]P_{M-1}$$

$$\cdot$$
$$\cdot$$
$$\cdot$$

$$2\lambda P_{N-2} + M\mu P_N = (M\mu + \lambda)P_{N-1}$$

$$\lambda P_{N-1} = M\mu P_N$$

Additionally,

$$\sum_{n=0}^{N} P_n = 1.$$

Solving these balance equations gives

$$P_0 = \left(\sum_{n=0}^{N} C_n \right)^{-1} \tag{11.38}$$

where

$$C_n = \begin{cases} \dfrac{N!}{(N-n)!\, n!} \left(\dfrac{\lambda}{\mu} \right)^n & \text{if } n = 0, 1, 2, \ldots, M \\[2ex] \dfrac{N!}{(N-n)!\, M!\, M^{n-M}} \left(\dfrac{\lambda}{\mu} \right)^n & \text{if } n = M+1, M+2, \ldots, N \end{cases} \tag{11.39}$$

Equations 11.38 and 11.39 can now be used to find the steady-state probability of n failed units as $P_n = P_0 C_n$ for $n = 0, 1, 2, \ldots, N$.

For example, assume that a finite population of 10 units exists, with each unit having a mean time between failure of 32 hours. Also assume that the mean time to service an item in a single service channel is 8 hours. From these data, the arrival rate and the service rate is found to be

$$\lambda = 1/32 \quad \text{and} \quad \mu = 1/8$$

from which

$$\frac{\lambda}{\mu} = 0.25.$$

First, compute C_n for $n = 0$ and 1 using the first expression in Equation 11.39 as

$$C_0 = \frac{10!\, (0.25)^0}{10!\, 0!} = 1$$

$$C_1 = \frac{10!\, (0.25)^1}{9!\, 1!} = 2.5$$

Continue computing C_n for $n = 2, 3, \ldots, 10$ using the second expression in Equation 11.39.

$$C_2 = \frac{10!\, (0.25)^2}{8!\, 1!\, 1^1} = 5.625$$

$$C_3 = \frac{10!\, (0.25)^3}{7!\, 1!\, 1^2} = 11.250$$

$$C_4 = \frac{10!\, (0.25)^4}{6!\, 1!\, 1^3} = 19.6875$$

$$C_5 = \frac{10!\, (0.25)^5}{5!\, 1!\, 1^4} = 29.53125$$

$$C_6 = \frac{10! \, (0.25)^6}{4! \, 1! \, 1^5} = 36.9140625$$

$$C_7 = \frac{10! \, (0.25)^7}{3! \, 1! \, 1^6} = 36.9140625$$

$$C_8 = \frac{10! \, (0.25)^8}{2! \, 1! \, 1^7} = 27.685547$$

$$C_9 = \frac{10! \, (0.25)^9}{1! \, 1! \, 1^8} = 13.842773$$

$$C_{10} = \frac{10! \, (0.25)^{10}}{0! \, 1! \, 1^9} = 3.460693$$

$$\sum_{n=0}^{10} C_n = 188.410888$$

From Equation 11.38 use

$$P_0 = \frac{1}{\sum\limits_{n=0}^{10} C_n} = \frac{1}{188.410888} = 0.0053076.$$

P_n for $n = 0, 1, 2, \ldots, N$ can now be computed from $P_n = P_0 C_n = 0.0053076$ (C_n) as follows.

$$P_0 = 0.0053076 \times 1 \qquad\qquad = 0.0053076$$
$$P_1 = 0.0053076 \times 2.5 \qquad\qquad = 0.0013269$$
$$P_2 = 0.0053076 \times 5.625 \qquad\qquad = 0.0298553$$
$$P_3 = 0.0053076 \times 11.250 \qquad\qquad = 0.0597105$$
$$P_4 = 0.0053076 \times 19.6875 \qquad\quad = 0.1044934$$
$$P_5 = 0.0053076 \times 29.53125 \qquad\quad = 0.1567400$$
$$P_6 = 0.0053976 \times 36.9140625 = 0.1959251$$
$$P_7 = 0.0053076 \times 36.9140625 = 0.1959251$$
$$P_8 = 0.0053976 \times 27.685547 \quad = 0.1469438$$
$$P_9 = 0.0053076 \times 13.842773 \quad = 0.0734719$$
$$P_{10} = 0.0053076 \times 3.460693 \quad = 0.0183680$$

The steady-state probabilities of n failed units calculated above can now be used to find the mean number of failed units from

$$\sum_{n=0}^{10} n \times P_n$$

as

$$0 \times 0.0053076 + 1 \times 0.0013269 + \cdots + 10 \times 0.183680 = 6.023 \text{ units.}$$

Finite Queuing Tables[1]

The finite queuing derivations in the previous section and the appropriate numerical applications provide complete probabilities for n units having failed, P_n. From these probabilities, the expected number of failed units was obtained.

In this section an approach based on the Finite Queuing Tables is presented. For convenience, the notation used in these tables will be adopted. Let

T = mean service time

U = mean time between calls for service

H = mean number of units being serviced

L = mean number of units waiting for service

J = mean number of units running or productive

Appendix F gives a portion of the *Finite Queuing Tables* (for populations of 10, 20, and 30 units). Each set of values is indexed by N, the number of units in the population. Within each set, data are classified by X, the service factor, and M, the number of service channels. Two values are listed for each value of N, X, and M. The first is D, the probability of a delay, expressing the probability that an arrival will have to wait. The second is F, an efficiency factor needed in the calculation of H, L, and J.

The service factor is a function of the mean service time and the mean time between calls for service,

$$X = \frac{T}{T + U}. \tag{11.40}$$

The mean number of units being serviced is a function of the efficiency factor, the number of units in the population, and the service factor,

$$H = FNX. \tag{11.41}$$

The mean number of units waiting for service is a function of the number of units in the population and the efficiency factor,

$$L = N(1 - F). \tag{11.42}$$

Finally, the mean number of units running or productive is a function of the number of units in the population, the efficiency factor, and the service factor,

$$J = NF(1 - X). \tag{11.43}$$

A knowledge of N, T, and U for the waiting-line system under study, the expressions given above, and a set of *Finite Queuing Tables* makes it easy to find mean

[1] L. G. Peck, and R. N. Hazelwood, *Finite Queuing Tables* (New York: John Wiley, 1958).

values for important queuing parameters. For example, the mean number of failed units calculated in the previous section to be 6.023 can be found from Equations 11.40, 11.41, and 11.42 as

$$X = \frac{8}{8 + 32} = 0.20$$

$$H = 0.497(10)(0.20) = 0.994$$

$$L = 10(1 - 0.497) = 5.03.$$

This gives $H + L = 6.024$ as the mean number failed. Other examples are given in the sections which follow.

Number of Service Channels Under Control

Assume that a population of 20 units exists, with each unit having a mean time between required service of 32 minutes. Each service channel provided will have a mean service time of 8 minutes. Both the time between arrivals and the service interval are distributed exponentially. The number of channels to be provided is under management control. The cost of providing one channel with a mean service-time capacity of 8 minutes is $10 per hour. The cost of waiting is $5 per unit per hour. The service factor for this system is

$$X = \frac{T}{T + U} = \frac{8}{8 + 32} = 0.20.$$

Table 11.3 provides a systematic means for finding the minimum-cost number of service channels. The values in columns A and B are entered from Appendix F, Table F.2, with $N = 20$ and $X = 0.20$. The mean number of units being serviced is found from Equation 11.41 and entered in column C. The mean number of units waiting for service is found from Equation 11.42 and is entered in column D. The mean number of units waiting in queue and in service is given in column E. The data of columns A and E may be multiplied by their respective costs to give the total system cost.

First, multiplying $5 per unit per hour by the mean number of units waiting gives the waiting cost per hour in column F. Second, multiplying $10 per channel

TABLE 11.3   COST AS A FUNCTION OF THE NUMBER OF SERVICE CHANNELS

M (A)	F (B)	H (C)	L (D)	$H + L$ (E)	WAITING COST (F)	SERVICE COST (G)	TOTAL COST (H)
8	0.999	4.00	0.02	4.02	$20.10	$80	$100.00
7	0.997	3.99	0.06	4.05	20.25	70	90.25
6	0.988	3.95	0.24	4.19	20.95	60	80.95
5	0.963	3.85	0.74	4.59	22.95	50	72.95
4	0.895	3.58	2.10	5.68	28.40	40	68.40
3	0.736	2.94	5.28	8.22	41.10	30	71.10
2	0.500	2.00	10.00	12.00	60.00	20	80.00

per hour by the number of channels gives the service cost per hour in column G. Finally, adding the expected waiting cost and the service cost gives the expected total system cost in column H. The minimum cost number of channels is found to be four.

In this example, the cost of waiting was taken to be $5 per unit per hour. If this is due to lost profit, resulting from unproductive units, the same solution may be obtained by maximizing profit. As before, the values in columns A and B of Table 11.4 are entered from Appendix F, Table F.2, with $N = 20$ and $X = 0.200$. The mean number of units running or productive is found from Equation 11.43 and is entered in column C.

The profit per hour in column D is found by multiplying the mean number of productive units by $5 profit per productive unit per hour. The cost of service per hour in column E is obtained by multiplying the number of channels by $10 per channel per hour. Finally, the net profit in column F is found by subtracting the service cost per hour from the profit per hour. As before, the number of channels that should be used is four. This example illustrates that either the minimum-cost or the maximum-profit approach may be used with the same results.

TABLE 11.4   PROFIT AS A FUNCTION OF THE NUMBER OF SERVICE CHANNELS

M (A)	F (B)	J (C)	GROSS PROFIT (D)	SERVICE COST (E)	NET PROFIT (F)
8	0.999	15.98	$79.90	$80	$-0.10
7	0.997	15.95	79.75	70	9.75
6	0.988	15.81	79.05	60	19.05
5	0.963	15.41	77.05	50	27.05
4	0.895	14.32	71.60	40	31.60
3	0.736	11.78	58.90	30	28.90
2	0.500	8.00	40.00	20	20.00

Mean Service Time Under Control

Assume that a population of 10 units is to be served by a single service channel. The mean time between calls for service is 30 mins. If the mean service rate is 60 units per hour, the service cost will be $100 per hour. The service cost per hour is inversely proportional to the time in minutes to service one unit, expressed as $100/T. Both the time between calls for service and the service duration are distributed exponentially. Lost profit due to units waiting in the system is $15 per hour.

Column A in Table 11.5 gives the capacity of the channel expressed as the mean service time in minutes per unit processed. The service factor for each service time is found from Equation 11.38 and entered in column B. The efficiency factors in column C are found by interpolation in Appendix F, Table F.1, for $N = 10$ and the respective service factors of column B. The mean number of units running is found from Equation 11.41 and entered in column D. The data given in column A and column D may now be used to find the service capacity that results in a maximum net profit.

TABLE 11.5   PROFIT AS A FUNCTION OF THE MEAN SERVICE TIME

T (A)	X (B)	F (C)	J (D)	GROSS PROFIT (E)	SERVICE COST (F)	NET PROFIT (G)
1	0.032	0.988	9.56	$143.20	$100.00	$43.20
2	0.062	0.945	8.86	132.90	50.00	82.90
3	0.091	0.864	7.85	117.75	33.33	84.42
4	0.118	0.763	6.73	101.00	25.00	76.00
5	0.143	0.674	5.77	86.51	20.00	66.51

The expected profit per hour in column E is found by multiplying the mean number of units running by $15 per hour. The cost of service per hour is found by dividing $100 by the value for T in column A. These costs are entered in column F. By subtracting the cost of service per hour from the expected gross profit per hour, the expected net profit per hour in column G is found. The mean service time resulting in an expected maximum profit is three periods.

Service Factor Under Control

Suppose that a population of production equipment is under study with the objective of deriving minimum cost maintenance policy. It is assumed that both the time between calls for maintenance for a unit of the population and the service times are distributed exponentially. Two parameters of the system are subject to management control: (1) by increasing the repair capability (reducing T) the average machine downtime will be reduced; and (2) alternative policies of preventive maintenance will alter the mean time between breakdowns, U. Therefore, the problem of machine maintenance reduces to one of determining the service factor, X, that will result in a minimum-cost operation. This section will present methods for establishing and controlling the service factor in maintenance operations.

As machines break down, they become unproductive with a resulting economic loss. This loss may be reduced by reducing the service factor. But a decrease in the service factor requires either a reduction in the repair time or a more expensive policy of preventive maintenance, or both. Therefore, the objective is to find an economic balance between the cost of unproductive machines and the cost of maintaining a specific service factor.

The analysis of this situation is facilitated by developing curves giving the percentage of machines not running as a function of the service factor. Figure 11.9 gives curves for selected populations when one service channel is provided. Each curve is developed from Equation 11.41 and the *Finite Queuing Tables*. As was expected, the percentage of machines not running increases as the service factor increases.

As an example of the determination of the minimum-cost service factor, suppose that eight machines are maintained by a mechanic and his helper. Each machine produces a profit of $22.00 per hour while it is running. The mechanic and his assistant cost the company $28.80 per hour. Three policies of preventive maintenance

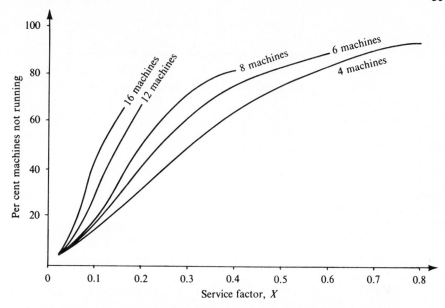

Figure 11.9   Percent machines not running as a function of the service factor.

are under consideration. The first will cost $80 per hour. After considering the increase in the mean time between breakdowns and the effect on service time, it is estimated that the resulting service factor will be 0.04. The second policy of preventive maintenance will cost only $42 per hour but will result in a service factor of 0.10. The third alternative involves no preventive maintenance at all; hence, it will cost nothing, but a service factor in excess of 0.2 will result. The time between calls for service and the service times are distributed exponentially. By reference to Figure 11.9, the results of Table 11.6 are developed. From the last column, it is evident that the second alternative should be adopted.

TABLE 11.6   THREE POLICIES OF PREVENTIVE MAINTENANCE

Maintenance Policy	Service Factor, X	Machines Not Running	Cost of Lost Profit	Cost of Maintenance	Cost of Mechanic	Total Cost
1	0.04	0.5	$11.00	$80	$28.80	$119.80
2	0.10	1.6	35.20	42	28.80	106.00
3	0.20	4.1	90.20	0	28.80	119.00

SELECTED REFERENCES

(1) Drake, A. W., *Fundamentals of Applied Probability Theory*. New York: McGraw-Hill, 1967.

(2) Fabrycky, W. J., Ghare, P. M., and P. E. Torgersen, *Applied Operations Research and Management Science*. Englewood Cliffs, N. J.: Prentice-Hall, 1984.

(3) Gross, D., and C. M. Harris, *Fundamentals of Queuing Theory*. New York: John Wiley, 1985.

(4) Hillier, F. S., and G. J. Lieberman, *Operations Research*. San Francisco: Holden-Day, 1986.

(5) Peck, L. G., and R. N. Hazelwood, *Finite Queuing Tables*. New York: John Wiley, 1958.

(6) Taha, H. A., *Operations Research*. New York: Macmillian, 1987.

(7) White, J. A., J. W. Schmidt, and G. K. Bennett, *Analysis of Queuing Systems*. New York: Academic Press, 1975.

QUESTIONS AND PROBLEMS

1. Name and describe (a) an infinite population and (b) a finite population.

2. Under what conditions is it mandatory that Monte Carlo analysis be used in the study of a queuing system?

3. Why is it essential that μ be greater than λ in a probabilistic waiting-line process?

4. Suppose that the time between arrivals is distributed exponentially with a mean of 6 mins and that service time is constant and equal to 5 mins. Service is provided on a first-come, first served basis. No units are in the system at 8:00 A.M. Use Monte Carlo analysis to estimate the total number of unit minutes of waiting between 8:00 A.M. and 12:00 noon.

5. Use Monte Carlo analysis to verify Equation 11.9 if the number of arrivals per period has a Poisson distribution with a mean of 0.10 and if the service duration is distributed exponentially with a mean of four periods. Plot the histogram of the number of units in the system and compare with Figure 11.6.

6. Suppose that arrivals are distributed according to the Poisson distribution with a mean of 0.125 unit per period and that the service duration is distributed exponentially with a mean of five periods. Develop the probability distribution of n units in the system. What is the probability of there being more than 4 units in the system?

7. The arrival rate for a certain waiting-line system obeys a Poisson distribution with a mean of 0.5 unit per period. It is required that the probability of one or more units in the system not exceed 0.20. What is the minimum service rate that must be provided if the service duration will be distributed exponentially?

8. What is the expected number of units in the system and the expected waiting time for the conditions of Problem 6?

9. The expected number of units in a waiting-line system experiencing Poisson arrivals with a rate of 0.4 unit per period must not exceed 8. What is the minimum service rate that must be provided if the service duration will be distributed exponentially? What will be the expected waiting time?

10. If the cost of waiting in Problem 7 is $5.00 per unit per period, and the service facility costs $2.50 per unit served, what is the total system cost? If the service rate is doubled at a total cost of $3.50 per unit served, what is the total system cost?

11. Plot the mean number of units in the system, and the mean waiting time, as a function of λ/μ, if the number of arrivals per period obey the Poisson distribution and if the service duration is distributed exponentially. What is the significance of this illustration?

12. The number of arrivals per period is distributed according to the Poisson with an expected value of 0.75 unit per period. The cost of waiting per unit per period is $3.20. The facility cost for serving one unit per period is $5.15. What expected service rate should be established if the service duration will be distributed exponentially? What is the expected total system cost?

13. The expected waiting time in a waiting-line system with Poisson arrivals at a rate of 1.5 units per day must not exceed 6 days. What is the minimum constant service rate that must be provided? What is the expected number in the system?

14. Trucks arrive at a loading dock in a Poisson manner at the rate of 3.5 per day. The cost of waiting per truck per day is $168. A three-person crew that can load at a constant rate of four trucks per day costs $224 per day. Compute the total system cost for the operation.

15. Customers arrive at a bank at the rate of 0.2 per period and the service rate is 0.4 customers per period. If there is one teller, what is the average length of the queue? If the arrival rate increases to 0.6 per period and the number of tellers is increased to two, how does the average length of the queue change. Assume Poisson arrivals and an exponential service time distribution.

16. Cars arrive at a toll plaza at the rate of 60 cars per hour. There are four booths each capable of servicing 30 cars per hour. Assuming a Poisson distribution for arrivals and an exponential distribution for the service time, determine (a) the probability of no cars in the toll plaza, (b) the average length of the queue, and (c) the average time spent by a car in the plaza.

17. In Problem 16, if it costs $17.20 per hour to operate a toll booth and the waiting cost per car is $0.25 per minute spent at the plaza, how many booths should be operated to minimize total cost?

18. In a three-channel queuing system the arrival rate is 1 unit per time period and the service rate is 0.6 unit per time period. Determine the probability that an arriving unit does not have to wait.

19. Use Equation 11.39 to extend the example in Section 11.6 to the case where there are two service channels. Compare the mean number waiting with the single channel assumption.

20. Each truck in a fleet of 30 delivery trucks will return to a warehouse for reloading at an average interval of 160 mins. An average of 20 mins is required by the driver and one warehouseman to load the next shipment. If the warehouse personnel are busy loading previous arrivals, the driver and his unit must wait in queue. Both the time between arrivals and the loading time are distributed exponentially. The cost of waiting in the system is $18.20 per hour per truck and the total cost per warehouseman is $10.65 per hour. Find the minimum cost number of warehousemen to employ.

21. A population of 10 cargo aircraft each produces a profit of $1,360 per 24-hour period when not waiting to be unloaded. The time between arrivals is distributed exponentially with a mean of 144 hours. The unloading time at the ramp is distributed exponentially with a mean of 18 hours. It costs $42 per hour to lease a ramp with this unloading capacity. How many ramps should be leased?

22. Each unit in a ten unit population returns for service at an average interval of 20 minutes. If the cost of waiting is $10 per hour per unit and the cost of service is $15/$T$ per unit per hour, find the optimum service time if there are two service channels. Assume Poisson arrivals and exponential service times.

23. A population of 30 chemical processing units is to be unloaded and loaded by a single crew. The mean time between calls for this operation is 68 mins. If the mean service rate is one unit per minute, the cost of the crew and equipment will be $16 per minute. This cost will decrease to $12, $9, and $7 for service intervals of two, three, and four minutes, respectively. The time between calls for service and the service duration are distributed exponentially. If it costs $28 per hour for each unit that is idle, find the minimum cost service interval.

24. Plot the percentage of machines not running as a function of the service factor for a population of 20 machines with exponential arrivals and services (a) if a single channel is employed and (b) if two channels are employed.

25. A population of ten production cells is to be served by an unknown number of service channels. The mean time between service calls is 15 minutes. The cost per service channel is $60 per hour at a mean service rate of 60 cells per hour and the service cost per channel per hour is inversely proportional to the service rate. Both the time between calls for service and the service times are distributed exponentially. If the profit from each production cell is $10 per hour, determine the optimum mix of service channels and service time to provide.

26. A group of 12 machines are repaired by a single repairperson when they break down. Each machine yields a profit of $5.20 per hour while running. The mechanic costs the firm $20.40 per hour, including overhead. At the present time no preventive maintenance is utilized. It is proposed that one technician be employed to perform certain routine maintenance and adjustment tasks. This will cost $16.60 per hour, but will reduce the service factor from 0.12 to 0.08. What is the economic advantage of implementing preventive maintenance?

12

Control Concepts
and Techniques

Most systems are deployed and then operate in an environment that changes over time. Except for static systems, a changing environment can lead to system instability unless control action is applied. The value of dynamic systems may be enhanced through control action applied during operations. Control of portions or all of a system can help maintain system performance within specified tolerances, or to increase the worth of system output.

The concern for control began with physical systems and servo theory. Engineers who turned their attention to large-scale systems in the 1950s were trained in servomechanisms and other control devices. A large-scale man-made system is one in which there are many states, control variables, and constraints. There may be several control loops and different measures of performance for each subsystem. Classical control theory cannot be used directly to select measures of effectiveness and to optimize outputs. Nevertheless, the concepts can be used to provide a basis for structuring system description and internal relationships so that feedback and adaptive phenomena are incorporated in system design.

12.1. SOME CONTROL CONCEPTS

A control problem may arise from the need to regulate speed, temperature, quality, quantity, or to determine the trajectory of an aircraft or space vehicle. In each case, the allocation of scarce resources over time is involved. Control variables must be related in some way to state variables describing the system characteristic or condition being controlled.

The classical example of control is the determination of a missile trajectory. Control variables are the amount, timing, and direction of the thrust applied to the missile, where the thrust available is constrained by fuel availability. State variables describe the trajectory incorporating mass, position, and velocity. The thrusts and the state variables are related by differential equations in the light of the effectiveness function to be optimized. The objective is usually to maximize the payload that can be delivered to a given destination.

Control problems also arise in large-scale person-equipment systems of activity. In air traffic control, state variables are the number of aircraft enroute, the number of aircraft waiting to land at available airports, the number of aircraft waiting on the ground, the available number of communication channels, and so on. Control variables in air traffic control are the instructions given by the traffic controllers.

The Elements of a Control System

Every control system has four basic elements. These elements always occur in the same sequence and have the same relationship to each other. They are

1. A controlled characteristic or condition.
2. A sensory device or method for measuring the characteristic or condition.
3. A control device that will compare measured performance with planned performance.
4. An activating device that will alter the system to bring about a change in the output characteristic or condition being controlled.

Figure 12.1 illustrates the relationships among the four elements of a control system. The output of the system (Figure 12.1, Block 1) is the characteristic or condition to be measured. This output may be speed, temperature, quality, or any other characteristic or condition of the system under consideration.

The sensory device (Figure 12.1, Block 2) measures output performance. System design must incorporate such devices as tachometers, thermocouples, thermometers, transducers, inspectors, and other physical and/or human sensors. Information gathered from a sensory device is essential to the operation of a control system.

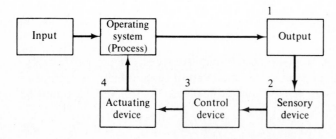

Figure 12.1 Control system-elements and relationships.

The third element in a control system is the control device (Figure 12.1, Block 3). This element exists to determine the need for control action based on the information provided by the sensory device. It may be a mini- or microcomputer, but need be no more than a visual or hand-calculated comparison depending upon the situation. It is important that this element in the control system be able to detect significant differences between planned output and actual output.

Any out-of-control signal received from the control device must be acted upon by the actuating device (Figure 12.1, Block 4). This device is the fourth element in a control system. Its role is that of implementation. Actuating devices may be mechanical, electromechanical, hydraulic, pneumatic, or they may be human, as in the case where a machine operator changes the setting on a machine to alter the dimension of a part being produced.

Information as the Medium of Control

It is the flow of measurement information which, when converted into corrective information, makes it possible for a characteristic or condition of a system to be controlled. To control every characteristic or condition of a system's output is usually not possible. Accordingly, it is important to make a wise choice of the characteristic to be measured. It is important that the successful control of the characteristic chosen lead to successful control of the system. This requirement is usually met if the controlled characteristic is stated in the language to be used in the corrective loop. Before information is gathered, one must determine what is to be transmitted.

Information to be compared with the control standard must be expressed in the same terms as the standard in order to facilitate a decision regarding the control status of the system. This information can often be secured by sampling the output. Of course, a sample is only an estimate of a group or population. Statistical sampling based on probability theory can, however, provide valid control information subject to some error.

The sensor element in Figure 12.1 provides information that becomes the basis for control action. Output information is compared with the characteristic or condition to be controlled. Significant deviations are noted and corrective information is provided to the activating device.

Information received from the control device causes the actuating device to respond. However, every control system has a finite delay in its action time which can cause problems of over correction. This difficulty can be overcome by reducing the time lag between output measurement and input correction or by introducing a lead time in the control system.

Types of Control Systems

There are two basic types of control systems, open-loop and closed-loop. Each has its own characteristic. The basic difference is whether or not the control device is an integral part of the system it controls.

In *open-loop control,* the optimal control action is completely specified at the initial time in the system cycle. In *closed-loop control,* the optimal control action is determined both by the initial conditions and by the current state-variable values. Whereas in open-loop control all decisions are made in advance, in closed-loop control decisions are revised in the light of new information received about the system state.

Common examples of open-loop and closed-loop control are the laundry dryer and the home thermostat. In the laundry dryer the length of the drying cycle is set once by human action. A home thermostat, on the other hand, provides control signals to the furnace whenever temperature falls below a predetermined level. When a system is regulated by a person, as in the case of a laundry dryer, attention must be paid to see that the desired output is obtained. In this open-loop example, the drying time must be determined by experience and be properly set into the timer.

If control is exercised in terms of system operation, as with a thermostat, closed-loop control is taking place. The significant difference in this type of control system is that the control device is an element of the heating system it serves. In closed-loop control the four elements of a control system all act together.

An essential part of a closed-loop system is feedback. *Feedback* is a process in which the output of a system is measured continuously or at predetermined intervals as in sample-data control. Generally, costs are incurred in measuring a characteristic or condition. These costs may be in terms of added workload, energy, weight, and so on. Open-loop control requires only initial measurements and is preferred unless there are random elements either in measurement or in system operation.

12.2. STATISTICAL PROCESS CONTROL

Variation is inherent in the output characteristic or condition of most processes. Patterns of statistical variation exist in the number of units processed through maintenance per time period, the dimension of a machined part, the procurement lead time for an item, and the number of defects per unit quantity of product produced. Statistical process control is a methodology for testing to see if an operating system is in control.

A statistical control chart is a control device (Figure 12.1, Block 3) that uses measurement information from a sensory device (Figure 12.1, Block 2) and provides system change information to an actuating device (Figure 12.1, Block 4). Control charts utilize samples taken from system output and control limits to determine if an operation is in control. When a sample value falls outside the control limits, a stable system of statistical variation probably no longer exists. Action may then be taken to find the cause of the "out-of-control" condition so that control may again be established, or compensation made for the change that has taken place.

Patterns of Statistical Variation

A stable or probabilistic steady-state pattern of variation exists when the parameters of the statistical distribution describing an operation remain constant over time. Steady-state variation of this type is normally an exception rather than the rule. Many operations will produce a probabilistic non-steady-state pattern of variation over time. When this is the case, the mean and/or the variance of the distribution describing the pattern change with time. Sometimes, the form of the distribution will also undergo change.

Control limits may often be placed about an initial stable pattern of variation to detect subsequent changes from that pattern, a statistical inference is made when a sample value is considered. If the sample falls within the control limits, the process under study is said to be *in control*. If the sample falls outside the limits, the process is deemed to have changed and is said to be *out of control*.

Figure 12.2 illustrates control limits placed at a distance $k\sigma$ from the mean of the initial stable pattern of variation. If a sample value falls outside the control limits, and the process has not changed, a Type I error of probability α has been made. On the other hand, if a sample value falls within these limits and the process has changed, a Type II error of probability β has been made. In the first case, the null hypothesis has been rejected in error. In the second, the null hypothesis has not been rejected, with a resulting error.

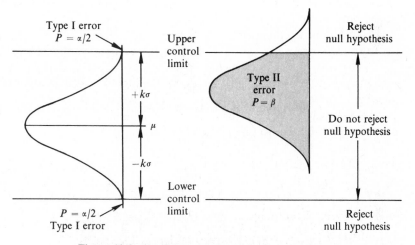

Figure 12.2   A stable and a changed pattern of variation.

If the control limits are set relatively far apart, it is unlikely that a Type I error will be made. The control model, however, is then not likely to detect small shifts in the parameter. In this case the probability, β, of a Type II error is large. If the other approach is taken, and the limits are placed relatively close to the initial stable pattern of variation, the value of α will increase. The advantage gained is greater sensitivity of the model for the shifts that may occur. The ultimate criterion will be the costs associated with the making of Type I and Type II errors. The limits should be established to minimize the sum of these two costs.

Control Charts for Variables

A *control chart* is a graphical representation of a mathematical model used to monitor a random variable process in order to detect changes in a parameter of that process. Charting statistical data is a test of the null hypothesis that the process from which the sample came has not changed. A control chart is employed to distinguish between the existence of a stable pattern of variation and the occurrence of an unstable pattern. If an unstable pattern of variation is detected, action may be initiated to discover the cause of the instability. Removal of the assignable cause should permit the process to return to the stable state.

Control charts for variables are used for continuous operations. Two charts are available for operations of this type. The $\bar{X}$ *chart* is a plot over time of sample means taken from a process. It is primarily employed to detect changes in the mean of the process from which the samples came. The *R chart* is a plot over time of the ranges of these same samples. It is employed to detect changes in the dispersion of the process. These charts are often employed together in control operations.

Constructing the $\bar{X}$ chart.   The $\bar{X}$ chart receives its input as the mean of a sample taken from the process under study. Usually, the sample will contain four or five observations, a number sufficient to make the Central Limit Theorem applicable.[1] Accepting an approximately normal distribution of the sample means allows the establishment of control limits with a predetermined knowledge of the probability of making a Type I error. It is not necessary to know the form of the distribution of the process.

The first step in constructing an $\bar{X}$ chart is to estimate the process mean, μ, and the process variance, σ^2. This requires taking m samples each of size n and calculating the mean, $\bar{X}$, and the range, R, for each sample. Table 12.1 illustrates the format that may be used in the calculations. The mean of the sample means, $\bar{\bar{X}}$, is used as an estimate of μ and is calculated as

$$\bar{\bar{X}} = \frac{\sum\limits_{i=1}^{m} \bar{X}}{m} \tag{12.1}$$

and the mean of the sample ranges, $\bar{R}$, is calculated as

$$\bar{R} = \frac{\sum\limits_{i=1}^{m} R}{m}. \tag{12.2}$$

TABLE 12.1   COMPUTATIONAL FORMAT FOR DETERMINING $\bar{X}$ AND R

SAMPLE NUMBER	SAMPLE VALUES	MEAN $\bar{X}$	RANGE R
1	$x_{11}, x_{12}, \ldots, x_{1n}$	$\bar{X}_1$	R_1
2	$x_{21}, x_{22}, \ldots, x_{2n}$	$\bar{X}_2$	R_2
$\vdots$	$\vdots$ $\vdots$ $\vdots$	$\vdots$	$\vdots$
m	$x_{m1}, x_{m2}, \ldots, x_{mn}$	$\bar{X}_m$	R_m

[1] The *Central Limit Theorem* states that the distribution of sample means approaches normality as the sample size approaches infinity, provided that the sample values are independent.

The expected ratio between the average range, $\bar{R}$, and the standard deviation of the process has been computed for various sample sizes, n. This ratio is designated d_2 and is expressed as

$$d_2 = \frac{\bar{R}}{\sigma}.$$

Therefore, σ can be estimated from the sample statistic $\bar{R}$ as

$$\sigma = \frac{\bar{R}}{d_2}. \tag{12.3}$$

Values of d_2 as a function of the sample size may be found in Table 12.2.

TABLE 12.2 FACTORS FOR THE CONSTRUCTION OF $\bar{X}$ AND R CHARTS

SAMPLE SIZE, n	$\bar{X}$ CHART		R CHART	
	d_2	A_2	D_3	D_4
2	1.128	1.880	0	3.267
3	1.693	1.023	0	2.575
4	2.059	0.729	0	2.282
5	2.326	0.577	0	2.115
6	2.534	0.482	0	2.004
7	2.704	0.419	0.076	1.924
8	2.847	0.373	0.136	1.864
9	2.970	0.337	0.184	1.816
10	3.078	3.308	0.223	1.777

Factors are taken from the *A.S.T.M. Manual on Quality Control of Materials* (1951), #STP 15C, with permission.

The mean of the $\bar{X}$ chart is set at $\bar{\bar{X}}$. The control limits are normally set at $\pm 3\sigma_{\bar{x}}$, which results in the probability of making a Type I error of 0.0027. Since

$$\sigma_{\bar{x}} = \frac{\sigma}{\sqrt{n}},$$

substitution into Equation (12.3) gives

$$\sigma_{\bar{x}} = \frac{\bar{R}}{d_2\sqrt{n}}$$

and

$$3\sigma_{\bar{x}} = \frac{3\bar{R}}{d_2\sqrt{n}}. \tag{12.4}$$

The factor $3/d_2\sqrt{n}$ has been tabulated as A_2 in Table 12.2. Therefore, the upper and lower control limits for the $\bar{X}$ chart may be specified as

$$UCL_{\bar{x}} = \bar{\bar{X}} + A_2\bar{R} \tag{12.5}$$

$$LCL_{\bar{x}} = \bar{\bar{X}} - A_2\bar{R} \tag{12.6}$$

Constructing the R chart.   The R chart is constructed in a manner similar to the $\bar{X}$ chart. If the $\bar{X}$ chart has already been completed, $\bar{R}$ has been calculated from Equation (12.2). Tabular values of 3σ control limits for the range have been compiled for varying sample sizes and are included in Table 12.2. The upper and lower control limits for the R chart are then specified as

$$UCL_R = D_4\bar{R} \tag{12.7}$$

$$LCL_R = D_3\bar{R} \tag{12.8}$$

Since $D_3 = 0$ for sample size of $n \le 6$ in Table 12.2, the $LCL_R = 0$. Actually, 3σ limits yield a negative lower control limit which is recorded as zero. This means that with samples of six or fewer, it will be impossible for a value on the R chart to fall outside the lower limit. Thus the R chart will not be capable of detecting reductions in the dispersion of the process output.

Application of the $\bar{X}$ and R charts.   Once the control limits have been specified for each chart, the data used in constructing the limits are plotted. Should all values fall within both sets of limits, the charts are ready for use. Should one or more values fall outside one set of limits, however, further inquiry is needed. A value outside the limits on the $\bar{X}$ chart indicates that the process may have undergone some change in regard to its central tendency. A value outside the limits on the R chart is evidence that the process variability may now be out of control. In either case, one should search for the source of the change in process behavior. If one or two values fall outside the limits and an assignable cause can be found, then these one or two values may be discarded and revised control limits calculated. If the revised limits contain all the remaining values, the control chart is ready for implementation. If they do not, the procedure may be repeated before using the control chart.

Assume that control charts are to be established to monitor the weight in ounces of the contents of containers being filled on an assembly line. The containers should hold at least 10 oz. and to guarantee this weight, the process must be set to deliver slightly more. Samples of five have been taken every 30 mins. The sample data, together with the sample means and sample ranges, are given in Table 12.3.

An R chart is first constructed from these data. Using Equation 12.2, $\bar{R}$ is calculated as 2.05. The control limits are then determined from Equations 12.7 and 12.8 as

$$UCL_R = 2.115(2.05) = 4.34$$

$$LCL_R = 0(2.05) = 0$$

These limits are used to construct the R chart of Figure 12.3. Since all values fall within the control limits, the R chart is accepted as a means of assessing subsequent process variation. Had a point fallen outside the calculated limits, that point would have had to be discarded and limits recalculated.

Attention should next be directed to the $\bar{X}$ chart. The mean of the sample means, $\bar{\bar{X}}$, is found from Equation 12.1 to be 12.19. The mean of the sample ranges,

TABLE 12.3   WEIGHT OF THE CONTENTS OF CONTAINERS (OUNCES)

SAMPLE NUMBER	SAMPLE VALUES					MEAN $\bar{X}$	RANGE R
1	11.3	10.5	12.4	12.2	12.0	11.7	1.9
2	9.6	11.7	13.0	11.4	12.8	11.7	3.4
3	11.4	12.4	11.7	11.4	12.4	11.9	1.0
4	12.0	11.9	13.2	11.9	12.2	12.2	1.3
5	12.4	11.9	11.7	11.6	10.5	11.6	1.9
6	13.8	12.5	13.9	11.9	11.4	12.7	2.5
7	13.3	11.6	13.2	10.7	11.4	12.0	2.6
8	11.1	11.3	13.2	12.8	12.0	12.1	2.1
9	12.5	11.9	13.8	11.6	13.0	12.6	2.2
10	12.1	11.7	12.0	11.7	12.9	12.1	1.2
11	11.7	12.6	12.3	11.2	10.8	11.7	1.8
12	13.8	12.3	12.4	14.1	11.3	12.8	2.8
13	10.6	11.8	13.1	12.8	11.7	12.0	2.5
14	12.0	11.2	12.1	11.7	12.1	11.8	0.9
15	11.5	13.1	13.9	11.9	10.7	12.2	3.2
16	13.4	12.6	12.4	11.9	11.8	12.4	1.6
17	12.1	13.1	14.1	11.4	12.3	12.6	2.7
18	11.5	13.2	12.4	12.6	12.2	12.4	1.7
19	13.8	14.2	13.5	13.2	12.8	13.5	1.4
20	11.5	11.4	13.1	11.6	10.8	11.7	2.3

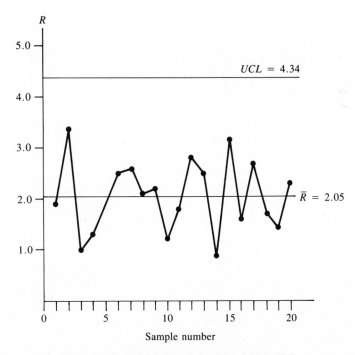

Figure 12.3   An R chart for the data of Table 12.3.

$\bar{R}$, has already been calculated as 2.05. Preliminary control limits for the $\bar{X}$ chart can now be calculated from Equations 12.5 and 12.6 as

$$UCL_{\bar{x}} = 12.19 + 0.577(2.05) = 13.37$$

$$LCL_{\bar{x}} = 12.19 - 0.577(2.05) = 11.01$$

These limits are used to construct the $\bar{X}$ chart shown in Figure 12.4.

The 20 sample means may now be plotted. It is noted that the mean of sample 19 exceeds the upper control limit. This would indicate that at this point in time, the universe from which this sample was selected was not exhibiting a stable pattern of variation. Some change occurred between the time of selecting sample 18 and sample 19. It is further noted that after sample 19, the process returned to its original state. One may base action on the assumption that these statements are true, particularly if one thinks that some recognized assignable cause effected the change. Actually, the mean of sample 19 might have exceeded the control limit by chance, and a Type I error might have been made. Alternatively, the pattern of variation might have shifted some time before sample 18 and/or not returned after this time. Then a Type II error would have been made at these other points in time.

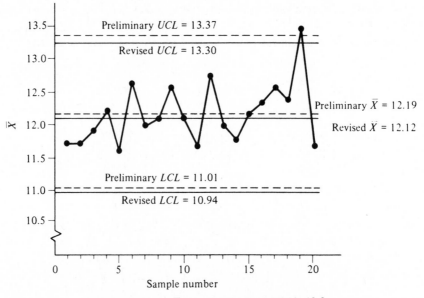

Figure 12.4 An $\bar{X}$ chart for the data of Table 12.3.

The data of sample 19 should now be discarded and the control limits recalculated for the remaining pattern of variation. The mean of the remaining sample means is 12.12. The range is not recalculated because it is assumed that the process variation did not change when sample 19 was selected. The control limits are now revised to

$$UCL_{\bar{x}} = 12.12 + 0.577(2.05) = 13.30$$

$$LCL_{\bar{x}} = 12.12 - 0.577(2.05) = 10.94$$

These are indicated on the control chart of Figure 12.4 and again checked to determine that no sample means exceed these limits. It is noted that no further changes are necessary, and the control chart of Figure 12.4, with the revised limits, and the chart of Figure 12.3 may now be employed to monitor the process. This control model may be implemented to test future process variation to see that it does not change from that used to construct the chart.

Control Charts for Attributes

A system characteristic or condition is sometimes expressible in a two-value manner. A defense system may be operationally ready or it may be down for maintenance. The surface finish of a workpiece may be acceptable or not. A maintenance crew may be working or it may be idle. When a two-value classification is made, the proportion of observations falling in one class during a predetermined time period may be monitored over time with a p chart.

Often observations yield a multivalue, but still discrete classification. An employee may suffer none, one, two, or more lost-time accidents during a given time period. The number of demands upon a supply system will be a discrete number during a specified time period. A maintenance mechanic may perform an assigned task perfectly, or may have made one or more errors. When a discrete classification is made, a system characteristic or condition can be monitored over time with a c chart.

The p chart.   When a characteristic or condition is sampled and then placed into one of two defined classes, the proportion of units falling into one class may be controlled over time or from one sample to another with a p chart. The applicable probability distribution is the binomial. The mean of this distribution and its standard deviation may be expressed as given in Equation (10.4).

$$\mu = np$$
$$\sigma = \sqrt{np(1 - p)}.$$

These parameters may be expressed as proportions by dividing by the sample size, n. If $\overline{p}$ is then defined as an estimate of the proportion parameter μ/n, and s_p as an estimate of σ/n, then these statistics can be expressed as

$$\overline{p} = \frac{\text{total number in the class}}{\text{total number of observations}} \tag{12.9}$$

$$s_p = \sqrt{\frac{\overline{p}(1 - \overline{p})}{n}}. \tag{12.10}$$

The application of the p chart will be illustrated with an example of an activity-sampling study. This activity measurement technique is used to obtain information about the activities of workers or machines, usually in less time and at lower cost than by conventional means. Random and instantaneous observations are taken by classifying the activity at a point in time into one and only one category. In the most simple form, the categories *idle* and *busy* are used. The control chart is useful in work-sampling studies in that the observed proportions can be verified as in control and following a stable pattern of variation or out of control with an unstable pattern. In the latter case, a search can be undertaken for an assignable cause.

Consider the case of an activity-sampling study involving computer terminals in an office. The objective was to determine the proportion of time the terminals were in use as opposed to the time they were idle. One hundred observations were taken each day over all working days in a month. The number of times the terminals were in use, and the proportion for each day are given in Table 12.4. From this table and Equation 12.9, $\bar{p}$ is established as

$$\bar{p} = \frac{546}{(21)(100)} = 0.260.$$

The standard deviation of the data is calculated with Equation 12.11 as

$$s_p = \sqrt{\frac{(0.26)(0.74)}{100}} = 0.044.$$

TABLE 12.4 NUMBER OF TIMES A DAY THE TERMINALS ARE IN USE

WORKING DAY	TIMES IN USE	PROPORTION	WORKING DAY	TIMES IN USE	PROPORTION
1	22	0.22	12	46	0.46
2	33	0.33	13	31	0.31
3	24	0.24	14	24	0.24
4	20	0.20	15	22	0.22
5	18	0.18	16	22	0.22
6	24	0.24	17	29	0.29
7	24	0.24	18	31	0.31
8	29	0.29	19	21	0.21
9	18	0.18	20	26	0.26
10	27	0.27	21	24	0.24
11	31	0.31			
			Total	546	

With $\bar{p} = 0.26$ as the best estimate of the population proportion, a control chart may now be constructed. Control limits in activity sampling are usually established at $\bar{p} \pm 2s_p$, and these limits will not vary from one day to another in this example, since a constant sample size has been maintained for each day. With these control limits, the probability of making a Type I error may be defined as $\alpha = 0.0456$ if the normal distribution is used as an approximation. Such an approximation is realistic for this example since the binomial will approximate the normal distribution if $n > 50$ and $0.20 < p < 0.80$. If these requirements were not met or a more accurate estimate were needed, the binomial distribution would have had to be employed. The control limits for the p chart are defined and calculated as

$$UCL_p = \overline{p} + 2s_p$$

$$= 0.260 + 2(0.044) = 0.348 \qquad (12.11)$$

$$LCL_p = \overline{p} - 2s_p$$

$$= 0.260 - 2(0.044) = 0.172. \qquad (12.12)$$

The p chart is constructed as shown in Figure 12.5. A plot of the data indicates that day 12 was not typical of the pattern of use established by the rest of the month. Subsequent investigation reveals that personnel from another office were also using the terminals that one day because their equipment had preceded them in a move to another building. As this is an atypical situation, the sample is discarded and a revised mean and standard deviation are calculated as

$$\overline{p} = \frac{500}{(20)(100)} = 0.250$$

$$s_p = \sqrt{\frac{(0.25)(0.75)}{100}} = 0.043.$$

Revised control limits are now calculated and placed on the same control chart as

$$UCL_p = \overline{p} + 2s_p$$

$$= 0.250 + 2(0.043) = 0.336$$

$$LCL_p = \overline{p} - 2s_p$$

$$= 0.250 - 2(0.043) = 0.164.$$

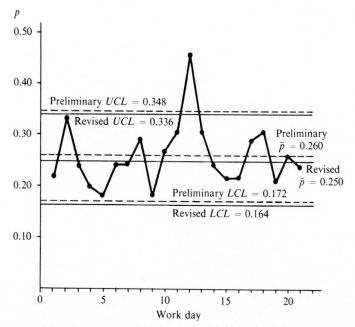

Figure 12.5   A p chart of daily computer terminal usage.

It is noted that no further days fall outside these limits and that the chart and data, without day 12, can be taken as a stable pattern of variation of terminal usage. Deviations from this in-control condition can be noted and appropriate control action taken.

The c chart.   Some systems exhibit characteristics expressible numerically. For example, the number of arrivals per hour seeking service at a toll booth is of interest when deciding upon the level of service capability to provide. If the number of arrivals per hour deviates from the stable pattern of variation, it may be necessary to compensate by either opening or closing certain toll booths.

The Poisson distribution is usually used to describe the number of arrivals per time period. Here, the opportunity for the occurrence of an event, n, is large but the probability of each occurrence, p, is quite small. The mean and the variance of the Poisson distribution are equal and were given in Table 10.1 as $\mu = \sigma^2 = np$. These parameters can be estimated from the statistics with $\bar{c}$ and s_c^2 defined as these estimates. In many applications, values for n and p cannot be determined, but their product np can be established. Then the mean and variance can be estimated as

$$\bar{c} = s_c^2 = \frac{\Sigma\,(np)}{m}. \tag{12.13}$$

Consider the application of a c chart to the arrival process previously described. Data for the past 20 hours have been collected and are presented in Table 12.5. The mean of the arrival population may be estimated from $\bar{c}$ with Equation 12.13 as

$$\bar{c} = \frac{84}{20} = 4.20$$

TABLE 12.5 NUMBER OF ARRIVALS PER HOUR DEMANDING SERVICE AT A TOLL BOOTH

HOUR NUMBER	np NUMBER OF ARRIVALS	HOUR NUMBER	np NUMBER OF ARRIVALS
1	6	11	6
2	4	12	4
3	3	13	2
4	5	14	2
5	4	15	4
6	6	16	8
7	5	17	2
8	4	18	3
9	2	19	5
10	5	20	4
		Total = 84	

and the standard deviation may be estimated as

$$s_c = \sqrt{\bar{c}}$$
$$= \sqrt{4.20} = 2.05.$$

With these estimates, the control chart may now be constructed. Control limits may be established as $\bar{c} \pm 3s_c$. The probability of making a Type I error can be determined from the cumulative values of Appendix E, Table E.1. For the example under consideration,

$$UCL_c = \bar{c} + 3s_c$$
$$= 4.2 + 3(2.05) = 10.35 \tag{12.14}$$

$$LCL_c = \bar{c} - 3s_c$$
$$= 4.2 - 3(2.05) < 0. \tag{12.15}$$

There is no lower control limit. The probability of making a Type I error is the probability of 11 or more arrivals in a given hour from a population with $\bar{c} = 4.2$. This is $1 - P(10 \text{ or less})$ or $1 - 0.994 = 0.006$. Alternatively, the control limit could have been defined as a probability limit. If it were thought desirable to define $\alpha \leq 0.01$, the control limit would have been specified as 9.5. Under this control policy the probability of detecting 10 or more arrivals would satisfy $\alpha \leq 0.01$.

In the c chart of Figure 12.6, all values fell within the control limits and the chart can be utilized as originally formulated. Action will be initiated to alter the service capability to meet the new demand pattern. At the same time, a new chart based upon the recent data will be constructed and implemented. In this way the decision maker is informed of conditions in the system that require modification of operating policy.

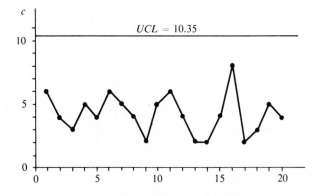

Hour

Figure 12.6   A c chart of the number of arrivals per hour.

12.3. OPTIMUM POLICY CONTROL

A statistical control chart may be used to monitor the optimum solution obtained from a decision model. Most decision models require monitoring because they do not automatically adapt themselves to the instability of the decision environment. These models are usually formulated at a point in time for a specific set of input parameters. A solution derived from a decision model will be optimal only as long as the input parameters retain the values initially established.[2]

Optimum policy control will be illustrated for the case where an optimum service factor has been determined for a maintenance situation.[3] Once the service factor resulting in a minimum-cost operation has been established, it might be desirable to implement a control model for detecting the effects of a shift in the arrival rate or the service rate or both. This may be accomplished by constructing a control chart for the number of machines not running. The statistical control charts presented cannot be applied directly to this variable, since its distribution is badly skewed. The expected form of the distribution is known in advance; however, a factor usually missing in other control applications.

As an example of the application of control charts in waiting-line operations, consider a one worker/N machine situation.[4] Suppose that 20 automatic machines are to be run by one worker, and that the minimum-cost service factor is found to be 0.03. It is assumed that the machines require service at randomly distributed times, that the service times are distributed exponentially, and that machines are serviced on a first-come, first-served basis. The probability of n machines in the queue and in service (not running) is given by

$$P_n = \frac{N!}{(N - n)!} \rho^n P_0 \tag{12.6}$$

where

P_n = probability that n machines are not running at any point in time
P_0 = probability that all machines are running at any point in time
ρ = ratio of λ/μ
N = number of machines assigned
n = number of machines not running at any point in time

Values for P_n are given in column C of Table 12.6 for $0 \leq n \leq N$. The first entry is found by dividing the first entry in column B by 2.30005387. The second results from dividing the second entry in column B by the same value, and so on. The results for P_n are plotted in Figure 12.7. The distribution of n is seen to be extremely skewed, being completely convex. Unlike some control chart applications, the distribution of the variable to be controlled is known. It may be used to determine the control limits.

[2] In terms of $E = f(X, Y)$, this means that E will remain optimal for specific values of X only as long as Y values retain their value initially established.

[3] This example follows the computational procedure of Section 11.6.

[4] Adapted from R. W. Llewellyn, "Control Charts for Queuing Applications," *Journal of Industrial Engineering*, vol. XI, no. 4 (July-August 1960).

TABLE 12.6   CALCULATION OF P_n AND ΣP_n

n (A)	$\dfrac{P_n}{P_0}$ (B)	P_n (C)	ΣP_n (D)
0	1.000000000	0.434772422	0.434772422
1	0.600000000	0.260863453	0.695635875
2	0.342000000	0.148692168	0.844328043
3	0.184680000	0.080293771	0.924621814
4	0.094186800	0.040949823	0.965571637
5	0.045209664	0.019655915	0.985227552
6	0.020344349	0.008845162	0.994072714
7	0.008544627	0.003714968	0.997787682
8	0.003332405	0.001448838	0.999236520
9	0.001199666	0.000521582	0.999758102
10	0.000395890	0.000172122	0.999930224
11	0.000118767	0.000051637	0.999981861
12	0.000032067	0.000013942	0.999995803
13	0.000007696	0.000003346	0.999999149
14	0.000001616	0.000000703	0.999999852
15	0.000000291	0.000000127	0.999999979
16	0.000000044	0.000000019	0.999999998
17	0.000000005	0.000000002	1.000000000
18	0.000000000	0.000000000	1.000000000
19	0.000000000	0.000000000	1.000000000
20	0.000000000	0.000000000	1.000000000

2.300053887

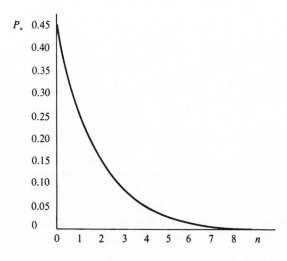

Figure 12.7 Probability distribution of the number of machines not running.

Since the probability that all machines are running is approximately 0.435, the lower control limit is obviously zero. The upper control limit is all that needs to be determined, since the variable n can go out of control only at the top. Thus, the entire critical range will be at the upper end of the distribution. It will be desirable to set this limit so that it will not be violated too frequently. Since n is an integer, the magnitude of the critical range can be observed as a function of n from column D in Table 12.6. If six machines not running is chosen as a point in control and seven is a point out of control, the probability of designating the system out of control when it is really in control is

$$1 - \sum_{n=0}^{6} P_n = 0.0059.$$

Therefore, a control limit set at 6.5 should be satisfactory. The control chart applied to a period of operation might appear as in Figure 12.8. It would be concluded that the service factor has not changed during this period of observation.

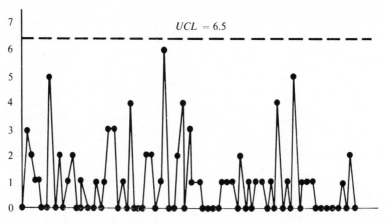

Figure 12.8   Control chart for the number of machines not running.

In order for the application of this control model to be valid, two conditions must be met. First, the observation must be made at random times. Second, the observations must be spaced far enough apart, so that the results are independent. For example, if the service time averages 5 minutes, and if six machines are idle in an observation, then an observation taken 10 minutes later would probably indicate at least four idle machines. The readings should be far enough apart for a waiting line to be dissipated. To ensure both randomness and spacing, all numbers from 45 to 90 could be taken from a random number table with their order preserved; numbers from 00 to 44 and from 91 to 99 would be dropped. These numbers can then be used to space the observations. This would yield an average of eight random observations per day with a minimum spacing of 45 mins and a maximum spacing of 90 mins.

If the control chart indicates that n is no longer in control, corrective action must be taken. This will require investigation to determine whether the service factor has changed because of a change in the mean time between calls or a change in the mean service time, or both. Specific items that might be studied are the policy of preventive maintenance, the age of the machines, the capability of the operator, or material characteristics. Once the assignable cause for the out-of-control condition is located, it may be corrected so that the optimal operating condition may be restored.

12.4. PROJECT CONTROL WITH CPM AND PERT

A project is composed of a series of activities directed to the accomplishment of a desired objective. Most projects are nonrepetitive and confront the decision maker with a situation in which prior experience and information for control is nonexistent. Therefore, techniques for project control should meet two conditions: (1) during the planning phase of a project they should make possible the determination of a logical, preferably optimal, project plan; and (2) during the execution of a project they should make possible the evaluation of a project progress against the plan.

The planning and control of a large-scale project may be accomplished by utilizing an activity network as the planning and control medium. Project planning and control techniques may be classified as deterministic and designated CPM (critical path method) or probabilistic and designated PERT (program evaluation and review technique).

CPM and PERT are similar in their logical structure. Projects are represented by sets of required activities, with the principal difference being the treatment of estimated time for the completion of each activity. In PERT, activity time is considered to be a random variable with an assumed probability distribution.

Critical Path Methods—CPM

Many engineering projects involving design, development, and construction operations are nonrepetitive in nature. Although a particular project is to be executed only once, many interdependent activities are required. Some must be executed simultaneously. Each will require personnel and equipment, the cost of which will depend upon the resource commitment. The total project duration and its overall cost depend upon the activities on the *critical path* and the resources allocated to them. Critical path methods (CPM) for dealing with this situation are presented in this section.

Activity and event networks.   A network is the basic structural entity behind all critical path methods. It is used to portray the interrelationships among the activities associated with a project. Figure 12.9 illustrates a network that represents the activities and events in connection with the preparation of a foundation. Four events and three activities are shown.

An event in CPM is represented by a circle. It indicates the completion of an activity. In the case of an initial event it indicates project start or the initiation of the associated activities. Events *A* and *C* in Figure 12.9 indicate the start of excavation and the start of equipment move activities. Event *D* represents the completion of foundation preparation.

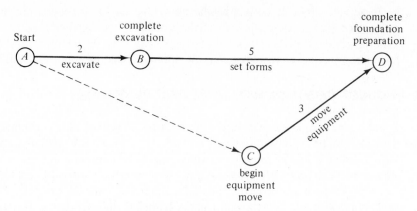

Figure 12.9   Activity-event network for preparing a foundation.

Activities in CPM are represented by arrows that interconnect events. These arrows symbolize the effort required to complete an event in units of time. Their direction specifies the order in which the events must occur. For example, activities *BD* (set forms) and *CD* (equipment move-in) are required in order for event *D* (complete preparation) to be achieved. The precedence relationships between event *D* and event *B* and between event *D* and event *C* are established by the arrows. Events *B* and *C* must be realized before the foundation is prepared (event *D*).

In some network representations it is necessary or useful to create dummy activities to clarify precedence relationships. A dummy activity is illustrated in Figure 12.9 by a broken arrow that connects activity *A* with activity *C*, establishing that event *C* is preceded by event *A*. Even though dummy activities do not require time and do not consume resources, they may alter the completion time of a project by establishing the order in which events are performed.

Associated with each activity is an elapsed time estimate. This estimate is usually shown above the arrow in the network. The length of the arrows need not be proportional to the duration of the activity. In Figure 12.9 the time estimates are weeks.

Two important absolute times for each event must be identified. The first is the earliest time, T_E, which indicates the calender time at which an event can take place, provided all previous activities were completed at their earliest possible times. The second is the latest time, T_L, for each event, defined as the latest calendar time at which an event can be completed without delaying initiating the following activities and completion of the project.

The difference between the latest and the earliest times for each event is called the *slack time* for that event. Positive slack for an event indicates the maximum amount of time that it can be delayed without delaying subsequent events and the overall project. In the example, event *B* has no slack since the latest time for completion of excavation and the earliest time for setting the forms is the same point in time. Any delay in realizing event *B* would cause a delay in the project. From this it is evident that the activities of excavation and setting the forms are critical. The activity of equipment move-in is noncritical to the realization of event *D* since the start of equipment move-in can occur at any time on or before 3 weeks before the completion of the project. Thus, 4 weeks is the slack time for event C.

Finding the critical path.   In CPM all critical activities must be identified and given special attention. This is necessary because any delay in performing these activities may lead to a delay in subsequent activities and the project as a whole. Often it is possible to identify the critical path by tracing the activities along a sequence of events which have zero slack time. The sum of the activity times along this critical path gives the shortest possible elapsed time for project completion. This is 7 weeks in the foundation example.

As an example of a systematic procedure for finding the critical path, consider a project requiring the construction of a certain structure. Five major events must be realized through seven activities as illustrated in Figure 12.10. The number appearing above each arrow is the estimated duration of the activity in months, with the arrow indicating the precedence relationship.

The procedure for finding the critical path starts by determining the earliest time and the latest time for each event. By subtracting T_E and T_L for each event, the slack time is found. To facilitate this process each event can be labeled (T_E, T_L, S), where S represents the slack time for the event. First, T_Es are determined for each event and are labeled $(T_E,\ ,\ )$. Next, T_Ls are found and the labels become $(T_E, T_L,\ )$. Finally, S is found by subtracting T_E from T_L and completing the label (T_E, T_L, S) for each event.

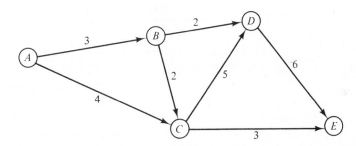

Figure 12.10   A network for the construction of a certain structure.

The earliest time for each event is determined by adding the earliest time for the event immediately preceding to the activity duration connecting the two events. When there is more than one preceding event, the largest value of the sum is chosen as the earliest time for the event under consideration. The process for determining these earliest times proceeds forward from the initial event, which is assumed to have an earliest time of zero. This process is illustrated in Figure 12.11 and Table 12.7 for the example under consideration.

The latest time for each event can be determined in a manner similar to that used in finding the earliest time except that the process starts with the final event and proceeds backward. The latest time for this final event is the same as its earliest time, for, if it were not, the project would be delayed. For each event the latest time is determined by subtracting the duration time of an activity connecting the event

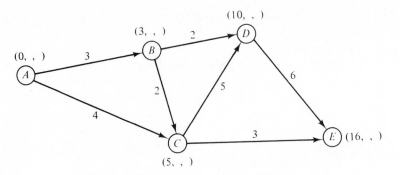

Figure 12.11   The earliest time for each event.

TABLE 12.7   COMPUTATION OF EARLIEST TIMES

EVENT	IMMEDIATE PREDECESSOR	T_E FOR IMMEDIATE PREDE-CESSOR PLUS ACTIVITY TIME	LARGEST SUM = THE EARLIEST TIME
A	None	$0 + 0 = 0$	0
B	A	$0 + 3 = 3$	3
C	A	$0 + 4 = 4$	
	B	$3 + 2 = 5$ ⟵	5
D	B	$3 + 2 = 5$	
	C	$5 + 5 = 10$ ⟵	10
E	C	$5 + 3 = 8$	
	D	$10 + 6 = 16$ ⟵	16

and an event immediately succeeding from the latest time of the event immediately succeeding. If there is more than one event immediately succeeding, the smallest value of the difference is chosen to be the latest time for the event under consideration. A slack time for each event is then found by computing the difference between T_L and T_E as illustrated in Figure 12.12 and Table 12.8.

The critical path can now be identified with the T_L, T_E, and S values entered in Figure 12.12. This is done by tracing all zero slack events and viewing them as a

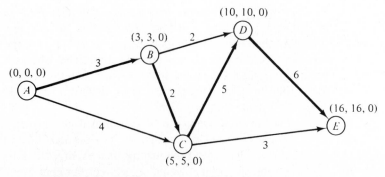

Figure 12.12   Labeling of events and identification of critical path.

TABLE 12.8 COMPUTATION OF LATEST TIMES

EVENT	IMMEDIATE SUCCESSOR	T_L FOR IMMEDIATE SUCCESSOR MINUS ACTIVITY TIME	SMALLEST DIFFERENCE = THE LATEST TIME
E	None	$16 - 0 = 16$	16
D	E	$16 - 6 = 10$	10
C	D	$10 - 5 = 5$ ⟵	5
	E	$16 - 3 = 13$	
B	C	$5 - 2 = 3$ ⟵	3
	D	$10 - 2 = 8$	
A	B	$3 - 3 = 0$ ⟵	0
	C	$5 - 4 = 1$	

path. In this example the critical path is $A \rightarrow B \rightarrow C \rightarrow D \rightarrow E$ with a minimum project duration of 16 months.

Economic aspects of CPM.   The CPM example above was presented under the assumption that the activities are performed in normal time with a normal allocation of resources. This is called a *normal schedule*. When one or more activities are performed with additional resources to shorten their time durations, a *crash schedule* is said to exist.

Two costs are associated with a project. Direct costs exist for each activity. These costs increase with an increase in the resources allocated for the purpose of decreasing the activity time. Indirect costs exist which are associated with the overall project duration. These increase in direct proportion to an increase in the total time required for project completion. This relationship should be self-evident, for overhead costs continue independent of activity levels and they depend upon the passage of time.

In most projects it is of interest to investigate the economic aspects of increasing direct costs as activities are expedited in the light of decreasing overhead costs. An optimum allocation of resources to each activity is sought so that the sum of direct and indirect costs will be a minimum.

The fist step in finding the optimum project schedule is to find the critical path under normal resource conditions. This is the starting point. Next, the activity on the critical path which has the least impact on direct cost is shortened. When shortening this activity, other critical paths may appear. If they do, it is necessary to reduce the duration of activities on these paths an equal number of time units for each path. No further activity time reduction should be attempted either when a limit has been reached or when the indirect costs saved are not greater than the extra expenditure of direct resources. This process is repeated for all activities on the critical path or paths.

Consider the example project shown in Figure 12.13. Indirect costs of $1,000 can be saved for each day removed from the total project duration. Event times shown are those that would occur under normal conditions. The label for each event represents T_E, T_L, and S as defined earlier. The critical path is $A \rightarrow B \rightarrow D \rightarrow E$.

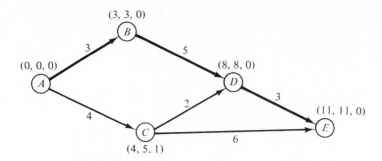

Figure 12.13   A CPM network with normal activities.

Data on activity durations under normal and crash conditions are given in Table 12.9. These data are daily costs computed from the assumption that each additional worker adds a direct cost of $50 per day. Additional equipment costs are those estimated to arise for each day reduced from the normal activity duration.

The project depicted in Figure 12.13 will take 11 days to complete under the normal activity times shown. No crash cost is incurred and therefore there is no saving in indirect cost.

A 10-day schedule can be achieved by reducing BD by one day. From Table 12.9 a crash cost of $700 is found. A saving of $1,000 in indirect cost will result. No further reduction in activity BD should be attempted because the critical path will change. This 10-day schedule is shown in Figure 12.14.

Two critical paths now exist: the original path and a new path $A \rightarrow C \rightarrow E$. Reducing activity BD one more day yields a nine-day schedule. This leads to a crash cost of $900 which is less than the crash cost for shortening BD one day and AB one day. The duration of one activity in the critical path $A \rightarrow C \rightarrow E$ should also be shortened by one day at the same time. A one-day reduction of AC costs $750 which is less than shortening CE by one day. A saving of $2,000 in indirect cost will result. This nine-day schedule is shown in Figure 12.15.

In the nine-day schedule, path $A \rightarrow B \rightarrow D \rightarrow E$ and path $A \rightarrow C \rightarrow E$ remain critical as in the ten-day schedule. Further equal time reduction should be sought in these critical paths.

An eight-day schedule is obtained by reducing AB by one day at a crash cost of $800 and BD by two days at a crash cost of $900 for a total of $1,700. This is better than a three-day reduction in activity BD. It is necessary to reduce CE and AC by one day each at a crash cost of $950 plus $750 or $1,700. This is better than a two-day reduction in CE. A saving of $3,000 is possible by this schedule. Three critical paths result: $A \rightarrow B \rightarrow D \rightarrow E$, $A \rightarrow C \rightarrow D \rightarrow E$, and $A \rightarrow C \rightarrow E$, each with a duration of 8 days. No further reductions are possible. The resulting schedule is shown in Figure 12.16.

Finally, it is necessary to find the optimum schedule. This is accomplished by summarizing the crash costs and the savings in Table 12.10. A maximum net saving occurs for a schedule of nine days.

TABLE 12.9 COMPUTATION OF CRASH COSTS

| ACTIVITY | NUMBER OF DAYS REDUCTION | ADDITIONAL LABOR COST | | | ADDITIONAL EQUIPMENT COST | TOTAL CRASH COST |
		ADDITIONAL WORKERS	WORKING DAYS	ADDITIONAL COSTS		
AB	1	2	2	$200	$ 600	$ 800
AC	1	3	3	450	300	750
BD	1	1	4	200	500	700
	2	3	3	450	450	900
	3	3	2	300	1,700	2,000
CD	0	0	2	0	0	0
CE	1	0	5	0	950	950
	2	4	4	800	1,110	1,900
DE	0	0	3	0	0	0

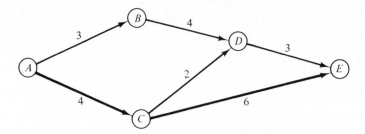

Figure 12.14   A ten-day schedule.

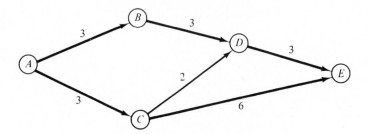

Figure 12.15   A nine-day schedule.

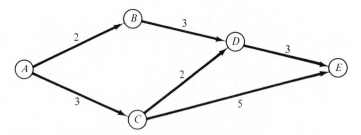

Figure 12.16   An eight-day schedule.

TABLE 12.10   NET SAVINGS FOR CRASH SCHEDULES

CRASH SCHEDULE	CRASH COST	INDIRECT COST SAVING	NET SAVING
10-day	$ 700	$1,000	$ 300
9-day	1,650	2,000	350
8-day	3,400	3,000	−400

Program Evaluation and Review Technique—PERT

PERT is ideally suited to early project planning where precise time estimates are not readily available. The advantages of PERT is that it offers a way of dealing with random variation, making it possible to allow for chance in the scheduling of activities. PERT may be used as a basis for computing the probability that the project will be completed on or before its scheduled date.

PERT network calculations.   A PERT network is illustrated in Figure 12.17. One starts with an end objective (e.g., event H in Figure 12.17) and works backward until event A is identified. Each event is labeled, coded, and checked in terms of program time frame. Activities are then identified and checked to ensure that they are properly sequenced. Some activities can be performed on a concurrent basis, and others must be accomplished in series. For each completed network, there is one beginning event and one ending event, and all activities must lead to the ending event.

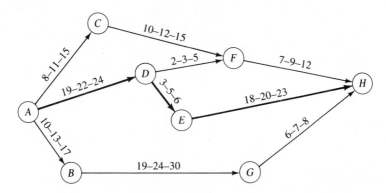

Figure 12.17   A PERT network with events and activity times.

The next step in developing a PERT network is to estimate activity times and to relate these times to the probability of their occurrences. An example of calculations for a typical PERT network is presented in Table 12.11. The following steps are required:

1. Column 1 lists each event, starting from the last event and working downward to the start.
2. Column 2 lists all previous events that have been indicated prior to the event listed in column 1.
3. Columns 3 to 5 give the optimistic (t_a), the most likely (t_b), and the pessimistic (t_c) times, in weeks or months, for each activity. Optimistic time means that there is very little chance that the activity can be completed before this time, while pessimistic time means that there is little likelihood that the activity will take longer. Time value (t) is defined as a random variable with a beta distribution and with range from a to b and mode m as shown in Figure 12.18.
4. Column 6 gives the expected or mean time (t_e) from

$$t_e = \frac{t_a + 4t_b + t_c}{6}. \qquad (12.7)$$

5. Column 7 gives the variance, σ^2, from

$$\sigma^2 = \left(\frac{t_c - t_a}{6}\right)^2. \qquad (12.8)$$

TABLE 12.11 EXAMPLE OF PERT CALCULATIONS

EVENT NUMBER (1)	PREVIOUS EVENTS (2)	t_a (3)	t_b (4)	t_c (5)	t_e (6)	σ^2 (7)	T_E (8)	T_L (9)	S (10)	PT (11)	PROB. (12)
H	G	6	7	8	7.0	0.108	46.8	46.8	0.0	46.0	0.266
	F	7	9	12	9.2	0.693					
	E	18	20	23	20.2	0.693					
G	B	19	24	30	24.2	3.349	37.7	39.8	2.1		
F	D	2	3	5	3.2	0.250	23.4	37.6	14.2		
	C	10	12	15	12.2	0.693					
E	D	3	5	6	4.8	0.250	26.6	26.6	0.0		
D	A	19	22	24	21.8	0.693	21.8	21.8	0.0		
C	A	8	11	15	11.2	1.369	11.2	25.4	14.2		
B	A	10	13	17	13.5	1.369	13.5	15.6	2.1		

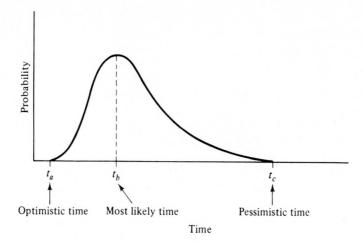

Figure 12.18   Beta distribution of activity times.

6. Column 8 gives the earliest expected time (T_E) as the sum of all times (t_e) for each activity and the cumulative total of the expected times through the preceding event, remaining on the same path throughout the network. When several activities lead to an event, the highest time value (t_e) is used. For example, path $A \rightarrow D \rightarrow E \rightarrow H$ in Figure 12.17 totals 46.8.

7. Column 9 gives the latest allowable time (T_L) calculated by starting with the latest time for the last event, which equals T_E (where T_E equals 46.8). Then one works backward, subtracting the value in column 6 for each activity, staying in the same path.

8. Column 10 gives the slack time (S) as the difference between the latest allowable time (T_L) and the earliest expected time (T_E) from

$$S = T_L - T_E.$$

9. Columns 11 and 12 give the required project time (PT) for the network. Assuming that PT is 46.0, it is necessary to determine the probability of meeting this requirement. This probability is determined from

$$Z = \frac{PT - T_L}{\sqrt{\Sigma \text{ variances}}}.$$

Z is the area under the normal distribution curve and the variance is the sum of the individual variances applicable to the critical path activities in Figure 12.17 (path $A \rightarrow D \rightarrow E \rightarrow H$). For this example,

$$Z = \frac{46.0 - 46.8}{\sqrt{0.693 + 0.250 + 0.693}} = -0.625.$$

From Appendix E, Table E.3, the calculated value of -0.625 represents an area of approximately 0.266; that is, the probability of meeting the scheduled time of 46 weeks is 0.266.

When evaluating the resultant probability value (column 12 of Table 12.11), management must decide on the range of factors allowable in terms of risk. If the probability factor is too low, additional resources may be applied to the project to reduce the activity schedule times and improve the probability of success. On the other hand, if the probability factor is too high, this may indicate that excess resources are being applied, some of which should be diverted to other projects.

PERT-cost.   An extension of PERT to include economic considerations brings cost factors into project control decisions. A cost network can be superimposed upon a PERT network by estimating the total cost and cost slope for each activity time.

When implementing PERT-cost, there is always the time-cost option which enables the decision maker to evaluate alternatives relative to the allocation of resources for activity accomplishment. In many instances time can be saved by applying more resources or, conversely, cost may be reduced by extending the time to complete an activity. The time-cost option can be attained by applying the following steps:

1. Determine alternative time and cost estimates for all activities and select the lowest-cost alternative in each instance.
2. Calculate the critical path time for the network. If the calculated value is equal to or less than the total time permitted, select the lowest-cost option for each activity and check to ensure that the total of the incremental activity times does not exceed the allowable overall program completion time. If the calculated value exceeds the program time permitted, review the activities along the critical path, select the alternative with the lowest cost slope, and then reduce the time value to be compatible with the program requirement.
3. After the critical path has been established in terms of the lowest-cost option, review all the network paths with slack time and shift activities to extend the times and reduce costs wherever possible. Activities with the steepest cost-time slopes should be addressed first.

The time-cost option can be illustrated by referring to two paths of the network presented in Figure 12.17. Data associated with each path are given in Table 12.12.

TABLE 12.12 TIME–COST OPTIONS

PATH $A \to D \to E \to H$				PATH $A \to C \to F \to H$			
Event Number	Previous Event	t_e	Estimated Activity Cost (\$)	Event Number	Previous Event	t_e	Estimated Activity Cost (\$)
H	E	20.2	3,500	H	F	9.2	1,050
E	D	4.8	940	F	C	12.2	2,350
D	A	21.8	4,800	C	A	11.2	1,870
Total		46.8	\$9,240	Total		32.6	\$5,370

In this example, path $A \rightarrow D \rightarrow E \rightarrow H$ is critical, and the alternative path ($A \rightarrow C \rightarrow F \rightarrow H$) has a slack of 14.2. If one is concerned with the critical path (since it is the governing factor in the overall network relative to program success or failure), the shifting of resources from the alternative path to critical-path activities should be considered. For instance, assume that $200 is reallocated from activity HF to activity HE. As a result, activity HF will now require 12.2 units of time, and activity HE will require 18.7 units of time. The total time along the critical path is reduced, and the project is completed in 45.3 units of time. On the other hand, the time along the alternative path is increased to 35.6 units, and the slack is reduced to 9.7 units. The entire network must now be reevaluated to determine the effects of the changes on the critical path. This process may be continued by trading off resources against time until the most acceptable result is obtained.

12.5. TOTAL QUALITY CONTROL

Total quality control (TQC) is a management concept being stimulated by the need to compete in global markets where higher quality, lower cost, and more rapid development are the keys to market leadership. Management for total quality is rapidly emerging as a means for improving the quality-cost characteristics of products, processes, and services.

Approaches now being promulgated emphasize that conformance to design specifications is necessary, but not sufficient. Quality losses begin to accumulate whenever a product parameter deviates from its nominal or optimal value. Accordingly, it is being recognized that quality must be designed into products and processes. Furthermore, optimal values for product and process parameters must be established and controlled throughout the life cycle.

Statistical Process Control

Statistical process control (SPC) was introduced in Section 12.2 and related to the process or operating system being controlled in Figure 12.1. Measurement is the key to evaluation in SPC. This measurement is applied to system output for the purpose of detecting a difference between actual values and ideal or target values.

By applying statistical control to a process properly centered on product specifications, considerable faith can be placed in the quality of the output product. Statistical control permits producers to identify and eliminate assignable causes of variation, while reducing random or unassignable causes. Furthermore, products produced under statistical control generally need not be subjected to sampling inspection. Sampling inspection seeks to find and eliminate defective items, a practice which is proving to be too disruptive and costly.

Statistical process control is based on the assumption of a stable process. If stable, one can measure a product characteristic that reflects the behavior of the process. The measured characteristic will have a steady-state distribution (in a probabilistic sense); that is, each sample of the product from the process will have the identical statistical distribution.

A stable process that yields product which satisfies specifications may later become unstable. Instability may arise from special causes. Continued monitoring of the process using statistical process control can detect the transition to an unstable state and the presence of special causes. A process that passes the statistical test is not always stable. However, the probability is very small that an unstable process will continue to produce output that passes the statistical tests (refer to Figure 12.2).

Charts are used in statistical process control for conducting stability tests. When the value of a characteristic can be measured, $\overline{X}$ and R charts are used; when the fraction of a two-valued characteristic is being measured, a p chart is appropriate; when the overall number of defects is being measured, the c chart is applicable. These were presented in Section 12.2 and are attributed to the work of Walter Shewhart in the 1920s.

The Shewhart approach to variation focuses more on process stability than on conformance to specifications. However, process control must not ignore specifications. Engineers should define specifications and statistical tolerances as part of the design process with the process capability in mind. A production process is said to have a process capability index, I, defined as

$$I = \frac{z}{3\sigma} \tag{12.9}$$

where $z = \min [(\overline{\overline{X}} - \text{lower specification limit})(\text{upper tolerance limit} - \overline{\overline{X}})]$ and $\overline{\overline{X}}$ is the grand average (from Equation 12.1). The process capability index is a measure of the ability of a process to produce quality product.

Although process control methods were first applied to manufacturing, their use is not restricted to production. Statistical process control is also used to evaluate and improve engineering processes. It is a means to obtain useful information to feed back into product and production system design. In addition to control charts, Pareto curves, cause and effect diagrams, CPM networks, and PERT are used to evaluate and improve engineering and production processes.

As real-time information becomes more readily available in design, production, and support, practitioners are beginning to measure not only product parameters but also changes in process control parameters. Such applications are closely related to control engineering and use many of its basic principles.

Experimental Design

Experimental design (ED) was first introduced in the 1920s by Sir Ronald Fisher. Its subsequent development led to gains in the efficiency of experimentation by changing factors, not one at a time, but together in a factorial design. Fisher introduced the concept of randomization (so that trends due to unknown disturbing factors would not bias results), the idea that a valid estimate of experimental error could be obtained from the design, and blocking to eliminate systematic differences introduced by using different lots.

Experimental designs were introduced into industry in the 1930s. At that time, the Industrial and Agricultural Section of the Royal Statistical Society was inaugurated in London and papers on applications to glass manufacturing, light bulb production, textile spinning, and the like were presented and discussed. This led to new statistical methods such as variance component analysis to reduce variation in textile production.

During the Second World War, the need for designs that could screen large numbers of factors led to the introduction of fractional factorial designs and other orthogonal arrays. These designs have been widely applied in industry and many successful industrial examples are described in papers and books dating from the 1950s. Also in the early 1950s, George Box developed response surface methods for the improvement and optimization of industrial processes by experimentation.

Modern engineering research and development involves considerable experimentation and analysis. Statistically designed experiments enable the engineer to achieve development objectives in less time with less resources. It is not unusual for development cycle time and cost to be reduced by one-half when experimental design methods are integrated into a concurrent engineering development process. When product quality and process design are addressed early in the life cycle, many decisions that affect product life, reliability, and producibility can be evaluated before the product is produced.

Parameter Design

Parameter design (PD) is based on the idea of controllables and uncontrollables in design and operations as introduced in Chapter 7 (refer to Section 7.3). Parameter optimization involves the selection of controllable values in the product and the processes of manufacture and maintenance, so that some measure of merit will be improved (or optimized if possible).

Parameter optimization is traditionally used when the goal is to improve some performance measure. It is now being applied to variability reduction for increased quality and lower cost. Although parameter design to reduce variability may be accomplished by a variety of methods, there is one technique which is gaining recognition. This technique is attributable to Genichi Taguchi.

Taguchi uses the term *Quality Engineering* (QE) to describe an approach to achieving both improved quality and low cost. He identifies two methods of dealing with variability: parameter optimization and tolerance design. The first attempts to minimize the affects of variation and the second seeks to remove its causes.

There have been many examples of the successful application of parameter optimization. The most widely reported is the case of tile manufacturing in Japan. Similar improvements have been demonstrated in automobile manufacturing, electronic component production, computer operations, IC chip bonding processes, ultrasonic welding, and the design of disc brakes and engines.

The concept of a quality loss function is used in parameter optimization to capture the effect of variability. This function represents the loss to society resulting from variability of function as well as from harmful side effects during the production and use of a product. Loss is calculated from the view of society with an emphasis on the viewpoint of the customer. For example, the loss to an automobile owner when a part fails includes the cost of repair, the cost of lost wages, and the inconvenience of not having use of the car.

In many applications the quality loss can be approximated by a quadratic function of the important design parameter. The function is assumed to be analytic, and so it can be expanded in a Taylor series about some known optimal value T. Then the first two terms of the expansion can be assumed to be zero and the principal term becomes

$$L(y) = (y - T)^2 \frac{L''(T)}{2!}$$

or

$$L(y) = k(y - T)^2$$

where

> $L(y)$ is the loss as a function of y
> T is the optimum or target value for parameter y
> $L''(T)$ is the second derivative of L evaluated at T
> k is the loss coefficient for the particular application.

Taguchi used the loss function to measure the effect of variation whether or not the specification is satisfied. He also used the orthogonal experimental design for efficiently evaluating the effect of individual parameter settings in the face of noise.

Quality Engineering

Quality engineering as promulgated by Taguchi involves the following activities:

1. *Plan an experiment*—Identify the main function, side effects, and failure modes. Identify noise factors and testing conditions for evaluating quality loss, the quality characteristic to be observed, the quality function to be optimized, and the controllable parameters and their most likely settings.
2. *Perform the experiment*—Conduct a limited number of experiments that simultaneously vary all of the parameters according to the pattern of an appropriate orthogonal matrix. Collect data on the results of the experiments.
3. *Analyze and verify the results*—Analyze the data, determine the optimum parameter settings and predict the performance under these settings. Conduct a verification experiment to confirm the results obtained at the optimum settings.

As with statistical process control, the Taguchi method relies on a hypotheses. Part of the evaluation includes a series of tests to verify that the hypothesis should not be rejected based on experimental evidence. To compare the effects of changing the values of different parameters, Taguchi introduces the signal-to-noise ratio measure embracing a simple addition of the contribution of individual parameter settings. This additive property is the hypothesis to be tested.

The relationship of Taguchi parameter design and the design dependent parameter approach introduced in Equation 7.2 is illustrated in Figure 12.19. A summary of the relationship follows.

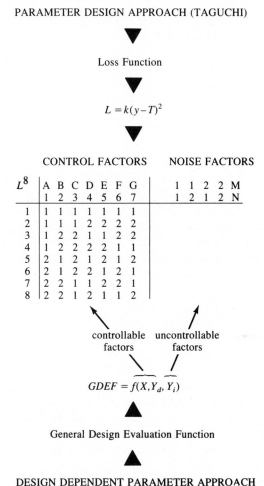

Figure 12.19   Parameter design and design dependent parameter relationships.

1. Parameter Design Approach (Taguchi).

    a. *Loss function*—Quadratic in the deviation of actual parameter value from target value and related to the specific design by k.

    b. *Experimental design*—A factorial experiment relating control factors at two levels to noise factors at two or more levels.

2. Design Dependent Parameter Approach (Equation 7.2).

   a. *Design evaluation function*—A life cycle cost function optimized on design variables for each set of design dependent parameters.
   b. *Indirect experimentation*—The use of modeling, Monte Carlo sampling, and sensitivity analysis to evaluate the LCC penalty for various combinations of design dependent parameters in the face of variation in design independent parameters.

Design dependent parameter approaches, as introduced in Chapter 7 and subsequently illustrated, and parameter design (Taguchi style) have much in common. Taguchi would call the concurrent design process and the design evaluation function approach *system design*. System design involves innovation and knowledge from the science and engineering fields. It includes the selection of materials, parts, and product parameter values in the product design stage (Figure 2.3, top life cycle) and the selection of production equipment and values for process factors in the process design stage (Figure 2.3, middle life cycle).

Tentative nominal values for design dependent parameters are then tested over specified ranges. This step is Taguchi's parameter design. Its purpose it to determine the best combination of parameter levels or values. Parameter design determines product parameter values that are least sensitive to joint changes in design independent parameter noise and environmental noise factors.

Indirect experimentation through modeling and simulated sampling is a means of determining the best combination of levels for design dependent parameters. The loss function best utilized is the design evaluation function of Equation 7.2. Taguchi parameter design is an approach yet to be adopted to conceptual and preliminary design. Since the product and the production process does not yet exist, *ED* approaches are not feasible except on surrogate systems.

Tolerance design, as defined by Taguchi, is expensive and, therefore, is to be minimized. It is in this area that SPC can be most helpful. The objective, however, is to define design dependent parameters in such a way that production, operations, and support capabilities are not adversely affected by design dependent parameter drift. The objective of the Taguchi and the design dependent parameter approaches is to achieve robustness against noise factors, so that life cycle cost will be minimized.

Total Quality Management

An overarching goal of the systems engineering process is to create quality systems and products in response to established needs. In the defense sector, this means deploying weapon systems which are competitive in both performance and life-cycle cost. In the nondefense sector, this means creating products that will compete in the world marketplace. Total quality management (TQM) may be the key.

The advent of computer-aided design technologies has led to the emergence of computer-aided concurrent engineering design (CACED) as a means by which a competitive advantage in defense and the world market may be retained. This engineering approach uses computer-aided methods to continually assess the anticipated quality of a product during the early design stage and throughout its entire life cycle.

The intrinsic quality of a deployed system or product is traceable to design features established during the design process. These features manifest themselves as design dependent parameters which have desired target values. Computer-aided concurrent engineering design focuses of these parameters and provides a basis for the development of products and systems which are robust to quality loss. By incorporating parameter design ideas attributed to Taguchi and process control approaches originated by Shewhart, the goal of total quality may be advanced.

SELECTED REFERENCES

(1) Banks, J., *Principles of Quality Control.* New York: John Wiley, 1989.

(2) Box, G. E., and N. R. Draper, *Empirical Model Building and Response Surfaces.* New York: John Wiley, 1987.

(3) Deming, W. E., *Out of the Crisis.* Cambridge: MIT Press, 1986.

(4) Fabrycky, W. J., Ghare, P. M., and P. E. Torgersen, *Applied Operations Research and Management Science.* Englewood Cliffs, N.J.: Prentice-Hall, 1984.

(5) Glasser, A., *Research and Development Management.* Englewood Cliffs, N.J.: Prentice-Hall, 1982.

(6) Juran, J. M., *Quality Control Handbook.* New York: McGraw-Hill, 1962.

(7) Montgomery, D. C., *Design and Analysis of Experiments.* New York: John Wiley, 1984.

(8) Singh, M. G., ed., *Systems and Control Encyclopedia: Theory, Technology, Applications.* Elmsford, N.Y.: Pergamon Press, 1989.

(9) Taguchi, G., E. A. Elsayed, and T. C. Hsiang, *Quality Engineering in Production Systems.* New York: McGraw-Hill, 1989.

QUESTIONS AND PROBLEMS

1. Speed is one characteristic or condition of an automobile which must be controlled. Discuss the role of each element of a control system for speed control.

2. Sketch a diagram such as Figure 12.1 which would apply to a thermostatically controlled heating system. Discuss each block.

3. Give an example of open-loop control; closed-loop control.

4. What is the relationship among an unstable pattern of variation, control limits, and a Type I error? a Type II error?

5. Samples of $n = 10$ were taken from a process for a period of time. The process average was estimated to be $\bar{X} = 0.0250$ in. and the process range was estimated as $\bar{R} = 0.0020$ in. Specify the control limits for an X chart and for an R chart.

6. Control charts by variables are to be established on the tensile strength in pounds of a yarn. Samples of five have been taken each hour for the past 20 hours. These were recorded as shown in the table.

									HOUR										
1	2	3	4	5	6	7	8	9	10	11	12	13	14	15	16	17	18	19	20
50	44	44	48	47	47	44	52	44	43	47	49	47	43	44	45	45	50	46	48
51	46	44	52	46	44	46	46	46	44	44	48	51	46	43	47	45	49	47	44
49	50	44	49	46	43	46	45	46	49	44	41	50	46	40	51	47	45	48	49
42	47	47	49	48	40	48	42	46	47	42	46	48	48	40	48	47	49	46	50
43	48	48	46	50	45	46	55	43	45	50	46	42	46	46	46	46	48	45	46

(a) Construct an $\bar{X}$ chart based on these data.
(b) Construct an R chart based on these data.

7. A lower specification limit of 42 lbs is required for the condition of Problem 6. Sketch the relationship between the specification limit and the control limits. What proportion, if any, of the yarn will be defective?

8. The total number of accidents during the long weekends and the number of fatalities for a 10-year period are given in the table. Assuming that the process is in control with regard to the proportion of fatalities, construct a p chart and record the data of the last eight weekends.

WEEKEND	NUMBER OF ACCIDENTS	NUMBER OF FATALITIES	WEEKEND	NUMBER OF ACCIDENTS	NUMBER OF FATALITIES
1	2,378	426	16	3,943	523
2	3,375	511	17	3,950	557
3	3,108	498	18	4,358	536
4	3,756	525	19	4,217	533
5	3,947	564	20	3,959	547
6	2,953	475	21	4,108	554
7	3,075	490	22	4,379	579
8	3,173	504	23	4,455	598
9	3,479	528	24	4,753	585
10	3,545	555	25	4,276	543
11	3,865	537	26	3,868	507
12	3,747	529	27	3,947	523
13	4,011	569	28	3,665	575
14	3,108	470	29	4,078	569
15	3,207	510	30	4,025	578

9. During a four-week inspection period, the number of defects listed in the table were found in a sample of 400 electronic components. Construct a c chart for these data. Does it appear as though there existed an assignable cause of variation during the inspection period?

DATE	NUMBER OF DEFECTS	DATE	NUMBER OF DEFECTS
1	7	11	6
2	8	12	8
3	9	13	16
4	8	14	2
5	3	15	4
6	9	16	2
7	5	17	6
8	6	18	5
9	15	19	3
10	9	20	7

10. A survey during a safety month showed the number of defective cars on a highway as given in the table. A sample of 100 cars was taken during each day. Construct a c chart for the data. Is there any assignable cause of variation during this period?

DATE	DEFECTIVE CARS	DATE	DEFECTIVE CARS	DATE	DEFECTIVE CARS
1	12	11	9	21	9
2	15	12	7	22	8
3	13	13	13	23	17
4	9	14	12	24	16
5	14	15	6	25	13
6	17	16	18	26	18
7	8	17	15	27	12
8	21	18	5	28	9
9	12	19	14	29	11
10	14	20	12	30	19

11. One operator is assigned to run 14 automatic machines. The minimum-cost service factor is 0.2. The machines require attention at random, and the service duration is distributed exponentially. Machines receive service on a first-come, first-served basis. Specify the upper control limit if the probability of designating the system out of control when it is really in control must not exceed 0.05.

12. Find the critical path in the network shown.

13. Eleven activities and eight events constitute a certain research and development project. The activities and their expected completion times are as follows:

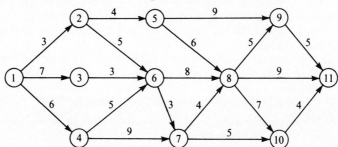

ACTIVITY	COMPLETION TIME IN WEEKS
$A \rightarrow B$	5
$A \rightarrow C$	6
$A \rightarrow D$	3
$B \rightarrow E$	10
$B \rightarrow F$	7
$C \rightarrow E$	8
$D \rightarrow E$	2
$E \rightarrow F$	1
$E \rightarrow G$	2
$F \rightarrow H$	5
$G \rightarrow H$	6

(a) Represent the project in the form of an activity-event network.

(b) Calculate T_E and T_L for each event and label each with T_E, T_L, and S.

(c) Identify the critical path and calculate the shortest possible time for project completion.

14. In Problem 13, assume that event C must occur before event D occurs. Modify the activity-event network for this assumption and show how the critical path would be changed.

15. Nine activities and six events are required to execute and complete a certain construction program. The activities and their completion times in weeks under normal and under expedited conditions are as follows:

	NORMAL		EXPEDITED	
Activity	Duration	Cost	Duration	Cost
AB	10	$3,000	8	$ 5,000
AC	5	2,500	4	3,600
AD	2	1,100	1	1,200
BC	6	3,000	3	12.000
BE	4	8,500	4	8,500
CE	7	9,800	5	10,200
CF	3	2,700	1	3,500
DF	2	9,200	2	9,200
EF	4	300	1	4,600

A linear crash cost-time relationship exists between the normal and expedited conditions. Overhead of $11,000 per week can be saved if the program is completed earlier than the normal schedule.

(a) Represent the program in the form of an activity-event network.

(b) Find the earliest time within which the program can be completed under normal conditions.

(c) Find the minimum cost schedule for the program.

16. Eight maintenance activities, their normal durations in weeks, and the crew sizes for the normal condition are as follows:

ACTIVITY	DURATION IN WEEKS	CREW SIZE
AC	6	8
BC	6	7
BE	2	6
CD	3	2
CE	4	1
DF	7	4
EF	4	6
FG	5	10

(a) Represent the maintenance project in the form of an activity-event network.

(b) Identify the critical path and find the minimum time for project completion.

(c) Extra crew members can be used to expedite activities *BC*, *CE*, *DF*, and *EF* at a cost given below:

EXTRA CREW	WEEKS SAVED	CRASH COST PER WEEK
1	1	$100
2	2	120
3	2	200
4	3	250

If there is a penalty cost of $1,250 per week beyond the minimum maintenance time of 17 weeks, recommend the minimum cost schedule.

17. Determine the critical path of the PERT network shown. What is the second most critical path?

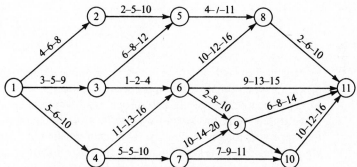

18. Calculate the probability of meeting a scheduled time of 50 units for the PERT network shown.

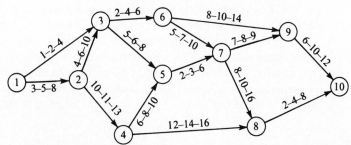

PART IV: DESIGN FOR OPERATIONAL FEASIBILITY

13

Design for Reliability

System design and development is accomplished through the systems engineering process, which requires the appropriate application of scientific and engineering efforts to ensure that the ultimate product is operationally feasible. Operational feasibility implies that the system will perform as intended in an effective and efficient manner for as long as necessary. The accomplishment of such requires the proper integration of design-related specialties such as reliability, maintainability, human factors, and so on, into the total engineering design effort. These design-related specialties are presented in this and the next chapters.

The first and one of the most significant design specialties requiring attention is *reliability*. Many systems (or products) in use today are highly sophisticated and will fulfill most expectations when operating. However, experience has indicated that these systems are inoperative much of the time, requiring extensive maintenance and the expenditure of scarce support resources. Unreliable systems are unable to fulfill the mission for which they were designed. In an environment of scarce resources, it is essential that reliability be considered as a major system parameter during the design process. Reliability is a characteristic inherent in design.

13.1. DEFINITION AND EXPLANATION OF RELIABILITY

Reliability can be defined simply as the probability that a system or product will perform in a satisfactory manner for a given period of time when used under specified operating conditions. This definition stresses the elements of *probability, satisfactory performance, time,* and *specified operating conditions.* These four elements are very important, since each plays a significant role in determining system or product reliability.

Probability, the first element in the reliability definition, is usually stated as a quantitative expression representing a fraction or a percent specifying the number of times that one can expect an event to occur in a total number of trials. For instance, a statement that the probability of survival of an item for 80 hours is 0.75 (or 75%) indicates that one can expect the item to function properly for at least 80 hrs 75 times out of 100. When there are a number of supposedly identical items operating under similar conditions, it can be expected that failures will occur at different points in time; thus, failures are described in probabilistic terms. The fundamental definition of reliability is heavily dependent on the concepts derived from probability theory.

The second element in the reliability definition is *satisfactory performance,* indicating that specific criteria must be established which describe what is considered to be satisfactory. A combination of qualitative and quantitative factors defining the functions that the system or product are to accomplish, usually presented in the context of a system specification, are required. These factors cover system operational requirements as identified in Section 3.1.

The third element, *time,* is one of the most important because it represents a measure against which the degree of system performance can be related. One must know the time parameter in order to assess the probability of completing a mission or a given function as scheduled. Of particular interest is the ability to predict the probability of an item surviving (without failure) for a designated period of time. Also, reliability is frequently defined in terms of mean time between failure (MTBF) or mean time to failure (MTTF), making time critical in reliability measurement.

The *specified operating conditions* under which a system or product is expected to function constitute the fourth significant element of the reliability definition. These conditions include environmental factors, such as the geographical location where the system is expected to operate, the operational profile, temperature cycles, humidity, vibration, shock, and so on. Such factors must not only address the conditions during the period when the system or product is operating, but during the periods when the system (or a portion thereof) is in a storage mode or being transported from one location to the next. Experience has indicated that the transportation, handling, and storage modes are sometimes more critical from a reliability standpoint than are the conditions experienced during actual system operational use.

The four elements discussed above are critical in determining the reliability of a system or product. As reliability is an inherent characteristic of design, it is essential that these elements be adequately considered at program inception. Thus, reliability must be a major factor throughout the systems engineering process, particularly during conceptual design.

13.2. RELIABILITY IN THE SYSTEM LIFE CYCLE

Reliability, as an inherent characteristic of design, must be an integral part of the overall systems engineering process illustrated in Figure 2.4. Reliability requirements are defined in conceptual design, reliability analyses and predictions are accomplished throughout preliminary and detail system or product design, reliability is considered in formal design reviews, and reliability testing is accomplished as part of system test and evaluation. Thus, reliability (along with other major design parameters) is considered throughout the system life cycle, and is particularly relevant during the early phases of system design and development. Specific reliability functions as related to the system life cycle are identified in Figure 13.1.

Referring to the figure, reliability quantitative and qualitative requirements are initially defined as part of conceptual design in Block 1. Such requirements may be specified in forms of a probability of survival (or probability of success), mean time between failure (MTBF), failure rate (λ), or some equivalent figure of merit. Qualitative goals relating to the selection and application of component parts, environmental stress conditions, and so on, may also be specified. These reliability requirements are integrated with the definition of operational requirements and the maintenance concept for the overall system, as is described in Chapter 3.

Concurrently, reliability planning must be accomplished as an integral part of the overall system planning effort. Specific program functions are identified and scheduled, the organization and implementation of reliability tasks are accomplished, and the results are evaluated through normal program review functions. The objective is to plan a program effort that will assure reliability involvement throughout all aspects of system design and development, production or construction, and system utilization. A reliability program plan is usually prepared at program inception and may be included as part of the system engineering management plan discussed in Part Five. A typical reliability plan should consider the following topics:

1. Reliability quantitative and qualitative requirements for the system (as related to system operational requirements and the maintenance concept discussed in Chapter 3).
2. Allocation or apportionment of reliability requirements to the subsystem level and beyond (as appropriate).
3. Design techniques and practices (i.e., component parts selection and control, derating, redundancy, etc.).
4. Reliability analysis to include the generation of block diagrams, mathematical models, stress-strength analysis, worst-case analysis, sneak-circuit analysis, and so on.
5. Failure mode, effect, and criticality analysis (FMECA).
6. Reliability prediction(s) and assessment.
7. Effects of storage, packaging, transportation, handling, and maintenance.

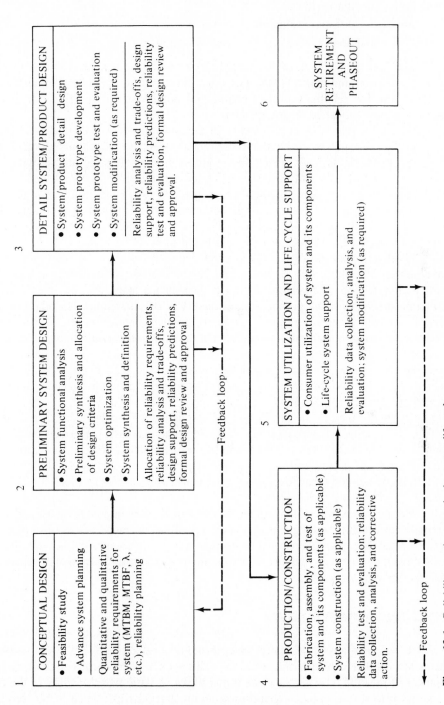

Figure 13.1 Reliability requirements in the system life cycle.

8. Formal design review(s).

9. Reliability test and evaluation.

10. Data collection, analysis, and corrective action.

The reliability plan must also include the essential management aspects of project organization and control. Not only must internal procedures be thoroughly delineated, but the relationships and contractual provisions involving suppliers (and supplier control) must be clearly defined.

Given that reliability requirements for the system have been defined, a series of activities are initiated to ensure that these requirements are indeed incorporated into the system design. Basically, reliability requirements are allocated and design criteria are established; reliability analyses and trade-off studies are accomplished to support major design decisions; predictions are made to assess the design configuration at various stages in the development process; and test and evaluation is conducted to measure the results of the design effort. These activities are represented in Blocks 2 and 3 of Figure 13.1 and are described in detail in Sections 13.4 and 13.5.

If reliability is successfully incorporated into the design, one may intuitively assume that all will be well. However, this can be a dangerous assumption, since the best design can be significantly degraded through the subsequent production/construction processes if the proper level of quality is not maintained. In other words, the quality control function must be properly structured and emphasized throughout production and/or construction to ensure that the inherent system reliability, demonstrated in qualification testing, is maintained (i.e., each item produced should possess the same reliability characteristics as verified through reliability qualification). Thus, reliability and quality control functions must be carefully planned and integrated at this stage.

Of significance in the production process is the accomplishment of reliability acceptance testing. This may include periodic sequential tests performed on a sampling basis, individual component reliability tests involving critical items, or some equivalent form of test. In any event, the objective is to ensure that the specified MTBF is maintained throughout production/construction (refer to Block 4 of Figure 13.1).

Finally, when the system becomes operational through consumer use, it is essential that a "true" assessment of system reliability be made. This assessment of system reliability in an operational environment is best accomplished through the establishment of an effective data collection, analysis, and system evaluation capability as described in Section 6.4.

13.3. MEASURES OF RELIABILITY

The evaluation of any system or product in terms of reliability is based on precisely defined reliability concepts and measures. This section is concerned with the development of selected reliability measures and terms. A basic understanding of these is required prior to discussing reliability program functions as related to the system/ product design. Terms such as the reliability function, failure rate, probability density function, reliability model, and so on, are briefly defined below.

The Reliability Function

The *reliability function,* also known as the *survival function,* is determined from the probability that a system (or product) will be successful at least for some specified time t. The reliability function, $R(t)$, is defined as

$$R(t) = 1 - F(t) \tag{13.1}$$

where $F(t)$ is the probability that the system will fail by time t. $F(t)$ is basically the *failure distribution function* or *unreliability function.* If the random variable t has a density function of $f(t)$, the expression for reliability is

$$R(t) = 1 - F(t) = \int_t^\infty f(t) \, dt. \tag{13.2}$$

If the time to failure is described by an exponential density function, then, from Chapter 10, Equation 10.7

$$f(t) = \frac{1}{\theta} e^{-t/\theta} \tag{13.3}$$

where θ is the mean life and t the time period of interest. The reliability at time t is

$$R(t) = \int_t^\infty \frac{1}{\theta} e^{-t/\theta} \, dt = e^{-t/\theta}. \tag{13.4}$$

Mean life (θ) is the arithmetic average of the lifetimes of all items considered, which for the exponential function is MTBF. Thus,

$$R(t) = e^{-t/M} = e^{-\lambda t} \tag{13.5}$$

where λ is the instantaneous failure rate and M the MTBF.

If an item has a constant failure rate, the reliability of that item at its mean life is approximately 0.37. Thus, there is a 37% probability that a system will survive its mean life without failure. Mean life and failure rate are related as

$$\lambda = \frac{1}{\theta}. \tag{13.6}$$

Figure 13.2 illustrates the exponential reliability function where time is given in units of t/M. The illustration presented focuses on the reliability function for the exponential distribution, which is commonly used in many applications.

Actually, the failure characteristics of different items are not necessarily the same. There are a number of well-known probability distribution functions which have been found in practice to describe the failure characteristics of different equipments. These include the binomial, exponential, normal, Poisson, gamma, and Weibull distributions. Thus, one should not assume that any one distribution is applicable in all instances.

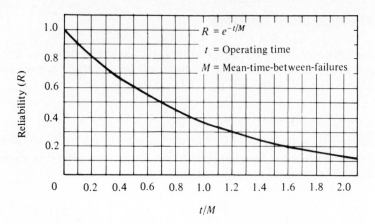

Figure 13.2   Reliability curve for the exponential distribution.

The Failure Rate

The rate at which failures occur in a specified time interval is called the *failure rate* for that interval. The failure rate per hour is expressed as

$$\lambda = \frac{\text{number of failures}}{\text{total operating hours}}. \tag{13.7}$$

The failure rate may be expressed in terms of failures per hour, percent failures per 1,000 hrs, or failures per million hours. As an example, suppose that 10 components were tested for 600 hours under specified operating conditions. The components (which are not repairable) failed as follows: component 1 failed after 75 hours, component 2 failed after 125 hours, component 3 failed after 130 hours, component 4 failed after 325 hrs, and component 5 failed after 525 hours. Thus, there were 5 failures and the total operating time was 4,180 hrs. Using Equation 13.7, the calculated failure rate per hour is

$$\lambda = \frac{5}{4,180} = 0.001196.$$

As a second example, suppose that the operating cycle for a given system is 169 hrs as illustrated in Figure 13.3. During that time, 6 failures occur at the points indicated. A failure is defined as an instance when the system is not operating within a specified set of parameters. The failure rate, or corrective maintenance frequency, per hour is

$$\lambda = \frac{\text{number of failures}}{\text{total mission time}} = \frac{6}{142} = 0.04225.$$

Assuming an exponential distribution, the system mean life or the mean time between failure (MTBF) is

$$\text{MTBF} = \frac{1}{\lambda} = \frac{1}{0.04225} = 23.6686 \text{ hrs.}$$

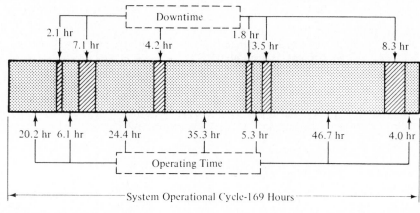

Figure 13.3 A system operational cycle.

Figure 13.4 presents a reliability nomograph (for the exponential failure distribution) which facilitates calculations of MTBF, λ, $R(t)$, and operating time. If the MTBF is 200 hours ($\lambda = 0.005$), and the operating time is 2 hours, the nomograph gives a reliability value of 0.99.

When determining the failure rate, particularly with regard to estimating corrective maintenance actions (i.e., the frequency of corrective maintenance), one must address all system failures to include failures due to primary defects, failures due to manufacturing defects, failures due to operator errors, and so on. A combined failure rate is presented in Table 13.1.

When assuming the negative exponential distribution, the failure rate is considered to be relatively constant during normal system operation if the system design is mature. That is, when equipment is produced and the system is initially distributed for operational use, there are usually a higher number of failures due to component variations and mismatches, manufacturing processes, and so on. The initial failure rate is higher than anticipated, but gradually decreases and levels off, as illustrated in Figure 13.5, during the *debugging* period. Similarly, when the system reaches a certain age, there is a *wear-out* period when the failure rate increases. The relatively level portion of the curve in Figure 13.5 is the constant failure-rate region, where the exponential failure law applies.

Figure 13.5 illustrates certain relative relationships. Actually, the curve may vary considerably depending on the type of system and its operational profile. Further, if the system is continually being modified for one reason or another, the failure rate may not be constant. In any event, the illustration does provide a good basis for considering relative failure-rate trends.

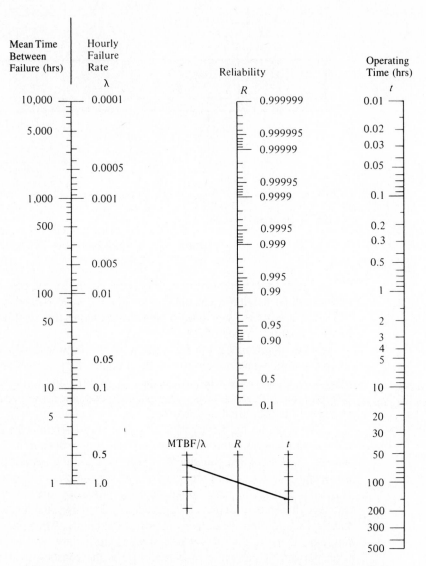

Figure 13.4 Reliability nomograph for the exponential failure distribution. Given equipment mean time to failure or hourly failure rate and operating time, solve for reliability. Connect MTBF and *t* values with straight line. Read *R*. *Source:* NAVAIR 00-65-502/NAVORD OD 41146, *Reliability Engineering Handbook,* Naval Air Systems Command and Naval Ordnance Systems Command, Revised March 1968.

TABLE 13.1 DETERMINING THE COMBINED FAILURE RATE

CONSIDERATION	ASSUMED FACTOR (INSTANCES/HR)
a. Inherent reliability failure rate	0.000392
b. Manufacturing defects	0.000002
c. Wear-out rate	0.000000
d. Dependent failure rate	0.000072
e. Operator-induced failure rate	0.000003
f. Maintenance-induced failure rate	0.000012
g. Equipment damage rate	0.000005
Total combined factor	0.000486

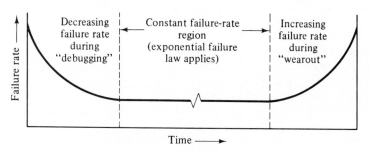

Figure 13.5   Typical failure-rate curve.

Reliability Component Relationships

Given the basic reliability function and the measures associated with failure rate, it is appropriate to consider their application in series networks, parallel networks, and combinations thereof. These networks are used in reliability block diagrams and in static models employed for reliability prediction and analysis.[1]

   Series networks.   The series relationship is probably the most commonly used and is the simplest to analyze. It is illustrated in Figure 13.6. In a *series network* all components must operate in a satisfactory manner if the system is to function properly. Assuming that a system includes subsystem *A*, subsystem *B*, and subsystem *C*, the reliability of the system is the product of the reliabilities for the individual subsystems expressed as

$$R = (R_A)(R_B)(R_C). \tag{13.8}$$

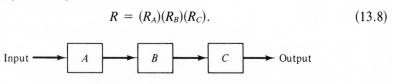

Figure 13.6   A series network.

[1] The development of reliability block diagrams should evolve directly from the functional flow diagrams for the system as described in Section 4.1 (refer to Figure 4.7).

As an example, suppose that an electronic system includes a transmitter, a receiver, and a power supply. The transmitter reliability is 0.8521, the receiver reliability is 0.9712, and the power supply reliability is 0.9357. The overall reliability for the electronic system is

$$R_s = (0.8521)(0.9712)(0.9357) = 0.7743.$$

If a series system is expected to operate for a specified time period, its required overall reliability can be derived. Substituting Equation 13.5 into Equation 13.8 gives

$$R_s = (e^{-\lambda_1 t})(e^{-\lambda_2 t}) \cdots (e^{-\lambda_n t})$$

for a series with n components. This may be expressed as

$$R_s = e^{-(\lambda_1 + \lambda_2 + \cdots + \lambda_n)t}. \tag{13.9}$$

Suppose that a series system consists of four subsystems and is expected to operate for 1,000 hours. The four subsystems have the following MTBFs: subsystem A, MTBF = 6,000 hours; subsystem B, MTBF = 4,500 hours; subsystem C, MTBF = 10,500 hours; subsystem D, MTBF = 3,200 hours. The objective is to determine the overall reliability of the series network, where

$$\lambda_A = \frac{1}{6,000} = 0.000167 \text{ failure/hr}$$

$$\lambda_B = \frac{1}{4,500} = 0.000222 \text{ failure/hr}$$

$$\lambda_C = \frac{1}{10,500} = 0.000095 \text{ failure/hr}$$

$$\lambda_D = \frac{1}{3,200} = 0.000313 \text{ failure/hr}$$

The overall reliability of the series network is found from Equation 13.9 as

$$R = e^{-(0.000797)(1,000)} = 0.4507.$$

This means that the probability of the system surviving (i.e., reliability) for 1,000 hrs is 45.1%. If the requirement were reduced to 500 hrs, the reliability would increase to about 67%.

Parallel networks.   A pure *parallel network* is one where a number of the same components are in parallel and where all the components must fail in order to cause total system failure. A parallel network with two components is illustrated in Figure 13.7. Assuming that components A and B are identical, the system will function if either A or B, or both, are working. The reliability is expressed as

$$R = R_A + R_B - (R_A)(R_B). \tag{13.10}$$

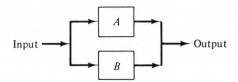

Figure 13.7   A parallel network.

Consider next a network with three components in parallel, as shown in Figure 13.8. The network reliability is expressed as

$$R = 1 - (1 - R_A)(1 - R_B)(1 - R_C). \tag{13.11}$$

If components A, B, and C are identical, the reliability expression can be simplified to

$$R = 1 - (1 - R)^3$$

for a system with three parallel components. For a system with n identical components, the reliability is

$$R = 1 - (1 - R)^n. \tag{13.12}$$

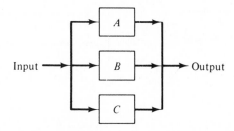

Figure 13.8   Parallel network with three components.

Parallel redundant networks are used primarily to improve system reliability as Equations 13.11 and 13.12 indicate mathematically. Consider the following example. A system includes two identical subsystems in parallel and the reliability of each subsystem is 0.95. The reliability of the system is found from Equation 13.10 as

$$R = 0.95 + 0.95 - (0.95)(0.95) = 0.9975.$$

Suppose that the reliability of the system above needs improvement beyond 0.9975. By adding a third identical subsystem in parallel the system reliability is found from Equation 13.11 to be

$$R = 1 - (1 - 0.95)^3 = 0.999875.$$

Note that there is a reliability improvement of 0.002375 over the previous configuration.

If the subsystems are not identical, Equation 13.10 can be used. For example, a parallel redundant network with two subsystems, $R_A = 0.75$ and $R_B = 0.82$, gives a system reliability of

$$R = 0.75 + 0.82 - (0.75)(0.82) = 0.955.$$

Combined Series-Parallel Networks. Various levels of reliability can be achieved through the application of a combination of series and parallel networks. Consider the three examples illustrated in Figure 13.9.

The reliability of the first network in Figure 13.9 is given by

$$R = R_A(R_B + R_C - R_B R_C). \tag{13.13}$$

For the second network the reliability is given by

$$R = [1 - (1 - R_A)(1 - R_B)][1 - (1 - R_C)(1 - R_D)]. \tag{13.14}$$

And for the third network the reliability is given by

$$R = [1 - (1 - R_A)(1 - R_B)(1 - R_C)] \times [R_D] \times [R_E + R_F - (R_E)(R_F)]. \tag{13.15}$$

Combined series-parallel networks such as those in Figure 13.9 require that the analyst first evaluate the redundant elements to obtain unit reliability. Overall system reliability is then determined by finding the product of all unit reliabilities.

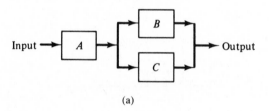

(a)

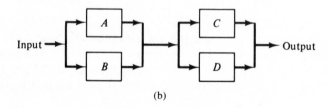

(b)

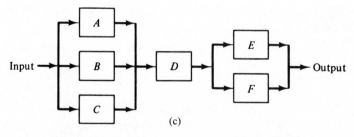

(c)

Figure 13.9 Some combined series-parallel networks.

Related Factors

There are several factors which may not be directly categorized as reliability figures-of-merit, but which are closely related. Some of these are availability, effectiveness, mean time between maintenance, and others, as described in the following paragraphs.

Availability (A).   Availability may be expressed and defined in three ways.

1. *Inherent Availability* (A_i)—Inherent availability is "the probability that a system or equipment, when used under stated conditions in an *ideal* support environment (i.e., readily available tools, spares, maintenance personnel, etc.), will operate satisfactorily at any point in time as required." It excludes preventive or scheduled maintenance actions, logistics delay time, and administrative delay time, and is expressed as

$$A_i = \frac{\text{MTBF}}{\text{MTBF} + \overline{\text{Mct}}} \qquad (13.16)$$

where MTBF is the mean time between failure and $\overline{\text{Mct}}$ the mean corrective maintenance time. The term $\overline{\text{Mct}}$ is equivalent to the mean time to repair (MTTR).

2. *Achieved Availability* (A_a)—Achieved availability is "the probability that a system or equipment, when used under stated conditions in an *ideal* support environment (i.e., readily available tools, spares, personnel, etc.), will operate satisfactorily at any point in time," This definition is similar to the definition for A_i except that preventive (i.e., scheduled) maintenance is included. It excludes logistics delay time and administrative delay time, and is expressed as

$$A_a = \frac{\text{MTBM}}{\text{MTBM} + \overline{\overline{\text{M}}}} \qquad (13.17)$$

where MTBM is the mean time between maintenance and $\overline{\text{M}}$ the mean active maintenance time. MTBM and $\overline{\text{M}}$ are functions of corrective (unscheduled) and preventive (scheduled) maintenance actions and times, respectively.

3. *Operational Availability* (A_0)—Operational availability is the "probability that a system or equipment, when used under stated conditions in an *actual* operational environment, will operate satisfactorily when called upon." It is expressed as

$$A_0 = \frac{\text{MTBM}}{\text{MTBM} + \text{MDT}} \qquad (13.18)$$

where MDT is the mean maintenance down time. The reciprocal of MTBM is the frequency of maintenance, which in turn is a significant factor in determining logistic support requirements. MDT includes active maintenance time ($\overline{M}$), logistics delay time, and administrative delay time.[2]

The term *availability* is used differently in different situations. If one is to impose an availability figure-of-merit as a *design requirement* for a given equipment supplier and the supplier has no control over the operational environment in which that equipment is to function, then A_a or A_i might be appropriate figures-of-merit against which the supplier's equipment can be properly assessed. On the other hand, if one is to assess a system or an equipment in a realistic operational environment, then A_0 is a preferred figure-of-merit to employ for assessment purposes. Further, availability may be applied at any time in the overall mission profile representing a point estimate or may be more appropriately related to a specific segment of the mission where the requirements are different from other segments. Thus, one must define precisely what is meant by "availability" and how it is to be applied. In any event, reliability is a major factor in the determination of system availability.

System Effectiveness (SE). System effectiveness may be defined as "the probability that a system can successfully meet an overall operational demand within a given time when operated under specified conditions" or "the ability of a system to do the job for which it was intended." System effectiveness, like operational availability, is a term used in a broad context to reflect the technical characteristics of a system (i.e., performance, availability, supportability, dependability, etc.) and may be expressed differently depending on the specific mission application. Sometimes a single figure-of-merit is used to express system effectiveness and sometimes multiple figures-of-merit are employed. The objective is to reflect system design attributes and logistic support elements as is illustrated in Figure 5.1.

Cost Effectiveness (CE). Cost effectiveness relates to the measure of a system in terms of mission fulfillment (system effectiveness) and total life-cycle cost, and can be expressed in various ways, depending on the specific mission or system parameters that one wishes to evaluate. In essence, cost effectiveness includes the elements illustrated in Figure 5.1 and can be expressed as indicated in Equations 5.1 through 5.5. Reliability is a major factor in determining the cost effectiveness of a system.

13.4. RELIABILITY IN SYSTEM OR PRODUCT DESIGN

Reliability in system design commences with the establishment of requirements as part of the conceptual design phase (refer to Figure 13.1, Block 1). Through the accomplishment of reliability analyses, these requirements are allocated to various elements of the system and design techniques are used as appropriate to help assure that the design is in compliance with allocated requirements (refer to Figure 13.1, Blocks 2 and 3). When a

---

[2] MTBM, $\overline{M}$ct, $\overline{M}$, and MDT and their applications are discussed further in Chapter 14.

prototype of the system is developed, test and evaluation is accomplished to assess the actual reliability of the system at that stage. The results reflect a measurement of the system just prior to entering the production/construction phase (refer to Figure 13.1, Block 3). The major reliability activities applicable to system or product design are covered in the sections which follow.

Reliability Requirements

Every system is developed in response to a need to fulfill some anticipated function. The effectiveness with which the system fulfills this function is the ultimate measure of its utility and its value to the consumer. This effectiveness is a composite of performance, reliability, and other factors. Reliability is a major factor in determining the usefulness of a system.

Reliability requirements for a system, specified in both quantitative and qualitative terms, are defined as part of the overall system operational requirements and the maintenance concept described in Section 3.1. Specifically, this includes

1. Definition of system performance factors, the mission profile, and system requirements (use conditions, duty cycles, and how the system is to be operated).
2. Definition of the operational life cycle (the anticipated time that the system will be in the inventory and in operational use).
3. Definition of the environment in which the system is expected to operate and be maintained (temperature, humidity, vibration, etc.). This should include a range of values as applicable and should cover all transportation, handling, maintenance, and storage modes.

The basic question at this point is: What reliability should the system have in order to successfully accomplish its intended mission, throughout the specified life cycle, and in the environment defined? If the operational requirements specify that the system must function 24 hours a day for 360 days a year without failure, the system reliability requirements may be rather stringent. On the other hand, if the system is only required to operate 2 hours per day for 260 days/year, the specified requirements may be quite different. In any event, quantitative and qualitative reliability requirements must be defined for the system based on the foregoing considerations. Quantitative requirements are usually expressed in terms of $R(t)$, MTBM, MTBF, λ, successful operational cycles per period, or a combination thereof.

In specifying reliability factors, one must first identify the major system requirements from the top-level functional flow diagram (as illustrated in Figure 4.3). The functions in the top-level diagram are expanded as necessary for the purposes of system definition, and each block in the functional flow diagram(s) is analyzed in terms of input and output requirements. A reliability block diagram is then developed from the functional flow diagram. The reliability diagram should show the series-parallel relationships of the subsystems (and to lower indenture levels of the subsystems, if necessary) required for the performance of individual system functions. Figure 13.10 illustrates a simplified reliability block diagram and the progressive expansion of such from the system level down as design detail becomes known. Generally, levels I and II are available through conceptual design activity, while levels III and on are defined in preliminary system design.

Referring to the figure, the reliability requirement for the system (e.g., R, MTBF) is specified for the entire network identified in level I, and an individual requirement is specified for each individual block in the network. For instance, the reliability of Block 3, function X, may be expressed as a probability of survival of 0.95 for a four-hour period of operation at level I. Similar requirements are specified for Blocks 1, 2, 4, and 5. These, when combined, will indicate the system reliability,

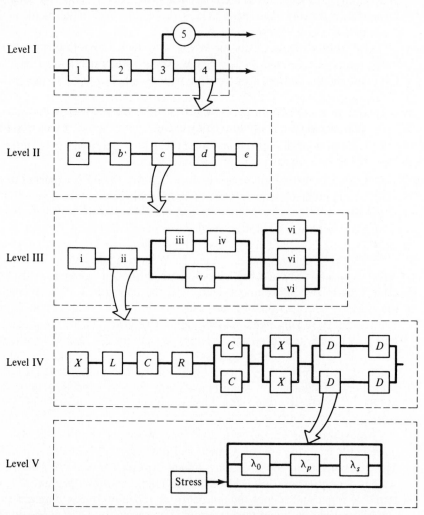

Figure 13.10 Progressive expansion of reliability block diagram. *Source:* NAVAIR 00-65-502/NAVORD OD 41146, *Reliability Engineering Handbook*, Naval Air Systems Command and Naval Ordnance Systems Command, Revised March 1968.

which in turn is evaluated in terms of the overall requirement. This constitutes an iterative process of requirements definition. In situations where a high degree of complexity exists, a reliability computer model may be developed to facilitate the requirements identification task.

Reliability Allocation

After an acceptable reliability figure of merit has been established for the system (e.g., R, MTBF), this requirement must then be allocated among the various subsystems, units, assemblies, and so on. This is accomplished as part of the system requirements allocation process described in Section 4.2, and is facilitated through the development of reliability block diagrams.

Block diagrams are generated to cover each major system function. Success criteria (go/no-go parameters) are established and failure rates (λ) are estimated for each block, the combining of which provides an overall factor for a series of blocks constituting a function or subfunction. Depending on the function, one or more of these diagrams can be related to a physical entity such as unit A in Figure 4.4, or an assembly of unit A. The failure-rate information provided at the unit or assembly level represents a reliability design goal. This, in turn, represents the anticipated frequency of corrective maintenance that is employed in the determination of logistics resource requirements.

The approach used in determining failure rates may vary depending on the maturity of system definition. Failure rates may be derived from direct field and/or test experience covering like items, reliability prediction reports covering items that are similar in nature, and/or engineering estimates based on judgment. In some instances, weighting factors are used to compensate for system complexity and environmental stresses.

When accomplishing reliability allocation, the following steps are considered appropriate:

1. Evaluate the system functional flow diagram(s) and identify areas where design is known and failure-rate information is available or can be readily assessed. Assign the appropriate factors and determine their contribution to the top-level system reliability requirement. The difference constitutes the portion of the reliability requirement that can be allocated to the other areas.
2. Identify the areas that are new and where design information is not available. Assign complexity weighting factors to each functional block. Complexity factors may be based on an estimate of the number and relationship of parts, the equipment duty cycle, whether an item will be subjected to temperature extremes, and so on. That portion of the system reliability requirement that is not already allocated to the areas of known design is allocated using the assigned weighting factors.

The end result should constitute a series of lower-level values that can be combined to represent the system reliability requirement initially specified. The combining of these values is facilitated through the application of a reliability mathematical model.

A reliability mathematical model is developed to relate individual block reliability to the reliabilities of its constitutent block elements. The procedure simply consists of determining a mathematical expression that represents the probability of survival for a small portion of the proposed configuration. Multiple applications of this process will eventually reduce the original complex system to an equivalent serial configuration. It is then possible to represent the system with a single probability statement. In the development of mathematical models, it is necessary to understand the basic series-parallel relationships described in Section 13.2.

When allocating a system-level requirement, one should construct a simplified

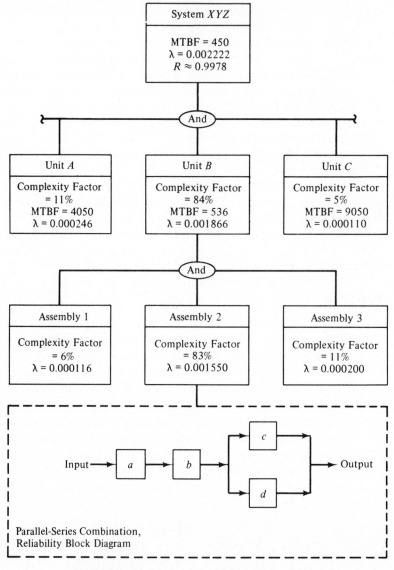

Figure 13.11   Allocation of reliability requirements.

functional breakdown as illustrated in Figure 13.11. The figure indicates a top-level system requirement supported by factors established at the unit level and on down. Unless otherwise specified, the requirements at the unit-level may be altered, or traded off, as long as the combined unit level requirements support the system objective. In other words, the failure rate of unit B may be higher and the failure rate of unit A may be lower than indicated without affecting the requirement of 0.002222 for the system.

The reliability factors established for the various items identified in Figure 13.11 serve as design criteria. For instance, the engineer responsible for unit B should design unit B such that the failure rate (λ) will not exceed 0.001866. As design progresses, reliability predictions are accomplished and the predicted value is compared against the requirement of 0.001866. If the predicted value does not meet the requirement (i.e., higher failure rate or lower MTBF), the design configuration must be reviewed for reliability improvement and design changes are implemented as appropriate.

Part Selection and Application

The reliability of a system depends largely on the reliability of its component parts, and the selection of parts must be compatible with the requirements of the particular application of those parts. The process of procuring what is advertised as being a reliable part is not adequate in itself and does not guarantee a reliable system. The specific application of the component part is of prime importance, particularly when considering factors such as part tolerances and drift characteristics, electrical and environmental stresses, and so on. In electrical and electronic systems, part tolerances, drift characteristics, electrical stresses, and environmental stresses can have a major impact on system reliability and on the individual failure modes of various system components. For mechanical items the same concern prevails when considering part tolerances, component stress-strength factors, and material fatigue characteristics. Consequently, a fundamental approach in attaining a high level of reliability is to select and apply those components and materials of known reliabilities and capable of meeting system requirements. Major emphasis in the design for reliability should take into account

1. The selection of *standardized* components and materials to the extent feasible.
2. The evaluation of all components and materials prior to design acceptance. This includes studying the operating features, tolerances, stresses, and other characteristics of the component(s) in its intended application.
3. The use of only those component parts capable of meeting reliability objectives, and not substitute parts that appear to meet specification requirements but whose quality is degraded through manufacture.

When considering component application(s) in system design, there are trade-offs involving the use of components in series and in parallel (or combinations thereof) which will allow for the achievement of the required system reliability. Figures 13.12 and 13.13 illustrate this fact. Figure 13.12 shows the relationship among

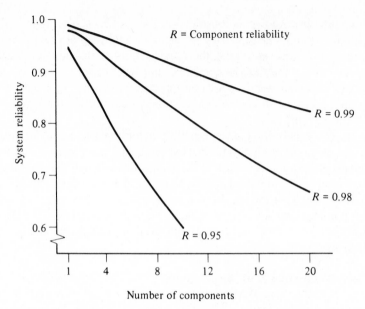

Figure 13.12   Series system reliability. *Source:* K. C. Kapur and L. R. Lamberson, *Reliability in Engineering Design.* New York: John Wiley, 1977.

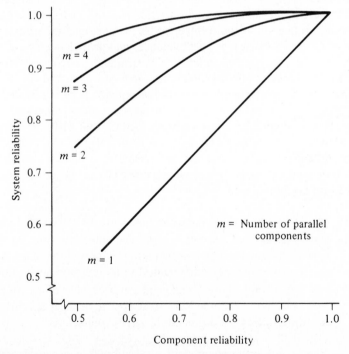

Figure 13.13   Reliability improvement by parallel components. *Source:* K. C. Kapur and L. R. Lamberson, *Reliability in Engineering Design.* New York: John Wiley, 1977.

system reliability, individual component reliability, and the quantity of *series* components included in the system design.

As indicated, the reliability of a series network can be improved by decreasing the number of components, or increasing the individual component reliabilities. It is evident that the system reliability deteriorates rather rapidly as the number of components increases.

Figure 13.13 illustrates the relationship among system reliability, individual component reliability, and the quantity of components in parallel. The paralleling of components is often considered as a means to improve system reliability. However, the overall gains may be marginal in terms of the added costs involved. It may be difficult to design parallel networks, particularly when dealing with mechanical devices. Also, it can be seen that the gain in system reliability by adding more than three parallel components is relatively small. However, the consideration of parallel components, in conjunction with series networks, should be accomplished in arriving at alternative ways to meet system reliability requirements.

Component Part Derating

Derating refers to the techniques of using a component part under stress conditions considerably below the rated value(s) in order to achieve a reliability margin in design. For example, if a 200-V capacitor is used in a 200-V application, then the part is operating at its full rating and it should function satisfactorily. However, if the capacitor is used in a 100-V application, it will in all probability last longer. Many electronic parts, such as resistors, transformers, transistors, and so on, tend to last longer when operated at lower power or voltage levels than at the rated values. Derating constitutes the application of parts in design operating at less than the rated values so as to improve system reliability.

Figure 13.14 illustrates a sample derating curve for a transistor. Similar curves are available for other components. Referring to the figure, the full rating is indicated along with the derated values presented as a function of ambient temperature. Given that the actual ambient temperature is expected to be 75°F, one can determine the derated value as

$$T_{normalized} = \frac{T_{actual} - T_{rated}}{T_{max} - T_{rated}}$$

$$= \frac{75 - 25}{150 - 25} = \frac{50}{125} = 0.4$$

derated value $= 1.0 - 0.4 = 0.6.$

In other words, the transistor should be incorporated into the system design and used at 0.6 of its fully rated value. This, in turn, can be related to a delta reduction in the component failure rate.

For many component parts, particularly in the case of electronic parts, data based on testing combined with actual experience have been converted into conve-

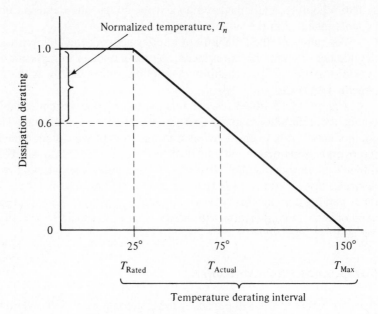

Figure 13.14   Dissipation derating curve.

nient tables that provide failure rates for various parts at derated values. These data are used for design and in reliability prediction.[3]

Redundancy in Design

Under certain conditions in system design it may be necessary to consider the use of redundancy to enhance system reliability by providing two or more functional paths (or channels of operation) in areas that are critical for successful mission accomplishment. But the application of redundancy per se will not necessarily solve all problems, since it usually implies increased weight and space, increased power consumption, greater complexity, and higher cost. On the other hand, the use of redundancy may be the only solution for reliability improvement in specific situations.

Redundancy can be applied at several levels, as illustrated in Figure 13.15. At the system level block G is redundant with the other blocks in the network and is at a different level than block C, which is redundant with block D. From the block diagram, the paths that will result in successful system operation are A, B, C, E; A, B, C, F; A, B, D, E; A, B, D, F; and G.

The probability of success, or the reliability, of each path may be calculated using the multiplication rule expressed in Equation 13.8. The calculation of system reliability (considering all paths combined) requires knowledge of the type of redundancy used and the individual block reliabilities. For operating redundancy, where all blocks are fully energized during an operational cycle, the appropriate equations

---

[3] One good source of reliability data is MIL-HDBK-217, *Reliability Prediction of Electronic Equipment,* Department of Defense, Washington, D.C.

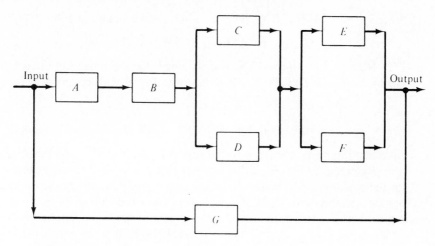

Figure 13.15   Reliability block diagram illustrating redundancy at system and subsystem levels.

in Section 13.3 may be used to calculate system reliability. Equations 13.10 through 13.12 and the related examples illustrate the gain in reliability that can be obtained through redundancy.

When design problems become somewhat more complex than the illustrations presented in Section 13.3, the number of possible events becomes greater. For example, in Figure 13.8, the following possibilities exist:

1. Subsystem A, B, and C are all operating.
2. Subsystems A and B are operating while C is failed.
3. Subsystems A and C are operating while B is failed.
4. Subsystems B and C are operating while A is failed.
5. Subsystem A is operating while B and C are failed.
6. Subsystem B is operating while A and C are failed.
7. Subsystem C is operating while A and B are failed.
8. Subsystems A, B, and C are all failed.

In the interest of simplicity, let R represent the reliability and $1 - R = Q$, the unreliability. Then

$$R^3 \quad \text{represents } A, B, \text{ and } C \text{ operating}$$

$$3R^2Q \begin{cases} \text{represents } A \text{ and } B \text{ operating, } C \text{ failed} \\ \text{represents } A \text{ and } C \text{ operating, } B \text{ failed} \\ \text{represents } B \text{ and } C \text{ operating, } A \text{ failed} \end{cases}$$

$$3RQ^2 \begin{cases} \text{represents } A \text{ operating, } B \text{ and } C \text{ failed} \\ \text{represents } B \text{ operating, } A \text{ and } C \text{ failed} \\ \text{represents } C \text{ operating, } A \text{ and } B \text{ failed} \end{cases}$$

$$Q^3 \quad \text{represents } A, B, \text{ and } C \text{ failed}$$

Since the sum of R^3, $3R^2Q$, $3RQ^2$, and Q^3 represents all possible events, then

$$R^3 + 3R^2Q + 3RQ^2 + Q^3 = 1. \tag{13.19}$$

Referring to Figure 13.8, it is assumed that the reliability of each block (i.e., blocks A, B, and C) is 0.95. Then the reliability of the network is determined as

$$R = R^3 + 3R^2Q + 3RQ^2$$
$$= (0.95)^3 + 3(0.95)^2(0.05) + 3(0.95)(0.05)^2 = 0.999875. \tag{13.20}$$

Note that this value checks with the results of Section 13.3 and Equation 13.11.

Referring to Figure 13.15, suppose that it is desired to calculate the reliability for the network as illustrated. The approach is to first calculate the reliability for the redundant subsystems C, D, E, and F; apply the product rule for A, B, and the resulting networks in this path (i.e., C, D, E, and F); and then determine the combined reliability of the two overall redundant paths (to include subsystem G). It is assumed that the reliability of each of the individual subsystems is as follows:

$$\text{Subsystem } A = 0.97$$

$$\text{Subsystem } B = 0.98$$

$$\text{Subsystem } C = 0.92$$

$$\text{Subsystem } D = 0.92$$

$$\text{Subsystem } E = 0.93$$

$$\text{Subsystem } F = 0.90$$

$$\text{Subsystem } G = 0.99$$

The reliability of the redundant network including subsystems C and D is

$$R_{CD} = R_C + R_D - (R_C)(R_D) = 0.9936.$$

The reliability of the redundant network including subsystems E and F is

$$R_{EF} = R_E + R_F - (R_E)(R_F) = 0.9930.$$

The reliability of the path is

$$R_{ABCDEF} = (R_A)(R_B)(R_{CD})(R_{EF}) = 0.9379.$$

The reliability of the combined network in Figure 13.15 is

$$R_{\text{system}} = R_{ABCDEF} + R_G - (R_{ABCDEF})(R_G) = 0.999379.$$

Thus far, the discussion has addressed only one form of redundancy, operating redundancy, where all subsystems are fully energized throughout the system operating cycle. In actual practice, however, it is often preferable to use standby redundancy. Figure 13.16 provides a simple illustration of standby redundancy where subsystem A is operating full time, and subsystem B is standing by to take over operation if subsystem A fails. The standby unit (i.e., subsystem B) is not operative until a failure-sensing device senses a failure in subsystem A and switches operation

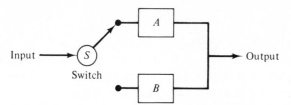

Figure 13.16 Standby redundant network.

to subsystem B, either automatically or through manual selection. Because of the fact that subsystem B is not operating unless a failure of subsystem A occurs, the system reliability for such a configuration is higher than for a comparable system where both subsystems A and B are operating continuously.

When determining the reliability of standby systems, the Poisson distribution may be used since standby systems display the constant λt characteristic of this distribution (refer to Chapter 10 for coverage of the Poisson process). In essence, the probability of no failure is represented by the first term, $e^{-\lambda t}$; the probability of one failure is $(\lambda t)(e^{-\lambda t})$; and so on.

Referring to Figure 13.16, where one operating subsystem and one standby subsystem are grouped together, one must consider the probability that no failure or one failure will occur (with one subsystem remaining in satisfactory condition). This combined probability is expressed as

$$P\,(\text{one standby}) = e^{-\lambda t} + (\lambda t)e^{-\lambda t} \qquad (13.21)$$

where λt is the expected quantity of failures. If one operating subsystem and two standby subsystems are grouped together, then the combined probability is

$$P\,(\text{two standbys}) = e^{-\lambda t} + (\lambda t)e^{-\lambda t} + \frac{(\lambda t)^2 e^{-\lambda t}}{2!}.$$

An additional term in the Poisson distribution is added for each subsystem in standby.

As an example, suppose that one must determine the system reliability for a configuration consisting of one operating subsystem and one identical standby operating for a period of 200 hours. This configuration is illustrated in Figure 13.16, and it is assumed that the reliability of the switch is 100%. The failure rate (λ) for each subsystem is 0.002 failure per hour. Using Equation 13.21, the reliability is

$$R = e^{-\lambda t}(1 + \lambda t) \text{ and } \lambda t = (0.002)(200) = 0.4$$

$$= e^{-0.4}(1 + 0.4)$$

$$= (0.67032)(1.4) = 0.9384.$$

To illustrate next the difference between operating redundancy and standby redundancy, assume that both of the subsystems in Figure 13.16 are operating throughout the mission. The reliability of the configuration is determined as

$$R = 1 - (1 - 0.67032)^2 = 0.8913.$$

As anticipated, the reliability of the standby system is higher (0.9384) than the reliability of the system using operating redundancy (0.8913).

Reliability Models

Throughout the system design process, reliability models are used to aid in the evaluation of alternatives and in the accomplishment of various types of analyses. There are series models, series-parallel models, parallel redundancy models, standby redundancy models, and so on. Models are used in the development of reliability block diagrams, the allocation of reliability requirements, stress-strength analyses, worst-case analyses, and in reliability prediction. Of course, each application may be somewhat unique in itself and the model must be tailored accordingly.

To illustrate model development, refer to the reliability block diagram in Figure 13.10. The block diagram enables one to visualize physical relationships and specific subsystem configurations. A mathematical model is used to relate the individual block reliabilities to the reliabilities of its elements and the reliability of the system as an entity. For instance, in progressing from level I to level V,

$$R_{\text{system}} = (R_1)(R_2)(R_3)(R_4)$$

where

$$R_4 = (R_a)(R_b)(R_c)(R_d)(R_e)$$

where

$$R_c = (R_i)(R_{ii})[1 - (1 - R_v)(1 - R_{iii}R_{iv})][1 - (1 - R_{vi})^3]$$

where

$$R_{ii} = (R_X)(R_L)(R_C)(R_R)[1 - (1 - R_C)^2][1 - (1 - R_X)^2][1 - (1 - R_D)^2]^2$$

where

$$R_D = e^{-(\lambda o + \lambda p + \lambda s)t}.$$

As one progresses, the notation often becomes quite complex and a computerized model can be effectively employed to facilitate calculations through the incorporation of simplified terms and equivalencies. Given the reliabilities (or failure rates) for each block in the diagram, one can readily calculate the reliability for the overall system.

The utilization of a computerized reliability model in this application provides additional benefits as follows:

1. It can be used to expeditiously evaluate alternative design approaches where different combinations of series-parallel relationships are being considered.
2. It can be used to expeditiously evaluate the reliability of the system (or various elements of the system) when different component parts are being considered for a specific application. This applies to evaluating component stress-strength characteristics, the effects of component-part derating, and so on.

The computerized reliability model is an excellent tool that can be used when (1) initially allocating reliability requirements to the subsystem level and below, (2) predicting the reliability for a newly proposed design configuration, and (3) when assessing the reliability of a system in operational use.[4]

Stress-Strength Analysis

Of major concern in the design for system reliability are the stress-strength characteristics of its components. Component parts are designed and manufactured to operate in a specified manner when utilized under nominal conditions. If additional stresses are imposed due to electrical loads, temperature, vibration, shock, humidity, and so on, then unexpected failures will occur and the reliability of the system will be less than anticipated. Also, if materials are utilized in a manner where nominal strength characteristics are exceeded, fatigue occurs and the materials may fail much earlier than expected. In any event overstress conditions will result in reliability degradation and understress conditions may be costly as a result of overdesign; that is, incorporating more than what is actually necessary to do the job.

A stress-strength analysis is often undertaken to evaluate the probability of identifying a situation(s) where the value of stress is much larger than (or the strength much less than) the nominal value. Such an analysis may be accomplished through the following steps:

1. For selected components, determine nominal stresses as a function of loads, temperature, vibration, shock, physical properties, and time.
2. Identify factors affecting maximum stress, such as stress concentration factors, static and dynamic load factors, stresses as a result of manufacturing and heat treating, environmental stress factors, and so on.
3. Identify critical stress components and calculate critical mean stresses (e.g., maximum tensile stress, shear stress).
4. Determine critical stress distributions for the specified useful life. Analyze the distribution parameters and identify component safety margins. Applicable distributions may include normal, Poisson, Gamma, Weibull, log-normal, or variations thereof.
5. For those components that are critical and where the design safety margins are inadequate, corrective action must be initiated. This may constitute component-part substitution or a complete redesign of the system element in question.

---

[4]Throughout the industrial and government sectors, there are many different computerized models being developed and used to fulfill specific reliability (and maintainability) program requirements. In general, these models are being employed on a stand-alone basis, and the results are not yet integrated into the CAD process in support of day-to-day design decisions. One of the objectives in systems engineering is to foster the integration of these various computerized tools into a CAD workstation, helping to ensure the availability of reliability evaluation on a concurrent basis (not after-the-fact). This objective is also being pursued by the DOD through such efforts as RAMCAD (Reliability and Maintainability in Computer-Aided Design) and CALS. A framework for these initiatives was presented in Section 5.4.

Reliability computerized models are used to facilitate the stress-strength analysis process. On the basis of such an analysis, component-part failure rates (λ) are adjusted as appropriate to reflect the effects of the stresses of the parts involved.

Critical-Useful-Life Analysis

A critical-useful-life item is one which, because of its short life, is incapable of satisfying the functional requirements imposed by its application unless corrective or preventive maintenance is performed. During the design phase, critical items are listed along with their expected life in terms of calendar time, operating cycles, or system operating hours. This listing specifies the requirement for maintenance, personnel support, and spare/repair parts. In the interest of equipment design for supportability, all critical items should be eliminated if at all possible.

Failure Mode and Effect Analysis (FMEA)

The FMEA, or FMECA (failure mode, effect, and criticality analysis), is performed during the early stages of preliminary system design to identify possible problems that could develop as a result of system failures. The FMEA is oriented to equipment only and does not generally cover the effects of human actions on equipment. The objective is to determine the ways in which equipment can fail, and the effects of such failures on other elements of the system. The FMEA includes the following information:

1. *Item identification*—Identify each significant system component that is likely to fail. This can be accomplished in conjunction with the functional analysis described in Section 4.1 (see Figure 4.7), where functional flow diagrams are employed to identify subsystems, units, assemblies, and components.
2. *Description of failure modes*—Define the most probable modes of failure for each identified item. Failure modes may be classified as ruptures, fractures or cracks, voids, short and open circuits, physical separations, and so on, and are related to the operational modes that the system experiences through the performance of its designated mission. In other words, how will the equipment fail, and under what operating or environmental conditions is the equipment subjected to when the failure occurs?
3. *Cause of failure*—The anticipated cause of failure should be described for each instance. Typical causes might include abnormal equipment stresses during operation, aging and wear-out, poor workmanship, defective materials damage due to transportation and handling, or operator- and maintenance-induced faults.
4. *Possible effects of failure*—Describe the most probable effects as a result of each identified failure. Effects may range from complete system destruction to partial system operation. To what extent does the failure affect successful accomplishment of the mission of the system?

5. *Probability of occurrence*—Through statistical means, estimate the probability of failure occurrence. Probabilities of occurrence initially may evolve from experience factors or through reliability allocation and will be based on reliability prediction data as system development progresses.

6. *Criticality of failure*—Failures may be classified in terms of criticality in any one of four categories, depending upon the defined failure effects as follows.

   a. *Minor failure*—Any failure that does not degrade the overall performance and effectiveness of the system beyond acceptable limits.

   b. *Major failure*—Any failure that will degrade the system performance and effectiveness beyond acceptable limits but can be controlled.

   c. *Critical failure*—Any failure that will degrade the system beyond acceptable limits and could create a safety hazard if immediate corrective action is not initiated.

   d. *Catastrophic failure*—Any failure that could result in significant system damage, such as to preclude mission accomplishment, and could cause deaths and personnel injuries.

7. *Possible corrective action or preventive measures*—Describe the action(s) that can be initiated to reduce the probability of failure occurrence or to minimize the effects of failure.

The FMEA can be effectively utilized to provide a logical review of system failure modes, to ensure that the results of any given failure will cause minimal damage to equipment and will not create hazardous conditions for system operators and maintenance personnel. The FMEA and the hazard analysis, often conducted as part of a safety analysis, provide an excellent means to assure the incorporation of system/product safety characteristics in design. In addition, the FMECA can be effectively used to help identify critical maintenance actions in the development of a recommended preventive maintenance program.

Reliability Prediction

As engineering data become available, reliability prediction is accomplished as a check on design in terms of the system requirement and the factors specified through allocation. The predicted values of MTBM and/or MTBF are compared against the requirement, and areas of incompatibility are evaluated for possible design improvement.

Prediction is accomplished at different times in the equipment design process and will vary somewhat depending on the type of data available. Basic prediction techniques are summarized as follows:

1. *Prediction may be based on the analysis of similar equipment*—This technique should only be used when the lack of data prohibits the use of more sophisti-

cated techniques. The prediction uses MTBF values for similar equipments of similar degrees of complexity performing similar functions and having similar reliability characteristics. The reliability of the new equipment is assumed to be equal to that of the equipment which is most comparable in terms of performance and complexity. Part quantity and type, stresses, and environmental factors are not considered. This technique is easy to perform, but not very accurate.

2. *Prediction may be based on an estimate of active element groups (AEG)*—The AEG is the smallest functional building block that controls or converts energy. An AEG includes one active element (relay, transistor, pump, machine) and a number of passive elements. By estimating the number of AEGs and using complexity factors, one can predict MTBF.

3. *Prediction may be accomplished from an equipment parts count*—There are a variety of methods used that differ somewhat due to data source, the number of part type categories, and assumed stress levels. Basically, a design parts list is used and parts are classified in certain designated categories. Failure rates are assigned and combined to provide a predicted MTBF at the system level. A representative approach is illustrated in Table 13.2.

TABLE 13.2 RELIABILITY PREDICTION DATA SUMMARY

COMPONENT PART	λ/PART (%/1000 HOURS)	QUANTITY OF PARTS	(λ/PART) (QUANTITY)
Part *A*	0.161	10	1.610
Part *B*	0.102	130	13.260
Part *C*	0.021	72	1.512
Part *D*	0.084	91	7.644
Part *E*	0.452	53	23.956
Part *F*	0.191	3	0.573
Part *G*	0.022	20	0.440

Failure Rate (λ) = 48.995%/1000 Hours $\qquad\qquad$ Σ = 48.995%

$$\text{MTBF} = \frac{1000}{0.48995} = 2041 \text{ Hours}$$

Source: MIL-HDBK-217, Military Handbook, *Reliability and Failure Rate Data for Electronic Equipment*, Department of Defense, Washington, D.C.

4. *Prediction may be based on a stress analysis (discussed earlier)*—When detailed equipment design is relatively firm, the reliability prediction becomes more sophisticated. Part types and quantities are determined, failure rates are applied, and stress ratios and environmental factors are considered. The interaction effects between components are addressed. This approach is peculiar and varies somewhat with each particular system/product design. Computer methods are often used to facilitate the prediction process.

The figures derived through reliability prediction constitute a direct input to maintainability prediction data, logistic support analysis, and the determination of specific support requirements (test and support equipment, spare and repair parts, etc.). Reliability basically determines the frequency of corrective maintenance and the quantity of maintenance actions anticipated throughout the life cycle; thus, it is imperative that reliability prediction results be as accurate as possible.

Reliability Degradation

Another critical design task is to determine the reliability degradation (if any) on the system or product due to storage, packing, transportation, handling, and maintenance. The equipment is subjected to these environments when initially shipped from the factory and distributed to the consumer, stored as a spare, returned to the depot or supplier for maintenance, and so on. Sometimes these environments include extreme conditions of rain, sand or dust, salt spray, high and low temperatures, and high humidity. In the event of degradation, either additional design provisions are needed to compensate for the reduction in reliability or an increase in the quantity of maintenance actions will result. In either case, the impact or logistic support and cost is evident.

Quite often the reliability of a system is degraded through the performance of preventive and corrective maintenance actions. Unless extreme care is taken, maintenance-induced faults may be inadvertently introduced in the accomplishment of maintenance actions, or components may be partially damaged to the extent that subsequent system failures may occur sooner and more frequently than intially anticipated. This is primarily due to carelessness on the part of the individual performing maintenance, using the wrong tools and test equipment, not following the approved maintenance procedures, and so on. Thus, it is extremely important that the proper logistic support resources be applied in performing system/product maintenance if the reliability of the system is to be maintained.

Reliability Design Review

The review of reliability in system design is accomplished as an inherent part of the process described in Sections 3.4, 4.5, and 5.6. The characteristics of the system (and its elements) are evaluated in terms of compliance with the initially specified reliability requirements for the system. If the requirements appear to be fulfilled, the design is approved as is. If not, the appropriate changes are initiated for corrective action.

In support of conducting reliability reviews, one may wish to develop a design review checklist, including questions such as those listed below.

1. Have reliability quantitative and qualitative requirements for the system been adequately defined and specified?
2. Are the reliability requirements compatible with other system requirements? Are they realistic?

3. Has system or product design complexity been minimized?

4. Have system failure modes and effects been identified?

5. Has the system or product wear-out period been defined?

6. Are system, subsystem, unit, and component-part failure rates known?

7. Have component parts with excessive failure rates been identified?

8. Has the system mean life been determined?

9. Have adequate derating factors been established and adhered to where appropriate?

10. Is protection against secondary failures (resulting from primary failures) incorporated where possible?

11. Have all critical-useful-life items been eliminated from system design?

12. Has the use of adjustable components been minimized?

13. Have cooling provisions been incorporated in design "hot-spot" areas?

14. Have all hazardous conditions been eliminated?

15. Have all system reliability requirements been met?

The items covered certainly are not intended to be all-inclusive, but merely represent a sample of possible interest areas (it should be noted that the answer to those questions that are applicable should be *yes*). The response to these questions is primarily dependent on the results of the reliability analyses and predictions discussed earlier.[5]

13.5. RELIABILITY TEST AND EVALUATION

Reliability testing is conducted as part of the system test and evaluation effort discussed in Chapter 6. Specifically, reliability testing is accomplished under the categories of Type 2 and Type 3 testing described in Section 6.2.

As is required for any category of testing, there is a planning phase, a test preparation phase, the actual test and evaluation itself, the data collection and analysis phase, and test reporting. The general requirements for each are discussed in Chapter 6. Only those procedures applicable to reliability test methods are covered in this chapter.

The ultimate objective of reliability testing is to determine whether the system (or product) under test meets the specified MTBF requirements. To accomplish this, the system is operated in a prescribed manner for a designated period and failures are recorded and evaluated as the test progresses. System acceptance is based on demonstrating a minimum acceptable life.

There are a number of different test methods and statistical procedures in practice today which are designed to measure system reliability. Most of these assume that the exponential distribution is applicable. The criteria for system acceptance (or rejection) are based on statistical assumptions involving test sample size, consumer–producer decision risk factors, test data confidence limits, and so on. These assump-

[5] Appendix B provides a more extensive design review checklist.

tions often vary from one application to the next, depending on the type of system, the mission that the system is expected to perform, and whether new design techniques are used in system development reflecting a potential high-risk area. In view of this, it is impossible to cover all facets of reliability testing within the confines of this chapter. However, in order to provide some understanding of the approaches used, it is intended to briefly cover reliability sequential qualification testing, reliability acceptance testing, and life testing.

Reliability Sequential Qualification Testing[6]

Reliability qualification testing is conducted to provide an evaluation of system development progress, as well as the assurance that specified requirements have been met prior to proceeding to the next phase (i.e., the production or construction phase of life cycle). Initially, a reliability MTBF is established for the system, followed by allocation and the definition of design criteria. System design is accomplished and reliability analyses and predictions are made to evaluate the design configuration (on an analytical basis) relative to compliance with system requirements. If the predictions indicate compliance, the design progresses to the construction of a prototype, or preproduction model, of the system and qualification testing commences.

When a reliability sequential test is conducted, there are three possible decisions: (1) accept the system, (2) reject the system, or (3) continue to test. Figure 13.17 illustrates a typical sequential test plan. The system under test is operated in a

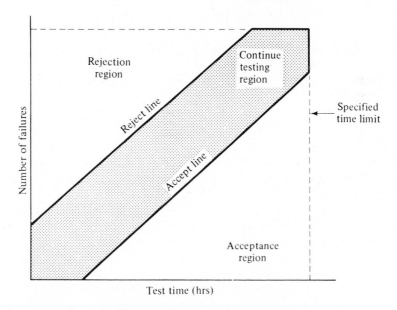

Figure 13.17   Sequential test plan.

[6] Reliability sequential testing is covered rather extensively in MIL-STD-781, *Reliability Design Qualification and Production Acceptance Tests: Exponential Distribution,* Department of Defense, Washington, D.C.

manner reflecting actual consumer utilization in a realistic environment. The objective is to simulate (to the extent possible) a mission profile, or at least to subject the system under test to conditions similar to those that will be present when the system is in use by the consumer. The accomplishment of this objective generally involves an environmental testing chamber and the operation of equipment through different duty cycles and environmental conditions. Figure 13.18 illustrates a sample duty cycle. The system is operated through a series of these duty cycles for a designated period of time.

Referring to the sequential test plan in Figure 13.17, system operating hours are accumulated along the abscissa and failures are plotted against the ordinate. If enough operating time is acquired without too many failures, a decision is made to accept the system and testing is discontinued. On the other hand, if there are a

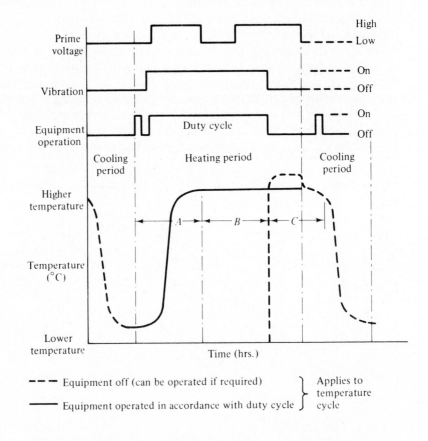

- - - Equipment off (can be operated if required) ⎫ Applies to
——— Equipment operated in accordance with duty cycle ⎬ temperature
                                                       ⎭ cycle

A. Time for chamber to reach stabilization at higher temperature

B. Time of equipment operation at higher temperature

C. Optional hot soak and hot start-up checkout

Figure 13.18 Sample environmental test cycle. *Source:* MIL-STD-781, *Reliability Design Qualification and Production Acceptance Tests: Exponential Distribution,* Department of Defense, Washington, D.C.

significant quantity of failures occurring early in test operations, then a reject decision may prevail and the system is unacceptable as is. A marginal condition results in a decision to continue testing until a designated point in time. In essence, the sequential test plan allows for an early decision. Highly reliable systems will be accepted with a minimum amount of required testing. If the system is unreliable, this will also be readily evident at an early point in time. In this respect, sequential testing is extremely beneficial. On the other hand, if the system inherent reliability is marginal, the amount of test time involved can be rather extensive and costly. An actual example of sequential testing is presented later.

Sequential test plans are highly influenced by the risks that both the producer and consumer are willing to accept in connection with decisions made as a result of testing. These risks are defined as

1. *Producer's risk* (α)—the probability of rejecting a system when the measured MTBF is equal to or better than the specified MTBF. In other words, this refers to the probability of rejecting a system when it really should be accepted, which constitutes a risk to the system manufacturer or producer (also known as a Type I error).

2. *User's or consumer's risk* (β)—the probability of accepting a system when the measured MTBF is less than the specified MTBF. This refers to the probability of accepting a system that actually should be rejected, which constitutes a risk to the user (also known as a Type II error).

The probability of making an incorrect decision on the basis of test results must be addressed in a manner similar to hypothesis testing. An assumption is made and a test is accomplished to support (or disprove) that assumption. In other words, a null hypothesis (H_0) is established, that is, a statement or conjecture about a parameter such as the true MTBF is equal to 100. The alternative hypothesis (H_1) is "the MTBF is not equal to 100." When testing an item (representing a sample of a total population), the question arises as to whether to accept H_0. The desired result is to accept when the null hypothesis is true and reject when false, or to minimize the chances of making an incorrect decision. The relationship of risks in sample testing is illustrated in Table 13.3. In selecting a sequential test plan, it is necesary to first identify two values of MTBF:

TABLE 13.3   RISKS IN SAMPLE TESTING

TRUE STATE OF AFFAIRS	ACCEPT H_0 AND REJECT H_1 (i.e., MTBF = 100)	REJECT H_0 AND ACCEPT H_1 (i.e., MTBF ≠ 100)
H_0 is true (i.e., MTBF = 100)	High probability $1 - \alpha$ (i.e., 0.90)	Low probability Error, α (i.e., 0.10)
H_0 is false and H_1 is true (i.e., MTBF ≠ 100)	Low probability Error, β (i.e., 0.10)	High probability $1 - \beta$ (i.e., 0.90)

1. Specified MTBF representing the system requirement (assume that θ_0 represents this value).

2. Minimum MTBF that is considered to be acceptable based on the results of testing (assume that θ_1 represents this value).

Given θ_0 and θ_1, one must next decide on the values of producer's risk (α) and consumer's risk (β). Most test plans in use today accept risk values ranging between 5 and 25%. For example, a test where $\alpha = 0.10$ means that in 10 of 100 instances the test plan will reject items that should have been accepted. Risk values are generally negotiated in the test planning phase.

Referring to Figure 13.17, the design of the sequential test plan is based on the values of θ_0, θ_1, α and β. For instance, the *accept line* slope is based on the following expressions.[7]

$$t_1 = \frac{\ln (\theta_0/\theta_1)}{1/\theta_1 - 1/\theta_0}(r) - \frac{\ln [\beta/(1 - \alpha)]}{1/\theta_1 - 1/\theta_0} \qquad (13.22)$$

where r is the number of failures (the accept line is plotted for assumed values of r). The ratio of θ_0 to θ_1 is known as the *discrimination ratio,* a standard test parameter. The *reject line* slope is based on Equation 13.23.

$$t_2 = \frac{\ln (\theta_0/\theta_1)}{1/\theta_1 - 1/\theta_0}(r) - \frac{\ln [(1 - \beta)/\alpha]}{1/\theta_1 - 1/\theta_0}. \qquad (13.23)$$

Determination of the specified time limit in Figure 13.17 is generally based on a multiple of θ_1.

As indicated earlier, there are a number of different sequential test plans that have been used for reliability testing. One example is illustrated in Figure 13.19. The established α and β decision risks are 10% (each), the discrimination ratio agreed upon is 1.5 : 1, and the accept-reject criteria are as specified.

Another approach is illustrated in Figure 13.20, representing actual experience in testing a specific system (to be designated as system *XYZ*). Referring to the figure, the specified MTBF for the system is 400 hours, and the maximum designated testing time for the sequential test plan used is 4,000 hours, or 10 times the specified MTBF. The test approach involves the selection of a designated quantity of equipments, operating the equipment under certain performance conditions over an extended period of time, and monitoring the equipment for failure. Failures are noted as events, corrected through the appropriate maintenance actions, and the applicable equipment is returned to full operational status for continued testing. An analysis of each event should determine the cause of failure, and trends may be established if more than one failure is traceable to the same cause. This may be referred to as a pattern failure, and in such cases, a change should be initiated to eliminate the occurrence of future failures of the same type.

[7] S., Halpern, *The Assurance Sciences: An Introduction to Quality Control and Reliability* Englewood Cliffs, N.J.: Prentice-Hall, 1978, Chapter II.

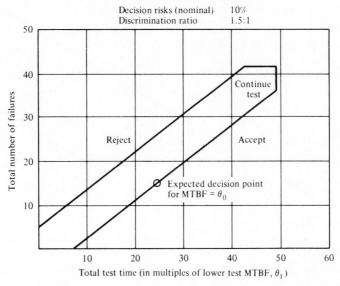

Decision risks (nominal)   10%
Discrimination ratio   1.5:1

Total test time (in multiples of lower test MTBF, θ_1)

	Total test time*				Total test time*	
Number of failures	Reject (equal or less)	Accept (equal or more)		Number of failures	Reject (equal or less)	Accept (equal or more)
0	N/A	6.60		21	18.92	32.15
1	N/A	7.82		22	20.13	33.36
2	N/A	9.03		23	21.35	34.58
3	N/A	10.25		24	22.56	35.79
4	N/A	11.46		25	23.78	37.01
5	N/A	12.68		26	24.99	38.22
6	0.68	13.91		27	26.21	39.44
7	1.89	15.12		28	27.44	40.67
8	3.11	16.34		29	28.65	41.88
9	4.32	17.55		30	29.85	43.10
10	5.54	18.77		31	31.08	44.31
11	6.75	19.98		32	32.30	45.53
12	7.97	21.20		33	33.51	46.74
13	9.18	22.41		34	34.73	47.96
14	10.40	23.63		35	35.94	49.17
15	11.61	24.84		36	37.16	49.50
16	12.83	26.06		37	38.37	49.50
17	14.06	27.29		38	39.59	49.50
18	15.27	28.50		39	40.82	49.50
19	16.49	29.72		40	42.03	49.50
20	17.70	30.93		41	49.50	N/A

*Total test time is total unit hours of equipment on time and is expressed in multiples of the lower test MTBF.

Figure 13.19   Accept-reject criteria for a sample test plan. *Source:* MIL-STD-781, *Reliability Design Qualification and Production Acceptance Tests: Exponential Distribution,* Department of Defense, Washington, D.C.

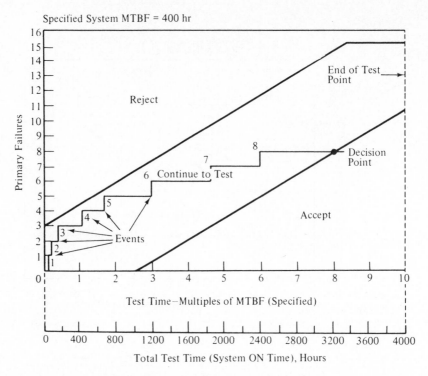

Figure 13.20   System *XYZ* reliability test plan.

Referring to Figure 13.20, the system is accepted after 3,200 hours of testing, and eight events are recorded. Thus, the specified MTBF requirement is fulfilled.

In this instance, the test sample is six pieces of equipment. If a single model is designated for test with a minimum of 4,000 test hours required, the length of the test program (assuming continuous testing) will obviously extend beyond six months. This may not be feasible in terms of the overall program schedule and the delivery of qualified equipment. On the other hand, the use of two or more models will result in a shorter test program. Sometimes the implementation of accelerated test conditions will result in an additional number of failures. These effects should be evaluated, and an optimum balance should be established between a smaller sample size and longer test times and a larger sample size and shorter test times.

Reliability Acceptance Testing

As indicated earlier, reliability testing may be accomplished as part of qualification testing prior to commencing with full-scale production (which has been discussed) and during full-scale production on a sampling basis. In order to determine the effects of the production process on system reliability, it may be feasible to select a sample number of equipments from each production lot and test them in the same manner as described above. The sample may be based on a percentage of the total equipments spread over the entire production period, or a set number of equipment(s) selected during a given calendar time period, (e.g., the two pieces of equipment per month throughout the production phase.) In any event, the selected equipment is tested and an assessed MTBF is derived from the test data. This value is compared against the specified MTBF and the measured value determined from earlier qualification testing. Growth MTBF trends (or negative trends) may be determined by plotting the resultant values as testing progresses.

Reliability Life Testing

There are two basic forms of life testing in use: (1) life tests based on a fixed-test time, and (2) life tests based on the occurrence of a predetermined number of failures.

     The first approach to life testing (based on time) assumes that a computed fixed test time will be planned and a specified quantity of failures will be predetermined. System acceptance occurs if the actual number of failures during test is equal to (or less than) the predetermined quantity of failures at the end of the scheduled test time.

     The second approach to life testing is (based on failures) assumes that a test plan is developed specifying a predetermined quantity of failures and a computed test time dependent on an expected system failure rate. Testing continues until the specified quantity of failures occurs. System acceptance takes place if the test time is equal to (or is greater than) the computed time at the point of last failure.

Operational Reliability Assessment

The activities discussed thus far have included requirements definition, allocation, analyses and predictions, and qualification and acceptance testing. The measure of system reliability has been based on a combination of analytical studies and the demonstration of prototype equipment. The opportunity to observe the system being utilized by the consumer in a realistic environment has not been possible until this point in time. Now, it is essential that a "true" assessment of the system reliability be made.

     This assessment of system reliability in an operational environment is best accomplished through the establishment of an effective data collection, analysis, and system evaluation capability as was described in Section 12.4. The purpose is twofold:

1. To provide ongoing data that can be analyzed to determine the true reliability of the system while performing its intended mission.
2. To provide historical data that can be beneficially used in the design and development of new systems and equipment having a similar function and nature. Such data will facilitate the accomplishment and accuracy of analyses and predictions in the future.

Figures 6.5 and 6.6 illustrated both the *success* and *maintenance* data elements that are considered important for the system as a whole. Of specific interest in the reliability assessment task are (1) the operational status and condition of the system at the time of failure, (2) the maintenance requirements necessary to restore the system to full operational status, and (3) the details associated with the actual cause of the failure and the effects on other elements of the system.

The first two items of concern are adequately covered through the data elements included in the *system operational information report* (Figure 6.5) and the *maintenance event report* (Figure 6.6). However, the details associated with the cause and effects of failures are not sufficiently covered. Thus, these two reports should be supplemented with a *failure analysis report*.

A failure analysis report should be prepared on each system failure and should provide information such as the time of failure, symptom of failure, effects of failure on system operation, effects of the failure on other elements of the system, the actual cause of system failure, and the specific failure mode. Enough data are required to determine *what* happened, *how* the failure occurred, and *why* did it occur.

These data should be collected throughout the system operational life cycle (to the maximum extent possible) and analyzed to determine trends and inherent weaknesses in the system. Major areas of deficiency should be corrected through the initiation of changes and the appropriate system modification(s).

SELECTED REFERENCES

(1) Amstadter, B. L., *Reliability Mathematics: Fundamentals, Practices, Procedures*. New York: McGraw-Hill 1971.

(2) Bazovsky, I., *Reliability Theory and Practice*. Englewood Cliffs, N.J.: Prentice-Hall, 1961.

(3) Blanchard, B. S., *Logistics Engineering and Management*. Englewood Cliffs, N.J.: Prentice-Hall, 1986.

(4) Calabro, S.R., *Reliability Principles and Practice*. New York: McGraw-Hill, 1962,

(5) Halpern, S., *The Assurance Sciences—An Introduction to Quality Control and Reliability*. Englewood Cliffs, N.J.: Prentice-Hall, 1978.

(6) Ireson, W. G., and C. F. Combs (Ed), *Handbook of Reliability Engineering and Management*. New York: McGraw-Hill, 1988.

(7) Juran, J.M., ed., *Quality Control Handbook*. New York: McGraw-Hill, 1979.

(8) Kapur, K. C., and L. R. Lamberson, *Reliability in Engineering Design*. New York: John Wiley, & 1977.

(9) Lloyd, D. K., and M. Lipow, *Reliability Management, Methods, and Mathematics*. Redondo Beach, Calif.: published by authors, Defense and Space Systems Group, TRW Systems and Energy, 1977.

(10) MIL-HDBK-217, Military Handbook, *Reliability Prediction of Electronic Equipment*. Department of Defense, Washington, D.C.

(11) MIL-STD-781, Military Standard, *Reliability Design Qualification and Production Acceptance Tests: Exponential Distribution*. Department of Defense, Washington, D.C.

(12) MIL-STD-785, Military Standard, *Reliability Program for Systems and Equipment Development and Production*. Department of Defense, Washington, D.C.

(13) Siewiorek, D. P., and R. B. Swarz, *The Theory and Practice of Reliable System Design*. Bedford, Mass.: Digital Press, 1982.

(14) Shooman, M. L., *Probabilistic Reliability: An Engineering Approach*. New York: Mc-Graw-Hill, 1968.

(15) Von Alven, W. H., ed., *Reliability Engineering*. Englewood Cliffs, N.J.: Prentice-Hall, 1964.

QUESTIONS AND PROBLEMS

1. Identify and discuss the four major elements in the reliability definition.
2. Describe how reliability relates to system or product design.
3. Describe how reliability ties in with the systems engineering process.
4. What are the quantitative measures of reliability?
5. What is the purpose of reliability allocation? Reliability models? Stress-strength analysis? Derating? FMEA? Critical useful life analysis? Reliability prediction? Reliability testing?
6. Define operational availability, achieved availability, and inherent availability.
7. A system consists of four subassemblies connected in series. The individual subassembly reliabilities are as follows:

$$\text{Subassembly } A = 0.98$$

$$\text{Subassembly } B = 0.85$$

$$\text{Subassembly } C = 0.90$$

$$\text{Subassembly } D = 0.88$$

Determine the overall system reliability.
8. When determining the overall system failure rate, what factors must be considered?
9. A system consists of three subsystems in parallel (assume operating redundancy). The individual subsystem reliabilities are as follows:

$$\text{Subsystem } A = 0.98$$

$$\text{Subsystem } B = 0.85$$

$$\text{Subsystem } C = 0.88$$

Determine the overall system reliability.
10. What information should be included in a reliability program plan?
11. Refer to Figure 13.15. Determine the overall network reliability if the individual reliabilities of the subsystems are as follows:

$$\text{Subsystem } A = 0.95$$

$$\text{Subsystem } B = 0.97$$

$$\text{Subsystem } C = 0.92$$

$$\text{Subsystem } D = 0.94$$

$$\text{Subsystem } E = 0.90$$

$$\text{Subsystem } F = 0.88$$

$$\text{Subsystem } G = 0.98$$

12. The failure rate (λ) of a device is 22 failures per million hours. Two standby units are added to the system. Determine the MTBF of the system.

13. Calculate the reliability of a system consisting of one operating unit and one identical standby unit operating for a period of 200 hrs. The failure rate (λ) of each unit is 0.003 failure/hr and the failure sensing switch reliability is 1.0.

14. Assume that you are assigned to a given system with the objective of improving the reliability of that system. Describe how you would approach the problem. What design methods would you consider?

15. A system consists of five subsystems with the following MTBFs:

$$\text{Subsystem } A: \text{MTBF} = 10{,}540 \text{ hrs}$$

$$\text{Subsystem } B: \text{MTBF} = 16{,}220 \text{ hrs}$$

$$\text{Subsystem } C: \text{MTBF} = 9{,}500 \text{ hrs}$$

$$\text{Subsystem } D: \text{MTBF} = 12{,}100 \text{ hrs}$$

$$\text{Subsystem } E: \text{MTBF} = 3{,}600 \text{ hrs}$$

The five subsystems are connected in series. Determine the probability of survival for an operating period of 1,000 hrs.

16. Define producer's risk and consumer's risk.

17. Describe in your own words how the effects of maintenance can affect system reliability.

18. Name some of the advantages of reliability sequential testing. Identify some of the disadvantages.

19. Define what is meant by discrimination ratio. How is it used in reliability testing?

20. Why is failure analysis so important in testing and evaluation?

21. What is the probability that a system will operate without a failure if the expected number of failures during its operational period is 0.65?

22. If the system operating time is 300 hr and the MTBF is 300 hrs, what is the probability of system success (assuming exponential distribution)? What is the probability of survival if the operating time is extended to 500 hr? Reduced to 200 hrs?

14

Design for Maintainability

The systems engineering process requires an appropriate application of scientific and engineering efforts to ensure that the final product is operationally feasible. One of the more important design specialties in the system engineering process is *maintainability*. Maintainability is a design characteristic dealing with the ease, accuracy, safety, and economy in the performance of maintenance functions. It can be specified, controlled and measured, and maintainability methods and techniques are available to facilitate this process. Maintainability must be considered along with performance, reliability, human factors, producibility, supportability, life-cycle cost, and other factors in systems design.

Systems (or products) in use today are highly sophisticated and will fulfill most expectations when operating. However, experience has indicated that the reliability of many of these systems is marginal and that these systems are inoperative much of the time. Unreliable systems are unable to fulfill the mission for which they were designed and require extensive maintenance. In an environment where resources are becoming scarcer, it is essential that system maintenance requirements be minimized and that maintenance costs be reduced. Therefore, it is essential that maintainability be considered as a major system parameter in the design process.

14.1. DEFINITION AND EXPLANATION OF MAINTAINABILITY

Maintainability is an inherent design characteristic of a system or product. It pertains to the ease, accuracy, safety, and economy in the performance of maintenance actions. Systems engineers must be concerned with the design and development of a system that can be maintained in the least amount of time, at the least cost, and with a minimum expenditure of support resources (e.g., personnel, materials, facilities, test equipment) without adversely affecting the mission of that system. Maintainability is the *ability* of an item to be maintained, whereas maintenance constitutes a series of actions to be taken to restore or retain an item in an effective operational state. Maintainability is a design parameter. Maintenance is a result of design.

Maintainability can also be defined as a characteristic in design that can be expressed in terms of maintenance frequency factors, maintenance times (i.e., elapsed times and labor hours), and maintenance cost. These terms may be presented as different figures-of-merit; therefore, maintainability may be defined on the basis of a combination of factors such as

1. A characteristic of design and installation which is expressed as the probability that an item will be retained in or restored to a specified condition within a given period of time, when maintenance is performed in accordance with prescribed procedures and resources.[1]

2. A characteristic of design and installation which is expressed as the probability that maintenance will not be required more than x times in a given period, when the system is operated in accordance with prescribed procedures.

4. A characteristic of design and installation which is expressed as the probability that the maintenance cost for a system will not exceed y dollars per designated period of time, when the system is operated and maintained in accordance with prescribed procedures.

Consideration must be given to a combination of factors involving all aspects of the system or product if the objective of maintainability in design is to be achieved. The degree to which a system can be adequately supported during its operational phase is closely related to its maintainability, and maintainability is a design characteristic.

14.2. MAINTAINABILITY IN THE SYSTEM LIFE CYCLE

The systems engineering process, in evolving functional detail and design requirements, has as one of its major goals the achievement of the proper integration and balance of various system parameters, such as performance, reliability, maintainability, human factors, supportability, producibility, economic feasibility, and other factors. This process of design evolution is illustrated in Figure 2.4. Maintainability, as an inherent characteristic of design, must be an integral part of this overall process.

[1] MIL-STD-721, *Definitions of Effectiveness Terms for Reliability, Maintainability, Human Factors, and Safety*, Department of Defense, Washington, D.C.

Maintainability requirements are defined in conceptual design as part of system operational requirements and the maintenance concept. Maintainability analysis is conducted as part of systems analysis. Maintainability predictions are accomplished throughout preliminary and detail system/product design, maintainability is considered in formal design reviews, and maintainability demonstration is accomplished as part of system test and evaluation. In essence, maintainability (along with reliability and other major design parameters) is considered throughout the system life cycle. It is particularly relevant during the early phases of system design and development. Specific maintainability functions as related to the system life cycle are identified in Figure 14.1.

The material in this chapter will amplify the functions noted in Figure 14.1 and will cover maintainability requirements, maintainability analysis, design techniques and practices, maintenance engineering analysis, maintainability demonstration, and data analysis and corrective action. These functions are oriented to the various phases of the life cycle (i.e., conceptual design, preliminary and detail design, production/construction, system utilization and life-cycle support, and system retirement and phase-out).

Maintainability Planning

Maintainability planning must be accomplished as an integral part of the overall systems planning effort. Specific program functions are identified and scheduled, the organization and implementation of maintainability tasks are accomplished, and the results are evaluated through normal program review functions. The objective is to plan a program effort that will assure maintainability involvement throughout all aspects of system design and development, production or construction, and system utilization. A maintainability program plan is usually prepared at program inception and may be included as part of the system engineering management plan discussed in Part Five.

A typical maintainability plan should consider the following topic areas:

1. Maintainability quantitative and qualitative requirements for the system (as related to system operational requirements and the maintenance concept discussed in Chapter 3).
2. Allocation or apportionment of maintainability requirements to the subsystem level and beyond (as appropriate).
3. Design techniques and practices (i.e., equipment packaging schemes, self-test and diagnostic provisions, component parts standardization and mounting, modularization and accessibility, etc.).
4. Maintainability analysis (i.e., the evaluation of alternative design concepts, level of repair analysis, the generation of logic trouble-shooting flow diagrams, etc.).
5. Maintainability prediction(s) and assessment.

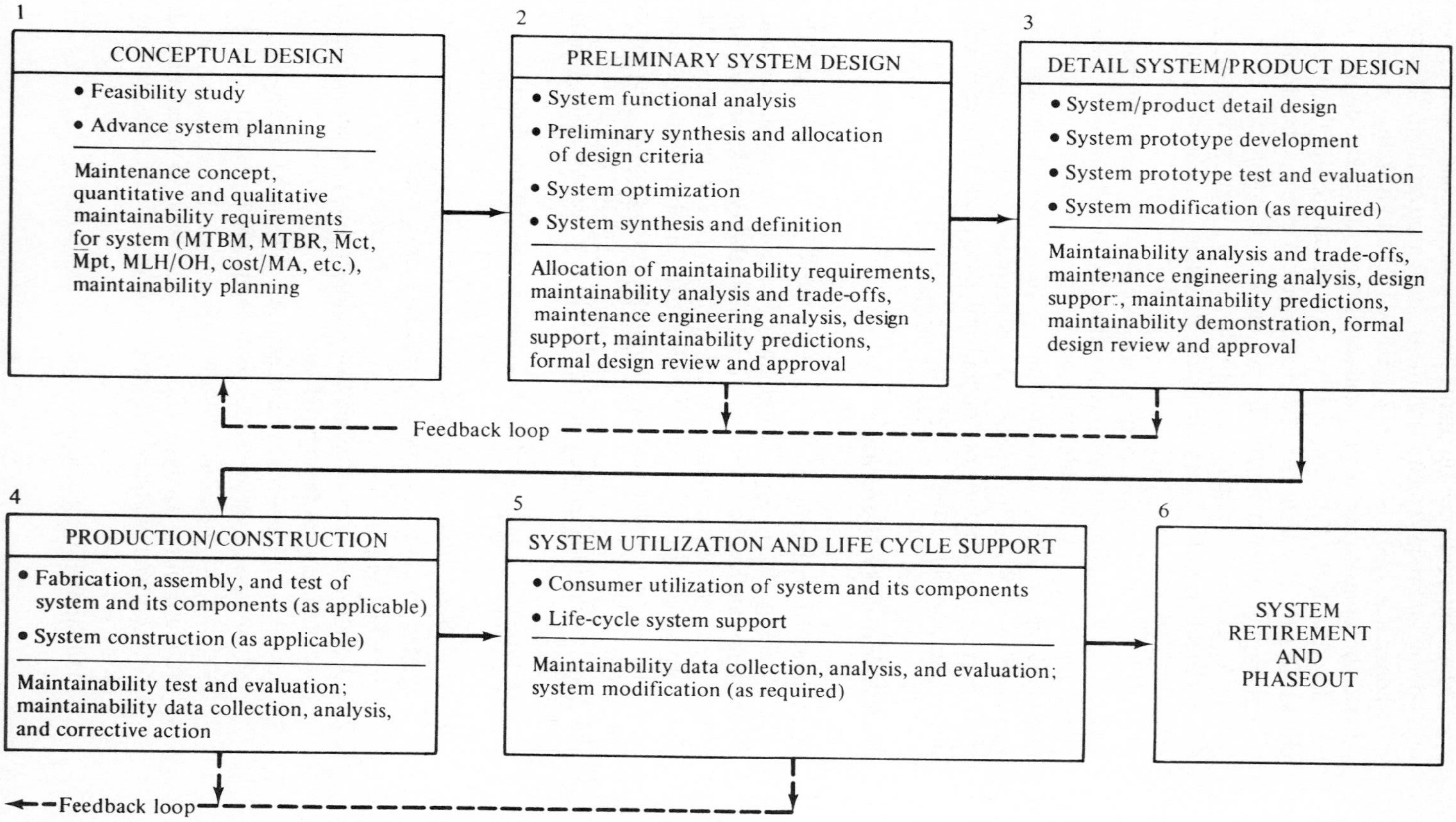

Figure 14.1 Maintainability requirements in the system life cycle.

CONCEPTUAL DESIGN

- Feasibility study
- Advance system planning

Maintenance concept, quantitative and qualitative maintainability requirements for system (MTBM, MTBR, Mct, Mpt, MLH/OH, cost/MA, etc.), maintainability planning

PRELIMINARY SYSTEM DESIGN

- System functional analysis
- Preliminary synthesis and allocation of design criteria
- System optimization
- System synthesis and definition

Allocation of maintainability requirements, maintainability analysis and trade-offs, maintenance engineering analysis, design support, maintainability predictions, formal design review and approval

DETAIL SYSTEM/PRODUCT DESIGN

- System/product detail design
- System prototype development
- System prototype test and evaluation
- System modification (as required)

Maintainability analysis and trade-offs, maintenance engineering analysis, design suppor;, maintainability predictions, maintainability demonstration, formal design review and approval

PRODUCTION/CONSTRUCTION

- Fabrication, assembly, and test of system and its components (as applicable)
- System construction (as applicable)

Maintainability test and evaluation; maintainability data collection, analysis, and corrective action

SYSTEM UTILIZATION AND LIFE CYCLE SUPPORT

- Consumer utilization of system and its components
- Life-cycle system support

Maintainability data collection, analysis, and evaluation; system modification (as required)

SYSTEM RETIREMENT AND PHASEOUT

Feedback loop

6. Maintenance engineering analysis (and its relationship to the logistic support analysis discussed in Chapter 16).

7. Formal design review(s).

8. Maintainability demonstration.

9. Data collection, analysis, and corrective action.

Finally, the maintainability plan must include the essential management aspects of project organization and control. Not only must internal procedures be thoroughly delineated, but the relationships and contractual provisions involving suppliers (and supplier control) must be clearly defined.

14.3. MEASURES OF MAINTAINABILITY

Maintainability, defined in the broadest sense, can be measured in terms of a combination of elapsed times, personnel labor hour rates, maintenance frequencies, maintenance cost, and related logistic support factors. The measures most commonly used are described in this section.

Maintenance Elapsed-Time Factors

Maintenance can be classified in two categories:

1. *Corrective maintenance*—the unscheduled actions accomplished, as a result of failure, to *restore* a system to a specified level of performance.

2. *Preventive maintenance*—the scheduled actions accomplished to *retain* a system at a specified level of performance by providing systematic inspection, detection, and/or prevention of impending failures.

Maintenance constitutes the act of diagnosing and repairing, or preventing, system failures. Maintenance time is made up of the individual task times associated with the required corrective and preventive maintenance actions for a given system or product. Maintainability is a measure of the ease and rapidity with which a system can be maintained, and is measured in terms of the time required to perform maintenance tasks. A few of the more commonly used maintainability time measures are defined below.

Mean corrective maintenance time ($\overline{M}ct$).   Each time that a system fails, a series of steps is required to repair or restore the system to its full operational status. These steps include failure detection, fault isolation, disassembly to gain access to the faulty item, repair, and so on, as illustrated in Figure 14.2. Completion of these steps for a given failure constitute a corrective maintenance cycle.

Throughout the system use phase, there will usually be a number of individual

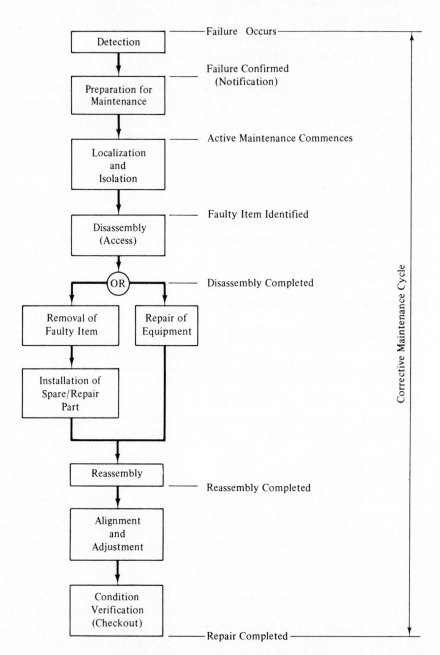

Figure 14.2 Corrective maintenance cycle.

maintenance actions involving the series of steps illustrated in Figure 14.2. The mean corrective maintenance time ($\overline{M}ct$), or the mean time to repair (MTTR) which is equivalent, is a composite value representing the arithmetic average of these individual maintenance cycle times.

For the purposes of illustration, Table 14.1 includes data covering a sample of 50 corrective maintenance repair actions on a typical equipment item. Each of the times indicated represents the completion of one corrective maintenance cycle illustrated in Figure 14.2. Based on the set of raw data presented, which constitutes a random sample, a frequency distribution table and frequency histogram may be prepared, as illustrated in Table 14.2 and Figure 14.3, respectively.

TABLE 14.1   CORRECTIVE MAINTENANCE TIMES (MINUTES)

40	58	43	45	63	83	75	66	93	92
71	52	55	64	37	62	72	97	76	75
75	64	48	39	69	71	46	59	68	64
67	41	54	30	53	48	83	33	50	63
86	74	51	72	87	37	57	59	65	63

TABLE 14.2   FREQUENCY DISTRIBUTION

CLASS INTERVAL	FREQUENCY	CUMULATIVE FREQUENCY
29.5–39.5	5	5
39.5–49.5	7	12
49.5–59.5	10	22
59.5–69.5	12	34
69.5–79.5	9	43
79.5–89.5	4	47
89.5–99.5	3	50

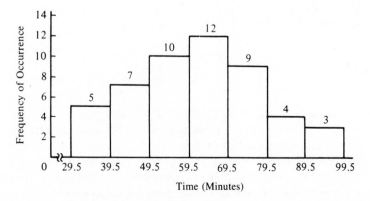

Figure 14.3   Histogram of maintenance times.

Referring to Table 14.2, the range of observations is between 97 min and 30 min, or a total of 67 min. This range can be divided into class intervals, with a class interval width of 10 assumed for convenience. A logical starting point is to select class intervals of 20–29, 30–39, and so on. In such instances, it is necessary to establish the dividing point between two adjacent intervals, such as 29.5, 39.5, and so on.

Given the frequency distribution of repair times, one can plot a histogram showing time values in minutes and the frequency of occurrence as in Figure 14.3. By determining the midpoint of each class interval, a frequency polygon can be developed as illustrated in Figure 14.4. This provides an indication of the form of the probability distribution applicable to repair times for this particular system.

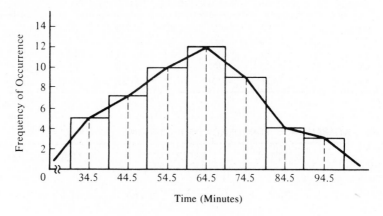

Figure 14.4   Frequency polygon.

As additional corrective maintenance actions occur and data points are plotted for the system in question, the curve illustrated in Figure 14.4 may take the shape of the normal distribution. The normal curve is defined by the arithmetic mean ($\overline{\text{Mct}}$) and the standard deviation (σ). From the maintenance repair times presented in Table 14.1, the arithmetic mean is determined from:

$$\overline{\text{Mct}} = \frac{\sum\limits_{i=1}^{n} \text{Mct}_i}{n} = \frac{3{,}095}{50} = 61.9 \quad \text{(assume 62)}$$

where Mct_i is the total active corrective maintenance cycle time for each maintenance action and n is the sample size. Thus, the average value for the sample of 50 maintenance actions is 62 minutes.

The standard deviation (σ) measures the dispersion of maintenance time values. When a standard deviation is calculated, it is convenient to generate a table giving the deviation of each task time from the mean of 62. Table 14.3 illustrates this for only four individual task times, although all 50 tasks should be treated. The total value of 13,013 does cover all 50 tasks.

TABLE 14.3 VARIANCE DATA

TOTAL	$\mathrm{Mct}_i - \overline{\mathrm{Mct}}$	$(\mathrm{Mct}_i - \overline{\mathrm{Mct}})^2$
40	-22	484
71	$+9$	81
75	$+13$	169
67	$+5$	25
etc.	etc.	etc.
Total		13,013

The standard deviation of the sample normal distribution curve can now be determined as

$$\sigma = \sqrt{\frac{\sum\limits_{i=1}^{n} (\mathrm{Mct}_i - \overline{\mathrm{Mct}})^2}{n-1}} = \sqrt{\frac{13,013}{49}} = 16.3 \text{ min} \quad (\text{assume } 16).$$

One may wish to know what percentage of the total repair actions falls within the range 46 to 78 minutes. This can be calculated by converting the range to multiples of standard deviation. In this case, the interval of 46 to 78 equals $\overline{\mathrm{Mct}} \pm 1\sigma$, or 62 minutes plus and minus the standard deviation of 16 minutes. When normality is assumed, it can be stated that 68% of the total population sampled falls within the range 46 to 78 min and that 99.7% of the sample population lies within the range $\overline{\mathrm{Mct}} \pm 3\sigma$, or 14 to 110 minutes.

As a typical application, it may be desirable to determine the percent of total population repair times that lies between 40 and 50 minutes. Graphically, this is represented in Figure 14.5. The problem is to find the percent represented by the shaded area. This can be calculated as follows:

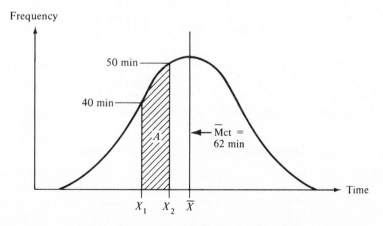

Figure 14.5 Normal distribution of repair times.

1. Convert maintenance times of 40 and 50 min into standard values (Z) or the number of standard deviations above and below the mean of 62 minutes.

$$Z \text{ for 40 min} = \frac{X_1 - \overline{X}}{\sigma} = \frac{40 - 62}{16} = -1.37$$

$$Z \text{ for 50 min} = \frac{X_2 - \overline{X}}{\sigma} = \frac{50 - 62}{16} = -0.75$$

The maintenance times of 40 and 50 minutes represent -1.37 and -0.75 standard deviations below the mean, since the values are negative.

2. Point X_1 $(Z = -1.37)$ represents an area of 0.0853 and point X_2 $(Z = -0.75)$ represents an area of 0.2266, as given in Appendix B, Table B.1.

3. The shaded area A in Figure 14.5 represents the difference in area or, area $X_2 - X_1 = 0.2266 - 0.0853 = 0.1413$. Thus, 14.13% of the population of maintenance times are estimated to lie between 40 and 50 minutes.

Next, confidence limits should be determined. Since the 50 maintenance tasks represent only a sample of all maintenance actions on the equipment being evaluated, it is possible that another sample of 50 maintenance actions on the same equipment could have a mean value either greater or less than 62 minutes. The original 50 tasks were selected at random, however, and statistically represent the entire population. Using the standard deviation, an upper and lower limit can be placed on the mean value $(\overline{M}ct)$ of the population. For instance, if one is willing to accept a chance of being wrong 15% of the time (85% confidence limit), then

$$\text{upper limit} = \overline{M}ct + Z\left(\frac{\sigma}{\sqrt{n}}\right) \tag{14.1}$$

where $\sigma/\sqrt{n}$ represents the standard error factor.

The Z value is obtained from Appendix E, where 0.1492 is close to 15% and reflects a Z of 1.04. Thus,

$$\text{upper limit} = 62 + (1.04)\left(\frac{16}{\sqrt{50}}\right) = 64.35 \text{ mins.}$$

This means that the upper limit is 64.4 minutes at a confidence level of 85%, or that there is an 85% chance that $\overline{M}ct$ will be less than 64.4. Variations in risk and upper limit values are shown in Table 14.4. If a specified $\overline{M}ct$ limit is established for the design of equipment (based on mission and operational requirements) and it is

TABLE 14.4 RISK-UPPER LIMIT VARIATIONS

RISK	CONFIDENCE	Z	UPPER LIMIT
5%	95%	1.65	65.72 minutes
10%	90%	1.28	64.89 minutes
15%	85%	1.04	64.35 minutes
20%	80%	0.84	63.89 minutes

known (or assumed) that maintenance times are normally distributed, then one would have to compare the results of predictions and/or measurements (e.g., 64.35 minutes) accomplished during the development process with the specified value to determine the degree of compliance.

When considering probability distributions in general, the time dependency be-- tween probability of repair and the time allocated for repair can usually be expected to produce a probability density function in one of three common distribution forms (normal, exponential, and log-normal) as illustrated in Figure 14.6.

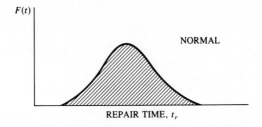

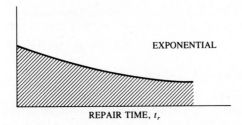

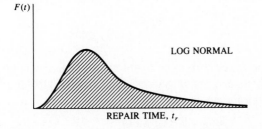

Figure 14.6   Repair time distributions.

1. The *normal* distribution applies to the relatively straightforward maintenance tasks and repair actions (simple removal and replacement tasks) which consistently require a fixed amount of time to complete with very little variation.

2. The *exponential* distribution applies to maintenance tasks involving part substitution methods of failure isolation in large systems that result in a constant repair rate.

3. The *log-normal* distribution applies to most maintenance tasks and repair actions comprised of several subsidiary tasks of unequal frequency and time du-

ration. Experience has shown that in almost all cases the distribution of maintenance times for complex equipment and systems is log-normal.

As indicated, the maintenance task times for many systems and equipments do not always fit within the normal curve. There may be a few representative maintenance actions where repair times are extensive, causing a skew to the right. This is particularly true for electronic equipment items, where the distribution of repair times often follows a log-normal curve as shown in Figure 14.7. Derivation of the specific distribution curve for a set of maintenance task times is accomplished using the same procedure as given in the preceding paragraphs. A frequency table is generated and a histogram is plotted.

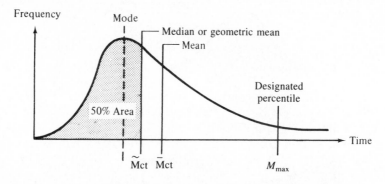

Figure 14.7   Log-normal distribution.

A sample of 24 corrective maintenance repair actions for a typical electronic equipment item is presented in Table 14.5. Using the data in the table, the mean is determined as

$$\overline{\text{Mct}} = \frac{\sum\limits_{i=1}^{n} \text{Mct}_i}{n} = \frac{1{,}637}{24} = 68.21 \text{ mins.}$$

When determining the mean corrective maintenance time ($\overline{\text{Mct}}$) for a specific sample population of maintenance repair actions, the use of Equation 14.2, which has wider application, is appropriate and is

$$\overline{\text{Mct}} = \frac{\sum (\lambda_i)(\text{Mct}_i)}{\sum \lambda_i} \tag{14.2}$$

where λ_i is the failure rate of the individual (ith) element of the item being mea-

TABLE 14.5   CORRECTIVE MAINTENANCE REPAIR
TIMES (MINUTES)

55	28	125	47	58	53	36	88
51	110	40	75	64	115	48	52
60	72	87	105	55	82	66	65

sured, usually expressed in failures per equipment operating hour. Equation 14.2 calculates $\overline{\text{M}}$ct as a weighted average using reliability factors.

It should be noted that $\overline{\text{M}}$ct considers only *active* maintenance time or that time which is spent working directly on the system. Logistics delay time and administrative delay time are not included. Although all elements of time are important, the $\overline{\text{M}}$ct factor is primarily oriented to a measure of the supportability characteristics in equipment design.

Mean preventive maintenance time ($\overline{\text{M}}$pt).   Preventive maintenance refers to the actions required to retain a system at a specified level of performance and may include such functions as periodic inspection, servicing, scheduled replacement of critical items, calibration, overhaul, and so on. $\overline{\text{M}}$pt is the mean (or average) elapsed time to perform preventive or scheduled maintenance on an item, and is expressed as

$$\overline{\text{M}}\text{pt} = \frac{\Sigma \, (\text{fpt}_i)(\text{Mpt}_i)}{\Sigma \, \text{fpt}_i} \tag{14.3}$$

where fpt_i is the frequency of the individual (ith) preventive maintenance action in actions per system operating hour, and Mpt_i is the elapsed time required for the ith preventive maintenance action.

Preventive maintenance may be accomplished while the system is in full operation or could result in downtime. In this instance, the concern is for preventive maintenance actions that result in system downtime. Again, $\overline{\text{M}}$pt includes only *active* system maintenance time, not logistic delay and administrative delay times.

Median active corrective maintenance time ($\widetilde{\text{M}}$ct).   The median maintenance time is that value which divides all of the downtime values so that 50% are equal to or less than the median and 50% are equal to or greater than the median. The median will usually give the best average location of the data sample. The median for a normal distribution is the same as the mean, while the median in a lognormal distribution is the same as the geometric mean, illustrated in Figure 14.7. $\widetilde{\text{M}}$ct is calculated as

$$\widetilde{\text{M}}\text{ct} = \text{antilog} \, \frac{\sum_{i=1}^{n} \log \text{Mct}_i}{n} = \text{antilog} \, \frac{\Sigma \, (\lambda_i)(\log \text{Mct}_i)}{\Sigma \, \lambda_i}. \tag{14.4}$$

For illustrative purposes, the maintenance time values in Table 14.5 are presented in the format illustrated in Table 14.6. The median is computed as

$$\widetilde{\text{M}}\text{ct} = \text{antilog} \, \frac{\sum_{1}^{24} \log \text{Mct}_i}{24}$$

$$= \text{antilog} \, \frac{43.315}{24} = \text{antilog} \, 1.805 = 63.8 \text{ mins.}$$

TABLE 14.6 CALCULATION FOR $\tilde{M}$ct

Mct_i	LOG Mct_i	$(LOG\ Mct_i)^2$	Mct_i	LOG Mct_i	$(LOG\ Mct_i)^2$
55	1.740	3.028	64	1.806	3.262
28	1.447	2.094	115	2.061	4.248
125	2.097	4.397	48	1.681	2.826
47	1.672	2.796	52	1.716	2.945
58	1.763	3.108	60	1.778	3.161
53	1.724	2.972	72	1.857	3.448
36	1.556	2.421	87	1.939	3.760
88	1.945	3.783	105	2.021	4.084
51	1.708	2.917	55	1.740	3.028
110	2.041	4.166	82	1.914	3.663
40	1.602	2.566	66	1.819	3.309
75	1.875	3.516	65	1.813	3.287
TOTAL				43.315	78.785

Median active preventive maintenance time ($\tilde{M}$pt). The median active preventive maintenance time is determined using the same approach as for calculating $\tilde{M}$ct. $\tilde{M}$pt is expressed as

$$\tilde{M}pt = antilog\ \frac{\Sigma\ (fpt_i)(log\ Mpt_i)}{\Sigma\ fpt_i}.$$ (14.5)

Mean active maintenance time ($\overline{M}$). $\overline{M}$ is the mean or average elapsed time required to perform scheduled (preventive) and unscheduled (corrective) maintenance. It excludes logistics delay time and administrative delay time, and is expressed as

$$\overline{M} = \frac{(\lambda)(\overline{M}ct) + (fpt)(\overline{M}pt)}{\lambda + fpt}$$ (14.6)

where λ is the corrective maintenance rate or failure rate, and fpt is the preventive maintenance rate.

Maximum active corrective maintenance time (M_{max}). M_{max} can be defined as that value of maintenance downtime below which a specified percent of all maintenance actions can be expected to be completed. M_{max} is related primarily to the log-normal distribution, and the 90th or 95th percentile point is generally taken as the specified value as shown in Figure 14.7. It is expressed as

$$M_{max} = antilog\ (\overline{log\ Mct} + Z\sigma_{log\ Mct_i})$$ (14.7)

where $\overline{\log \text{Mct}}$ is the mean of the logarithms of Mct_i, Z the value corresponding to the specific percentage point at which M_{max} is defined (see Table 14.4, -1.65 for 95%), and

$$\sigma_{\log \text{Mct}_i} = \sqrt{\frac{\sum_{i=1}^{n} (\log \text{Mct}_i)^2 - (\sum_{i=1}^{n} \log \text{Mct}_i)^2/n}{n-1}} \qquad (14.8)$$

or the standard deviation of the sample logarithms of average repair times, Mct_i.

For example, determining M_{max} at the 95th percentile for the data sample in Table 14.5 is accomplished as

$$M_{max} = \text{antilog } [\log \widetilde{\text{Mct}} + (1.65)\sigma_{\log \text{Mct}_i}] \qquad (14.9)$$

where, referring to Equation (14.8) and Table 14.6,

$$\sigma_{\log \text{Mct}_i} = \sqrt{\frac{78.785 - (43.315)^2/24}{23}} = 0.163.$$

Substituting the standard deviation factor and the mean value into Equation 14.9, gives

$$M_{max} = \text{antilog } [\log \widetilde{\text{Mct}} + (1.65)(0.163)]$$

$$= \text{antilog } (1.805 + 0.269) = 119 \text{ mins.}$$

If maintenance times are distributed log-normally, M_{max} cannot be derived directly by using the observed maintenance values. However, by taking the logarithm of each repair value, the resulting distribution becomes normal, facilitating usage of the data in the manner identical to the normal case.

Logistics delay time (LDT).   Logistics delay time refers to that maintenance downtime which is expended as a result of waiting for a spare part to become available, waiting for the availability of an item of test equipment in order to perform maintenance, waiting for transportation, waiting to use a facility required for maintenance, and so on. LDT does not include active maintenance time, but does constitute a major element of total maintenance downtime (MDT).

Administrative delay time (ADT).   Administrative delay time refers to that portion of downtime during which maintenance is delayed for reasons of an administrative nature (e.g., personnel assignment priority, labor strike, organizational constraint, etc.). ADT does not include active maintenance time, but often constitutes a significant element of total maintenance downtime (MDT).

Maintenance downtime (MDT).   Maintenance downtime constitutes the total elapsed time required (when the system is not operational) to repair and restore a system to full operating status, and/or to retain a system in that condition. MDT includes mean active maintenance time ($\overline{M}$), logistics delay time (LDT), and administrative delay time (ADT). The mean or average value is calculated from the elapsed times for each function and the associated frequencies (similar to the approach used in determining $\overline{M}$).

Maintenance Labor-Hour Factors

The maintainability factors covered in the previous paragraphs relate to elapsed times. Although elapsed times are extremely important in the performance of maintenance, one must also consider the maintenance labor-hours expended in the process. Elapsed times can be reduced (in many instances) by applying additional human resources in the accomplishment of specific tasks. However, this may turn out to be an expensive trade-off, particularly when high skill levels are required to perform tasks which result in less overall clock time. In other words, maintainability is concerned with the *ease* and *economy* in the performance of maintenance. As such, an objective is to obtain the proper balance among elapsed time, labor time, and personnel skills at a minimum maintenance cost.

When considering measures of maintainability, it is not only important to address such factors as $\overline{Mct}$ and MDT, but it is also necessary to consider the labor-time element. Thus, some additional measures must be employed, such as the following:

1. Maintenance labor-hours per system operating hour (MLH/OH).
2. Maintenance labor-hours per cycle of system operation (MLH/cycle).
3. Maintenance labor-hours per month (MLH/month).
4. Maintenance labor-hours per maintenance action (MLH/MA).

Any of these factors can be specified in terms of mean values. For example, $\overline{\text{MLH}}_c$ is the mean corrective maintenance labor-hours, expressed as

$$\overline{\text{MLH}}_c = \frac{(\Sigma\ \lambda_i)(\text{MLH}_i)}{\Sigma\ \lambda_i} \qquad (14.10)$$

where λ_i is the failure rate of the ith item (failures/hour), and MLH_i is the average maintenance labor-hours necessary to complete repair of the ith item.

Additionally, the values for mean preventive maintenance labor-hours and mean total maintenance labor-hours (to include preventive and corrective maintenance) can be calculated on a similar basis. These values can be predicted for each echelon or level of maintenance and are employed in determining specific support requirements and associated cost.

Maintenance Frequency Factors

Section 13.3 covers the measures of reliability, with MTBF and λ being key factors. Based on the discussion thus far, it is obvious that reliability and maintainability are very closely related. The reliability factors, MTBF and λ, are the basis for determining the frequency of corrective maintenance. Maintainability deals with the characteristics in system design pertaining to minimizing the corrective maintenance requirements for the system when it assumes operational status. Thus, in this area, reliability and maintainability requirements for a given system must be compatible and mutually supportive.

In addition to the corrective maintenance aspect of system support, maintainability also deals with the characteristics of design which minimize (if not eliminate) preventive maintenance requirements for that system. Sometimes, preventive maintenance requirements are added with the objective of improving system reliability (e.g., reducing failures by specifying selected component replacements at designated times). However, the introduction of preventive maintenance can turn out to be quite costly if not carefully controlled. Further, the accomplishment of too much preventive maintenance (particularly for complex systems or products) often has a degrading effect on system reliability as failures are frequently induced in the process. Hence, an objective of maintainability is to provide the proper balance between corrective maintenance and preventive maintenance at least overall cost.

Mean time between maintenance (MTBM).   MTBM is the mean or average time between all maintenance actions (corrective and preventive) and can be calculated as

$$\text{MTBM} = \frac{1}{1/\text{MTBM}_u + 1/\text{MTBM}_s} \qquad (14.11)$$

where MTBM_u is the mean interval of unscheduled (corrective) maintenance and MTBM_s is the mean interval of scheduled (preventive) maintenance. The reciprocals of MTBM_u and MTBM_s constitute the maintenance rates in terms of maintenance actions per hour of system operation. MTBM_u should approximate MTBF, assuming that a combined failure rate is used which includes the consideration of primary inherent failures, dependent failures, manufacturing defects, operator and maintenance induced failures, and so on. The maintenance frequency factor, MTBM, is a major parameter in determining system achieved availability, Equation 13.17, and operational availability, Equation 13.18, as described in Section 13.3.

Mean time between replacement (MTBR).   MTBR, a factor of MTBM, refers to the mean time between item replacement and is a major parameter in determining spare part requirements. On many occasions, corrective and preventive maintenance actions are accomplished without generating the requirement for the replacement of a component part. In other instances, item replacements are required, which in turn necessitates the availability of a spare part and an inventory requirement. Additionally, higher levels of maintenance support (i.e., intermediate and depot levels) may also be required.

In essence, MTBR is a significant factor, applicable in both corrective and preventive maintenance activities involving item replacement, and is a key parameter in determining logistic support requirements. A maintainability objective in system design is to minimize MTBR where feasible.

Maintenance Cost Factors

For many systems/products, maintenance cost constitutes a major segment of total life-cycle cost. Further, experience has indicated that maintenance costs are significantly affected by design decisions made throughout the early stages of system development. Thus, it is essential that total life-cycle cost be considered as a major design parameter beginning with the definition of system requirements (refer to Chapters 3 and 4).

Of particular interest in this chapter is the aspect of *economy* in the performance of maintenance actions. In other words, maintainability is directly concerned with the characteristics of system design that will ultimately result in the accomplishment of maintenance at minimum overall cost.

When considering maintenance cost, the following cost-related indices may be appropriate as criteria in system design:

1. Cost per maintenance action ($/month).
2. Maintenance cost per system operating hour ($/OH).
3. Maintenance cost per month ($/month).
4. Maintenance cost per mission or mission segment ($/mission).
5. The ratio of maintenance cost to total life-cycle cost.

Related Maintenance Factors

It is evident from the system maintenance concept in Section 3.1 that there are a number of additional factors that are closely related to and highly dependent on the maintainability measures described. These include various logistics factors, such as

1. Supply responsiveness or the probability of having a spare part available when needed, supply lead times for given items, levels of inventory, and so on.
2. Test and support equipment effectiveness (reliability and availability of test equipment), test equipment utilization, system test thoroughness, and so on.
3. Maintenance facility availability and use.
4. Transportation times between maintenance facilities.
5. Maintenance organizational effectiveness and personnel efficiency.

There are numerous other logistics factors that should also be specified, measured, and controlled if the ultimate mission of the system is to be fulfilled. For instance, specifying a 15-min $\overline{M}ct$ requirement is highly questionable if there is a low probability of having a spare part available when required (resulting in a long logistics delay-time possibility); specifying specific maintenance labor-hour requirements may not be appropriate if the maintenance organization is not properly staffed or available to perform the required function(s); specifying system test-time requirements may be inappropriate if the predicted reliability of the test equipment is less than the reliability of the item being tested; and so on.

There are many examples where the interactions between the prime system and its elements of support are critical, and both areas must be considered in the establishment of system requirements during conceptual design. Maintainability, as a characteristic in design, is closely related to the area of system support since the results of maintainability directly affect maintenance requirements. Thus, when specifying maintainability factors, one should also address the qualitative and quantitative requirements for system support in order to determine the effects of one area on another. Logistic support factors are discussed further in Chapter 16.

14.4. *MAINTAINABILITY IN SYSTEM OR PRODUCT DESIGN*

Maintainability in system or product design begins with the establishment of requirements as part of the conceptual design phase, as is shown in Figure 14.1, Block 1. Through maintainability analyses, these requirements are allocated to various elements of the system. Design techniques are then used as appropriate to assure that the design is in compliance with allocated requirements (refer to Figure 14.1, Blocks 2 and 3). When a prototype of the system is developed, a demonstration is conducted to assess the actual maintainability of the system at that stage. The results reflect a measurement of the system just prior to entering the production/construction phase (refer to Figure 14.1, Block 3). The major maintainability activities applicable to system/product design are covered in this section.

Maintainability Requirements

Every system is developed in response to a need or to fulfill some anticipated function. The effectiveness with which the system fulfills this function is the ultimate measure of its utility and its value to the consumer. This effectiveness is a composite of performance, maintainability, and other factors. In essence, maintainability constitutes a major factor in determining the usefulness of the system.

Maintainability requirements for a system, specified in both quantitative and qualitative terms, are defined as part of the overall system operational requirements and the maintenance concept described in Section 3.1. Of particular interest is the following information:

1. Definition of system performance factors, the mission profile, and system utilization requirements (use conditions, duty cycles, and how the system is to be operated).
2. Definition of the operational life cycle (i.e., the anticipated time that the system will be in the inventory and in operational use).
3. Definition of the basic system support concept [i.e., the anticipated levels of maintenance, maintenance responsibilities, major functions at each level, and the prime elements of logistic support at each level (type of spares, test equipment, personnel skills, facilities, etc.)].
4. Definition of the environment in which the system is expected to operate and be maintained (e.g., temperature, humidity, vibration, etc.). This should include a range of values as applicable and should cover all transportation, handling, and storage modes.

Given the foregoing information, one must determine the extent to which maintainability characteristics should be incorporated in the system design. If the operational requirements specify that the system must function 90% of the time and the estimated reliability is low, then the system maintainability requirements may be rather stringent in order to maintain the overall 90% availability. On the other hand, if the estimated reliability is high (resulting in very few failures), the specified maintainability requirements may be quite different. Further, if the maintenance concept dictates that only the organization and depot levels of maintenance are allowed, then the maintainability requirements may be quite different than what would likely be specified with three levels of maintenance planned (to include intermediate maintenance). In any event, quantitative and qualitative maintainability requirements must be defined for the system based on the foregoing considerations. Quantitative requirements are usually expressed in terms of MTBM, MTBR, $\overline{M}$, $\overline{M}$ct, $\overline{M}$pt, MLH/OH, \$/MA, or a combination thereof.

In specifying maintainability, one must first identify the major system requirements from the top-level functional flow diagram (as illustrated in Figure 4.3). The functions in the top-level diagram are expanded as necessary for the purposes of system definition, and a functional packaging scheme is identified as illustrated in Figure 4.4. Maintenance functional diagrams are developed as shown in Figure 4.6, significant maintenance functions are defined, and maintainability requirements are established for the appropriate maintenance functions and the various functional packages of the system. This is an iterative process of requirements definition.

Maintainability Allocation

Once requirements for the system have been established, it is then necessary to translate these requirements into lower-level design criteria through maintainability allocation. This is accomplished as part of the system requirements allocation process described in Section 4.2.

For the purpose of illustration, it is assumed that System *XYZ* in Figure 4.9 must be designed to meet an inherent availability requirement of 0.9989, a MTBF of 450, and a MLH/OH (for corrective maintenance) of 0.2, and a need exists to allocate $\overline{M}$ct and MLH/OH to the assembly level.[2] From Equation 13.16, $\overline{M}$ct is

$$\overline{M}\text{ct} = \frac{\text{MTBF}(1 - A_i)}{A_i} \tag{14.12}$$

or

$$\overline{M}\text{ct} = \frac{450(1 - 0.9989)}{0.9989} = 0.5 \text{ hr.}$$

Thus, the system's $\overline{M}$ct requirement is 0.5 hour, and this requirement must be allocated to units *A*, *B*, *C*, and the assemblies within each unit. The allocation process is facilitated through the use of a format similar to that illustrated in Table 14.7.

Referring to Table 14.7, each item type and the quantity (*Q*) of items per system are indicated. Allocated reliability factors are specified in column 3 (refer to Figure 13.11) and the degree to which the failure rate of each unit contributes to the overall failure rate (represented by C_f) is entered in column 4. The average correc-

---

[2] MTBM and $\overline{M}$pt may be allocated on a comparable basis with MTBF and $\overline{M}$ct, respectively.

TABLE 14.7 MAINTAINABILITY ALLOCATION FOR SYSTEM ZYZ

1	2	3	4	5	6	7
ITEM	QUANTITY OF ITEMS PER SYSTEM (Q)	FAILURE RATE $(\lambda) \times 1000$ HR	CONTRIBUTION OF TOTAL FAILURES $C_f = (Q)(\lambda)$	PERCENT CONTRIBUTION $C_p = C_f \Sigma C_f \times 100$	AVERAGE CORRECTIVE MAINT. TIME $\overline{M}ct(hr)$	CONTRIBUTION OF TOTAL CORRECTIVE MAINT. TIME $C_t = (C_f)(\overline{M}ct)$
1. Unit A	1	0.246	0.246	11%	0.9	0.221
2. Unit B	1	1.866	1.866	84%	0.4	0.746
3. Unit C	1	0.110	0.110	5%	1.0	0.110
Total			$\Sigma C_f = 2.222$	100%		$\Sigma C_t = 1.077$

$\overline{M}ct$ for System $XYZ = \dfrac{\Sigma C_t}{\Sigma C_f} = \dfrac{1.077}{2.222} = 0.485$ Hour (Requirement: 0.5 Hour)

tive maintenance time for each unit is estimated and entered in column 6. These times are ultimately based on the inherent characteristics of equipment design, which are not known at this point in the system life cycle. Thus, corrective maintenance times are initially derived using a complexity factor, which is indicated by failure rate. As a goal, the item that contributes the highest percentage to the anticipated total failures (unit B in this instance) should require a low $\overline{M}ct$, and those with low contributions may require a higher $\overline{M}ct$. On certain occasions, however, the design costs associated with obtaining a low $\overline{M}ct$ for a complex item may lead to a modified approach, which is feasible as long as the end result ($\overline{M}ct$ at the system level) falls within the quantitative requirement.[3]

The estimated value of C_t for each unit is entered in column 7, and the sum of the contributions for all units can be used to determine the overall system's $\overline{M}ct$ as

$$\overline{M}ct = \frac{\Sigma\ C_t}{\Sigma\ C_f} = \frac{1.077}{2.222} = 0.485\ hr. \qquad (14.13)$$

In Table 14.7, the calculated $\overline{M}ct$ for the system is within the requirement of 0.5 hour. The $\overline{M}ct$ values for the units provide corrective maintenance downtime criteria for design, and the values are included in equipment design specifications.

Once allocation is accomplished at the unit level, the resultant $\overline{M}ct$ values can be allocated to the next-lower equipment indenture item. For instance, the 0.4 hour $\overline{M}ct$ value for unit B can be allocated to assemblies 1, 2, and 3, and the procedure for allocation is the same as employed in Equation (14.13). An example of allocated values for the assemblies of unit B is included in Table 14.8.

TABLE 14.8 UNIT B ALLOCATION

1	2	3	4	5	6	7
Assembly 1	1	0.116	0.116	6%	0.5	0.058
Assembly 2	1	1.550	1.550	83%	0.4	0.620
Assembly 3	1	0.200	0.200	11%	0.3	0.060
Total			1.866	100%		0.738

$$\overline{M}ct\ for\ Unit\ B = \frac{\Sigma\ C_t}{\Sigma\ C_f} = \frac{0.738}{1.866} = 0.395\ Hour\ (Requirement:\ 0.4\ Hour)$$

The $\overline{M}ct$ value covers the aspect of *elapsed* or *clock time* for restoration actions. Sometimes this factor, when combined with a reliability requirement, is sufficient to establish the necessary maintainability characteristics in design. On other occasions, specifying $\overline{M}ct$ by itself is not adequate, since there may be a number of design approaches that will meet the $\overline{M}ct$ requirement but not necessarily in a cost-effective manner. Meeting a $\overline{M}ct$ requirement may result in an increase in the

---

[3] Note that, in any event, the maintainability parameters are dependent upon the reliability parameters. Also, it will frequently occur that reliability allocations are incompatible with maintainability allocations (or vice versa). Hence, a close feedback relationship between these activities is mandatory.

skill levels of personnel accomplishing maintenance actions, increasing the quantity of personnel for given maintenance functions, and/or incorporating automation for manual operations. In each instance there are costs involved; thus, one may wish to specify additional constraints, such as the skill level of personnel at each maintenance level and the maintenance labor-hours per operating hour (MLH/OH) for significant equipment items. In other words, a requirement may dictate that an item be designed such that it can be repaired within a specified elapsed time with a given quantity of personnel possessing skills of a certain level. This will influence design in terms of accessibility, packaging schemes, handling requirements, diagnostic provisions, and so on, and is perhaps more meaningful overall.

The factor MLH/OH is a function of task complexity and the frequency of maintenance. The system-level requirement is allocated on the basis of system operating hours, the anticipated quantity of maintenance actions, and an estimate of the number of labor-hours per maintenance action. Experience data are used where possible.

Following the completion of quantitative allocations for each indenture level of equipment, all values are included in the functional breakdown illustrated in Figure 4.9. The figure provides an overview of major system design requirements.

Maintainability Analysis

The maintainability analysis constitutes the iterative process of requirements allocation, synthesis, optimization, and definition and is an inherent part of the overall systems analysis effort described throughout Chapters 3, 4, and 5. Specifically, the maintainability analysis involves the ongoing evaluation of various possible design and support alternatives, following the steps illustrated in Figure 3.7, with the ultimate objective of developing a system configuration that will fulfill the stated requirements.

The maintainability analysis per se pertains primarily to the evaluation of alternative system/product design configurations involving maintainability options, alternative repair policies, alternative logistic support plans, and so on. The specific problems vary from system to system depending on the phase of the life cycle, the stage of development, and so on. To illustrate the types of problems addressed through maintainability analysis, two examples will be presented in the sections which follow.

Trade-off evaluation of reliability and maintainability.   Suppose that there is a requirement to replace an existing equipment item with a new item for the purpose of improving operational effectiveness. The current need specifies that the equipment must operate 8 hrs per day, 360 days per year, for 10 years. The existing equipment meets an availability of 0.961, a MTBF of 125 hours, and a $\overline{M}ct$ of 5 hours. The new system must meet an availability of 0.990, a MTBF greater than 300 hours, and a $\overline{M}ct$ not to exceed 5.0 hours. An anticipated quantity of 200 items of equipment is to be procured. Three different alternative design configurations are being considered to satisfy the requirement, and each configuration constitutes a modification of the existing equipment.

Figure 14.8 graphically shows the relationships among inherent availability (A_i), MTBF, and $\overline{M}ct$ and illustrates the allowable area for trade-off. The selected configuration must reflect the reliability and maintainability characteristics represented by the shaded area. Obviously, the existing design is not compatible with the new requirement. Three alternative design configurations are being considered. Each configuration meets the availability requirement, with configuration A having the highest estimated reliability MTBF and configuration C reflecting the best maintainability characteristics with the lowest $\overline{M}ct$ value. The objective is to select the best of the three configurations on the basis of cost.

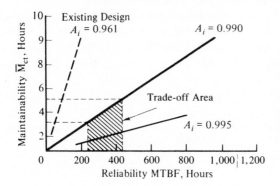

Conf.	A_i	MTBF	$\overline{M}$ct
Existing	0.961	125	5.0
Alt. A	0.991	450	4.0
Alt. B	0.990	375	3.5
Alt. C	0.991	320	2.8

Other systems are eligible for consideration as long as the effectiveness parameters fall within the trade-off areas

Figure 14.8   Reliability-maintainability trade-off.

When considering cost, there are costs associated with research and development (R&D) activity, investment or manufacturing costs, and operation and maintenance (O&M) costs.[4] For instance, improving reliability and/or maintainability characteristics in design will result in an increase in R&D and investment (manufacturing) cost. In addition, experience has indicated that such improvements will result in lower O&M cost, particularly in the areas of maintenance personnel and support cost and the cost of spare/repair parts. Thus, initially the analyst may look only at these categories. If the final decision is close, it may be appropriate to investigate other categories. A summary of partial cost data is presented in Table 14.9.

As shown in Table 14.9, the delta costs associated with the three alternative equipment configurations are included for R&D and investment. Maintenance personnel and support costs, included as part of the O&M cost, are based on estimated operating time for the 200 items of equipment throughout the required 10-year period of use (i.e., 200 items of equipment operating 8 hours per day, 360 days per year, for 10 years) and the reliability MTBF factor. Assuming that the average cost per maintenance action is $100, maintenance personnel and support costs are determined by multiplying this factor by the estimated quantity of maintenance actions, which is determined from total operating time divided by the MTBF value. Based on the delta values presented in Table 14.9, configuration A satisfies the system availability, reliability, and maintainability requirements with the least cost.

---

[4] Only those *delta costs* considered significant for this evaluation are included here.

TABLE 14.9 COST SUMMARY (PARTIAL COSTS)

CATEGORY	CONF. A	CONF. B	CONF. C	REMARKS
R and D cost				High reliability parts, packaging, accessibility
●Reliability design	$ 17,120	$ 15,227	$ 12,110	
●Maintainability design	2,109	4,898	7,115	
Investment cost manufacturing (200 Systems)	$3,422,400	$3,258,400	$3,022,200	$17,112/Equipment A; $16,292/Equipment B; $15,111/Equipment C
O&M cost				
●Maintenance personnel and support	$1,280,000	$1,536,000	$1,800,000	12,800 Maint. Action/Equipment A; 15,360 Maint. Action/Equipment B; 18,000 Maint. Action/Equipment C;
●Spare and Repair Parts	342,240	325,840	302,220	10% of manufacturing cost for spares
Total	$5,063,869	$5,140,365	$5,143,645	

Repair versus discard evaluation. In expanding the maintenance concept to establish criteria for equipment design, it is necessary to determine whether it is economically feasible to repair certain assemblies or to discard them when failures occur. If the decision is to accomplish repair, it is appropriate to determine the maintenance level at which the repair should be accomplished (i.e., intermediate maintenance or supplier/depot maintenance). This example illustrates a level-of-repair analysis based on life-cycle-cost criteria which may be accomplished during preliminary system design.[5]

Suppose that a computer system will be distributed in quantities of 65 throughout three major geographical areas. The system will be utilized to support both scientific and management functions within various industrial firms and government agencies. Although the actual system utilization will vary from one consumer organization to the next, an average utilization of 4 hours per day (for a 360-day year) is assumed.

The computer system is currently in the early development stage, should be in production in 18 months, and will be operational in 2 years. The full complement of 65 computer systems is expected to be in use in 4 years and will be available through the eighth year of the program before system phaseout commences. The system life cycle, for the purposes of the analysis, is 10 years.

Based on early design data, the computer system will be packaged in major units with a built-in test capability that will isolate faults to the unit level. Faulty units will be removed and replaced at the organizational level (i.e., consumer's facility), and sent to the intermediate maintenance shop for repair. Unit repair will be accomplished through assembly replacement, and assemblies will be either repaired or discarded. There is a total of 15 assemblies being considered, and the requirement is to justify the assembly repair or discard decision on the basis of life-cycle-cost criteria. The operational requirements, maintenance concept, and program plan are illustrated in Figure 14.9.

The stated problem primarily pertains to the analysis of 15 major assemblies of the given computer system configuration to determine whether the assemblies should be repaired or discarded when failures occur. In other words, the various assemblies will be individually evaluated in terms of (1) assembly repair at the intermediate level of maintenance, (2) assembly repair at the supplier or depot level of maintenance, and (3) disposing of the assembly. Life-cycle costs, as applicable to the assembly level, will be developed and employed in the alternative selection process. Total overall computer system costs have been determined at a higher level, and are not included in this example.

Given the information in the problem statement, the next step is to develop a cost breakdown structure (CBS) and to establish evaluation criteria. The evaluation criteria include consideration of all costs in each applicable category of the CBS, but the emphasis is on operation and support (O&S) costs as a function of acquisition cost. Thus, the research and development cost and the production cost are presented as one element, while various segments of O&S costs are identified individually. Figure 14.10 presents evaluation criteria, cost data, and a brief description and

---

[5] This example is taken from B. S. Blanchard, *Design and Manage to Life Cycle Cost* (Forest Grove, Ore.: M/A Press, 1978).

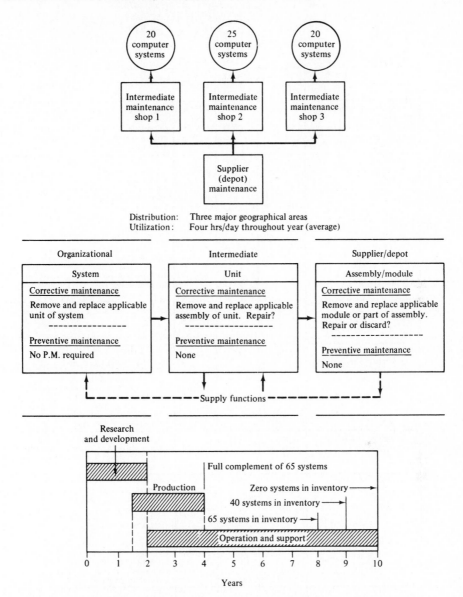

Figure 14.9 Basic system concepts.

justification supporting each category. The information shown in the figure covers only one of the 15 assemblies, but is typical for each case.

The data presented in Figure 14.10 represent assembly A-1, utilized in a manner illustrated in Figure 14.9. The next step is to employ the same criteria to determine the recommended repair-level decision for each of the other 14 assemblies (i.e., assemblies 2 through 15). Although acquisition costs, reliability and maintainability factors, and certain logistics requirements are different for each assembly, many of the cost-estimating relationships are the same. The objective is to be *consis-*

Evaluation criteria	Repair at intermediate cost ($)	Repair at supplier cost ($)	Discard at failure cost ($)	Description and justification
1. Estimated acquisition costs for assembly A-1 (to include R and D cost and production cost)	550/Assy. or 35,750	550/Assy. or 35,750	475/Assy. or 30,875	Acquisition cost includes all applicable costs allocated to assembly A-1 based on a requirement of 65 systems. Assembly design and production are simplified in the discard area.
2. Unscheduled maintenance costs	6,480	8,100	Not applicable	Based on the 8-year useful system life, 65 systems, a utilization of 4 hrs/day, a failure rate (λ) of 0.00045 for assembly A-1, and a $\overline{M}ct$ of 2 hrs, the expected number of maintenance actions is 270. When repair is accomplished, one technician is required on a full-time basis. The labor rates are $12/hr for intermediate maintenance and $15/hr for supplier maintenance.
3. Supply support spare assemblies	3,300	4,950	128,250	For intermediate maintenance 6 spare assemblies are required to compensate for transportation time, the maintenance queue, TAT, etc. For supplier/depot maintenance 9 spare assemblies are required. 100% spares are required in the discard case.
4. Supply support spare modules or parts for assembly repair	6,750	6,750	Not applicable	Assume $25 for materials per repair action.
5. Supply support inventory management	2,010	2,340	25,650	Assume 20% of the inventory value (spare assemblies, modules, and parts).
6. Test and support equipment	5,001	1,667	Not applicable	Special test equipment is required in the repair case. The acquisition and support cost is $25,000 per installation. The allocation for assembly A-1 per installation is $1,667. No special test equipment is required in the discard case.
7. Transportation and handling	Not applicable	4,725	Not applicable	Transportation costs at the intermediate level are negligible. For supplier maintenance, assume 540 one-way trips at $175/100 pounds. One assembly weighs 5 pounds.
8. Maintenance training	260	90	Not applicable	Delta training cost to cover maintenance of the assembly is based on the following: Intermediate – 26 students, 2 hrs each, $200/student week. Supplier - 9 students, 2 hrs each, $200/student week.
9. Maintenance facilities	594	810	Not applicable	From experience, a cost estimating relationship of $0.55 per direct maintenance laborhour is assumed for the intermediate level, and $0.75 is assumed for the supplier level.
10. Technical data	1,250	1,250	Not applicable	Assume 5 pages for diagrams and text covering assembly repair at $250/page.
11. Disposal	270	270	2,700	Assume $10/assembly and $1/module or part as the cost of disposal
Total estimated cost	61,665	66,702	187,475	

Figure 14.10   Repair versus discard evaluation (Assembly A-1).

tent in analysis approach and in the use of input cost factors to the maximum extent possible and where appropriate. The summary results for all 15 assemblies are presented in Table 14.10.

Referring to Table 14.10, note that the decision for assembly A-1 favors repair at the intermediate level; the decision for assembly A-2 is repair at the supplier or depot level; the decision for assembly A-3 is not to accomplish repair at all but to discard the assembly when a failure occurs; and so on. The figure reflects recommended policies for each individual assembly. In addition, the overall policy decision, when addressing all 15 assemblies as an integral package, favors repair at the supplier.

TABLE 14.10   SUMMARY OF REPAIR-LEVEL COSTS

	MAINTENANCE STATUS			
Assembly Number	Repair at Intermediate Cost($)	Repair at Supplier Cost ($)	Discard at Failure Cost ($)	Decision
A-1	61,665	66,702	187,475	Repair—intermediate
A-2	58,149	51,341	122,611	Repair—supplier
A-3	85,115	81,544	73,932	Discard
A-4	85,778	78,972	65,071	Discard
A-5	66,679	61,724	95,108	Repair—supplier
A-6	65,101	72,988	89,216	Repair—intermediate
A-7	72,223	75,591	92,114	Repair—intermediate
A-8	89,348	78,204	76,222	Discard
A-9	78,762	71,444	89,875	Repair—supplier
A-10	63,915	67,805	97,212	Repair—intermediate
A-11	67,001	66,158	64,229	Discard
A-12	69,212	71,575	82,109	Repair—intermediate
A-13	77,101	65,555	83,219	Repair—supplier
A-14	59,299	62,515	62,005	Repair—intermediate
A-15	71,919	65,244	63,050	Discard
Policy cost	1,071,267	1,037,362	1,343,449	Repair—supplier

Prior to arriving at a final conclusion, the analyst should reevaluate each situation where the decision is close. Referring to Figure 14.10, it is clearly uneconomical to accept the discard decision; however, the two repair alternatives are relatively close. Based on the results of the various individual analyses, the analyst knows that repair-level decisions are highly dependent on the unit acquisition cost of each assembly and the total estimated number of replacements over the expected life cycle (i.e., maintenance actions based on assembly reliability). The trends are illustrated in Figure 14.11, where the decision tends to shift from discard to repair at the intermediate level as the unit acquisition cost increases and the number of replacements increases (or the reliability decreases).[6]

[6] The curves projected in Figure 14.11 are characteristic for this particular analysis and will vary with changes in operational requirements, system utilization, the maintenance concept, production requirements, and so on.

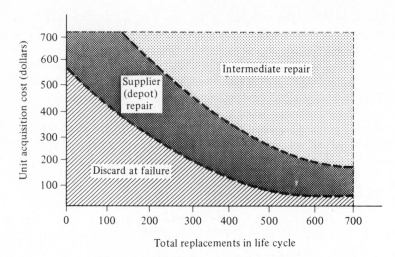

Figure 14.11   Economic screening criteria.

In instances where the individual analysis result lies close to the crossover lines in Figure 14.11, the analyst may wish to review the input data, the assumptions, and accomplish a sensitivity analysis involving the high-cost contributors. The purpose is to assess the risk involved and verify the decision. This is the situation for assembly A-1, where the decision is close relative to repair at the intermediate level versus repair at the supplier's facility (refer to Figure 14.10).

After reviewing the individual analyses of the 15 assemblies to ensure that the best possible decision is reached, the results in Table 14.10 are updated as required. Assuming that the decisions remain basically as indicated, the analyst may proceed in either of two ways. First, the decisions in Table 14.10 may be accepted without change, supporting a *mixed* policy with some assemblies being repaired at each level of maintenance and other assemblies being discarded at failure. With this approach, the analyst should review the interaction effects that could occur (i.e., the effects on spares, utilization of test and support equipment, maintenance personnel utilization, etc.). In essence, each assembly is evaluated individually based on certain assumptions, the results are reviewed in the context of the whole, and possible feedback effects are assessed to ensure that there is no significant impact on the decision.

A second approach is to select the overall least-cost policy for all 15 assemblies treated as an entity (i.e., assembly repair at the supplier or depot level of maintenance). In this case, all assemblies are designated as being repaired at the supplier's facility, and each individual analysis is reviewed in terms of the criteria in Figure 14.11 to determine the possible interaction effects associated with the single policy. The result may indicate some changes to the values in Table 14.10.

Finally, the output of the repair-level analysis must be reviewed to ensure compatibility with the initially specified system maintenance concept. The analysis data may either directly support and be an expansion of the maintenance concept, or the maintenance concept will require change as a consequence of the analysis. If the latter occurs, other facets of system design may be significantly impacted. The consequences of such maintenance concept changes must be thoroughly evaluated prior to arriving at a final repair-level decision. The basic level of repair analysis procedure is illustrated in Figure 14.12.

Maintainability Prediction

Maintainability prediction involves an early assessment of the maintainability characteristics in system design, and is accomplished periodically at different stages in the design process. Through the review of design data, predictions of the MTBM, $\overline{M}ct$, $\overline{M}pt$, MLH/OH, and so on, are made and are compared with the initially specified requirements identified in the maintainability allocation process. Areas of noncompliance are evaluated for possible design improvement.

Maintainability prediction, in the broadest sense, includes the early quantitative estimation of maintenance elapsed-time factors, maintenance labor-hour factors, maintenance frequency factors, and maintenance cost factors. The accomplishment of this requires that the engineer or analyst review design drawings, layouts, component-part lists, reliability factors, and supporting data with the intent of identifying anticipated maintenance tasks and the resources required for task completion. Maintainability prediction not only requires the estimation of task times and frequencies, but requires a qualitative assessment of the design characteristics for supportability.

Prediction of $\overline{M}ct$.   Prediction of mean corrective maintenance ($\overline{M}ct$) may be accomplished using a system functional-level breakdown, as illustrated in Figure 4.9, and determining maintenance tasks and the associated elapsed times in progressing from one function to another. The functional breakdown covers subsystems, units, assemblies, subassemblies, and parts. Maintainability characteristics such as localization, isolation, accessibility, repair, and checkout as incorporated in the design are evaluated and identified with one of the functional levels and down to each component part. Time applicable to each part (assuming that every part will fail at some point) are combined to provide factors for the next-higher level. A sample data format for an assembly is presented in Table 14.11.

Referring to Table 14.11, the frequency factors are represented by the failure rate (λ) for each part determined from reliability prediction data and the elapsed times required for fault localization, isolation, disassembly in order to gain access, and so on. Maintenance task times are usually estimated from experience and data obtained on similar systems already in use. The summation of the various individual times constitutes a maintenance cycle time (i.e., Mct_i) as illustrated in Figure 14.2.

Similar data prepared on each assembly in the system are combined as illustrated in Table 14.12, and the factors are computed to arrive at the predicted $\overline{M}ct$.

Prediction of $\overline{M}pt$.   Prediction of preventive maintenance time may be accomplished using a method similar to the corrective maintenance approach discussed above. Preventive maintenance tasks are estimated along with frequency and task times. An example is presented in Table 14.13.

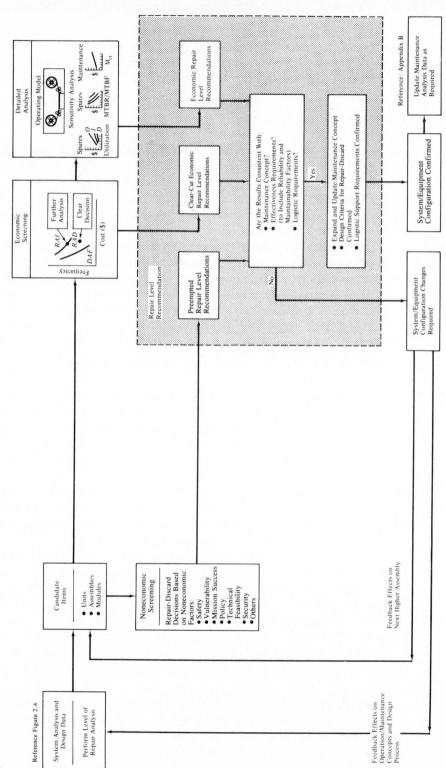

Figure 14.12 Level-of-repair analysis procedure.

TABLE 14.11 MAINTAINABILITY PREDICTION WORKSHEET (ASSEMBLY 4)

PART CATEGORY	λ	N	$(N)(\lambda)$	MAINTENANCE TIMES (HOURS)							
				Loc	Iso	Acc	Ali	Che	Int	Mct_i	$(N)(\lambda)(Mct_i)$
Part A	0.161	2	0.322	0.02	0.08	0.14	0.01	0.01	0.11	0.370	0.119
Part B	0.102	4	0.408	0.01	0.05	0.12	0.01	0.02	0.12	0.330	0.134
Part C	0.021	5	0.105	0.03	0.04	0.11	—	0.01	0.14	0.330	0.034
Part D	0.084	1	0.084	0.01	0.03	0.10	0.02	0.03	0.11	0.300	0.025
Part E	0.452	9	4.060	0.02	0.04	0.13	0.02	0.03	0.08	0.320	1.299
Part F	0.191	8	1.520	0.01	0.02	0.11	0.01	0.02	0.07	0.240	0.364
Part G	0.022	7	0.154	0.02	0.05	0.15	—	0.05	0.15	0.420	0.064
	Total		6.653							Total	2.039

N = Quantity of Parts   Iso = Isolation   Che = Check-out
λ = Failure Rate   Acc = Access   Int = Interchange
Loc = Localization   Ali = Alignment Mct_i = Maintenance Cycle Time
For determination of $\overline{MMH}c$, enter manhours for maintenance times.

TABLE 14.12 MAINTAINABILITY PREDICTION DATA SUMMARY

WORK SHEET NO.	ITEM DESIGNATION	WORK SHEET FACTOR	
		$\Sigma\,(N)(\lambda)$	$\Sigma\,(N)(\lambda)(\mathrm{Mct}_i)$
1	Assembly 1	7.776	3.021
2	Assembly 2	5.328	1.928
3	Assembly 3	8.411	2.891
4	Assembly 4	6.653	2.039
5	Assembly 5	5.112	2.576
$\vdots$	$\vdots$	$\vdots$	$\vdots$
13	Assembly 13	4.798	3.112
Grand Total		86.476	33.118

$$\overline{\mathrm{Mct}} = \frac{\Sigma\,(N)(\lambda)\mathrm{Mct}_i}{\Sigma\,(N)(\lambda)} = \frac{33.118}{86.486} = 0.382 \text{ hours}$$

TABLE 14.13 PREVENTIVE MAINTENANCE DATA SUMMARY

DESCRIPTION OF PREVENTIVE MAINTENANCE TASK	TASK FREQUENCY $(\mathrm{fpt}_i)(N)$	TASK TIME (Mpt_i)	PRODUCT $(\mathrm{fpt}_i)(N)(\mathrm{Mpt}_i)$
1. Lubricate	0.115	5.511	0.060
2. Calibrate	0.542	4.234	0.220
$\vdots$	$\vdots$	$\vdots$	$\vdots$
31. Service	0.321	3.315	0.106
Grand Total	13.260		31.115

$$\overline{\mathrm{Mpt}} = \frac{\Sigma\,(\mathrm{fpt}_i)(N)(\mathrm{Mpt}_i)}{\Sigma\,(\mathrm{fpt}_i)(N)} = \frac{31.115}{13.260} = 2.346 \text{ hours}$$

Prediction of maintenance resource requirements. Maintenance resources in this instance include the personnel and training requirements, test and support equipment, supply support (e.g., spare and repair parts), transportation and handling requirements, facilities, and data needed in the accomplishment of maintenance actions. For instance, what test equipment is required to accomplish fault isolation? What personnel types and skill level are required to repair the faulty item? What type of handling equipment is necessary to transport an item to the intermediate maintenance shop? In essence, the elapsed-time element alone does not provide an adequate prediction of the maintainability characteristics in design. One needs to predict the time element and the resources required.

The prediction of maintenance resource requirements may be accomplished using a variety of techniques. One accepted approach, described as the *logistic support analysis,* will be presented in Chapter 16. The relationship between maintainability prediction and the logistic support analysis is extremely close, and it is essential that both areas of activity be compatible and mutually supportive. This is particularly true since the objectives associated with the design for supportability and the design for maintainability are common in many areas.

Maintainability Review

The review of maintainability in system design is accomplished as an inherent part of the process described in Sections 3.4, 4.5, and 5.6. The characteristics of the system (and its elements) are evaluated in terms of the initially specified maintainability requirements for the system. If the requirements appear to have been met, the design is approved and the program enters the next phase. If not, the appropriate changes are initiated for corrective action.

In accomplishing a maintainability review, a checklist may be developed to facilitate the review process. An abbreviated representative list, not to be considered as being all inclusive, is presented below (it should be noted that the answer to those questions that are applicable should be *yes*).[7]

1. Have maintainability quantitative and qualitative requirements for the system been adequately defined and specified?

2. Are the maintainability requirements compatible with other system requirements? Are they realistic?

3. Are the maintainability requirements compatible with the system maintenance concept?

4. Has the proper level of accessibility been provided in the design to allow for the easy accomplishment of repair and/or item replacement? Are access requirements compatible with the frequency of maintenance? Accessibility for items requiring frequent maintenance should be greater than that for items requiring infrequent maintenance.

5. Is standardization incorporated to the maximum extent possible throughout the design? In the interest of developing an efficient supply support capability, the number of different types of spares should be held to a minimum.

6. Is functional packaging incorporated to the maximum extent possible? Interaction affects between modular packages should be minimized. It should be possible to limit maintenance to the removal of one module (the one containing the failed part) when a failure occurs and not require the removal of two, three, or four different modules.

7. Have the proper diagnostic test provisions been incorporated into the design? Is the extent or depth of testing compatible with the level of repair analysis?

8. Are modules and components having similar functions electrically, functionally, and physically interchangeable?

9. Are the handling provisions adequate for heavy items requiring transportation (e.g., hoist lugs, lifting provisions, handles, containers, etc.)?

---

[7] Appendix B provides a more extensive design review checklist.

10. Have quick-release fasteners been used on doors and access panels? Have the total number of fasteners been minimized? Have fasteners been selected based on the requirement for standard tools in lieu of special tools?

11. Have adjustment, alignment, and calibration requirements been minimized (if not eliminated)?

12. Have servicing and lubrication requirements been held to a minimum (if not eliminated)?

13. Are assembly, subassembly, module, and component labeling requirements adequate? Are the labels permanently affixed and unlikely to come off during a maintenance action or as a result of environmental conditions?

14. Have all system maintainability requirements been met?

14.5. MAINTAINABILITY DEMONSTRATION

Maintainability demonstration is conducted as part of the system test and evaluation effort discussed in Chapter 6. Specifically, maintainability demonstration is accomplished as part of Type 2 testing to verify that qualitative and quantitative maintainability requirements have been achieved. It also provides for the assessment of various logistic support factors related to, and having an impact on, maintainability parameters and item downtime (e.g., test and support equipment, spare/repair parts, technical data, personnel, maintenance policies).

Maintainability demonstration is usually accomplished during the latter part of detail design, and should be conducted in an environment that simulates, as closely as practical, the operational and maintenance environment planned for the item. The maintainability demonstration may vary considerably depending on system requirements and the test objectives. Two representative approaches are described in this section to provide an idea of the steps involved.

Demonstration Method 1

Maintainability demonstration method 1 follows a sequential test approach that is similar to the reliability test described earlier. Two different sequential test plans are employed to demonstrate $\overline{M}ct$ and M_{max} (for corrective maintenance). An accept decision for the equipment under test is reached when that decision can be made for both test plans. The test plans assume that the underlying distribution of corrective maintenance task times is log-normal. The sequential test plan approach allows for a quick decision when the maintainability of the equipment under test is either far above or far below the specified values of $\overline{M}ct$ and M_{max}.

Maintainability testing is accomplished by simulating faults in the system and observing the task times and logistic resources required to correct the situation. It involves the following steps:

1. A failure is induced in the equipment without knowledge of the test team. The induced failure should not be evident in any respect other than that normally

resulting from the simulated mode of failure. In other words, the technician(s) scheduled to perform the maintenance demonstration shall not be given any hints (through visual or other evidence) as to where the failure is induced.

2. The maintenance technician will be called upon to operationally check out the equipment. At some point in the checkout procedure, a symptom of malfunction is detected.

3. Once a malfunction has been detected, the maintenance technician(s) will proceed to accomplish the necessary corrective maintenance tasks (i.e., fault localization, isolation, disassembly, remove and replace, repair, reassembly, adjustment and alignment, and system checkout). In the performance of each step, the technician should follow approved maintenance procedures and use the proper test and support equipment. The maintenance tasks performed must be consistent with the maintenance concept and the specified levels of maintenance appropriate to the demonstration. Replacement parts required to perform repair actions being demonstrated shall be compatible with the spare and repair parts recommended for operational support.

4. While the maintenance technician is performing the corrective tasks (commencing with the identification of a malfunction and continuing until the equipment has been returned to full operational status), a test recorder collects data on task sequences, areas of task difficulty, and task times. In addition, the adequacies and inadequacies of logistic support are noted. Was the right type of support provided? Were there test delays due to inadequacies? Was there an overabundance of certain items and a shortage of others? Did each specified element of support do the job in a satisfactory manner? Were the test procedures adequate? These and related questions should be in the mind of the data recorder during the observation of a test.

This maintainability test cycle is accomplished n times, where n is the selected sample size. For the sequential test, the number of demonstrations could possibly extend to 100, assuming that a continue-to-test decision prevails. Thus, in preparing for the test, a sample size of 100 demonstrations should be planned. The selected tasks should be representative and based on the expected percent contribution toward total maintenance requirements. Those items with high failure rates will fail more often and require more maintenance and logistic resources; hence, they should appear in the demonstration to a greater extent than items requiring less maintenance.

The task selection process is accomplished by proportionately distributing the 100 tasks among the major functional elements of a system. Assuming that three units compose a system, the 100 tasks may be allocated as illustrated in Table 14.14. Referring to the table, the elements of the system and the associated failure rates (from reliability data) are listed. The percent contribution of each item to the total anticipated corrective maintenance (column 5) is computed as

$$\text{item percent contribution} = \frac{Q\lambda}{\Sigma \; Q\lambda} \times 100. \qquad (14.14)$$

TABLE 14.14   CORRECTIVE MAINTENANCE TASK ALLOCATION

(1) ITEM	(2) QUANTITY OF ITEMS (Q)	(3) FAILURES/ITEM % 1000 HOURS (λ)	(4) TOTAL FAILURES $(Q)(\lambda)$	(5) % CONTRIBUTION	(6) ALLOCATED MAINT. TASKS FOR DEMONSTRATION
Unit A	1	0.48	0.48	21	21 Tasks
Unit B	1	1.71	1.71	76	76 Tasks
Unit C	1	0.06	0.06	3	3 Tasks
TOTAL			2.25	100%	100 Tasks

This factor is used to allocate the tasks proportionately to each unit. In a similar manner, the 21 tasks within unit A can be allocated to assemblies within that unit, and so on. When the allocation is completed, there may be one task assigned to a particular assembly and the assembly may contain a number of components, the failure of which reflects different failure modes (e.g., no output, erratic output, low output, etc.). Through a random process, one of the components in the assembly will be selected as the item where the failure is to be induced, and the method by which the failure is induced is specified.

With the tasks identified and listed in random order, the demonstration proceeds with the first task, then the second, third, and so on. The criteria for accept-reject decisions are illustrated in Figure 14.13. Task times (Mct_i) are measured and compared with the specified Mct and M_{max} values. When the demonstrated time exceeds the specified value, an event is noted along the ordinate of the graph and problem areas are described. Testing then continues until the event line either enters the reject region or the accept region.

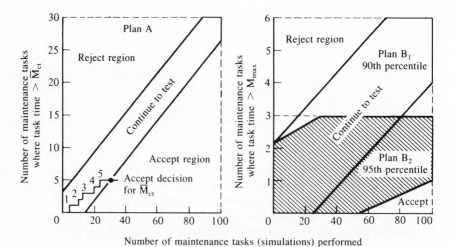

Figure 14.13 Graphical representation of maintainability demonstration plans. *Source:* MIL-STD-471, *Maintainability Verification, Demonstration, Evaluation,* Department of Defense, Washington, D.C.

An example of the demonstration test score sheet is illustrated in Table 14.15. The accept–reject numbers support the decision lines in Figure 14.13 (refer to the $\overline{M}$ct curve). In this instance, 29 tasks were completed before an accept decision was reached. The sequential test requires that both the $\overline{M}$ct criteria and the M_{max} criteria be met before the equipment is fully acceptable. M_{max} may be based on either the 90th or 95th percentile, depending on the specified system requirement and the test plan selected. If one test plan is completed with the event line crossing into the accept region, testing will continue until a decision is made in the other test plan.

TABLE 14.15   DEMONSTRATION SOURCE SHEET

REQMT: $\overline{M}$ct = 0.5 HOURS = 30 Min   Plan *A*

MAINT. TASK NO.	TASK TIME Mct_i	CUM NO. $Mct_i > \overline{M}ct$	ACCEPT WHEN CUM ≤ THAN	REJECT WHEN CUM > THAN
1	12 Min	0	—	—
2	6	0	—	—
3	18	0	—	—
4	32	1	—	—
5	19	1	—	5
6	27	1	—	6
7	108	2	—	6
8	6	2	—	6
9	14	2	—	7
10	47	3	—	7
11	28	3	—	7
12	19	3	0	7
13	4	3	0	8
14	24	3	0	8
15	78	4	1	8
⋮	⋮	⋮	⋮	⋮
24	20	4	3	11
25	127	5	4	11
26	21	5	4	12
27	13	5	4	12
28	28	5	4	12
29	8	5	5	12

Accept for $\overline{M}$ct

Referring to Figure 14.13, the criteria for sequential testing specifies that the minimum number of tasks possible for a quick decision for $\overline{M}$ct is 12 (test plan A). For M_{max} at the 90th percentile, the least number of tasks possible is 26 (test plan B_1) while the figure is 57 for the 95th percentile (test plan B_2). Thus, if the maintainability of an item is exceptionally good, demonstrating the complete sample of 100 tasks may not be necessary, thus saving time and cost. On the other hand, if the maintainability of an item is marginal and a continue-to-test decision prevails, the test program may require the demonstration of all 100 tasks. If truncation is reached, the equipment is acceptable for $\overline{M}$ct if 29 or less tasks exceed the specified $\overline{M}$ct value. Comparable factors for M_{max} are 5 or less for the 90th percentile and 2 or less for the 95th percentile.

Demonstration Method 2

This method is applicable to the demonstration of $\overline{M}ct$, $\overline{M}pt$, and $\overline{M}$. The underlying distribution of maintenance times is not restricted (no prior assumptions), and the sample size constitutes 50 corrective maintenance tasks for $\overline{M}ct$ and 50 preventive maintenance tasks for $\overline{M}pt$. $\overline{M}$ is determined analytically from the test results for $\overline{M}ct$ and $\overline{M}pt$. $\overline{M}_{max}$ can also be determined if the underlying distribution is assumed to be log-normal. This method offers the advantage of a fixed sample size, which facilitates the estimation of test costs.

The method involves the selection and performance of maintenance tasks in a similar manner as described for demonstration method 1. Tasks are selected based on their anticipated contribution to the total maintenance picture, and each task is performed and evaluated in terms of maintenance times and required logistic resources. Illustration of this method is best accomplished through an example. It is assumed that a system is designed to meet the following requirements and must be demonstrated accordingly:

$$\overline{M} = 75 \text{ min}$$

$$\overline{M}ct = 65 \text{ min}$$

$$\overline{M}pt = 110 \text{ min}$$

$$M_{max} = 120 \text{ min}$$

$$\text{producer's risk } (\alpha) = 20\%$$

This test is accomplished and the data collected are presented in Table 14.16. The determination of $\overline{M}ct$ (upper confidence limit) is based on the expression

$$\text{upper limit} = \overline{M}ct + Z\left(\frac{\sigma}{\sqrt{n_c}}\right) \tag{14.15}$$

TABLE 14.16 MAINTENANCE TEST-TIME DATA

DEMONSTRATION TASK NUMBER	OBSERVED TIME Mct_i	$Mct_i\text{-}\overline{M}ct$ $(Mct_i - 62)$	$(Mct_i - \overline{M}ct)^2$
1	58	−4	16
2	72	+10	100
3	32	−30	900
⋮	⋮	⋮	⋮
50	48	−14	196
Total	3105		15,016

where

$$\overline{\text{Mct}} = \frac{\sum\limits_{i=1}^{n_c} \text{Mct}_i}{n_c} = \frac{3,105}{50} = 62.1 \quad \text{(assume 62)}$$

and $Z = 0.84$ from Table 14.4. Then,

$$\sigma = \sqrt{\frac{\sum\limits_{i=1}^{n_c} (\text{Mct}_i - \overline{\text{Mct}})^2}{n_c - 1}} = \sqrt{\frac{15,016}{49}} = 17.5$$

where n_c is the corrective maintenance sample size of 50

$$\text{upper limit} = 62 + \frac{(0.84)(17.5)}{\sqrt{50}} = 64.07 \text{ mins.}$$

The completed $\overline{\text{Mct}}$ statistic is compared to the corresponding accept–reject criteria, which is to accept if

$$\overline{\text{Mct}} + Z\left(\frac{\sigma}{\sqrt{n_c}}\right) \leq \overline{\text{Mct}} \text{ (specified)} \tag{14.16}$$

and reject if

$$\overline{\text{Mct}} + Z\left(\frac{\sigma}{\sqrt{n_c}}\right) > \overline{\text{Mct}} \text{ (specified).} \tag{14.17}$$

Applying demonstration test data, it can be seen that 64.07 minutes (the upper value of $\overline{\text{Mct}}$ derived by test) is less than the specified value of 65 minutes. Therefore, the system passes the $\overline{\text{Mct}}$ test and is accepted.

For preventive maintenance the same approach is used. Fifty (50) preventive maintenance tasks are demonstrated and task times (Mpt_i) are recorded. The sample mean preventive downtime is

$$\overline{\text{Mpt}} = \frac{\sum \text{Mpt}_i}{n_p}. \tag{14.18}$$

The accept–reject criteria is the same as stated in Equations 14.16 and 14.17, except that preventive maintenance factors are used. That is, accept if

$$\overline{\text{Mpt}} + Z\left(\frac{\sigma}{\sqrt{n_p}}\right) \leq 110 \text{ mins}$$

and reject if

$$\overline{\text{Mpt}} + Z\left(\frac{\sigma}{\sqrt{n_p}}\right) > 110 \text{ mins.}$$

Given the test values for $\overline{\text{Mct}}$ and $\overline{\text{Mpt}}$, the calculated mean maintenance time is

$$\overline{\text{M}} = \frac{(\lambda)(\overline{\text{Mct}}) + (\text{fpt})(\overline{\text{Mpt}})}{\lambda + \text{fpt}} \tag{14.19}$$

where

λ = corrective maintenance rate or the expected number of corrective maintenance tasks occurring in a designated period of time.

fpt = preventive maintenance rate or the expected number of preventive maintenance tasks occurring in the same time period.

Using test data, the resultant value of $\overline{M}$ should be equal to or less than 75 minutes. Finally, M_{max} is determined from

$$M_{max} = \text{antilog } (\overline{\log Mct} + Z\sigma_{\log Mct_i}). \tag{14.20}$$

The calculated value should be equal to or less than 120 for acceptance. An example of calculation for M_{max} is presented in Section 14.3. If all the demonstrated values are better than the specified values, following the criteria defined above, the system is accepted. If not, some retest and/or redesign may be required, depending on the seriousness of the problem.

Maintainability Assessment

The true assessment of system maintainability in an operational environment is accomplished through an effective data collection, analysis, and evaluation capability, as described in Section 6.5. The maintenance event report illustrated in Figure 6.6 is designed for the collection of maintenance data throughout the system operational life cycle when failures occur. Maintenance tasks, task sequences, task times, and logistics factors are recorded. Major areas of deficiency are noted and should be corrected through the initiation of changes and the appropriate system modification(s).

SELECTED REFERENCES

(1) AMCP 706–134, *Engineering Design Handbook—Maintainability Guide for Design*. Department of the Army, October 1972.

(2) Blanchard, B. S., *Logistics Engineering and Management* (3rd ed.). Englewood Cliffs, N.J.: Prentice-Hall, 1986.

(3) Blanchard, B. S., and E. E. Lowery, *Maintainability—Principles and Practices*. New York: McGraw-Hill, 1969.

(4) Cunningham, C. E., and W. Cox, *Applied Maintainability Engineering*. New York: John Wiley, 1972.

(5) Goldman, A., and T. Slattery, *Maintainability—A Major Element of System Effectiveness*. New York: John Wiley, 1967.

(6) Halpern, S., *The Assurance Sciences—An Introduction to Quality Control and Reliability*. Englewood Cliffs, N.J.: Prentice-Hall, 1978.

(7) Jardine, A. K. S., *Maintenance, Replacement, and Reliability*. New York: John Wiley (Halsted Press Book), 1973.

(8) Kelly, A., and M. J. Harris, *Management of Industrial Maintenance*. London: Newnes–Butterworths, 1978.

(9) MIL-STD-470, Military Specification, *Maintainability Program Requirements for Systems and Equipments*. Department of Defense, Washington, D.C.

(10) MIL-STD-471, Military Specification, *Maintainability Verification, Demonstration, Evaluation*. Department of Defense, Washington, D.C.

QUESTIONS AND PROBLEMS

1. Describe how maintainability relates to system or product design.
2. Describe how maintainability ties in with the system engineering process.
3. What are the quantitative measures of maintainability?
4. What is the significant difference between MTBF and MTBM? Between MTBF and MTBR?
5. Describe the basic differences between maintainability and maintenance.
6. Corrective-maintenance task times were observed as given in the table.

TASK TIME (MIN)	FREQUENCY	TASK TIME (MIN)	FREQUENCY
41	2	37	4
39	3	25	10
47	2	36	5
35	5	31	7
23	13	13	3
27	10	11	2
33	6	15	8
17	12	29	8
19	12	21	14

(a) What is the range of observations?
(b) Using a class interval width of 4, determine the number of class intervals. Plot the data and construct a curve. What type of distribution is indicated by the curve?
(c) What is the geometric mean of the repair times?
(d) What is the standard deviation of the sample data?
(e) What is the M_{max} value? Assume 90% confidence level.
(f) What is the $\overline{M}ct$?

7. Corrective-maintenance task times were observed as given in the table.

TASK TIME (MIN)	FREQUENCY	TASK TIME (MIN)	FREQUENCY
35	2	25	12
17	6	19	10
12	2	21	12
15	4	23	13
37	1	29	8
27	10	13	3
33	3	9	1
31	6	—	—

(a) What is the range of observations?

(b) Assuming 7 classes with a class interval width of 4, plot the data and construct a curve. What type of distribution is indicated by the curve?

(c) What is the mean repair time?

(d) What is the standard deviation of the sample data?

(e) The system is required to meet a mean repair time of 25 minutes at a stated confidence level of 95%. Do the data reveal that the specification requirements will be met? Why?

8. With a specified inherent availability of 0.990 and a calculated MTBF of 400 hrs, what is the $\overline{M}ct$?

9. Calculate as many of the following parameters as you can with the information given.

DETERMINE		GIVEN
A_i	MTBM	$\lambda = 0.004$
A_a	MTBF	Total operation time = 10,000 hrs
A_o	$\overline{M}$	Mean downtime = 50 hrs
$\overline{M}ct$	$MTTR_g$	Total number of maintenance actions = 50
M_{max}		Mean preventive maintenance time = 6 hrs
		Mean logistics plus administrative time = 30 hrs

10. Find the upper 90% confidence limit for the population mean repair time based on a normally distributed sample of 70 repair actions, given $\bar{x} = 80$ minutes and

$$\sum_1^N (x_1 - \bar{x})^2 = 18,000.$$

11. What is maintainability allocation? How is it accomplished?

12. Recommend either of the following maintenance concepts based on the data provided.

    —Concept 1. Discard all component-board assemblies upon failure.

    —Concept 2. Repair all component-board assemblies upon failure.

    —Concept 3. Discard component-board assemblies upon failure when the replacement cost is less than $100 each, and repair all other component boards.

(a) Quantity data

Total number of assemblies: 10,000

Number of items under $100 unit cost: 5,000

Number of items over $100 unit cost: 5,000 (average cost per unit is $300)

(b) Failure-rate data

The items under $100 unit cost will fail on the average of every 36 months. The failure rates, respectively, are 0.00048 and 0.00016 per item; or 2.40 and 0.80 per class.

(c) The service life of the system is 10 years. The system is operated 2,080 hours per year.

(d) An investment of $1,000,000 is necessary for equipment to repair any or all of the subassemblies. The equipment life is 10 years, and it will have no scrap value at the end of that time.

(e) Labor costs are $3.00 for each repair. Component spares cost $2.00 per repair.

13. What is the purpose of maintainability prediction? When are maintainability predictions accomplished in the life cycle?

14. Given the following data, calculate the achieved availability.
    (a) $\overline{Mct}$ = 0.5 hour.
    (b) $MTBM_u$ = 2.0 hours.
    (c) $\overline{Mpt}$ = 2.0 hours.
    (d) $MTBM_s$ = 1,000 hours.

15. A $\overline{Mct}$ of 1 hour has been assigned to system ABC. Allocate this time to the assembly level of the system using the data given in the table.

ASSEMBLY	QUANTITY	λ	$\overline{Mct}$
A	1	0.05	?
B	2	0.16	?
C	1	0.27	?
D	1	0.12	?

16. Is MDT in itself a true measure of equipment design for maintainability? If so, explain. If not, why not?

17. What is the purpose of maintainability demonstration? What elements are demonstrated?

18. When selecting tasks for maintainability demonstration, how does the selection process result in a representative sample of what is to be expected in an operational situation?

19. What criteria are used in the selection of personnel for the demonstration of maintenance tasks? Define the criteria for the selection of test and support equipment.

20. Refer to maintainability demonstration method 2 and determine whether the system will meet the $\overline{Mct}$ requirement if the assumed producer's risk is 5%.

21. The $\overline{Mct}$ requirement for an equipment item is 65 min and the established risk factor is 10%. A maintainability demonstration is accomplished and yields the results given in the table for the 50 tasks demonstrated. (Task times are in minutes.)

39	57	70	51	74	63	66	42	85	75
42	43	54	65	47	40	53	32	50	73
64	82	36	63	68	70	52	48	86	36
74	67	71	96	45	58	82	32	56	58
92	91	75	74	67	73	49	62	64	62

Did the equipment item pass the maintainability demonstration?

22. The $\overline{Mpt}$ requirement for an equipment is 120 minutes. A maintainability demonstration is accomplished and yields the results given in the table for the 50 tasks demonstrated. (Task times are in minutes.) Did the equipment item pass the maintainability demonstration? The risk factor is 10%.

150	120	133	92	89	115	122	69	172	161
144	133	121	101	114	112	181	78	112	91
82	131	122	159	135	108	95	67	118	103
78	93	144	152	136	86	113	102	65	115
113	101	94	129	148	118	102	106	117	115

15
Design for Manability

Most of the material presented thus far has been concerned with the hardware and software elements of a system. For the system design to be complete, one also needs to address the human element and the interface(s) between the human and the equipment. Optimum hardware design alone will not guarantee effective results. Consideration must be given to human physiological factors (e.g., reaction to environmental forces), psychological factors (e.g., need, expectation, attitude, motivation), anthropometric factors (e.g., human physical dimensions), and their interrelationships. Good system design requires the proper integration of human factors into the process illustrated in Figure 2.4, along with hardware, software, data, and other elements.

The purpose of this chapter is to briefly deal with the human being as an integral part of the system. In this context, the activities discussed will be presented under the broad category of human factors, which includes the determination of personnel requirements, the accomplishment of analyses and trade-offs dealing with human–machine relationships, equipment design functions, the determination of training requirements, and the test and evaluation of the human being in the system. These activities may also be included under such general categories as human engineering, ergonomics, engineering psychology, and systems psychology.

15.1. HUMAN FACTORS REQUIREMENTS

Human factors requirements in design are derived initially from system operational requirements and the system maintenance concept in Chapter 3. A description of the

434

system and the mission to be performed is an essential first step in the requirements definition process. Based on this information, a functional analysis is accomplished in which operator and maintenance functions are identified as is illustrated in Chapter 4. As one progresses through the analysis, individual functions are described and expressed in verbal form (i.e., acquire, perform, monitor, check, isolate, control, etc.). This provides a basic indication of what needs to occur (in series or in parallel) if the system is to fulfill its mission objective in a satisfactory manner.

The identification of functions is one of the first steps in systems analysis. Often, functions are defined in rather general terms as indicated in the earlier chapters. Although this may be sufficient in providing the necessary information for many facets of system design, the definition of human factors requirements often dictates a further breakdown of functions to job operation, duty, task, subtask, and task element. One needs to provide enough detail to make possible the prediction and assessment of individual human behavior as it pertains to the smallest logically definable set of perceptions, decisions, and responses associated with a given activity. Thus, while equipment is presented in a hierarchical form (i.e., subsystem, unit, assembly, subassembly, component part), one needs to establish a hierarchy of personnel-related activities. A brief description of this hierarchy is presented below.[1]

1. *Job operation*—Completion of a function normally includes a combination of duties and tasks. A job operation may involve one or more related groups of duties, and may require one or more individuals in its accomplishment (i.e., positions in a given specialty field). An example of a job operation is operate a motor vehicle or accomplish scheduled maintenance.

2. *Duty*—Defined as a set of related tasks within a given job operation. For instance, when considering the operating of a motor vehicle, a set of related tasks may include (a) driving the motor vehicle in traffic on a daily basis, (b) registering the motor vehicle yearly, (c) servicing the motor vehicle as required, and (d) accomplishing vehicle preventive maintenance on a periodic basis.

3. *Task*—Constitutes a composite of related activities (informational, decision, and control activities) performed by an individual in accomplishing a prescribed amount of work in a specified environment. A task may include a series of closely associated operations, maintenance inspections, and so on. Relative to driving a motor vehicle, examples of tasks are (a) applying the appropriate pressure on the accelerator in order to maintain the desired vehicle speed, (b) shifting gears as necessary in order to maintain engine rpm, or (c) turning the steering wheel as required to enable the motor vehicle to move in the desired direction.

4. *Subtask*—Depending on the complexity of the situation, a task may be broken down into subtasks to cover discrete actions of a limited nature. A subtask may

[1] A hierarchy of behavioral terms is discussed further in K. B. DeGreene, ed., *Systems Psychology* (New York: McGraw-Hill, 1970).

constitute the shifting of gears from first to second (in the motor vehicle example), a machine adjustment, or a similar act.

5. *Task element*—Task elements may be categorized as to the smallest logically definable facet of activity (based on perceptions, decisions, and control actions) that requires individual behavioral responses in completing a task or a subtask. For example, the identification of a specific signal level on a display, a decision pertaining to a single physical action, the actuation of a switch on a control panel, and the interpretation of a go/no-go signal might be classified as a task element. This is the lowest category of activity where job behavioral characteristics are identified and evaluated.

The hierarchy outlined above represents a breakdown from the top system-level functions described in Chapter 4, to the smallest element of activity involving human performance. In developing this breakdown, it is not always easy to separate functions from tasks, tasks from subtasks, and so on. However, one must proceed from the top-level function(s) down to the level necessary to establish the proper human–machine interface. As with an equipment hierarchy, this breakdown constitutes a logical division of human activity which serves as a basis in establishing human factors requirements.

With the functional analysis results and the basic hierarchy of human activity identified (at least, on a preliminary basis), one needs to address some of the environmental and personnel factors that will ultimately influence work accomplishment. Figure 15.1 illustrates the relationship of these elements in the process of defining the overall human factors requirements for the system. The environmetal considerations noted in Figure 15.1 constitute those factors which impose a constraint on the system development activity as an entity. These factors are introduced in Chapter 2. Personnel factors may be categorized in terms of anthropometric factors, human sensory factors, physiological factors, and psychological factors. These categories are briefly described below.

Anthropometric Factors

When establishing basic design requirements for the system, particularly with regard to human activities, one obviously must consider the physical dimensions of the human body. The weight, height, arm reach, hand size, and so on, are critical when designing operator stations, consoles, control panels, accesses for maintenance purposes, and the like. Further, body dimensions will vary somewhat from a static position to a dynamic condition. *Static* measurements pertain to the human subject in a rigid standardized position, while *dynamic* measurements are made with the human in various working positions and undergoing continuous movement. As movement takes place, body measurements will change. Thus, when considering various system design alternatives, the designer must be aware of the different human profiles that are likely to occur in the performance of operator and maintenance activities associated with the system.

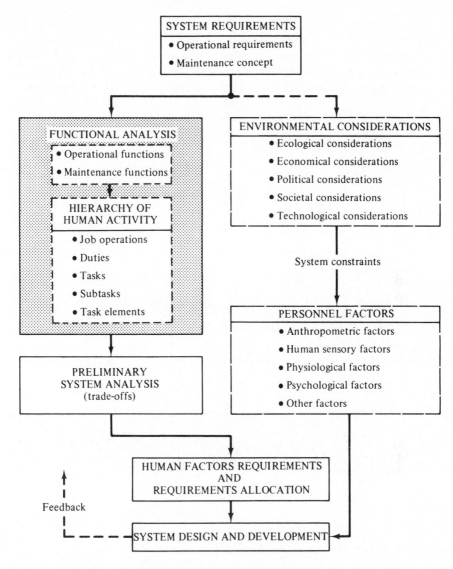

Figure 15.1 Human factors requirements.

To aid the designer in obtaining information on body measurements, there are two basic sources of data to consider: (1) anthropometric surveys, in which measurements of a sample of the population have been made; and (2) experimental data derived from simulating the operating conditions peculiar to the system being developed. The designer may have access to either source of data, or a combination of both. In many instances, however, the acquisition of meaningful experimental data is quite costly. Hence, static measurements are often used for the purposes of design guidance.

Figures 15.2 and 15.3 illustrate examples of static body measurements in the standing and sitting positions, respectively.[2] Anthropometric measurements are usually provided in percentiles, ranges, and means (or medians). The data presented represent the 5th and 95th percentiles for men and women for a given sample population. Although the designer will not be able to cover all possible sizes or profiles,

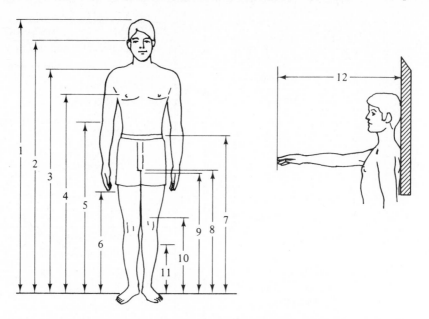

	Percentile values (cm)			
Factors	5th Percentile		95th Percentile	
	Men	Women	Men	Women
Weight (kg)	57.4	46.4	91.6	76.5
Standing body dimensions				
1. Stature	163.8	152.4	185.6	172.2
2. Eye height (standing)	152.1	142.7	173.3	160.1
3. Shoulder (acromiale) height	133.6	123.0	154.2	143.4
4. Chest (nipple) height		110.0		127.3
5. Elbow (radiale) height	104.8		120.0	
6. Fingertip (dactylion) height	61.5		73.2	
7. Waist height	97.5	93.1	115.2	110.1
8. Crotch height	76.3	66.4	91.8	81.4
9. Gluteal furrow height		66.2		79.4
10. Kneecap height	47.5		58.6	
11. Calf height	31.1		40.6	
12. Functional reach	72.7	67.7	90.9	80.4

Figure 15.2 Anthropometric data—standing body dimensions.

[2] The factors presented in Figures 15.2 and 15.3 were taken from MIL-STD-1472, Military Standard, *Human Engineering Design Criteria for Military Systems, Equipment and Facilities,* Department of Defense, Washington, D.C. Another good source for anthropometric data is Sanders, M. S. and E. J. McCormick, *Human Factors in Engineering Design,* 6th ed. (New York: McGraw-Hill, 1987).

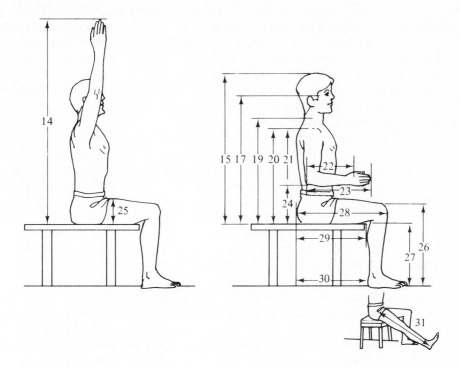

Factors	Percentile values (cm)			
	5th Percentile		95th Percentile	
	Men	Women	Men	Women
Seated body dimensions				
14. Vertical arm reach, sitting	128.6		147.8	
15. Sitting height, erect	84.5	78.4	96.9	90.9
16. Sitting height, relaxed	82.5	76.9	94.8	89.7
17. Eye height, sitting erect	72.8	68.7	84.6	78.8
18. Eye height, sitting relaxed	70.8	67.2	82.5	77.6
19. Mid-shoulder height	57.1	53.7	67.7	62.5
20. Shoulder height, sitting	54.2		65.4	
21. Shoulder-elbow length	33.8	30.2	40.2	36.2
22. Elbow-grip length	32.6		37.9	
23. Elbow-fingertip length	44.3	38.9	51.9	45.7
24. Elbow rest height	17.5	18.7	28.0	26.9
25. Thigh clearance height		10.4		14.6
26. Knee height, sitting	49.7	43.7	58.7	51.6
27. Popliteal height	40.6	38.0	50.0	44.1
28. Buttock-knee length	54.9	52.0	64.3	61.9
29. Buttock-popliteal length	45.8	43.4	54.5	52.6
30. Buttock-heel length	46.7		56.4	
31. Buttock-heel length (diagonal)	103.9		120.4	

Figure 15.3 Anthropometric data—seated body dimensions.

he or she can cover most situations as well as decide on where to cut off. The ultimate design configuration must accommodate the majority of the population if operator and maintenance activities are to be accomplished efficiently. Those individuals having extreme measurements (e.g., the small percentage of the population below the 5th percentile and above the 95th percentile) may be able to perform the necessary operator and/or maintenance functions; however, the results are likely to be inefficient due to the introduction of various human stresses causing early personal fatigue and possible system failure.

Work-space requirements must be established for both operator personnel functions and maintenance personnel functions. Figure 15.4 illustrates typical requirements for accomplishment of a variety of maintenance tasks. Body dimensions and work-space dimensions must be viewed on an integrated basis if optimum results are to be realized.

The application of anthropometric data in design involves many considerations. The human body and work-space dimensions are, of course, significant. However, one should consult the literature prior to proceeding further since the limited material in this section is provided for illustrative purposes only. There exist criteria covering many facets of console and control panel design, work-space design, seat design, work surface design, and so on. Also, associated with body dimensions are the aspects of force and weight-lifting capacity. A human being can exert more force, with less fatigue, if the system (i.e., the segment of the system where the interface exists) is designed properly. The amount of force that can be exerted is determined by the position of the body and the members applying the force, the direction of application, and the object to which the force is applied. Weight-lifting capacity, which is closely related to force, is highly dependent on the individual's weight, size, and position. The field of anthropometry deals primarily with the science and technique of human measurements; however, there are numerous other factors that are directly affected by design decisions.

Human Sensory Factors [3]

In dealing with the human–machine interface in system design, one must be cognizant of certain human sensory capacities. Although most of the human senses may be affected through system design, factors pertaining to vision (sight) and hearing (noise) are of particular significance.

Vision.   Vision or sight, as it pertains to system design, is usually limited to the desired vertical and horizontal fields, as illustrated in Figure 15.5. The designer should consider the specified degrees of eye and head rotation as the maximum allowable values in the design of operator consoles and control panels. Requirements outside of these recommended limits will result in operator inefficiency and system failures.

Within the broad field of view illustrated in Figure 15.5, the human eye may see different objects from different angles. Sight is stimulated by the electromag-

---

[3] NAVSHIPS 94324, *Maintainability Design Criteria Handbook for Designers of Shipboard Electronic Equipment*, Naval Ship Systems Command, Department of the Navy, 1964.

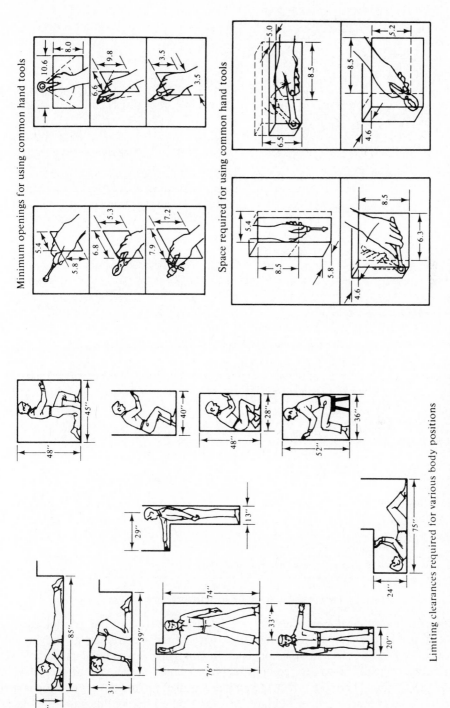

Minimum openings for using common hand tools

Space required for using common hand tools

Limiting clearances required for various body positions

Figure 15.4 Anthropometric data pertaining to work space requirements. *Source:* NAV-SHIPS 94324, *Maintainability Design Criteria Handbook for Designers of Shipboard Electronics Equipment,* Naval Ship Systems Command, U.S. Navy.

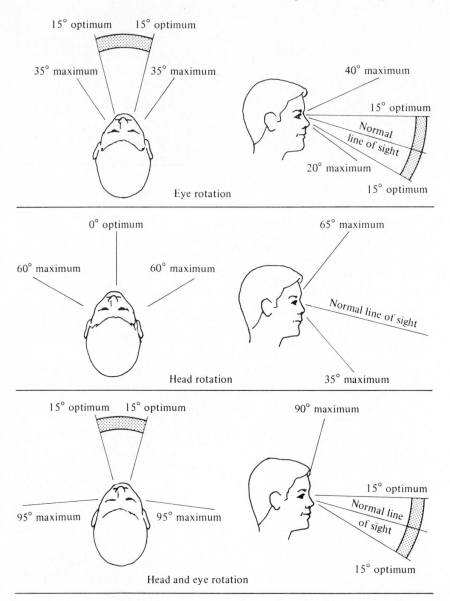

Figure 15.5   Vertical and horizontal visual field.

netic radiation of certain wavelengths, or the visible portion of the electromagnetic spectrum. The eye sees different lines (i.e., parts of the spectrum) with varying degrees of brightness. Relative to color, one can usually perceive all colors while looking straight ahead. However, color perception begins to decrease as the viewing angle decreases. Therefore, when designing consoles (or panels) with color-coded meters or color warning light displays, one must take into consideration the placement of such relative to the operator's field of view. Figure 15.6 illustrates the limits of color vision.

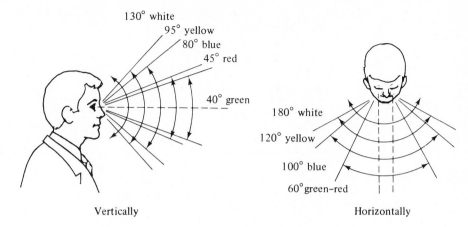

Figure 15.6   Approximate limits of color differentiation.

Finally, the satisfactory performance of tasks is highly dependent on the level of illumination. Not only must the designer be concerned with field of view and color, but the proper level of illumination is an obvious necessity. Illumination levels will vary somewhat depending on the task to be performed. Table 15.1 identifies suggested levels of illumination for designated items.

Hearing.   In human activities the designer needs to address both the requirements for oral communication and the aspects of noise. Of particular concern is the effect of noise on the performance of work. Noise is generally regarded as a distractor and a deterrent when considering efficiency of work accomplishment. As the noise level increases, a human being begins to experience discomfort, and both productivity and efficiency decrease. Further, oral communication becomes ineffectual or impossible. When the noise level approaches 120 dB, a human being will usually experience a physical sensation in some form; and at levels above 130 dB, pain can occur.

Another major factor in determining the extent to which noise is a distractor is its character—whether steady or intermittent. In situations where the noise is steady, an individual can adapt to it, and work efficiency may not be significantly compromised. Of course, if the noise level is too high (even though steady), the individual will probably experience permanent injury through loss of hearing. On the other hand, when the noise is intermittent, the individual is usually distracted regardless of the intensity. In this instance, a greater effort is required to maintain job efficiency and the possibility of early fatigue occurs.

The designer, when dealing with the human being as a component of the system, must identify the operator and maintenance tasks that are to be accomplished by people. These tasks must be evaluated in terms of the environment in which the tasks are to be performed, and the noise generated by the system (or generated externally) must be maintained at a level where human efficiency is maximized. The desired intensity level will likely fall within a range 50 to 80 dB; however, the designer should consult the available literature to establish the proper design criteria for the variety of applications that are likely to occur.

TABLE 15.1   SPECIFIC TASK ILLUMINATION REQUIREMENTS

ILLUMINATION LEVELS

	Foot-candles*	
Type of Task	*Recommended*	*Minimum*
1. Business machine operation	100	50
2. Console surface	50	30
3. Dials, gages, and meters	50	30
4. Factory assembly—general	50	30
5. Factory assembly—precise	300	200
6. Inspection tasks—extra fine	300	200
7. Inspection tasks—rough	50	30
8. Office work, general	70	50
9. Ordinary seeing tasks	50	30
10. Panels	50	30
11. Passageways and stairways	20	10
12. Reading, large print	30	10
13. Reading, small type	70	50
14. Repair work—general	50	30
15. Repair work—instrument	200	100
16. Testing—extra fine	200	100
17. Testing—rough	50	30
18. Transcribing and tabulation	100	50

*As measured at the task object or 30 in. above the floor.
This information (in part) was extracted from MIL-STD-1472, Military Standard, *Human Engineering Design Criteria for Military Systems, Equipment, and Facilities,* Department of Defense, Washington, D.C.

Physiological Factors[4]

The study of physiology is obviously well beyond the scope of this text; however, some recognition must be given to the effects of environmental stresses on the human body while performing system tasks. *Stress* refers to any aspect of external activity or the environment acting on the individual (who is performing a system task) in such a manner as to cause a degrading effect. Stress may result in physiological effects and psychological effects, and *strain* is often the consequence of stress measured in terms of one or more physical characteristics of the human body. Some of the causes of stress are:

1. *Temperature extremes*—Experience has indicated that certain temperature extremes are detrimental to work efficiency. As the temperature increases above the comfort zone (e.g., 55 to 75°F), mental processes slow down, motor response is slower, and the likelihood of error increases. On the other hand, as

[4] Additional coverage of physiological factors is presented in Sanders, M. S. and E. J. McCormick, *Human Factors in Engineering and Design,* 6th ed. (New York: McGraw-Hill, 1987).

the temperature is lowered (e.g., 50°F and lower), physical fatigue and stiffening of the extremities begins. For this reason, the designer needs to be aware of the environmental profiles for the system, particularly with regard to anticipated system use in the arctic or in the tropics.

2. *Humidity*—Heat and humidity are usually significant factors in causing a reduction in the operational efficiency of personnel. A human being can tolerate much higher temperatures with dry air than if the air were humid.

3. *Vibration*—High levels of vibration will often affect human proficiency. Figure 15.7 illustrates desired criteria for various values of frequency and acceleration. The designer must consider these factors to permit safe conditions for the accomplishment of system operational and maintenance activities.

4. *Noise*—The consequences of noise and its impact on human efficiency was discussed earlier in the context of human sensory factors (i.e., hearing). However, this factor is listed again, since high steady and/or intermittent levels of noise may have a highly degrading effect on human performance.

5. *Other factors*—A number of additional factors may (to varying degrees) cause stress on the human body. These include the effects of radiation, the effects of gas or toxic substances in the air, the effects of sand and dust, and so forth.

The external stress factors listed above will normally result in individual human strain. Strain may, in turn, have an impact on any one or more of the human body systems (i.e., circulatory system, digestive system, nervous system, respiratory system, etc.). Measures of strain include such common parameters as pulse rate, blood pressure, body temperature, oxygen consumption, and the like.

Psychological Factors

Psychological factors pertain to the human mind and the aggregate of emotions, traits, and behavior patterns as they relate to job performance. All other conditions may be optimum from the standpoint of performing a task in an efficient manner; however, if the individual operator (or maintenance technician) lacks initiative, motivation, dependability, self-confidence, communication skills, and so on, the probability of performing in an efficient manner is low.

Psychological factors may be highly influenced by physiological factors. In general, one's attitude, initiative, motivation, and so on, is dependent on the needs and expectations of the individual. Fulfillment of these needs and expectations is a function of the organizational environment within which the individual performs and the leadership characteristics of supervisory personnel in that organization. If the individual operator does not perceive the opportunity for on-the-job growth, or if the managerial style of his or her immediate supervisor is not directly supportive of personal goals and objectives, job performance may deteriorate accordingly.

The psychological aspects of human performance are many and varied, and the reader is advised to review the literature on behavioral science in order to gain some

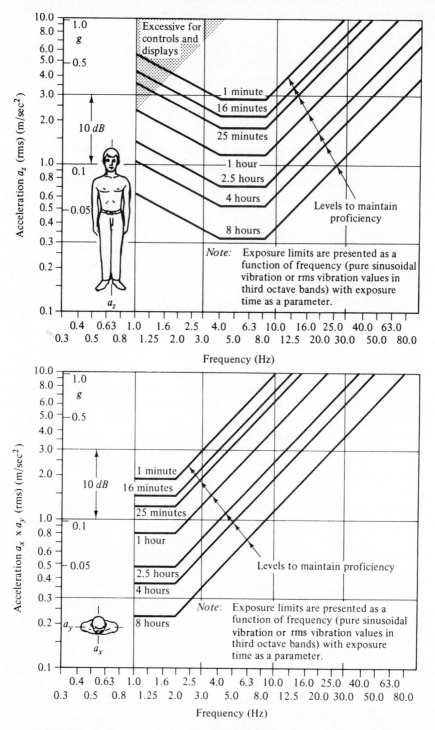

Figure 15.7 Vibration exposure criteria for longitudinal (upper curve) and transverse (lower curve) directions with respect to body axis. *Source:* MIL-STD-1472, Military Standard, *Human Engineering Design Criteria for Military Systems, Equipment, and Facilities,* Department of Defense, Washington, D.C.

insight as to the cause-and-effect relationships pertaining to good versus poor performance.[5]

15.2. HUMAN FACTORS IN THE SYSTEM LIFE CYCLE

As shown in Figure 15.1, human factors requirements are established through the logical breakdown of operation and maintenance functions to the extent feasible during early system design. These requirements are influenced by the environmental and personnel factors described. Within the limits provided, a preliminary system analysis is accomplished with the intent of identifying functions/tasks that will be (1) accomplished solely by the human being, (2) accomplished automatically by equipment, or (3) accomplished by a combination of the human being and equipment. Functions (or elements thereof) are evaluated relative to the work required in terms of human capabilities and cost. Trade-off analyses are conducted to support the allocation process. The results provide the identification of system activities and the allocation of human resources for these activities. Human factors requirements for the system will specify:

1. The number of persons required for system operation and maintenance throughout the life cycle. Personnel numbers should be justified through life-cycle-cost analyses.
2. The personnel job categories and skill levels necessary to perform the designated operation and maintenance functions/tasks. Skill level requirements should normally be minimized where possible.
3. The anthropometric, human sensory, physiological, and psychological parameters that are necessary to facilitate satisfactory human performance in the fulfillment of operation and maintenance functions/tasks. Qualitative and quantitative requirements covering human body dimensions, the environment, and the organization should be specified as applicable.

Human factors, as with any other engineering discipline, are applicable to all phases of the system life cycle. Figure 15.8 depicts the basic requirements by phase and these are discussed below.

Conceptual Design (Figure 15.8, Block 1)

In the early stages of conceptual design it is just as important to establish criteria covering system personnel requirements as it is to specify performance factors, reliability and maintainability factors, supportability factors, economic goals, and so

---

[5] An overview of organizational characteristics (i.e., managerial styles, leadership characteristics, needs of the individual, creativity, etc.) is presented in B. S. Blanchard, *Engineering Organization and Management* (Englewood Cliffs, N.J.: Prentice-Hall, 1976), Chapter 9. A more in-depth coverage of behavioral science may be found in B. M. Bass, *Organizational Psychology* (Boston: Allyn and Bacon, 1968), and W. G. Scott, *Organization Theory—A Behavioral Analysis for Management* (Homewood, Ill.: Richard D. Irwin, 1967).

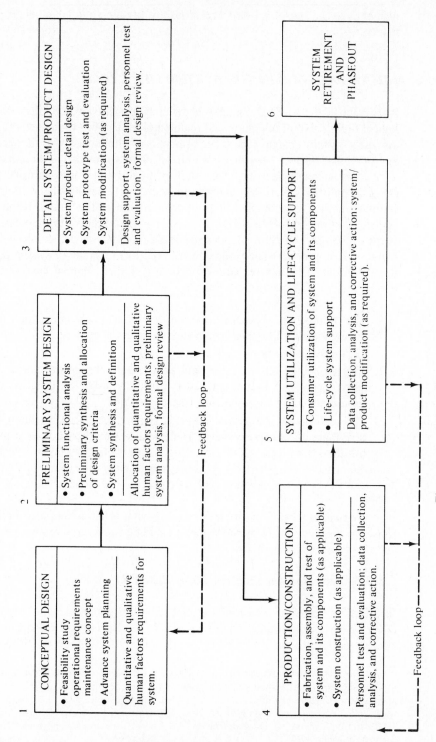

Figure 15.8 Human factors in the system life cycle.

on. As people often constitute a major element of the system, and personnel costs are usually relatively high (as compared to other categories), it is necessary to establish design guidelines. Of particular significance are the number of people required for system operation and maintenance support together with the job categories and skill levels of these personnel.

Relative to personnel numbers, an objective is to minimize the total number of personnel required to perform system operation and maintenance functions. Quantitative measures may be expressed in terms of direct operator labor hours per hour of system operation, direct maintenance labor-hours per hour of system operation, and/or in terms of x number of people.

Where personnel are required, a major objective is to identify a job category that relates to the function(s) to be performed within a current supply of personnel available to perform this function. It is highly desirable to keep the personnel entry-skill-level requirements to a minimum. For instance, an objective may be to design the system such that it can be successfully operated by an individual with a *basic* skill level, where a basic skill level constitutes a person with the following characteristics:

1. Age: 18 to 21 years.
2. Education level: high school graduate.
3. General reading/writing level: ninth grade.
4. Experience: no regular work experience prior to training.
5. General qualifications: after a limited amount of familiarization training on the system, this individual can accomplish normal operator functions, involving system actuation, manipulation of controls, communication, and the recording of data. The individual is able to follow clearly presented instructions where interpretation and decision making are not necessary. This individual will normally require close supervision.

The objective here is to establish system design criteria that will promote simplicity in operation (and simplicity in the performance of maintenance). The accomplishment of such will tend to minimize personnel training costs, reduce the probability of personnel-induced system failures, and minimize overall life-cycle cost.

In addition to personnel numbers and skill levels, there are many individual factors involving the human being which may be judiciously applied to both equipment and to the environment in which the system is to operate. These include anthropometric parameters, human sensory limitations, and the appropriate physiological and psychological factors as they pertain to the environment. These factors should be specified individually and in quantitative terms in the areas of system design considered critical from the standpoint of mission success, or in areas where personnel safety is of concern. On the other hand, for more general applications, it may be sufficient to merely reference criteria documentation for design guidance purposes. In any event, qualitative and quantitative requirements are specified in this program phase to the extent necessary to promote efficiency in human performance, and to optimize the human–machine interface.

Preliminary System Design (Figure 15.8, Block 2)

Qualitative and quantitative parameters specified for the overall system during conceptual design are allocated to various elements of the system to the extent necessary for design guidance purposes. The allocation process described in Chapter 4 is applicable here, with emphasis on the human element of the system. The allocation results constitute specific parameters to which the system should be designed.

Within the limitations and bounds established through the specified design criteria, there may be alternative approaches in responding to a given objective. Feasible alternatives are evaluated as part of the iterative system analysis process, with a preferred approach being justified on the basis of technical performance requirements and life-cycle cost. Inherent in this process is the ongoing requirement for trade-off studies dealing with the human–machine interface.

In arriving at a specific recommended approach, one must proceed down through the human hierarchy described in Section 15.1 (i.e., function, job operation, duty, task, etc.) to the extent necessary in determining actual personnel quantities and skill-level requirements for each alternative being considered. In this evaluation effort, the results may dictate a change from the initial allocation of human–machine requirements (e.g., a function that was initially designated as being accomplished by an individual is now automated to be accomplished by the machine, or vice versa). Changes of this nature are implemented through the feedback loop illustrated in Figure 15.8.

Detail System/Product Design (Figure 15.8, Block 3)

As the design progresses and data become available, a system analysis activity may be applied in the form of an ongoing design evaluation effort. Operational sequence diagrams are developed to depict information–decision–action flow in the system; detail operator and maintenance task analyses are accomplished to evaluate human–machine interactions at the appropriate level; and error analyses are performed to identify critical activities where operator-induced system failures are likely to occur. Thus, the design configuration being considered is evaluated from the standpoint of the human being in the system. Problem areas are noted and changes are initiated as required.

On a concurrent basis, the analysis activity serves as a data base for determining personnel numbers, job categories, and skill-level requirements. Equipment lists are developed, duty and task worksheets are prepared, and personnel requirements are identified. The data covering anticipated maintenance activities are based on the maintainability and logistic support analysis. The selected operator and maintenance personnel categories and skills are evaluated in terms of available human resources and training requirements are defined. The object is to convert available resources into qualified operator and maintenance staffing for the system through effective training programs. Training programs are designed to include technical program content, student entry-level requirements, instructor qualifications, training material needs, training aids, and so on.

Toward the end of the detail system/product design phase, prototype equipment is developed and tested for the purposes of verifying or demonstrating various system characteristics. Along with the demonstration of selected hardware and software components of the system, it may be necessary to verify certain operator and maintenance tasks to ensure complete human–equipment compatibility. The verification of human performance is usually accomplished through a formal personnel test and evaluation program.

Production/Construction, Use, and Support (Figure 15.8, Blocks 4 and 5)

During the later phases of the system life cycle, the primary activity constitutes data collection, analysis, and corrective action as required. The objective is to identify major operation and maintenance activities where human performance is marginal or inefficient, and where personnel costs are high. Problems are noted, cause-and-effect relationships are established, and recommendations for system or product improvement are initiated as appropriate.

15.3. HUMAN FACTORS ANALYSIS

Throughout system design, a human factors analysis is performed as an integral part of the overall system analysis effort. The human factors analysis constitutes a composite of individual program activities directed toward (1) the initial establishment of human factors requirements for system design, (2) the evaluation of system design to ensure that an optimum interface exists between the human and other elements of the system, and (3) the assessment of personnel number and skill-level requirements for a given system design configuration. The analysis effort employs a number of the analytical techniques discussed in Part Three and is closely related to the reliability analysis, maintainability analysis, logistic support analysis, and life-cycle cost analysis.

The human factors analysis begins with conceptual design when functions are identified, and trade-off studies are accomplished to determine whether these functions are to be performed manually using human resources, automatically with equipment, or by a combination thereof. Given the requirements for human resources, one must then ensure that these resources are used as efficiently as possible. Thus, the analysis continues through an iterative process of evaluation, system modifications for improvement, reevaluation, and so on. In support of this latter phase of the overall analysis process, there are a number of methods and techniques that can be employed for evaluation purposes. These include the generation of operational sequence diagrams, the accomplishment of detail task analyses, the performance of an error analysis, the preparation of duty and task worksheets, and so on.

Operational Sequence Diagrams

As part of the human factors analysis activity, one of the major tasks is the evaluation of the flow of information from the point in time when the operator first becomes involved with the system to completion of the mission. Information flow in this instance pertains to operator decisions, operator control activities, and the transmission of data.

There are a number of different techniques that can be employed to show information flow. The use of operational sequence diagrams is one.[6] Operational sequence diagrams are decision–action flow devices that integrate operational functions and equipment design. More specifically, these diagrams project different sequences of operation, showing:

1. Manual operations.
2. Automatic operations.
3. Operator decision points.
4. Operator control actuations or movements.
5. Transmitted information.
6. Received information using indicator displays, meter readouts, and so on.

Operational sequence diagrams are similar to industrial engineering work-flow process charts and time-line analyses, and are used to evaluate decision–action sequences and human–machine interfaces. The evaluation of operator control panel layouts and work-space design configurations are good examples of where these diagrams may be profitably used. Figure 15.9 presents an example of an operational sequence diagram.

Detail Task Analysis

This facet of analysis involves a systematic study of the human behavior characteristics associated with the completion of a system task(s). It provides data basic to human engineering design and to the determination of personnel types and skill-level requirements. Tasks may be classified as being discrete or continuous. Further, there are operator tasks and maintenance tasks. Thus, one may wish to divide the analysis effort into the detail operator task analysis and the detail maintenance task analysis. The portion of the analysis covering maintenance tasks may evolve directly from a combination of maintainability analysis and logistic support analysis.

In accomplishing a task analysis, there are varying degrees of emphasis and types of formats used. However, the following general steps apply in most instances:

1. Identify system operator and maintenance functions and establish a hierarchy of these functions in terms of job operations, duties, tasks, subtasks, and task elements, as described in Section 15.1.
2. Identify those functions (or duties, tasks, etc.) that are controlled by the human being and those functions that are automated and controlled by the machine.
3. For each function involving the human element, determine the specific infor-

---

[6] Operational sequence diagrams are discussed further and illustrated in: K. B. DeGreene, *Systems Psychology* (New York: McGraw-Hill, 1970), Chapter 8; and W. E. Woodson, *Human Factors Design Handbook*, New York: McGraw-Hill, 1981).

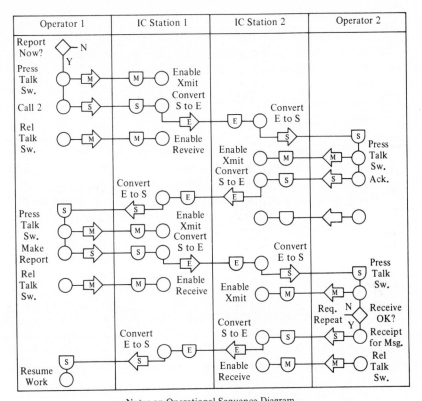

Operator 1	IC Station 1	IC Station 2	Operator 2

Notes on Operational Sequence Diagram

Symbols

◇ Decision
○ Operation
⇨ Transmission
⊍ Receipt
◻ Delay
▢ Inspect, Monitor
▽ Store

Links

M   Mechanical or Manual
E   Electrical
V   Visual
S   Sound
etc.

Stations or subsystems are shown by columns; sequential time progresses down the page.

Figure 15.9 Operational sequence diagram (example). *Source:* MIL-H-46855, Military Specification, *Human Engineering Requirements for Military Systems, Equipment and Facilities,* Department of Defense, Washington, D.C.

mation necessary for operator and/or maintenance personnel decisions. Such decisions may lead to the actuation of a control, the monitoring of a system condition, or the equivalent. Information required for decision making may be presented in the form of a visual display or an audio signal of some type.

4. For each action, determine the adequacy of the information fed back to the human being as a result of control activations, operational sequences, and so on.

5. Determine the impact of the environmental and personnel factors and constraints on the human activities identified in Figure 15.1.
6. Determine the time requirements, frequency of occurrence, accuracy requirements, and criticality of each action (or series of actions) accomplished by the human being.
7. Determine the human skill level requirements for all operator and maintenance personnel actions.

A task analysis is generated to ensure that each stimulus is tied to a response, and that each response is directly related to a stimulus. Further, individual human motions are analyzed on the basis of dexterity, mental and motor skill requirements, stress and strain characteristics of the human being performing the task, and so on. The purpose is (1) to identify those areas of system design where potential human–machine problems exist, and (2) identify the necessary personnel skill level requirements for operating and maintaining the system in the future.

Error Analysis

One of the major objectives in system design is to minimize (if not eliminate) the possibility of introducing human error in the performance of system operation functions and in the accomplishment of maintenance tasks. Errors may occur as a result of a combination of factors such as:

1. Failure to comply with system operating and maintenance procedures.
2. Failure to obtain and/or consider the proper input data for decision-making purposes.
3. Failure to read displays or operate controls properly.
4. Failure to monitor and respond to system changes.

An error analysis may be accomplished in conjunction with the detail task analysis, using a fault-tree approach or a failure mode and effect analysis (refer to the FMECA/FMEA discussion in Chapter 13). Given the requirement for human manipulation in the accomplishment of a task, one must address the questions: What else can happen? or How might a fault be introduced in the system? After recognizing, verifying, and classifying failures, one must determine the effects of these failures on the overall system and on the mission to be fulfilled. Verifying a failure at the operator level and determining its effect(s) at the system level is a prime objective of the reliability FMECA. In this instance, the primary source of failure is the human being.

Safety Analysis

The safety analysis (or the hazard analysis) is closely aligned with the task and error analyses. Safety, of course, pertains to both personnel and equipment, with personnel safety being emphasized herein. The safety analysis generally includes the following basic information:[7]

1. *Description of hazard*—System operational requirements and the system maintenance concept are reviewed to identify possible hazardous conditions. Past experience on similar systems or products utilized in comparable environments serves as a good starting point. Hazardous conditions may include acceleration and motion, electrical shock, chemical reactions, explosion and fire, heat and temperature, radiation, pressure, moisture, vibration and noise, and toxicity.

2. *Cause of hazard*—Possible causes should be described for each identified hazard. In other words, what events are likely to take place in creating the hazard?

3. *Identification of hazard effects*—Describe the effects of each identified hazard on both personnel and equipment. Personnel effects may include injuries such as cuts, bruises, broken bones, punctures, heat exhaustion, asphyxiation, trauma, and respiratory or circulatory damage.

4. *Hazard classification*—Hazards may be categorized according to their impact on personnel and equipment as follows:

   a. *Negligible hazard* (*category I*)—such conditions as environment, personnel error, characteristics in design, errors in procedures, or equipment failures that will not result in significant personnel injury or equipment damage.

   b. *Marginal hazard* (*category II*)—such conditions as environment, personnel error, characteristics in design, errors in procedures, or equipment failures that can be controlled without personnel injury or major system damage.

   c. *Critical hazard* (*category III*)—such conditions as environment, personnel error, characteristics in design, errors in procedures, or equipment failures that will cause personnel injury or major system damage, or that will require immediate corrective action for personnel or system survival.

   d. *Catastrophic hazard* (*category IV*)—such conditions as environment, personnel error, characteristics in design, errors in procedures, or equipment failures that will cause death or severe injury, and/or complete system loss.

5. *Anticipated probability of hazard occurrence*—Through statistical means, estimate the probability of occurrence of the anticipated hazard frequency in terms of calendar time, system operation cycles, equipment operating hours, or equivalent.

[7] Safety analysis and hazard classifications are identified in: W. P. Rodgers, *Introduction to System Safety Engineering* (New York: John Wiley, 1971); and H. E. Roland and B. Moriarty, *System Safety Engineering And Management* (New York: John Wiley, 1983).

6. *Corrective action or preventive measures*—Describe the action(s) that can be taken to eliminate or minimize (through control) the hazard. Hopefully, all hazardous conditions will be eliminated; however, in some instances it may only be possible to reduce the hazard level from category IV to one of the lesser critical categories.

The safety analysis serves as an aid in initially establishing design criteria and as an evaluation tool for the subsequent assessment of design for safety. Although the format may vary somewhat, a safety analysis may be applied in support of requirements for both industrial safety and system/product safety.

15.4. PERSONNEL AND TRAINING REQUIREMENTS

Personnel requirements for the system may be categorized in terms of *operator* personnel and *maintenance* personnel. Operator personnel requirements are derived through the human factors analysis, particularly through the generation of operational sequence diagrams and the detail operator task analysis. Maintenance personnel requirements evolve from the maintainability analysis, the logistic support analysis, and the detail maintenance task analysis.

In defining these requirements for the system, personnel numbers and skill levels are determined for all human activities by function, job operation, duty, task, and so on. These requirements, initially identified in small increments, are combined on the basis of similarities and complexity level. Individual position requirements are identified, and worksheets are prepared specifying the duties and tasks to be performed by each individual. From this information, personnel quantities and skill level requirements are identified for the system as an entity.

Given the personnel requirements (in terms of positions) for the system, the next step is to identify available resources and those individuals who can be employed to operate and maintain the system. These selected individuals are evaluated relative to their current skills as compared to the skill levels necessary for the system. The differences, of course, dictate the needs for training.

For the sake of illustration, let us assume that as a result of the human factors analysis the system will require x individuals with a *basic* skill, y individuals with an *intermediate* skill, and z individuals with a *high* skill, and that these skill level classifications are broadly defined as follows:[8]

Basic skill level. A basic skill level is assumed to require an individual between 18 and 21 years of age, a high school graduate with a ninth-grade general reading and writing level and having no regular work experience. After a limited amount of familiarization training on the system, this individual can accomplish simple operator functions involving system actuation, manipulation of

---

[8] As indicated, these classifications are defined in rather broad terms to illustrate a concept. In certain instances it may be necessary to elaborate further, particularly with regard to complex functions, in order to provide a good basis for determining specific training needs.

controls, communications, and the recording of data. The individual is able to follow clearly presented instructions where interpretation and decision making are not required. Close supervision of this individual is normally required.

Intermediate skill level. An intermediate skill level normally requires an individual over 21 years of age, with approximately two years of college or equivalent course work in a technical institute, has had some specialized training on similar systems in the field, and has had from two to five years of work experience. Personnel in this classification can perform relatively complex tasks, where the interpretation of data and some decision making may be required, and can accomplish simple on-site preventive maintenance tasks. This individual requires very little supervision.

High skill level. A high skill level normally requires an individual with two to four years of formal college or equivalent course work in a technical institute, who has taken a number of specialized training courses in various related fields and possesses ten years or more of related on-the-job experience. An individual in this classification may be assigned to train and supervise basic and intermediate-skill-level personnel, and is in the position to interpret procedures, accomplish complex tasks, and make major decisions affecting system operating policies. Further, this individual is qualified to accomplish (or supervise) all on-site preventive maintenance requirements for the system.

Given requirements for the system, it is assumed that all available resource personnel currently possess the necessary qualifications for entry into positions requiring a basic skill level; that is, with some general familiarization training on the system, they can usually perform the level of activity noted for an individual with a basic skill.

On the basis of these assumptions, it is necessary to develop a training program that will

1. Train entry-level personnel in the fundamentals of system operation to fill the positions where a *basic skill* level is required.
2. Train entry-level personnel in the performance of operator and maintenance functions to the extent necessary to satisfy the *intermediate skill-level* requirements for the system.
3. Train entry-level personnel in the performance of operator and maintenance functions to the extent necessary to satisfy the *high skill-level* requirements for the system.

Training may be accomplished through a combination of formal structured programs and on-the-job (OJT) training. Training requirements include both the training of personnel initially assigned to the system and the training of replacement personnel throughout the life cycle as required due to attrition. In addition, training must cover both operator activities and maintenance activities.

In designing training programs, the specific requirements may vary considerably depending on the complexity of the human functions to be performed in the operation and maintenance of the system. The level of complexity is, of course, a function of system design as discussed in the earlier sections of this chapter. If there are many human functions of a highly complex nature, then the training requirements will likely be extensive and the associated cost will be high. On the other hand, if the functions to be performed by the human being are relatively simple, then training requirements (and associated cost) will be minimal. In any event, a training program should be designed and tailored to meet the personnel needs for the system. The program content must be at the appropriate level to provide future system operators and maintenance personnel with the tools necessary to enable the performance of their respective functions in an effective and efficient manner.

With this objective in mind, a formal training plan is often developed for the system during the detail system/product design phase. This plan should include a statement of training objectives, a description of the training program in terms of modules and content, a proposed training schedule, a description of required training aids and data, and an estimate of anticipated training cost(s). The training plan must be implemented in time to support system operation and maintenance activities as the system is distributed for use.

15.5. PERSONNEL TEST AND EVALUATION

Chapter 6 addressed the overall requirements for system test and evaluation in a fairly comprehensive manner. Included in the system test and evaluation effort is, of course, the human element of the system. As equipment is being evaluated for performance, reliability, maintainability, and so on, the operator and the maintenance technician should be evaluated in terms of human performance efficiency.

During the latter part of the detail system/product design phase and throughout the subsequent phases of the life cycle (refer to Figure 15.8), there is a personnel test and evaluation activity. The basic objective of this activity is to (1) monitor human performance in fulfilling system operator and maintenance requirements; (2) assess operator and maintenance tasks in terms of elapsed time, step sequences, dexterity requirements, areas of difficulty, and human error; (3) identify problem areas and initiate recommendations for corrective action; and (4) modify personnel selection and training requirements as necessary for compatibility with any system changes that may occur. Thus, the process described in Chapter 6 is followed, but the emphasis is on the human being.

SELECTED REFERENCES

(1) Blanchard, B. S., *Engineering Organization and Management*. Englewood Cliffs, N.J.: Prentice-Hall, 1976.

(2) Blum, M. L., and J. C. Naylor, *Industrial Psychology—Its Theoretical and Social Foundations*. New York: Harper & Row, 1968.

(3) DeGreene, K. B., ed., *Systems Psychology*. New York: McGraw-Hill, 1970.

(4) Meister, D., *Behavioral Analysis and Measurement Methods*. New York: John Wiley, 1985.

(5) MIL-STD-1472, Military Standard, *Human Engineering Design Criteria for Military Systems, Equipment and Facilities,* Department of Defense, Washington, D.C.

(6) Roland, H. E., and B. Moriarty, *System Safety Engineering and Management.* New York: John Wiley, 1983.

(7) Salvendy, G., ed., *Handbook of Human Factors.* New York: John Wiley, 1987.

(8) Sanders, M. S., and E. J. McCormick, *Human Factors in Engineering and Design,* (6th ed.). New York: McGraw-Hill, 1987.

(9) Van Cott, H. P., and R. G. Kinkade, eds., *Human Engineering Guide to Equipment Design.* U.S. Government Printing Office, Washington, D.C., 1972.

(10) Woodson, W. E., *Human Factors Design Handbook.* New York: McGraw-Hill, 1981.

QUESTIONS AND PROBLEMS

1. In the development of a new system, describe the steps that you would follow in identifying human functions versus equipment functions.

2. Define cybernetics. How does cybernetics relate to system design?

3. What is meant by anthropometry and how does it relate to system design? How do human physiology and psychology relate to system design? Are they important? Why?

4. Identify some of the measures that you would employ in assessing the physiological effects on human performance.

5. Describe the possible impact(s) of psychological factors on system operation and the successful completion of its defined mission. Provide some examples.

6. What physiological and/or psychological effects are likely to occur if the operator in performing his or her function is overtrained for the job? Undertrained? How would these effects influence system operation?

7. How are personnel quantities and skill-level requirements defined for the system? Describe the steps involved.

8. Why is it important to establish human factors requirements early in conceptual design?

9. What is the purpose of generating operational sequence diagrams? When in the system life cycle should they be developed? What information is conveyed through operational sequence diagrams?

10. What is the purpose of a detail operator task analysis and what information is conveyed? How does the operator task analysis relate to operational sequence diagrams?

11. What is the purpose of a detail maintenance task analysis and what information is conveyed? How might the detail maintenance task analysis relate to the maintainability analysis and the logistic support analysis?

12. What is the purpose of a human error analysis?

13. What is the purpose of a safety analysis? What information is conveyed? How does the safety analysis relate to the FMEA or FMECA?

14. How are training requirements for the system determined?

15. Review the analytical techniques in Part Three. Which techniques could be appropriately used in the performance of a human factors analysis? Identify some applications.

16. What is the purpose of personnel test and evaluation? What system parameters are measured?

17. Design for manability is essential. Why?

16
Design for Supportability

Major systems and products have been planned, designed and developed, produced, and delivered to the consumer with very little consideration given to the aspects of maintenance and logistic support. The primary area of emphasis has been placed on those facets of the system that deal directly with the fulfillment of the designated mission (i.e., performance), while the sustaining life-cycle support of the system has been addressed somewhat after the fact. This practice has been costly because (1) the costs of system support often constitute a major portion of the projected overall life-cycle cost, and these costs are rapidly rising; and (2) the costs of system support are largely influenced by decisions made in the early phases of the life cycle.

In recent years, systems and products have become more complex as technology advances. At the same time, the current economic dilemma of decreasing budgets combined with upward inflationary trends results in less money available for both the acquisition of new systems and the support of those systems already in being. Thus, there is an increasing need for more effective and efficient management of resources. The requirement to deal with all facets of the system as an integrated entity is readily apparent. This includes consideration for the aspect of logistic support, as well as the performance-related aspects of the system.

16.1. SYSTEM SUPPORT REQUIREMENTS

The material presented in earlier chapters primarily addresses areas that relate directly to system performance factors. This will now be augmented with the subject of logistic support and the topic of design for supportability. Based on cause-and-

460

effect relationships between system design and support, and the fact that logistics costs may assume major proportions, it is essential that logistic support be included in the early stages of system and product planning and design.

System support (defined herein) is viewed as the composite of all considerations needed to assure the effective and economic support of a system throughout its programmed life cycle. It is an integral part of all aspects of system planning, design and development, test and evaluation, production and/or construction, consumer use, and system retirement. The elements of support, which must be developed on an integrated basis with all other segments of the system, are described below:

1. *Maintenance planning* includes all planning and analysis associated with the establishment of requirements for the overall support of the system throughout its life cycle. Maintenance planning constitutes a sustaining level of activity beginning with the development of the maintenance concept (described in Chapter 3) and continuing through the accomplishment of logistic support analyses during design and development, the procurement and acquisition of support items, and through the consumer use phase when an ongoing system and product support capability is required to sustain operations. Maintenance planning is accomplished to integrate the various other facets of support.

2. *Supply support* includes all spares (units, assemblies, modules, etc.), repair parts, consumables, special supplies, and related inventories needed to support prime mission-oriented equipment, software, test and support equipment, transportation and handling equipment, training equipment, and facilities. Supply support also covers provisioning documentation, procurement functions, warehousing, distribution of material, and the personnel associated with the acquisition and maintenance of spare and repair part inventories at all support locations. Considerations include each maintenance level and each geographical location where spare and repair parts are distributed and stocked; spares demand rates and inventory levels; the distances between stockage points; procurement lead times; and the methods of material distribution.

3. *Test and support equipment* includes all tools, condition monitoring equipment, diagnostic and checkout equipment, metrology and calibration equipment, maintenance stands, and handling equipment required to support scheduled and unscheduled maintenance actions associated with the system or product. Test and support equipment requirements at each level of maintenance must be addressed as well as the overall requirements for test traceability to a secondary or primary standard of some type.

4. *Transportation and handling* includes all equipment, special provisions, containers (reusable and disposable), and supplies necessary to support packaging, preservation, storage, handling, and/or transportation of prime mission equipment, test and support equipment, spares and repair parts, personnel, technical data, and mobile facilities. This category basically covers the initial distribution of products and the transportation of personnel and materials for maintenance purposes. Primary transportation modes include air, highway, pipeline, railway, and waterway.

5. *Personnel and training* required for the installation, checkout, operation, handling, and sustaining maintenance of the system (or product) and its associated test and support equipment are included in this category. Maintenance personnel required at each level of maintenance are considered. Formal training includes both initial training for system and product familiarization and replenishment training to cover attrition and replacement personnel. Training is designed to upgrade assigned personnel to the skill levels defined for the system. Training data and equipment (e.g., simulators, mockups, special devices) are developed as required to support personnel training operations.

6. *Facilities* refer to all supporting facilities needed for the performance of maintenance functions at each level. Physical plant, portable buildings, housing, intermediate maintenance shops, calibration laboratories, and special depot repair and overhaul facilities must be considered. Capital equipment and utilities (heat, power, energy requirements, environmental controls, communications, etc.) are generally included as part of facilities.

7. *Data* includes system installation and checkout procedures, operating and maintenance instructions, inspection and calibration procedures, overhaul procedures, modification instructions, facilities information, drawings, and specifications that are necessary in the performance of system operation and maintenance functions are included. Such data not only cover the prime mission equipment but test and support equipment, transportation and handling equipment, training equipment, and facilities.

8. *Computer resources* (*software*) is a facet of support that refers to all computer programs, condition monitoring and diagnostic tapes, and so on, necessary in the performance of system maintenance functions.

For large-scale systems, the logistic support requirements throughout the life cycle may be significant. The prime mission-oriented segment of the system must be designed with support in mind, and the various elements of logistic support must be designed to be compatible with the prime mission equipment. Further, these different elements of logistic support interact with each other and the effects of these interactions must be reviewed and evaluated continually. A major decision or a change involving any one of these elements could significantly impact other elements and the system as a whole.

On the other hand, the logistics requirements for relatively small systems (or products) may entail only the functions of product distribution to the consumer and initial system installation and checkout, while the sustaining life-cycle maintenance support will be minimal. In this instance, the emphasis on logistic support (and the design for supportability) will not be as great as for large complex systems. The specific requirements must be tailored accordingly.

The objective is to develop a system reflecting the proper balance between the prime mission elements of that system and its related support. This balance deals with establishing the proper relationships between system performance characteristics, reliability and maintainability characteristics, supportability characteristics, economic factors, and so on. To attain such a balance requires the consideration of logistic support throughout the system life cycle, particularly in the early phases of planning and conceptual design when major decisions affecting the overall system configuration are made. This emphasis in the early phases of the life cycle is provided through design for supportability, an inherent part of the systems engineering process.

16.2. LOGISTIC SUPPORT IN THE SYSTEM LIFE CYCLE

Logistics involves the distribution, maintenance, and support of systems and products. The primary activities include material flow, transportation and distribution, maintenance of inventories, and customer service. Although many of these activities are directly oriented to the latter phases of the system life cycle, the ultimate impact on such activities stems from early planning and design functions. Thus, a life-cycle approach to logistics is assumed throughout this text, and major emphasis is directed toward the engineering design facet, or the design for supportability.[1]

Logistics is important in all phases of the life cycle, and the fulfillment of logistic support objectives can be accomplished through the concept of *integrated logistic support* (ILS). ILS is a management function providing the initial planning, funding, and controls which help to assure that the ultimate consumer will receive a system (or product) that will not only meet performance requirements, but one that can be expeditiously and economically supported throughout its programmed life cycle.

Figure 16.1 illustrates the system life cycle along with the major logistic support activities in each phase. The flow diagram is not intended to reflect specific levels of effort or organization relationships. Its purpose is to illustrate a process; that is, one must progress systematically through the steps of planning, design and analysis, test and evaluation, production, and so on, regardless of the type of system being considered. These steps (particularly those reflected by Blocks 1, 2, and 3 in the figure) directly tie in with the systems process illustrated in Figure 2.4. A brief discussion of the major steps in this process is provided below.

Logistic Support Planning

Logistic support planning begins during the early conceptual phase of a program through the development of a preliminary integrated logistic support plan (Figure 16.1, Block 1), and extends through the preparation of a formal integrated logistic support plan during preliminary design (Block 2). The preliminary plan leads directly into the final ILS plan. Although the level and depth of planning may vary

---

[1] Refer to B. S. Blanchard, *Logistics Engineering And Management*, 3rd ed., Prentice-Hall, New Jersey, 1986, for a more in-depth discussion of this subject and its orientation to an engineering life-cycle approach.

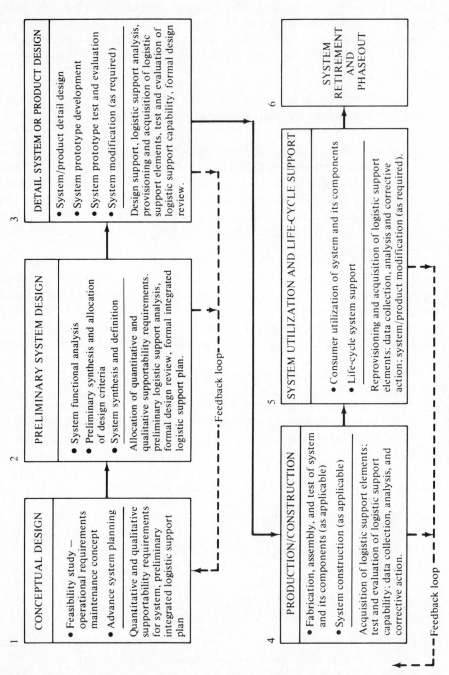

Figure 16.1 Logistic support in the system life cycle.

1 CONCEPTUAL DESIGN
- Feasibility study — operational requirements maintenance concept
- Advance system planning

Quantitative and qualitative supportability requirements for system, preliminary integrated logistic support plan

2 PRELIMINARY SYSTEM DESIGN
- System functional analysis
- Preliminary synthesis and allocation of design criteria
- System synthesis and definition

Allocation of quantitative and qualitative supportability requirements, preliminary logistic support analysis, formal design review, formal integrated logistic support plan.

3 DETAIL SYSTEM OR PRODUCT DESIGN
- System/product detail design
- System prototype development
- System prototype test and evaluation
- System modification (as required)

Design support, logistic support analysis, provisioning and acquisition of logistic support elements, test and evaluation of logistic support capability, formal design review.

4 PRODUCTION/CONSTRUCTION
- Fabrication, assembly, and test of system and its components (as applicable)
- System construction (as applicable)

Acquisition of logistic support elements; test and evaluation of logistic support capability; data collection, analysis, and corrective action.

5 SYSTEM UTILIZATION AND LIFE-CYCLE SUPPORT
- Consumer utilization of system and its components
- Life-cycle system support

Reprovisioning and acquisition of logistic support elements; data collection, analysis and corrective action; system/product modification (as required).

6 SYSTEM RETIREMENT AND PHASEOUT

Feedback loop

Feedback loop

Feedback loop

significantly depending on the type and size of the system, the planning approach proposed herein is still considered to be appropriate provided that the result is properly tailored to the need.

Preliminary integrated logistic support plan.   The preliminary ILS plan should address the overall approach to system support as well as the functions to be performed throughout system design as related to supportability. Specifically, this plan should cover the following:

1. Definition of system operational requirements and the system maintenance concept (logistic support requirements must be based on the mission profile, the anticipated geographical area where the system is to be located, the environment in which it is intended to operate, and the maintenance concept as defined in Chapter 3).

2. Quantitative and qualitative supportability design requirements for the system.

3. Functional analysis and the allocation of supportability requirements to the subsystem level and beyond as appropriate.

4. Logistic support analysis (i.e., maintenance analysis, level-of-repair analysis, life-cycle cost analysis, etc.).

5. Design of new logistic support elements as required (e.g., new test and support equipment, new facilities, training equipment, etc.).

6. Formal design reviews.

7. Early provisioning and the acquisition of selected logistic support elements (i.e., the plan for the procurement of long-lead-time items to include spares and repair parts, test and support equipment, etc., as necessary to meet program schedules).

8. Test and evaluation of logistic support elements.

The preliminary ILS plan covers those functions in the early phases of the life cycle that directly interface with detail day-to-day design activities, reliability activities, maintainability activities, and so on. This plan may be included as part of the system engineering management plan discussed in Part Five and must lead directly into the formal ILS plan.

Formal integrated logistic support plan (ILSP).   The formal ILS plan is initially prepared during the latter phases of preliminary design. It evolves directly from the preliminary ILS plan and covers logistic support activities that deal with the detailed requirements for the system and the provisioning, procurement, and acquisition, and the maintaining of a support capability for the system throughout its planned life cycle. This plan directly ties into the systems engineering process, since knowledge of the support requirements is acquired directly from the design process. The formal ILS plan usually includes the following individual sections:[2]

1. *Maintenance plan*—A detailed plan for maintenance is developed from the maintenance concept, the ongoing maintenance planning effort accomplished throughout system design, and the results of the logistic support analysis. This plan includes a description of the levels of maintenance required for sustaining system life-cycle support, the major functions to be performed at each level, the organizations responsible for performing the required maintenance, and a summary of the logistic support resources required. The maintenance plan is based on a given system design configuration, while the maintenance concept reflects a before-the-fact approach to system support. The information contained in this plan serves as a *technical baseline* for all other sections of the formal ILS plan.

2. *Supply support plan*—A supply support plan is developed to include:

   a. A summary listing of spares, repair parts, and consumables for each level of maintenance. This list is based on the results of the logistic support analysis.

   b. Procedures covering the procurement and handling (packaging and transportation) of spares, repair parts, and consumables.

   c. Definition of warehousing and the accountability functions associated with maintenance support. This includes initial cataloging, stocking, inventory maintenance and control, establishing procurement cycles, and the disposition of residual assets.

   d. Procedures for data collection, analysis, and the updating of spare and repair part demand factors necessary to improve procurement cycles and reduce waste.

3. *Test and support equipment plan*—A detailed plan is developed to cover the acquisition of test and support equipment, and should include:

   a. A summary listing of test and support equipment requirements for each level of maintenance. This listing is based on the results of the logistic support analysis.

   b. Procedures covering the acquisition of newly designed support equipment and the procurement of off-the-shelf items from supplier inventories.

   c. Procedures covering support equipment/prime equipment compatibility testing and the subsequent distribution of test and support equipment to the geographical point of need.

---

[2] While the formal ILS plan is initially prepared during the latter stages of preliminary design, it needs to be updated periodically if it is to reflect specific support requirements being defined on a continuing basis through detail design. For the purposes of this text, the ILS plan is to be considered as a living document. An ILSP may include a LSA plan and/or a Reliability and Maintainability plan depending on the program.

    d. Procedures for data collection, analysis, and the evaluation of test and support equipment to assess testing accuracy and thoroughness, test equipment utilization, reliability of the test equipment, and test equipment maintenance.

4. *Personnel and training plan*—A plan should be developed to cover:

    a. Operator training (training of system operators) to include the type and level of training, the proposed schedule for training, basic entry-skill requirements, and a brief course outline.

    b. Maintenance training for each level of maintenance to include the type and depth of training, the proposed schedule for training, basic entry requirements, and a brief course outline.

    c. Training equipment, devices, aids, and data required to support operator and maintenance training activity.

    d. Proposed approach for the accomplishment of periodic training for new personnel and/or for the recycling of existing personnel for upgrading purposes.

Personnel quantities and skill-level requirements, particularly with regard to the accomplishment of maintenance functions, are identified through the logistic support analysis. The differences between the required skills and the existing skills of currently available personnel will lead to the definition of training requirements (i.e., the process of upgrading individuals to higher-level skills in order to successfully operate and maintain the system).

5. *Facilities plan*—A facilities plan is developed to identify all real property and equipment to support system testing (Types 3 and 4), training, operation, and maintenance functions. The plan must contain sufficient qualitative and quantitative information to allow facility planners to:

    a. Initially assess and allocate requirements. The logistic support analysis identifies maintenance activities that will lead to the establishment of facility requirements for maintenance support.

    b. Analyze existing facilities to determine adequacies or deficiencies.

    c. Determine requirements for new facilities and/or modifications to existing facilities.

    d. Estimate the cost of construction (and capital equipment) or modification projects required to meet the needs of the system, and establish the appropriate facility development schedule.

6. *Data plan*—A data plan should be prepared to include

    a. A summary of technical data requirements (installation, operating, maintenance, servicing, inspection, calibration, overhaul instructions) for each level of maintenance for the system and its components.

  b. Technical data development schedule (major milestones).

  c. A procedure for data verification and validation.

7. *Computer resources plan*—A detailed listing of maintenance software programs, and the procedures for modifying such programs.

8. *Retirement plan*—Throughout the system life cycle, faulty nonrecoverable items will be phased out of the inventory as maintenance actions occur. Further, the entire system and its components will ultimately be phased out upon retirement after the system has fulfilled its intended purposes. In the past, the disposal of residual items has not been very well planned, system components have been inappropriately discarded and left to the whims of nature, and the results have had a costly impact on the environment. Thus, it is essential that one address the areas of disposability, possible candidates and procedures for item recycling, and methods for material decomposition. The accomplishment of material recycling and/or disposal often requires a relatively high level of logistic support. The intent of this plan is to identify the magnitude of effort required in this final phase of the life cycle, and to define the anticipated logistic support resources needed.

The formal ILS plan provides the necessary guidance for life-cycle system support. This guidance is developed through system design activities and the decisions pertaining to supportability. If the system is to be effectively and economically supported, it is essential that the logistic support elements covered in the plan be addressed at the early stages of the life cycle through development of the maintenance concept, and expanded subsequently through the iterative process involving the logistic support analysis. Thus, logistic support planning is a significant ongoing activity, and the early phases of this planning effort must be closely integrated with the system engineering management plan.

Design for Supportability

There are many contributing factors that must be considered in providing effective results. Obviously, the system must be able to perform certain designated functions, it must be reliable, it must be maintainable, it must be operable, it must be economically feasible, and so on. Relative to the aspect of logistic support, the major level of activity occurs during the system use phase, when the consumer is operating the system on a day-to-day basis. However, the greatest impact on system support is realized during the early stages of system design, when decisions are made that significantly affect the type and extent of support to be provided. Thus, it is essential that logistic support be a major consideration during the early system design and development process (i.e., design for supportability must take place). When design for supportability takes place, one must consider:

1. The design of the prime-mission-oriented elements of the system, to ensure that the inherent characteristics in the design reflect a configuration that can be easily and economically supported.

2. The design of the various elements of logistic support, to ensure that these elements are compatible with the prime mission elements of the system, compatible with each other, and reflect an overall economical approach to system support.

The design of the prime-mission-oriented elements of the system for supportability is accomplished largely through the reliability and maintainability efforts described in Chapters 13 and 14, while the design of the logistic support elements of the system is emphasized in this chapter. The interaction effects between these two major segments of the system can be significant. Dealing with one of these areas without considering the other on an integrated basis can prove to be costly. For instance, the following problems often arise when only prime mission equipment design is accomplished without the proper integration of elements of support (this list is not all-inclusive, but represents some of the areas of concern):

1. Individual items of test equipment are not compatible with the tasks scheduled to be accomplished at the maintenance level assigned. Items incorporating insufficient test capability will not allow for a positive verification of the system condition. This will often lead to the accomplishment of maintenance based on a "best guess" approach which, in turn, results in the consumption of valuable resources. On the other hand, items incorporating more test capability than what is necessary are costly, both from the standpoint of acquisition cost and from the standpoint of requiring more maintenance (in themselves) than what otherwise would be required.

2. Individual items of test equipment are not as reliable as the elements of the system requiring the test support. This, of course, tends to destroy the validity of the test being conducted, particularly in situations when a failure occurs and it is difficult (if not impossible) to determine whether the failure is in the system itself or in the test equipment. Such situations usually result in more total system downtime than desired.

3. Spares and repair parts stocked at each level of maintenance are not compatible with the removal and replacement requirements or the repair-discard decisions anticipated. This usually results in excessive system downtime, the possibility of costly inventory requirements, and/or the added high costs associated with the increase in transportation and handling requirements necessary to obtain the needed item(s).

4. The personnel quantities and skill levels are not adequate to perform the maintenance functions that are assigned at a given level. Personnel performing maintenance tasks without the necessary skills will likely induce problems into the system and cause an increase in the maintenance requirements overall. Personnel possessing skills greater than what are actually required will tend to

induce problems due to carelessness or seek employment elsewhere (resulting in cost to the organization).

5. The facilities are not adequate for the operating and maintenance tasks scheduled to be performed. This often leads to poor maintenance practices, the possible introduction of hazards due to unsafe conditions, excessive system downtime, or a combination thereof.

6. Maintenance procedures and supporting technical data do not track the tasks that are to be performed at a given maintenance level. Further, such procedures are not prepared to the correct skill levels of the personnel performing the tasks. Inadequate data coverage tends to lead to maintenance-induced faults in the system, since the assigned personnel are performing tasks without the necessary guidance. On the other hand, procedures that provide more information than what is actually required will encourage the accomplishment of maintenance beyond that which is considered to be economically feasible.

These six points are symptomatic of the problems that can occur, particularly for large systems, when inadequate consideration is given to logistic support. The decisions associated with system support often turn out to be quite significant in terms of life-cycle cost, and it may be necessary to make a change in one element of support in order to reach a cost-effective solution for the system as a whole. This, in turn, may necessitate a change in the prime mission equipment design, as well as requiring changes in the other elements of support for the purposes of compatibility. In essence, a change in any given area may require changes in many other areas, and the interactions that occur between the prime equipment and the elements of logistic support are numerous.

When designing for system supportability, it is essential that logistic support be considered at program inception. This is accomplished through the development of the maintenance concept as described in Chapter 3. The design must be compatible with the levels of maintenance identified, the basic functions to be performed at each level, the anticipated personnel skills at each level, the test and support equipment approach conveyed, and the quantitative requirements specified. These requirements, where appropriate, are allocated to various subsystems, units, or other parts of the system. As design progresses, alternative support approaches are evaluated and a preferred configuration is selected. This iterative process involving the evaluation of alternatives is accomplished through the logistic support analysis (LSA), which is directly tied into the overall systems analysis activity. The LSA constitutes the integration and application of various analytical techniques to define and optimize (to the extent possible) the logistic support requirements for a given system. The output of the LSA provides (1) an analytical assessment of the system in terms of its supportability, and (2) a data base for the identification, provisioning, and acquisition of those logistic support elements required for system support during the consumer use phase. The results of the LSA are used in the development of the formal ILS plan.

The logistic support activities in the system design process are highlighted in Figure 16.1, Blocks 1, 2, and 3. These activities are an integral part of the systems engineering process described in Part Two and are paramount in meeting the objective of design for supportability.

Logistics in the Production Phase

Logistics in the production phase includes two basic areas of activity. The first deals with the production process itself and the flow of materials from the producer's manufacturing facility to the consumer. This aspect of logistics, often classified under the term of *business logistics* or *industrial logistics,* covers the overall physical distribution of the prime mission equipment and consumer products. Specific functions include packaging, transportation and handling, warehousing and inventory control, and related activities involved in the transition of items from the producer to the consumer.

The second area relates to the production and distribution of the elements of logistics as described in Section 16.1 required for the sustaining support of the system throughout its useful life. This includes test and support equipment, spares and repair parts, training equipment, data, and the like. The specific requirements in this area are based on the logistic support analysis (LSA), specifications are prepared, and production begins.

As the production process evolves, it is essential that the item being produced is of high quality and performs satisfactorily. For instance, if the proper level of quality control is not maintained in the production of test equipment, the test equipment will, in all probability, not provide the desired accuracy in testing or be reliable. If the spares are not manufactured to the same specifications and quality standards as those comparable items in the prime equipment with the same part numbers, the system may not function properly when replacements occur as a result of maintenance actions. In essence, the elements of logistics must be treated in the production process using the same controls as those imposed for the prime equipment.

Logistic Support for Operating Systems

The major elements of logistics, applicable to the sustaining support of a system or product being utilized by the consumer in the field, are described in Section 16.1. These elements are required to support the ongoing scheduled and unscheduled maintenance activities that are necessary to keep the system or product in proper operating condition. The prime objective in this phase is to effectively and efficiently manage all logistic support resources such that the right type of support is available at the right location and in a timely manner. A major requirement toward meeting this objective is to be able to evaluate and assess the effectiveness of the overall logistic support capability, and to initiate corrective action in areas where problems exist.

The assessment process is accomplished through the development of a data collection and analysis capability. Data may be collected using the forms illustrated in Figures 6.5 and 6.6. When maintenance actions occur, maintenance and logistics factors are analyzed to determine the adequacy (or inadequacy) of the logistic support for the system.

16.3. MEASURES OF LOGISTIC SUPPORT

The quantitative factors of logistic support can vary significantly depending on the type of system, its mission, and the criteria employed in system evaluation. These factors pertain to supply, transportation, test equipment, maintenance organizations, facilities, and the like. Thus, one should deal with the measures associated with the elements defined in Section 16.1. A few of these measures are described below.

Supply Support Measures

Supply support includes the spare parts and the associated inventories necessary for the accomplishment of unscheduled and scheduled maintenance actions. At each maintenance level, one must determine the type of spare part (by manufacturing part number) and the quantity of items to be purchased and stocked. Also, it is necessary to know how often various items should be ordered and the number of items that should be procured in a given purchasing transaction.

Spare part requirements are intially based on the system maintenance concept and are subsequently defined and justified through logistic support analysis (LSA). Essentially, spare-part quantities are a function of demand rates and include consideration of:

1. Spares and repair parts covering actual item replacements occurring as a result of corrective and preventive maintenance actions. Spares are considered as being major replacement items which are repairable, while repair parts are nonrepairable smaller components.

2. An additional stock level of spares to compensate for repairable items in the process of undergoing maintenance. If there is a backup (lengthy queue) of items in the intermediate shop or at the depot awaiting repair, these items obviously will not be available as recycled spares for subsequent maintenance actions; thus, the inventory is further depleted (beyond expectations) and a stock-out condition may result. In addressing this problem, it becomes readily apparent that the test equipment capability, personnel, and facilities directly impact the maintenance turnaround times and the quantity of additional spare items needed.

3. An additional stock level of spares and repair parts to compensate for the procurement lead times required for item acquisition. For instance, prediction data may indicate that 10 maintenance actions requiring the replacement of a certain item will occur within a 6-month period and it takes 9 months to acquire replacements from the supplier. What additional repair parts will be necessary to cover the operational needs and yet compensate for the long supplier lead

time? The added quantities will, of course, vary depending on whether the item is designated as repairable or will be discarded at failure.

4. An additional stock level of spares to compensate for the condemnation or scrapage of repairable items. Repairable items returned to the intermediate shop or depot are sometimes condemned (i.e., not repaired) because, through inspection, it is decided that the item was not economically feasible to repair. Condemnation will vary, depending on equipment utilization, handling, environment, and organizational capability. An increase in the condemnation rate will generally result in an increase in spare part requirements.

In reviewing the foregoing considerations, of particular significance is the determination of spares requirements covering item replacements in the performance of corrective maintenance resulting from system failure. Major factors involved in this process are (1) the reliability of the item to be spared, (2) the quantity of items used, (3) the required probability that a spare will be available when needed, (4) the criticality of item application with regard to mission success, and (5) cost. Use of the reliability and probability factors are illustrated in the examples presented below.

Probability of success with spares availability considerations.
Assume that a single component with a reliability of 0.8 is used in a unique system application and that one backup spare component is purchased. Determine the probability of system success having a spare available in time, t, (given that failures occur randomly and are exponentially distributed).

This situation is analogous to the case of an operating component and a parallel component in standby (i.e., standby redundancy) discussed in Section 13.3. The applicable expression is Equation 13.21 stated as

$$P = e^{-\lambda t} + (\lambda t)e^{-\lambda t}. \tag{16.1}$$

With a component reliability of 0.8, the value of λt is 0.223. Substituting this value into Equation 16.1 gives a probability of success of

$$P = e^{-0.223} + (0.223)e^{-0.223}$$

$$= 0.8 + (0.223)(0.8) = 0.9784.$$

Assume next that the component is supported with two backup spares (where all three components are interchangeable). The probability of success during time, t, is determined from

$$P = e^{-\lambda t} + (\lambda t)e^{-\lambda t} + \frac{(\lambda t)^2 e^{-\lambda t}}{2!}$$

or

$$P = e^{-\lambda t}\left[1 + \lambda t + \frac{(\lambda t)^2}{2!}\right]. \tag{16.2}$$

With a component reliability of 0.8, and a value of λt of 0.223, the probability of success is

$$P = 0.8\left[1 + 0.223 + \frac{(0.223)^2}{(2)(1)}\right]$$

$$= 0.8(1.2479) = 0.9983.$$

Thus, adding another spare component results in one additional term in the Poisson expression. If two spare components are added, two additional terms are added, and so forth.

The probability of success for a configuration consisting of two operating components, backed by two spares, with all components being interchangeable can be found from the expression

$$P = e^{-2\lambda t}\left[1 + 2\lambda t + \frac{(2\lambda t)^2}{2!}\right]. \tag{16.3}$$

With a component reliability of 0.8 and $\lambda t = 0.223$,

$$P = e^{-0.446}\left[1 + 0.446 + \frac{(0.446)^2}{(2)(1)}\right]$$

$$= 0.6402(1 + 0.446 + 0.0995) = 0.9894.$$

These examples illustrate the computations used in determining system success with spare parts for three simple component configuration relationships. Various combinations of operating components and spares can be assumed, and the system success factors can be determined by using

$$1 = e^{-\lambda t} + (\lambda t)e^{-\lambda t} + \frac{(\lambda t)^2 e^{-\lambda t}}{2!} + \frac{(\lambda t)^3 e^{-\lambda t}}{3!} + \cdots + \frac{(\lambda t)^n e^{-\lambda t}}{n!}. \tag{16.4}$$

Equation 16.4 can be simplified into the general Poisson expression,

$$f(x) = \frac{(\lambda t)^x e^{-\lambda t}}{x!}. \tag{16.5}$$

The objective is to determine the probability of x failures occurring if an item is placed in operation for t hours, and each failure is corrected (through item replacement) as it occurs. With n items in the system, the number of failures in t hours will be $n\lambda t$, and the general Poisson expression becomes

$$f(x) = \frac{(n\lambda t)^x e^{-n\lambda t}}{x!}. \tag{16.6}$$

To facilitate calculations, a cumulative Poisson probability graph is presented in Figure 16.2, derived from Equation 16.6. The ordinate value can be viewed as a confidence factor. Several simple examples will be presented to illustrate the application of Figure 16.2.

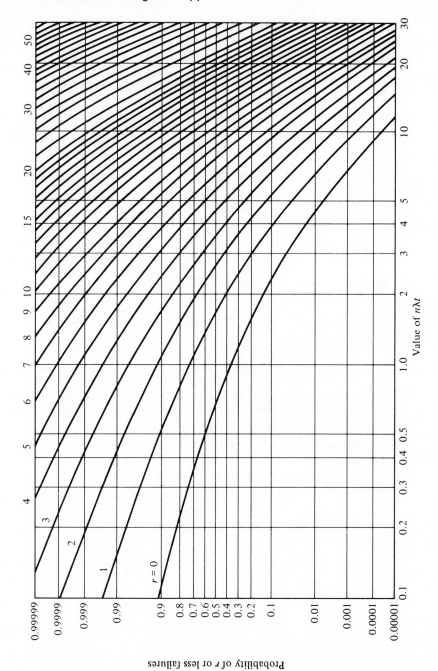

Figure 16.2   Spares determination using Poisson cumulative probabilities. *Source:* NAVAIR 00-65-502/NAVORD OD 41146, *Reliability Engineering Handbook,* Naval Air Systems Command and Naval Ordnance Systems Command, Revised March 1968.

Probability of mission completion. Suppose that one must determine the probability that a system will complete a 30-hour mission without a failure when the system has a known mean life of 100 hours. Let

$$\lambda = 1 \text{ failure}/100 \text{ hrs or } 0.01 \text{ failure/hr}$$

$$t = 30 \text{ hrs}$$

$$n = 1 \text{ system}$$

$$n\lambda t = (1)(0.01)(30) = 0.3$$

Enter Figure 16.2 where $n\lambda t$ is 0.3. Proceed to the intersection where r equals zero and read the ordinate scale, indicating a value of approximately 0.73. Thus, the probability of the system completing a 30-hour mission is 0.73.

Assume that the system identified above is installed in an aircraft and there are 10 aircraft scheduled for a 15-hour mission. Determine the probability that at least seven systems will operate for the duration of the mission without failure as

$$n\lambda t = (10)(0.01)(15) = 1.5$$

and

$$r = 3 \text{ failures or less (allowed)}$$

Enter Figure 16.2 where $n\lambda t$ equals 1.5. Proceed to the intersection where r equals 3, and read the ordinate scale, indicating a value of approximately 0.92. Thus, there is a 92% confidence that at least seven systems will operate successfully out of 10. If an 80% operational reliability is specified (i.e., eight systems must operate without failure), the confidence factor decreases to about 82%.

Although the graph in Figure 16.2 provides a simplified solution, the use of Equation 16.4 is preferable for more accurate results.

Spare-part-quantity determination. Spare-part-quantity determination is a function of the probability of having a spare part available when required, the reliability of the item in question, the quantity of items used in the system, and so on. An expression, derived from the Poisson distribution, useful for spare-part-quantity determination is

$$P = \sum_{n=0}^{n=s} \left[\frac{(R)[-(\ln R)^n]}{n!} \right] \tag{16.7}$$

where

P = probability of having a spare of a particular item available when required
S = number of spare parts carried in stock
R = composite reliability (probability of survival); $R = e^{-K\lambda t}$
K = quantity of parts used of a particular type
$\ln R$ = natural logarithm of R

In determining spare-part quantities, one should consider the level of protection desired (safety factor). The protection level is the P value in Equation 16.7. This is the probability of having a spare available when required. The higher the protection level, the greater the quantity of spares required. This results in a higher cost for item procurement and inventory maintenance. The protection level, or safety factor, is a hedge against the risk of a stock depletion.

When determining spare-part quantities, one should consider system operational requirements (e.g., system effectiveness, availability) and establish the appropriate level at each location where corrective maintenance is accomplished. Different levels of corrective maintenance may be appropriate for different items. For instance, spares required to support prime equipment components that are critical to the success of a mission may be based on one factor; high-value or high-cost items may be handled differently than low-cost items; and so on. In any event, an optimum balance between stock level and cost is required.

Figure 16.3 (sheets 1 and 2) presents a nomograph that simplifies the determination of spare-part quantities using Equation 16.7. The nomograph not only simplifies solutions for basic spare part availability questions, but provides information that can aid in the evaluation of alternative design approaches in terms of spares and in the determination of provisioning cycles. The following examples illustrate the use of this nomograph.[3]

Suppose that a piece of equipment contains 20 parts of a specific type with a failure rate (λ) of 0.1 failure per 1,000 hrs of operation. The equipment operates 24 hours a day and spares are procured and stocked at 3-month intervals. How many spares should be carried in inventory to ensure a 95% chance of having a spare part when required? Let

$$K = 20 \text{ parts}$$

$$\lambda = 0.1 \text{ failure}/1,000 \text{ hrs}$$

$$T = 3 \text{ months}$$

$$K\lambda T = (20)(0.0001)(24)(30)(3) = 4.32$$

$$P = 95\%$$

Using the nomograph in Figure 16.3 as illustrated, approximately eight spares are required.

As a second example, suppose that a particular part is used in three different items of equipment (A, B, C). Spares are procured every 180 days. The number of parts used, the part failure rate, and the equipment operating hours per day are given in Table 16.1.

The number of spares that should be carried in inventory to ensure a 90% chance of having a spare available when required is calculated as follows:

1. Determine the product of K, λ, and T as

$$A = (25)(0.0001)(180)(12) = 5.40$$

$$B = (28)(0.00007)(180)(15) = 5.29$$

$$C = (35)(0.00015)(180)(20) = 18.90$$

[3] NAVSHIPS 94324, *Maintainability Design Criteria Handbook for Designers of Shipboard Electronic Equipment*, Naval Ship Systems Command, Department of the Navy, 1964.

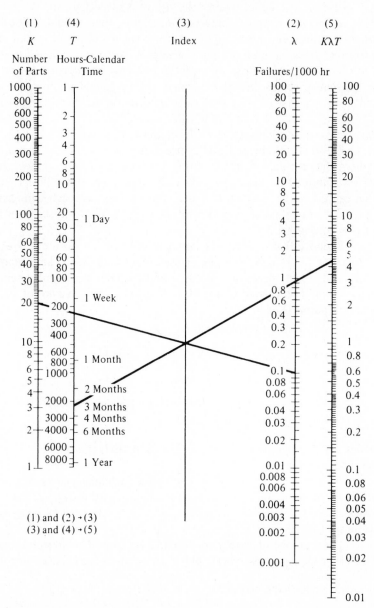

Figure 16.3 Spare part requirement nomograph. *Source:* NAVAIR 00-65-502/
NAVORD OD 41146, *Reliability Engineering Handbook,* Naval Air Systems Command and Naval Ordnance Systems Command, Revised March 1968.

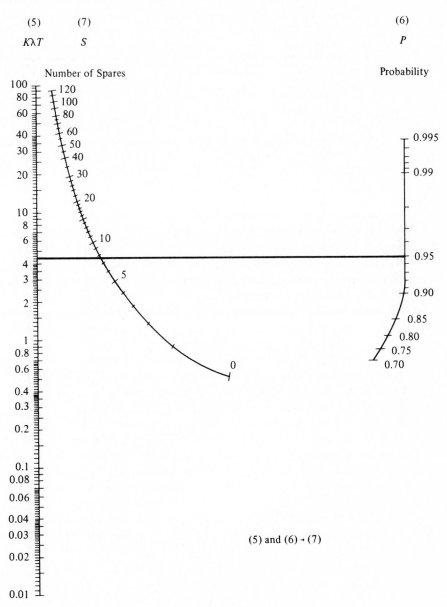

Figure 16.3 Continued.

TABLE 16.1 DATA FOR SPARES INVENTORY

ITEM	K	FAILURES PER 1000 HOURS	OPERATING HOURS PER DAY
A	25	0.10	12
B	28	0.07	15
C	35	0.15	20

2. Determine the sum of the $K\lambda T$ values as

$$\sum K\lambda T = 5.40 + 5.29 + 18.90 = 29.59$$

3. Using sheet 2 of the nomograph (Figure 16.3), construct a line from $K\lambda T$ value of 29.59 to the point where P is 0.90. The approximate number of spares required is 36.

Inventory system considerations. Overall inventory requirements for spares must be addressed in addition to evaluating specific demand situations. Too much inventory on hand may be very responsive to demand, but the cost will be high due to inventory holding cost. In addition, waste could occur if system design changes are introduced, rendering certain spares obsolete. On the other hand, too little inventory on hand increases the risk of a stock depletion and the cost, and/or disruption due to an inoperative system. An optimum balance must be sought among the inventory on hand, the procurement frequency, and the procurement quantity, as is modeled in Section 9.2.

Figure 9.8 shows the general deterministic inventory process (constant demand and constant procurement lead time). The illustration is a theoretical representation of an inventory cycle for a given item. Actually, demands are not always constant and quite often the reorder cycle changes with time.

Figure 16.4 presents a situation which is more realistic, where demand and/or procurement lead time is a random variable. The necessity for safety stock is evident from this illustration. Safety stock is provided to recognize that demand and procurement lead time will not be constant as assumed, but that demand and/or procurement lead time will probably vary. Terms identified are defined as follows:

1. *Operating level* is the quantity of material items required to support normal system operations in the interval between orders and the arrival of successive shipments.
2. *Safety level* is the additional stock required to compensate for unexpected demands, repair and recycle times, procurement lead time, and unforeseen delays.
3. *Procurement cycle* is interval of time between successive procurement orders.
4. *Procurement lead time* is the span of time from the date of procurement to receipt of the order in the inventory. This includes (a) administrative lead time from the date that a decision is made to initiate an order to the receipt of the order at the supplier, (b) production lead time or the time from receipt of the order by the supplier to completion of the manufacture of the item ordered,

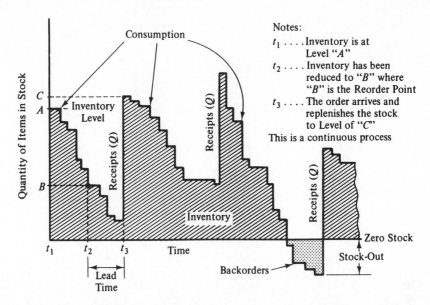

Figure 16.4   Representation of an actual inventory cycle.

and (c) delivery lead time from completion of manufacture to receipt of the item in the inventory.

5. *Procurement level* is the stock level when a replenishment order is initiated for additional stock of the spare/repair parts.

In Figure 16.4, the objective is to have the needed amount and type of supplies available for the lowest total cost. Procurement costs vary with the number of orders placed. The economic inventory principle involves a trade-off between the placing of many orders, resulting in high material acquisition costs and the placing of orders less frequently while maintaining a higher level of inventory, causing high inventory maintenance and carrying costs. The economic order principle equates the cost to order to the cost to hold, and the point at which the combined costs are at a minimum indicates the desired size of order.

The optimization model in Section 9.2 can be adapted to this simple inventory situation. To do so means that the derivation based on Figure 9.8 must be modified. This is accomplished by setting $R = \infty$ (instantaneous replenishment), $C_s = \infty$ (no shortages permitted), and adding safety stock. With these modifications, Figure 9.8 takes on the form of Figure 16.4.

Neglecting the cost of holding safety stock, the total cost per period (day) can be found from Equation 9.39, which is stated as

$$TC = C_i D + \frac{C_p D}{Q} + \frac{C_h}{2Q(1 - D/R)} [Q(1 - D/R) + L - DT]^2 + \frac{C_s(DT - L)^2}{2Q(1 - D/R)}.$$

With $R = \infty$ and $C_s = \infty$

$$TC = C_i D + \frac{C_p D}{Q} + \frac{C_h}{2Q}(Q + L - DT)^2. \qquad (16.8)$$

And since $L = DT$, plus safety stock which is neglected, Equation 16.8 reduces further to

$$TC = C_i D + \frac{C_p D}{Q} + \frac{C_h Q}{2}. \qquad (16.9)$$

Optimization is accomplished by taking the derivative of TC with respect to Q and setting the result equal to zero as

$$\frac{dTC}{dQ} = -\frac{C_p D}{Q^2} + \frac{C_h}{2} = 0.$$

Solving for the optimal value of Q gives

$$Q^* = \sqrt{\frac{2 C_p D}{C_h}}. \qquad (16.10)$$

This also could have been found from Equation 9.44 with R and C_s set equal to ∞.

The optimum value of the procurement level, L^*, is simply D times T plus safety stock, where the cost of holding safety stock is justified by the contribution it makes to reducing shortages. An approach to setting a level of safety stock was presented in Chapter 10, Section 10.6, where the distribution of lead time demand, LTD, is developed through Monte Carlo analysis.

As an example of the application of Equation 16.10, assume that the annual demand for a spare part is 1,000 units. The cost per unit is $6.00 delivered in the field. Procurement cost per order placed is $10.00 and the cost of holding one unit in inventory for one year is estimated to be $1.32.

The economic procurement quantity may be found by substituting the appropriate values into Equation 16.10 as

$$Q^* = \sqrt{\frac{2(\$10)(1,000)}{\$1.32}} = 123 \text{ units.}$$

Total cost may be expressed as a function of Q by substituting the costs and various values of Q into Equation 16.9. The result is shown in Table 16.2. The tabu-

TABLE 16.2   INVENTORY COST AS A FUNCTION OF Q

Q	$C_i D$	$\dfrac{C_p D}{Q}$	$\dfrac{C_h D}{2}$	TC
0	$6,000	$ \infty	$ 0	$ \infty
50	6,000	200	33	6,233
100	6,000	100	66	6,166
123	6,000	81	81	6,162
150	6,000	67	99	6,165
200	6,000	50	132	6,182
400	6,000	25	264	6,289
600	6,000	17	396	6,413

lated total cost value for $Q = 123$ is the minimum-cost quantity for the condition specified. Total cost as a function of Q is illustrated in Figure 16.5.[4]

The economic procurement model of Equation 16.10 is generally applicable in instances where there are relatively large quantities of common spares and repair parts. However, it may be feasible to employ other models of procurement for major high-value items and for those items considered to be particularly critical to mission success.

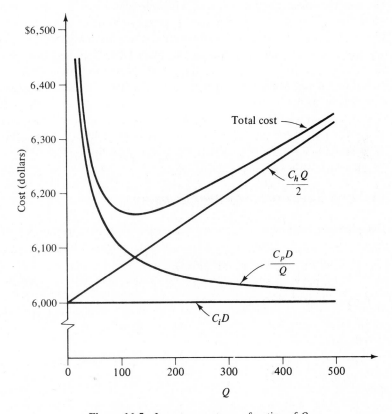

Figure 16.5   Inventory costs as a function of Q.

High-value items are those components with a high unit procurement cost that should be purchased on an individual basis. The dollar value of these components is usually significant and may even exceed the total value of the hundreds of other spares and repair parts in the inventory.

Often a relatively small number of items may represent a large percentage of the total inventory value. Because of this, it may be preferable to maintain a given quantity of these items in the inventory to compensate for repair and recycle times, pipeline and procurement lead times, and so on, and order new items on a one-for-

---

[4] The total cost found above does not include the cost of safety stock, which must be separately justified, as was indicated.

one basis as failures occur and as a spare is withdrawn from the inventory. Thus, where it may appear to be economically feasible to purchase many of these items in a given purchase transaction, only a small quantity of items may be actually procured, owing to the cost of tying up too much capital and the resultant high inventory maintenance cost.[5]

Another consideration in the spares acquisition process is that of *criticality*. Some items are considered more critical than others in terms of impact on mission success. For instance, the lack of a $10 item may cause a system to be inoperative, while the lack of a $1,000 item might not cause a major problem. The criticality of an item is generally based on its function in the system and not necessarily on its acquisition cost. The degree of criticality may be determined through failure mode, effect, and criticality analysis (FMECA), described in Chapter 13.

The spares acquisition process may also vary somewhat between items of a comparable nature if the usage rates (i.e., duty cycles) are significantly different. Fast-moving items may be procured locally near the point of usage (such as an intermediate maintenance shop), while slower-moving items stocked at the depot may be acquired from a remotely located supplier if the pipeline and procurement lead times are not as critical.

Test and Support Equipment Measures

The general category of test and support equipment may include a wide spectrum of items, such as precision electronic test equipment, mechanical test equipment, ground handling equipment, special jigs and fixtures, maintenance stands, and the like. These items, in varying configurations and mixes, may be assigned to different levels of maintenance and geographically dispersed throughout the country (or world). However, regardless of the nature and application, the objective is to provide the right item for the job intended, at the proper location, and in the quantity required.

Because of the likely diversification of the test and support equipment for any given system, it is difficult to specify quantitative measures that can be universally applied. Certain measures are appropriate for electronic test equipment, other measures are applicable to ground-handling equipment, and so on. Further, the specific location and application of a given item of test equipment may also result in different measures. For instance, an item of electronic test equipment used in support of on-site organizational maintenance may have different requirements than a similar item of test equipment used for intermediate maintenance accomplished in a remote shop.

Although all the test and support equipment requirements at each level of maintenance are considered to be important relative to successful system operation, the testers or test stations in the intermediate and depot maintenance facilities are of particular concern, since these items are likely to support a number of system ele-

---

[5] The classification of high-value items will vary with the program and may be established at a certain dollar value (i.e., all components whose initial unit cost exceeds x dollars are considered as high-value items).

ments at different consumer locations. That is, an intermediate maintenance facility may be assigned to provide the necessary corrective maintenance support for a large number of system elements dispersed throughout a wide geographical area. This means, of course, that a variety of items (all designated for intermediate-level maintenance) will arrive from different consumer sites at different times.

When determining the specific test equipment requirements for a shop, one must define (1) the type of items that will be returned to the shop for maintenance; (2) the test functions to be accomplished, including the performance parameters to be measured as well as the accuracies and tolerances required for each item; and (3) the anticipated frequency of test functions per unit of time. The type and frequency of item returns (i.e., shop arrivals) is based on the maintenance concept and system reliability data. The distribution of arrival times for a given item is often a negative exponential with the number of items arriving within a given time period following a Poisson distribution, as presented in Chapter 11. As items arrive in the shop, they may be processed immediately or there may be a waiting line, or queue, depending on the availability of the test equipment and the personnel to perform the required maintenance functions.

When evaluating the test process itself, one should calculate the anticipated test equipment use requirements (i.e., the total amount of on-station time required per day, month, or year). This can be estimated by considering the repair-time distributions for the various items arriving in the shop. However, the ultimate elapsed times may be influenced significantly, depending on whether manual, semiautomatic, or automatic test methods are employed.

Given the test equipment utilization needs (from the standpoint of total test station time required for processing shop arrivals), it is necessary to determine the anticipated availability of the test equipment configuration being considered for the application. As defined in Equation 13.18, availability is a function of reliability and maintainability. Thus, one must consider the MTBM and MDT values for the test equipment itself. Obviously, the test equipment configuration should be more reliable than the system component being tested. Also, in instances where the complexity of the test equipment is high, the logistic resources required for the support of the test equipment may be extensive (e.g., the frequent requirement to calibrate an item of test equipment against a secondary or primary standard in a *clean-room* environment). There may be a requirement to determine the time that the test equipment will be available to perform its intended function.

The final determination of the requirements for test equipment in a maintenance facility is accomplished through an analysis of various alternative combinations of arrival rates, queue length, test station process times, and/or quantity of test stations. In essence, one is dealing with a single-channel or multichannel queuing situation using the queuing models discussed in Chapter 11. As the maintenance configuration becomes more complex, involving many variables (some of which are probabilistic in nature), Monte Carlo analysis may be appropriate. In any event, there may be a number of feasible servicing alternatives, and a preferred approach is sought.

Organizational Measures

The measures associated with a maintenance organization are basically the same as those factors which are typical for any organization. Of particular interest relative to logistic support are:

1. The direct maintenance labor time for each personnel category, or skill level, expended in the performance of system maintenance activities. Labor time may be broken down to cover both unscheduled and scheduled maintenance individually. Labor time may be expressed in:

   a. Maintenance labor-hours per system operating hour (MLH/OH).
   b. Maintenance labor-hours per mission cycle (or segment of a mission).
   c. Maintenance labor-hours per month (MLH/Mo).
   d. Maintenance labor-hours per maintenance action (MLH/MA).

2. The indirect labor time required to support system maintenance activities (i.e., overhead factor).
3. The personnel attrition rate or turnover rate (in percent).
4. The personnel training rate or the worker-days of formal training per year of system operation and support.
5. The number of maintenance work orders processed per unit of time (e.g., week, month, or year), and the average time required for work-order processing.
6. The average administrative delay time, or the average time from when an item is initially received for maintenance to the point when active maintenance on that item actually begins.

When addressing the total spectrum of logistics (and the design for supportability), the organizational element is critical to the effective and successful life-cycle support of a system. The right personnel quantities and skills must be available when required, and the individuals assigned to the job must be properly trained and motivated. As in any organization, it is important to establish measures dealing with organizational effectiveness and productivity.

Transportation Measures

Transportation requirements include the movement of human and material resources, in support of both operational and maintenance activities, from one location to another. When evaluating the effectiveness of transportation, one must deal with such factors as:

1. Transportation route, both national and international (number of nationalities, customs requirements, distances, and so on).
2. Transportation capability or capacity (modes of transportation, volume of goods transported, ton-miles per year, quantity of loads).

3. Transportation time (short-haul versus long-haul time, mean delivery time, and so on).

4. Transportation cost (cost per shipment, cost of transportation per mile, cost of packaging and handling, and so on).

Transportation and handling are significant concerns in the design of a system for transportability and mobility.

Facility Measures

Facilities are required to support activities pertaining to the accomplishment of active maintenance tasks, providing warehousing functions for spares and repair parts, and providing housing for related administrative functions. Although the specific quantitative measures associated with facilities may vary significantly from one system to the next, the following factors are considered to be relevant in most instances:

1. Item process time or turnaround time (TAT); i.e., the elapsed time necessary to process an item for maintenance, returning it to full operational status.

2. Facility utilization; i.e., the ratio of the time utilized to the time available for use, percent utilization in terms of space occupancy, and so on.

3. Energy utilization in the performance of maintenance; i.e., unit consumption of energy per maintenance action, cost of energy consumption per increment of time or per maintenance action, and so on.

4. Total facility cost for system operation and support; i.e., total cost per month, cost per maintenance action, and so on.

In summary, the measures of logistic support are numerous and will vary depending on the system and the nature of its mission. For some systems, supply support may represent a major portion of the total logistics capability, and the measures associated with supply support are critical in the system design and development process. In other cases, facility support may turn out to be predominant, particularly if the system requires the design and construction of a unique maintenance building (or equivalent).

16.4. LOGISTIC SUPPORT ANALYSIS

The *logistic support analysis* (LSA) constitutes the integration and application of different analytical techniques to solve a wide variety of problems. The LSA, in its application, is the process employed on an iterative basis throughout system development that addresses the aspect of supportability in design.[6]

[6] The LSA is being implemented rather extensively by the Department of Defense on all major programs involving systems and large equipment. Two major references are MIL-STD-1388-1, Military Standard, *Logistic Support Analysis*. Department of Defense, Washington, D.C.; and MIL-STD-1388-2, Military Standard, *DOD Requirements for a Logistic Support Analysis Record*, Department of Defense, Washington, D.C.

The LSA, which is an inherent part of the overall system engineering effort, is used in the evaluation of a given or proposed configuration to determine (1) the direct impact of the prime mission segment of the system on total logistic support, and (2) the feedback effects of the planned elements of logistic support on the prime configuration itself. This interactive relationship, illustrated in Figure 16.6, forces the early consideration of supportability in design through analysis.

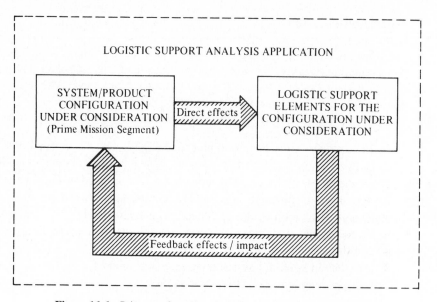

Figure 16.6   Prime configuration–logistic support interface relationship.

The LSA process, which follows the approach for systems analysis illustrated in Figure 3.7, can be applied to a broad spectrum of problems. More specifically,

1. The LSA aids in the evaluation of prime equipment design characteristics in terms of logistic support requirements. This includes consideration of alternative equipment packaging schemes, test approaches, accessibility features, transportation and handling provisions, and so on. Logistic support resource requirements are estimated for each alternative being considered and a preferred approach is selected.

2. The LSA aids in the evaluation of alternative repair policies allowable within the constraints dictated by the maintenance concept. For example, the LSA supports repair-discard decisions, and decisions pertaining to where repair is to be accomplished. In this context, the LSA is closely interrelated with the maintainability analysis described in Chapter 14.

3. The LSA aids in the evaluation of two or more off-the-shelf equipment items being considered for a single system application. Assuming that new design is

not appropriate and outside supplier sources are being considered, which item (when installed as part of the system) will reflect the least overall logistics burden? The LSA facilitates the evaluation effort which, in turn, affects procurement decisions in system acquisition.

4. Given an assumed or fixed design configuration for the prime mission segment of the system, the LSA aids (through reliability and maintainability analyses and predictions) in determining the specific logistic support resource requirements for that configuration. Once design data are available, it is possible to determine the type and quantity of test and support equipment, the type of quantity of spares and repair parts, personnel quantities and skill levels, transportation and handling requirements, facility requirements, and technical data needs.

5. During system test and evaluation and when the system is in operational use by the consumer, the LSA aids (through application of operations analysis and logistics modeling techniques) in the measurement of the overall effectiveness of the system and its associated support. Problem areas readily become apparent and the LSA can assist in the evaluation of alternatives for corrective action, including improvements in the use and applications of support resources.

The LSA is developed on an iterative basis throughout system definition and design. Basically, the analysis effort stems from the maintenance concept and is ultimately dependent on engineering data, reliability and maintainability analyses and predictions, cost data, and so on. The analyst, in developing LSA data, reviews engineering documentation with the objective of estimating future support requirements. Thus, the LSA includes coverage of:

1. All significant repairable items (i.e., system, subsystem, assembly, and subassembly). Items defined through the level-of-repair analysis as being "repairable" are evaluated to determine the type and extent of support necessary.

2. All maintenance requirements (i.e., troubleshooting, remove and replace, servicing, alignment and adjustment, functional test and checkout, inspection, calibration, overhaul, and the like).

LSA data are used to support design decisions and to serve as the basis for the subsequent provisioning and acquisition of specific logistic support elements needed for the operation and maintenance of the system. The accomplishment of these objectives requires the generation of certain qualitative and quantitative factors dealing with the various elements of logistics discussed earlier. Typical LSA data cover the factors identified in Table 16.3.[7]

[7] A maintenance analysis is often accomplished within the scope of LSA in addition to what might be required in a maintainability program (see Section 14.2). For a detailed description of the procedures for accomplishing maintenance analyses, refer to B. S. Blanchard, *Logistics Engineering and Management*, 3rd ed. (Englewood Cliffs, N.J.: Prentice-Hall, 1986).

TABLE 16.3   LOGISTIC SUPPORT ANALYSIS (LSA) DATA OUTPUT

1. Maintenance levels
2. Maintenance tasks and levels
   —Tasks sequence(s)
   —Task time
   —Task frequency
3. Supply support
   —Repair levels and location
   —Quantity and type of spares, repair parts, and consumables
   —Critical items
   —Replacement frequency
   —Condemnation rate
   —Wear-out rate
   —Inventory level and location
   —Safety stock level
   —High-value items
   —Shelf life
   —Spares and repair part availability
   —Procurement lead time
   —Inventory cost (order cost, material cost, holding cost)
   —Provisioning cycle
4. Test and support equipment
   —Quantity, type, and location
   —Utilization rate
   —Availability and reliability
   —Maintenance requirements
   —Cost (R&D, production, operation, maintenance)
5. Transportation and handling equipment
   —Quantity, type, and location
   —Containers (disposable and reusable)
   —Packing and shipping
   —Transportation cost
6. Personnel and training
   —Personnel quantity and skill level

      requirements and maintenance location
      —Attrition rate
      —Learning curve(s)
      —Indirect labor (overhead)
      —Initial training requirements
      —Replacement training requirements
      —Training aids (equipment and data)
      —Personnel costs (direct and indirect)
      —Training costs (direct and indirect)
7. Facilities
   —Maintenance and training facility requirements (space and layout)
   —Storage requirements
   —Capital equipment
   —Tooling and special handling equipment
   —Environmental requirements
   —Utility requirements (power, light, heat, water, telephone, etc.)
   —Facility utilization
   —Facilities cost (R&D, construction, operation, maintenance, taxes, energy utilization, etc.)
8. Technical data
   —Technical manual requirements (operating and maintenance instructions, overhaul procedures, etc.)
   —Logistics provisioning data
   —Maintenance reporting
9. Additional factors
   —Availability (A_o, A_a, A_i)
   —MTBM, MTBR, MTBF, R, λ, fpt
   —$\overline{M}$ct, $\overline{M}$pt, $\overline{M}$, MDT, M_{max}
   —MLH/OH, MLH/MA, MLH/year
   —Turnaround time (TAT)
   —Maintenance actions (MA/year)
   —Logistic support cost factors

The LSA data format (i.e., output data package) is by no means fixed. It varies from program to program and is tailored to the specific information desired and the time in the life cycle when needed. Early in the life cycle, major trade-offs are accomplished and include a wide variety of individual evaluations at a gross level. For instance, the analyst may wish to evaluate the effects of system fault isolation capability on spares cost. The basic relationship is illustrated in Figure 16.7 and the evaluation process requires prime equipment design data and supply support data. A second example is illustrated in Figure 16.8. This figure indicates the overall relationships between corrective maintenance and preventive maintenance. Quite often,

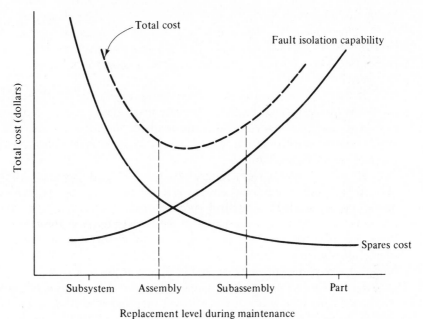

Figure 16.7   Cost-effectiveness of system diagnostic capability.

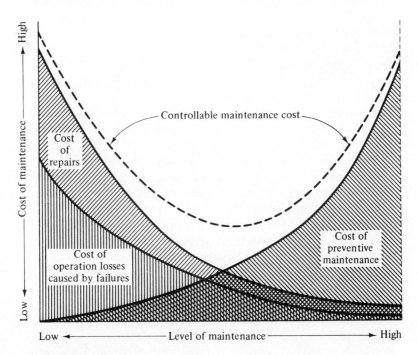

Figure 16.8   Cost-effectiveness evaluation: corrective maintenance versus preventive maintenance.

the question arises as to how much maintenance should be planned and the nature of such maintenance. To evaluate this requires knowledge of many of the factors listed in Table 16.3, although the level of detail required may not be extensive. In any event, the evaluation is accomplished, a preferred approach is selected, and the system design reflects the results.

In essence, the early analysis efforts must provide the right information expeditiously as the designer is required to make many decisions in a relatively short period of time. If this information is not available in a timely manner, the decisions will be made anyway and the results may not prove to be beneficial from the standpoint of system supportability.

As system development progresses, the LSA effort assumes a more comprehensive approach. Additional design data are available and the analyst is able to define support needs in a relatively precise manner. LSA data are used to evaluate the design configuration (as proposed at the time) in terms of the measures described in Section 16.3. Further, LSA data are used to identify specific logistic support resource requirements. The transition from the LSA data package to the identification of logistics requirements is illustrated in Figure 16.9.

Figure 16.9 shows the LSA output leading to the definition of support requirements. Actually, the process is somewhat more complex. For instance, one might refer to the measures of supply support discussed in Section 16.3. To determine spares requirements, the analyst must review design data and identify operating and maintenance functional requirements, equipment packaging schemes, repair levels, fault isolation capability, and so on. After identifying the specific items that are to be replaced during maintenance actions, it is necessary to estimate demand rates and the various quantitative parameters associated with the anticipated inventories. These factors are evaluated, an optimum approach is selected, and a formal spare and repair parts list is developed for the system. This, in turn, leads to the provisioning and acquisition of the required spares. The process is illustrated in Figure 16.10. Note that the LSA includes the analytical steps involved in determining supply support requirements, not the aspects of provisioning and acquisition.

Figure 16.11 covers the development process for test and support equipment in a manner similar to the illustration for supply support (see Figure 16.10). Referring to the logistics measures in Section 16.3, one needs to identify the items to be tested, the parameters to be measured for each item tested, and the frequency of test. Loading studies are accomplished and alternative configurations are evaluated. The results are presented in the form of a list of recommended test and support equipment. The LSA deals with the analytical aspects of this process.

Although only two of the elements of logistic support are covered in Figures 16.10 and 16.11, the same basic approach is applicable to the other elements. The LSA provides a data base, covering the factors in Table 16.3, that is used for requirements definition purposes. The analysis is iterative in nature, and is employed throughout the early phases of the life cycle to promote supportability in system design.

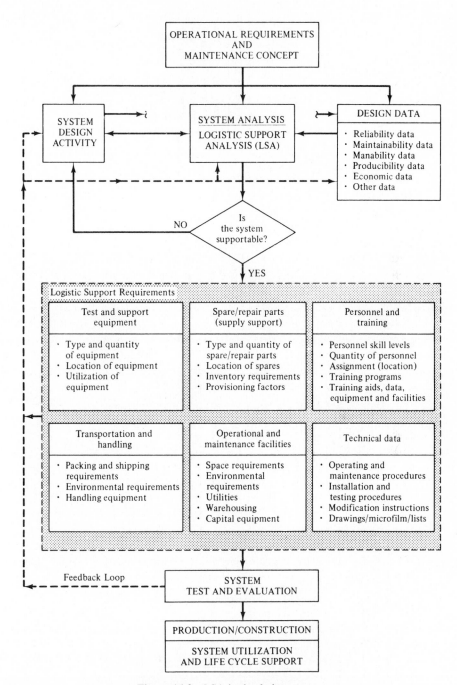

Figure 16.9 LSA in the design process.

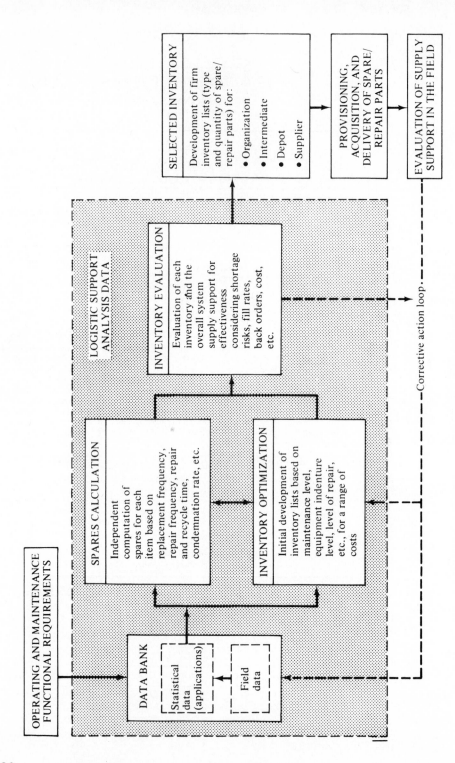

Figure 16.10 Spare and repair parts development process.

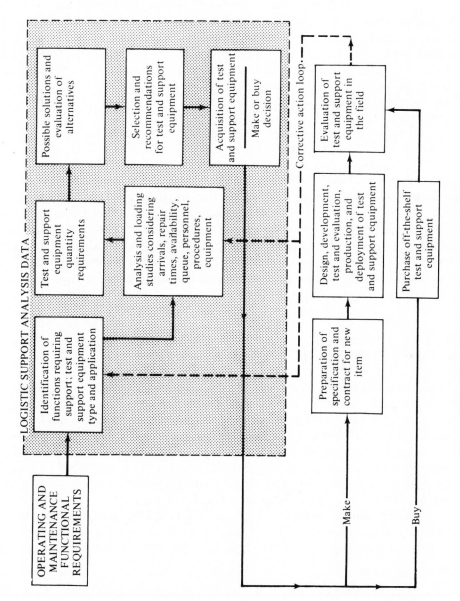

Figure 16.11 Test and support equipment development process.

16.5. *COMPUTER-AIDED ACQUISITION AND LOGISTIC SUPPORT (CALS)*

Inherent within the field of logistics is the requirement for an extensive amount of data (as described in Section 16.4), particularly for complex systems and large-scale programs. With the ever increasing need to address supportability in design, it becomes essential that such data (which provides an assessment of system design for supportability) be processed and made available to the design community in a timely manner. In other words, as the LSA activity evolves commencing during conceptual design, a design evaluation is accomplished, the results are recorded, and recommendations for possible corrective action need to be fed back to designers on a real-time basis.

In the past, design evaluation and feedback associated with logistic support have been almost nonexistent, primarily due to inabilities to collect, process, and evaluate appropriate data in a timely manner. In recognition of this deficiency, combined with the advent of computers and their use in the design process through computer-aided design (see Section 5.4), the Department of Defense is addressing this problem through the advent of Computer-Aided Acquisition and Logistic Support (CALS). Basically, the objective is to develop and convert existing data bases to a digital format and to expedite the processing of data within government organizations, within the supplier organizations, and between the government and its various contractors. Of further significance is the processing of information to enhance the design-for-supportability objectives discussed throughout this chapter.[8]

16.6. *LOGISTIC SUPPORT TEST AND EVALUATION*

The ultimate objective in designing for supportability is to ensure that the proper level of logistic support is provided for the system in a timely and economical manner throughout its life cycle. Specific quantitative and qualitative goals are established during conceptual design, analyses are completed throughout detail design and development, a fixed design configuration is assumed based on these analyses, a prototype system is produced, and test and evaluation is accomplished to compare the results with the initially specified requirements. System test and evaluation and its general requirements are discussed in Chapter 6.

The test and evaluation effort associated with logistic support is accomplished through a combination of individual demonstrations (Type 2 testing), an integrated systems test at a designated test site (Type 3 testing), and an operational test in a con-

---

[8] At the time that this text went to press, there was a great deal of effort being expended on CALS by the government and industrial sectors alike. Initially, there is an educational requirement (educating those associated with the logistics field to recognize the potential in this area), followed by technical growth in the areas of data base design and information system processing.

sumer environment (Type 4 testing). Obviously, the most information can be derived through the latter, while individual demonstrations provide limited results. However, if problems do occur, it is far preferable to make changes as early in the life cycle as possible. Thus, it is important to begin with the assessment of the logistic support capability through Type 2 testing when prototype equipment first becomes available.

Much of the early evaluation of the logistic support elements for the system can be accomplished in conjunction with other testing activities. For instance, the results of reliability testing as is described in Chapter 13 will provide some indication of the failures that are likely to occur, causing maintenance actions and the demand for logistics resources. To an even greater extent, the results of maintainability demonstration, as discussed in Chapter 14, will provide an initial assessment of:

1. The most likely maintenance tasks that will be accomplished throughout the life cycle, task sequences, and task times.
2. The personnel skill levels that will be required for maintenance task accomplishment, the labor-hours expended in task completion, complexity, dexterity factors, and error rates.
3. The removals and replacements required in accomplishing corrective and preventive maintenance (spare and repair part requirements).
4. The compatibility between the test and support equipment, the transportation and handling equipment, and the prime-mission-oriented equipment. (Will the item perform the intended function effectively and efficiently?)
5. The procedures that will be used in the accomplishment of corrective and preventive maintenance.

Maintainability demonstration is primarily designed to evaluate the supportability characteristics inherent in the prime mission equipment. As such, this requires the availability of the right elements of the logistic support capability for the system, and much valuable information can be acquired.

System test and evaluation, accomplished as Types 3 and 4 testing, provides the first real opportunity to look at the system as a whole. Prime equipment and all the required elements of logistic support are delivered to an operational test site, integrated, and scheduled through a series of mission profiles. The logistic support provided is in accordance with the requirements of the formal ILS plan. Throughout the accomplishment of the scheduled mission exercises, system operational, performance, and supportability characteristics are measured and evaluated. Areas of deficiency are noted and corrective action is initiated as appropriate. In essence, the same procedure as that discussed in Chapter 6 is employed, except that the major emphasis here is on the elements and measures of logistic support covered in Sections 16.1 and 16.3.

SELECTED REFERENCES

(1) *Annual Department of Defense Bibliography of Logistics Studies and Related Documents.* Defense Logistics Studies Information Exchange (DLSIE), U.S. Army Logistics Management Center, Fort Lee, Virginia 23801.

(2) Blanchard, B. S., *Logistics Engineering and Management* (3rd ed.). Englewood Cliffs, N.J.: Prentice-Hall, 1986.

(3) Bowersox, Donald, David Closs, and Omar Helferich, *Logistical Management.* New York: Macmillan, 1986.

(4) *Compendium of Authenticated System and Logistics Terms, Definitions, and Acronyms.* AU-AFIT-LS-3-81, U.S. Air Force Institute of Technology, Wright-Patterson AFB, Dayton, Ohio.

(5) Defense Systems Management College (DSMC), *Integrated Logistics Support Guide.* DSMC, Fort Belvoir, Virginia.

(6) Heskett, J. L., N. A. Glaskowsky, and R. M. Ivie, *Business Logistics* (2nd ed.). New York: Ronald, 1973.

(7) Hutchinson, Norman E., *An Integrated Approach to Logistics Management.* Englewood Cliffs, NJ: Prentice-Hall, 1987.

(8) Jones, J. V., *Integrated Logistics Support Handbook.* Tab Books, Inc., P. O. Box 40, Blue Ridge Summit, Pennsylvania, 1987.

(9) *Logistics Spectrum,* Journal of the Society of Logistics Engineers (SOLE), 125 West Park Loop, Suite 201, Huntsville, Alabama.

(10) MIL-STD-1388-1, Military Standard, *Logistic Support Analysis.* Department of Defense, Washington, D.C.

(11) MIL-STD-1388-2, Military Standard, *Department of Defense Requirements for a Logistic Support Analysis Record.* Department of Defense, Washington, D.C.

(12) TM 38-710, APF, NAVMAT P-4000, *Integrated Logistics Support Implementation Guide for DOD Systems and Equipment,* Department of Defense, Washington, D.C.

QUESTIONS AND PROBLEMS

1. Define supportability. How does it relate to reliability, maintainability, and manability?
2. Define logistics. Identify and describe the elements of logistic support.
3. How does logistic support fit into the system life cycle? Briefly describe the logistics activities in each phase of the life cycle.
4. How does logistics tie in with the systems engineering process?
5. Why is it important to consider logistics during the early planning and conceptual stages of system or product design?
6. What factors should be considered in defining supply support requirements? Test and support equipment requirements? Personnel and training requirements? Transportation and handling requirements? Facility requirements? Technical data requirements?
7. Assuming that a single component with a reliability of 0.85 is used in a unique application in the system and that one backup spare component is purchased, determine the probability of having a spare available in time, t, when required.

8. Assuming that the component in Problem 7 is supported with two backup spares, determine the probability of having a spare available when needed.

9. Determine the probability of having a spare available for a configuration consisting of two operating components backed by two spares (assume that the component reliability is 0.875).

10. There are 10 systems located at a site scheduled to perform a 20-hour mission. The system has an expected MTBF of 100 hours. What is the probability that at least eight of these systems will operate for the duration of the mission without failure?

11. An equipment item contains 30 parts of the same type. The part has a predicted mean failure frequency of 10,000 hours. The equipment operates 24 hours a day and spares are provisioned at 90-day intervals. How many spares should be carried in the inventory to ensure a 95% probability of having a spare available when required?

12. Determine the economic order quantity of an item for spares inventory replenishment where:
    (a) The cost per unit is $100.
    (b) The cost of preparing for a shipment and sending a truck to the warehouse is $25.
    (c) The estimated cost of holding the inventory, including capital tied up, is 25% of the inventory value.
    (d) The annual demand is 200 units. Assume that the cost per order and the inventory carrying charge are fixed.

13. Referring to Figure 16.4, what happens to Q when the demand increases? What happens when there are outstanding back orders? What factors are included in procurement lead time?

14. Some spare and repair parts are considered to have a higher priority (in importance) than others. What factor(s) determine this priority?

15. Assume that you are requested to set up a maintenance organization for system support. What factors would you consider? Identify the steps involved. How would you later evaluate this organization in terms of effectiveness? (Identify the measures that you would employ.)

16. For an operating system, how would you verify the adequacy of the following:
    (a) Spare and repair parts?
    (b) Test and support equipment?
    (c) Frequency distributions for unscheduled maintenance actions? Maintenance times?
    (d) Personnel quantities and skill levels?
    (e) Operating and maintenance procedures?

17. How could the maintenance requirements for test and support equipment affect prime equipment availability?

18. Quite often in system design preventive maintenance requirements are specified with the objective of improving system reliability. What adverse effects could occur as a result of this practice?

19. What is logistic support analysis (LSA), and how is it utilized to enhance supportability in system design? How does LSA relate to the systems analysis effort?

20. Refer to Part Three. Identify some of the analytical techniques presented and discuss their application in defining logistic support requirements.

21. Throughout the system life cycle, design changes may be initiated from time to time to improve system performance. How might a design change in the prime equipment affect

test and support equipment? Spare and repair parts? Facilities? Personnel and training requirements? Technical data? Why is change control important?

22. How are the elements of logistic support evaluated through testing?

23. How can the results of reliability and maintainability testing be employed to evaluate logistic support requirements?

24. Will the level of logistic support provided when a system or product is initially delivered for consumer use likely be the same as that required for subsequent periods of the life cycle? Why?

25. What is the purpose of the formal ILS plan? What is usually included in such a plan? How does the formal ILS plan affect production/construction operations? How does it affect system/product operations throughout the consumer use period?

17

Design for Economic Feasibility

Many systems and products have been planned, designed, produced, and operated with very little concern for their *life-cycle cost* (LCC). Although different facets of cost have been considered in the development of new systems, these costs have often been viewed in a fragmented manner. The costs associated with activities such as research, design, testing, production or construction, operations, consumer use, and support have been isolated and addressed at various stages in the system life cycle, and not viewed on an *integrated* basis.

Experience has indicated that a large portion of the total cost for many systems is the direct result of activities associated with the operation and support of these systems, while the commitment of these costs is based on decisions made in the early stages of the system life cycle. Further, the various costs associated with the different phases of the life cycle are all interrelated. Thus, in addressing the economic aspects of a system, one must look at total cost in the context of the overall life cycle, particularly during the early stages of conceptual design and advanced system planning. Life-cycle cost, when included as a parameter in the systems engineering process, provides the opportunity to design for economic feasibility.

17.1. INTRODUCTION TO LIFE-CYCLE COSTING

The recent combination of economic trends, rising inflation, cost growth experienced for many systems and products, the continuing reduction in buying power, budget limitations, increased competition, and so on, has created an awareness and interest in *total* system and product cost. Not only are the acquisition costs associated with new systems rising, but the costs of operating and maintaining systems already in use are increasing at alarming rates. This is due primarily to a combination of inflation and cost growth from causes such as:

1. Cost growth due to engineering changes occurring throughout the design and development of a system or product (for the purposes of improving performance, adding capability, etc.).
2. Cost growth due to changing suppliers in the procurement of system components.
3. Cost growth due to system production and/or construction changes.
4. Cost growth due to changes in the logistic support capability.
5. Cost growth due to initial estimating inaccuracies and changes in estimating procedures.
6. Cost growth due to unforeseen problems.

It has been noted on occasion that cost growth due to these various causes has ranged from 5 to 10 times the rate of inflation over the past several decades. Figure 17.1 illustrates basic cost growth trends for different systems. The projections relate

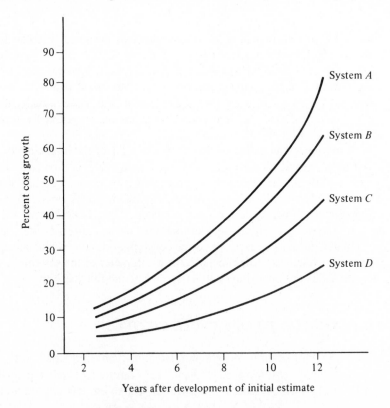

Figure 17.1 Some cost growth trends.

the initial estimate of system cost to subsequent cost estimates of the same system at later points in the life cycle.

At a time when considerable system cost growth is being experienced, budget allocations for many categories of systems are decreasing from year to year. The net result is that less money is available for acquiring and operating new systems or products and in maintaining and supporting systems that are already in being. The available funds for projects (i.e., buying power), when inflation and cost growth are considered, are decreasing at a rapid rate.

The current economic situation is further complicated by some additional problems related to the actual determination of system and/or product cost:

1. Total system cost is often not visible, particularly those costs associated with system operation and support. The cost visibility problem can be related to the "iceberg effect" illustrated in Figure 17.2. One must not only address system acquisition cost, but other costs as well.

2. In estimating cost, individual factors are often improperly applied. For instance, costs are identified and often included in the wrong category; variable costs are treated as fixed costs (and vice versa); indirect costs are treated as direct costs; and so on.

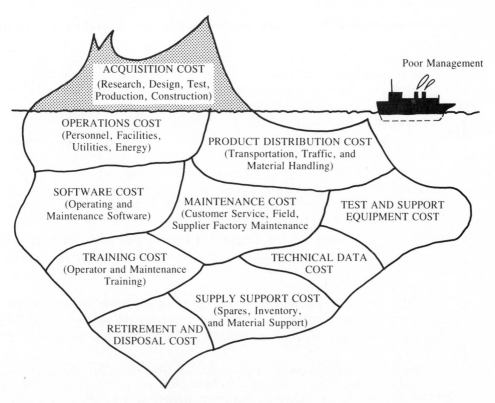

Figure 17.2 Total cost visibility.

3. Existing accounting procedures do not always permit a realistic and timely assessment of total cost. In addition, it is often difficult (if not impossible) to determine costs on a functional basis.

4. Budgeting practices are often inflexible regarding the shift in funds from one category to another, or from year to year, to facilitate improvements in system acquisition and utilization.

The current trends of inflation and cost growth, combined with these additional problems, have caused inefficiencies in the utilization of valuable resources. Systems and products have been developed that are not cost-effective. Further, it is anticipated that conditions will become worse unless an increased degree of cost consciousness is assumed in day-to-day activities. Design for economic feasibility must address all aspects of *life-cycle cost*, not just segments thereof.

Life-cycle cost refers to all costs associated with the system or product as applied to the defined life cycle. The life cycle and the major functions associated with each phase are illustrated in Figure 2.1.[1] The life cycle, tailored to the specific system being addressed, forms the basis for life-cycle costing. In general, life-cycle cost includes the following:

1. *Research and development cost*—initial planning; market analysis; feasibility studies; product research; engineering design; design documentation; software; test and evaluation of engineering models; and associated management functions.

2. *Production and construction cost*—industrial engineering and operations analysis; manufacturing (fabrication, assembly, and test); facility construction; process development; production operations; quality control; and *initial* logistic support requirements (e.g., initial consumer support, the manufacture of spare parts, the production of test and support equipment, etc.).

3. *Operation and support cost*—consumer or user operations of the system or product in the field; product distribution (marketing and sales, transportation, and traffic management); and sustaining logistic support throughout the system or product life cycle (e.g., customer service, maintenance activities, supply support, test and support equipment, transportation and handling, technical data, facilities, system modifications, etc.).

4. *Retirement and disposal cost*—disposal of nonrepairable items throughout the life cycle; system/product retirement; material recycling; and applicable logistic support requirements.

Life-cycle cost is determined by identifying the applicable functions in each phase of the life-cycle, costing these functions, applying the appropriate costs by function on a year-to-year schedule, and ultimately accumulating the costs for the

---

[1] Describing the system/product life cycle may appear to be rather elementary; however, experience has indicated that many different interpretations may exist of what constitutes the life cycle. Since this establishes the major reference point for life-cycle costing, it is essential that a *common understanding* prevail as to what is meant by the life cycle and what is included (or excluded).

entire span of the life cycle. Life-cycle cost includes all producer and consumer costs.[2]

The application of life-cycle costing methods in system and product design and development is realized through the accomplishment of life-cycle cost analyses. A *life-cycle cost analysis* may be defined as a systematic analytical process of evaluating various alternative courses of action with the objective of choosing the best way to employ scarce resources.

Life-cycle costing is employed in the evaluation of alternative system design configurations, alternative production schemes, alternative logistic support policies, and so on. The analysis constitutes a step-by-step approach employing life-cycle cost figures-of-merit as criteria to arrive at a cost-effective solution. The analysis process is iterative in nature and can be applied to any phase of the system or product life cycle.

17.2. COST EMPHASIS IN THE SYSTEM LIFE CYCLE[3]

Experience has shown that a major portion of the projected life-cycle cost for a given system or product stems from the consequences of decisions made during early planning and as part of system conceptual design. These decisions deal with system operational requirements, performance and effectiveness factors, the maintenance concept, the system configuration, quantity of items to be produced, consumer utilization factors, logistic support policies, and so on. Such decisions, made as a result of a market analysis or a design feasibility study, actually guide subsequent design and production activities, product distribution functions, and the various aspects of sustaining system support. Thus, if ultimate life-cycle costs are to be optimized in designing for economic feasibility, it is essential that a high degree of cost emphasis be applied in the early stages of system/product development.

Figure 17.3 reflects a characteristic life-cycle cost trend curve as related to actions occurring during the various phases of the life cycle. The system/product life-cycle phases presented in Figure 2.1 are translated to reflect emphasis on the early planning and design stages of a program. As illustrated, approximately 60% of the projected life-cycle cost is *committed* by the end of the system planning and conceptual design stage, even though actual project expenditures are relatively minimal at this point in time. This curve varies with the individual system; however, it does convey a trend relative to affects of decisions on ultimate life-cycle cost.

The initial step in a life-cycle cost analysis is to establish cost targets or firm goals [i.e., one or more quantitative figures-of-merit to which the system or product should be designed, produced (or constructed), and supported for a designated period of time]. Second, these cost targets are then allocated to specific subsystems or

---

[2] It should be noted that *all* life-cycle costs may be difficult (if not impossible) to predict and measure. For instance, some indirect costs caused by the interaction effects of one system on another, social costs, and so on, may be impossible to quantify. Thus, the emphasis should relate primarily to those costs that can be *directly* attributed to a given system or product.

[3] Much of this material was adapted from B. S. Blanchard, *Design and Manage to Life Cycle Cost* (Forest Grove, Ore.: M/A Press, 1978).

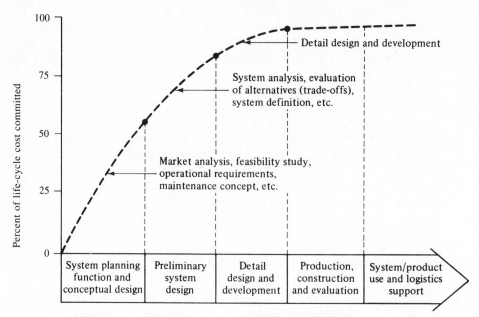

Figure 17.3 Actions affecting life-cycle cost.

elements as design constraints or criteria. With the progression of design, various alternative configurations are evaluated in terms of compliance with the allocated targets, and a preferred approach is selected. As the system or product continues to evolve through various stages of development, life-cycle cost estimates are made and the results are compared against the initially specified targets. Areas of noncompliance are noted and corrective action is initiated where appropriate. Cost emphasis throughout the system/product life cycle is illustrated in Figure 17.4 and discussed below.

Conceptual Design (Figure 17.4, Block 1)

In the early stages of planning and conceptual design, quantitative cost figures-of-merit should be established as requirements to which the system or product is to be designed, tested, produced (or constructed), and supported. A *design-to-cost* (DTC) concept may be adopted to establish life-cycle cost (LCC) as a system or product *design parameter*, along with performance, effectiveness, capacity, accuracy, size, weight, reliability, maintainability, supportability, and so on. Cost must be an *active* rather than a *resultant* factor throughout the design process.

Design-to-cost figures-of-merit are usually specified in terms of life-cycle cost at the system level. However, DTC parameters sometimes are established at a lower level to facilitate improved cost visibility and closer control throughout the life cycle:

1. *Design to unit acquisition cost*—a factor that includes only research and development cost and production or construction cost.

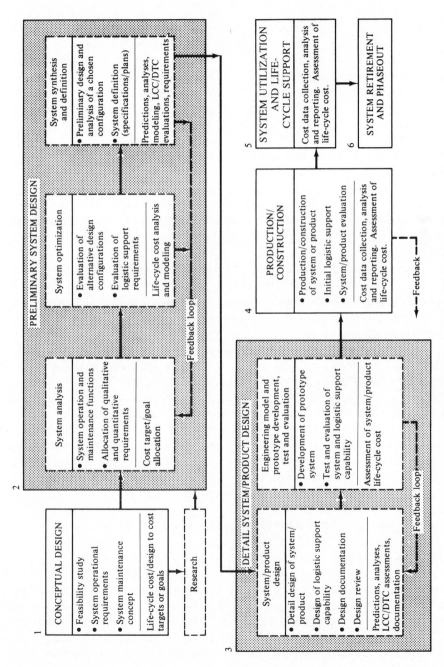

Figure 17.4 System or product life-cycle process.

2. *Design to unit operation and support cost*—a factor that includes only operation and maintenance support cost.

When sub-optimizing by considering only a single segment of life-cycle cost, one must be sure that decisions are not based on that one segment alone, without considering the overall effects on total life-cycle cost. For example, one can propose a given design configuration on the basis of a low unit acquisition cost, but the projected operation and support cost, and life-cycle cost, for that configuration may be considerably higher than necessary. Ideally, acquisition cost should not be addressed without consideration for operation and support cost (and vice versa), and both segments of cost must be viewed in terms of life-cycle cost.

Preliminary System Design (Figure 17.4, Block 2)

With the quantitative cost requirements established, the next step involves an iterative process (i.e., optimization, synthesis, and system/product definition). The criteria defined in block 1 are initially allocated, or apportioned, to various segments of the system to establish guidelines for the design and/or the procurement of the applicable element(s).

As illustrated in Figure 17.5, allocation is accomplished from the system level down to the level necessary to provide adequate control. The factors projected reflect the target cost per individual unit (i.e., a single system or product in a total population), and are based on system operational requirements, the system maintenance concept, and so on.

As the design process evolves, various alternative approaches are considered in arriving at a preferred system configuration. Life-cycle cost analyses are accomplished in evaluating each possible candidate with the objective of (1) ensuring that the candidate selected is compatible with the established cost targets, and (2) determining which of the various candidates being considered is preferred from an overall cost-effectiveness standpoint. Numerous trade-off studies are accomplished, using life-cycle cost analysis as an evaluation tool, until a preferred design configuration is chosen. Areas of compliance are justified, and noncompliant approaches are discarded. This is an iterative process with the necessary feedback and corrective action loop as illustrated by Block 2 of Figure 17.4.

Detail System or Product Design (Figure 17.4, Block 3)

As system or product design is further refined and design data become available, the life-cycle cost analysis effort involves the evaluation of specific design characteristics (as reflected by design documentation and engineering or prototype models), the prediction of cost-generating variables, the estimation of costs, and the projection of life-cycle cost as a cost profile. The results are compared with the initial requirement and corrective action is taken as necessary. Again, this is an iterative process, but at a lower level than what is accomplished during preliminary system design.

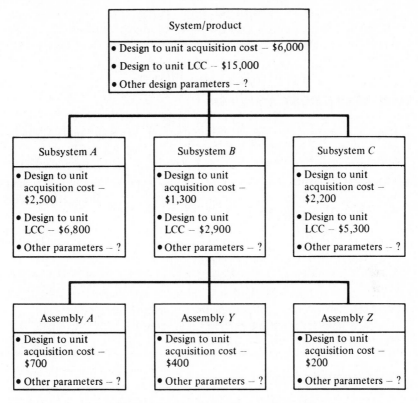

Figure 17.5   System/product cost allocation.

Production, Construction, Utilization, and Support (Figure 17.4, Blocks 4 and 5)

Cost concerns in these latter stages of the system or product life cycle involve a data collection, analysis, and assessment function. Hopefully, valuable information is gained and utilized for the purposes of product improvement and for the development of good historical data for future applications.

In summary, life-cycle costing is applicable in all phases of system design and development, production, construction, operational use and logistic support. Cost emphasis is created early in the life cycle by establishing quantitative cost factors as *requirements*. As the life cycle progresses, cost is employed as a major parameter in the evaluation of alternative design configurations and in the selection of a preferred approach. Subsequently, cost data are generated based on established design and production characteristics and used in the development of life-cycle cost projections. These projections, in turn, are compared with the initial requirements to determine degree of compliance and the ultimate necessity for corrective action. In essence, life-cycle costing evolves from a series of rough estimates to a relatively refined methodology, and is employed as a management tool for decision-making purposes.

17.3. LIFE-CYCLE COST ANALYSIS

Thoroughout the system/product life cycle, illustrated in Figure 17.4, there are many decisions required, of both a technical and a nontechnical nature. The majority of these decisions, particularly those at the earlier stages, have life-cycle implications and definitely affect life-cycle cost. It may appear initially that a specific decision will not directly affect life-cycle cost; however, the indirect effects on cost may turn out to be very significant.

For each specific problem where there are possible alternative solutions and a decision is required in the selection of a preferred approach, there is an overall analysis process that one usually follows, either intuitively or on a formal basis. This process is comparable to the engineering analysis process illustrated in Figure 3.7. Specifically, one should (1) define the need for analysis, (2) establish the analysis approach, (3) select a model to facilitate the evaluation process, (4) generate the appropriate data for each alternative being considered, (5) evaluate the alternatives, and (6) recommend a proposed solution in response to the problem at hand.

Cost Analysis Goals

The possible applications of life-cycle cost analysis are numerous. Specifically, life-cycle cost analysis may be employed in the evaluation of:

1. Alternative system/product operational use and environmental profiles.
2. Alternative system maintenance concepts and logistic support policies.
3. Alternative system/product design configurations such as packaging schemes, diagnostic routines, built-in test versus external test, manual functions versus automation, hardware versus software approaches, component selection and standardization, reliability versus maintainability, levels of repair versus discard decisions, and so on.
4. Alternative procurement source selection for a given item.
5. Alternative production approaches, such as continuous versus discontinuous production, quantity of production lines, number of inventory points and levels of inventory, levels of product quality, inspection and test alternatives, and so on.
6. Alternative product distribution channels, transportation and handling methods, warehouse locations, and so on.
7. Alternative logistic support plans, such as customer service levels, sustaining supply support levels, maintenance functions and tasks, and so on.

8. Alternative product disposal and recycling methods.
9. Alternative management policies and their impact on the system.

There are many different facets of a system that one can study, and it is relatively easy to become overwhelmed by undertaking too large an effort, or by proceeding in the wrong direction. Thus, an important initial step constitutes the clarification of analysis objectives, defining the issues of concern, and bounding the problem such that it can be studied in an efficient and timely manner. In many cases, the nature of the problem appears to be obvious, whereas the precise definition of the problem may be the most difficult part of the entire process. The problem at hand must be defined clearly, precisely, and presented in such a manner as to be easily understood by all concerned. Otherwise, it is highly doubtful whether an analysis of any type will be meaningful.

From the problem statement the cost analyst needs to identify specific goals. For instance, is the objective to evaluate two alternatives on the basis of life-cycle cost? Is there a requirement to determine the life-cycle cost of System *XYZ* for budgetary purposes? Does the evaluation need to show system performance in terms of design to unit acquisition cost? Is it necessary to evaluate supply support costs as a function of the equipment design packaging configuration?

Actually, there are many such questions that the decision maker might wish to address. There may be a single overall analysis goal (e.g., design to minimum life-cycle cost), and any number of subgoals. The primary question is: What is the purpose of the analysis, and what is to be learned through the analysis effort?

Identifying the goals of the analysis may seem elementary, but is extremely important. It is not uncommon to find instances where the analysis effort becomes the driving force and the original goals are lost in the process; or the goals have unintentionally shifted as a result of the analyst becoming too involved in the details and losing sight of the big picture. The analyst must be careful to ensure that realistic goals are established at the start of the analysis process, and that these goals remain in sight throughout the process.

Guidelines and Constraints

In support of the problem definition and the goals, the cost analyst must define the guidelines and constraints (or bounds) within which the analysis effort is to be accomplished. Guidelines are composed of information concerning such factors as the resources available for conducting the analysis (i.e., necessary labor skills, availability of a computer if required, etc.), the time schedule allowed for completion of the analysis, and/or related management policy or direction that in any way will effect the analysis. In many instances, a manager may not completely understand the problem or the analysis process and direct that certain tasks be accomplished in a prescribed manner and at a designated time, which in turn may not be compatible with the analysis objectives. On other occasions, a manager may have a preconceived idea as to a given decision outcome and direct that the analysis support the decision, whether realistic or not. Thus, at times there are external inhibiting factors that may impact the validity of the analysis effort. In such cases, the cost analyst should make every effort to alleviate the problem through education. Should any unresolved problems exist, the analyst should document them as part of the analysis report and relate their effects to the analysis results.

Relative to the technical characteristics of a system or product, the analysis output may be constrained by bounds (or limits) that are established through the definition of system performance features, operational requirements, the maintenance concept, and/or through advanced program planning. For instance, there may be a maximum weight requirement for a given product, a minimum reliability requirement, a maximum allowable unit life-cycle cost goal (e.g., the allocated value in Figure 17.5), a minimum capacity output for a plant, and so on. These various bounds, or constraints, must be defined in terms of the trade-off areas allowable in the evaluation of alternatives. Figure 4.12 illustrates this point. All alternative candidates that fall within the trade-off area are eligible for consideration, while those that fall outside this area are not—even though one of these alternatives may turn out to be more cost effective in the long run.

Identification of Feasible Alternatives

Section 4.3 describes the process relating to the identification and evaluation of alternatives. Within the established bounds and constraints defined, there may be any number of different approaches leading to a possible solution. All possible candidates should be initially considered, with the most likely candidates selected for further evaluation.

All possible alternatives are not necessarily attainable. Alternatives are frequently proposed for analysis even though there seems to be little likelihood that they will prove feasible. This is done with the thought that it is better to consider many alternatives than to overlook one that may be very good. Alternatives not considered cannot be adopted, no matter how desirable they may prove to be.

Development of Cost Breakdown Structure (CBS)

With feasible alternatives identified, the next step is to develop a cost breakdown structure (CBS) to provide a mechanism for initial cost allocation, cost categorization, and cost monitoring and control. An example of a top-level cost breakdown structure is presented in Figure 17.6. The CBS is used as a basis for assessing the life-cycle cost of each alternative being considered.

The cost breakdown structure links objectives and activities with resources, and constitutes a logical subdivision of cost by functional activity area, major element of a system, and/or one or more discrete classes of common or like items. The cost breakdown structure, which is usually adapted or tailored to meet the needs of each individual program, should exhibit the following characteristics:

1. All life-cycle costs should be considered and identified in the cost breakdown structure. This includes research and development cost, production and con-

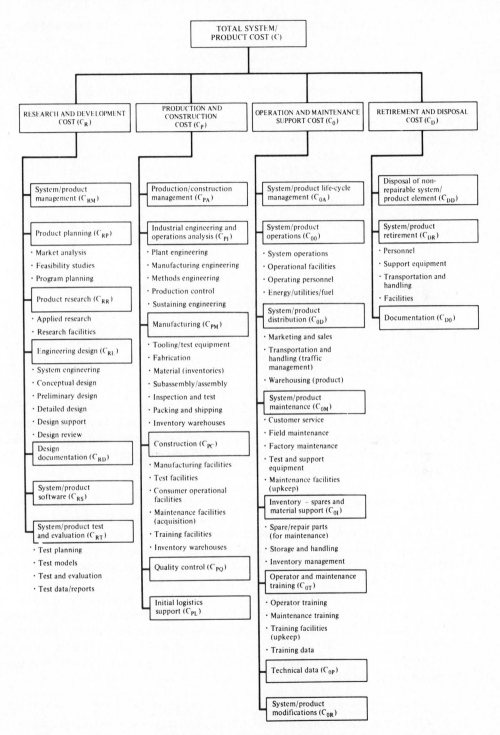

Figure 17.6 Cost breakdown structure.

struction cost, operation and system support cost, and retirement and disposal cost.[4]

2. Cost categories are generally identified with a significant level of activity or with a major item of material. Cost categories in the CBS must be well defined, and managers, engineers, accountants, and others must have the same understanding of what is included in a given cost category and what is not included. Cost omissions and doubling (i.e., counting the same cost in two or more categories) must be avoided.

3. Costs must be broken down to the level necessary to provide management with the visibility required in evaluating various facets of system design and development, production, operational use, and support. Management must be able to identify high-cost areas and cause-and-effect relationships.

4. The CBS and the categories defined should be coded in a manner to facilitate the analysis of specific areas of interest while virtually ignoring other areas. For example, the analyst may wish to investigate supply support costs as a function of engineering design or distribution costs as a function of manufacturing, independent of other aspects of the system. The CBS should be designed with this objective in mind.

5. The CBS and the categories defined should be coded in such a manner as to enable the separation of producer costs, supplier costs, and consumer costs in an expeditious manner.

6. When related to a particular program, the cost structure should be directly compatible (through cross-indexing, coding, etc.) with planning documentation, the work breakdown structure, work packages, the organization structure, PERT-CPM and PERT-COST scheduling networks, Gantt charts, and so on. Costs that are reported through various management information systems must be compatible and consistent with those comparable cost factors in the CBS.

Referring to Figure 17.6, the cost categories identified are obviously too broad to ensure any degree of accountability and control. The analyst can not readily determine what is and what is not included, nor can he or she validate that the proper relationships or parameters have been utilized in determining the specific cost factors that are inputted into the illustrated cost structure. The cost analyst requires much more information than is presented in Figure 17.6.

In response, the CBS illustrated in Figure 17.6 must be expanded to include a detail description of each cost category (in the order presented in the CBS), along with the symbology and quantitative relationships used to derive costs. The system design application in Chapter 19 of this text includes a sample cost breakdown structure (Figure 19.5), and Appendix C includes a complete breakdown of cost categories in a typical CBS.

---

[4] This does not mean to imply that all cost categories are relevant to all analyses. The objective is to include all life-cycle costs and then identify those categories that are considered significant relative to the problem at hand.

In developing the CBS, one needs to expand to the depth required to (1) provide the necessary information for a true and valid assessment of the system or product life-cycle cost, (2) identify high-cost contributors and enable determination of the cause-and-effect relationships, and (3) illustrate the various cost parameters and their application in the analysis. Traceability from the system level LCC figure-of-merit to the specific input factor is required.

Establishing the cost breakdown structure is one of the most significant steps in life-cycle costing. The CBS constitutes the framework for defining life-cycle costs and provides the communications link for cost reporting, analysis, and ultimate cost control. The CBS is the basic reference point for much of the material presented in subsequent sections of this chapter.

Selection of a Cost Model

After the establishment of the cost breakdown structure, it is necessary to develop a model of some type to facilitate the life-cycle cost evaluation process. The model may be a simple series of relationships or a complex set of computer subroutines, depending on the phase of the system life cycle and the nature of the problem at hand. In any case, the criteria for model selection and application discussed in Section 4.3 and in Chapter 7 are applicable.

Life-cycle costing itself includes a compilation of a variety of cost factors, reflecting the many different types of activities indicated by the CBS. The objective in using a model is to evaluate a system in terms of total life-cycle cost, as well as the various individual segments of cost. Figure 17.7 illustrates this point. Total system life-cycle cost is compiled through the use of accounting techniques in the life-cycle cost summary subroutine, whereas major areas of cost (which constitute input data to the LCC summary subroutine) can be viewed on an individual basis.

The subroutines that reflect segments of cost may be structured quite differently depending on the system element covered. For instance, supply support costs may be compiled through the application of spare and repair part demand factors and inventory techniques discussed in Chapter 16. This, in turn, may require factors derived through the use of a reliability model. On the other hand, engineering design costs may be extracted directly from a proposal or a set of engineering cost projections. The relationship here is analogous to the relationship of the models as was illustrated in Figure 4.13.

Cost Estimating Relationships[5]

Definition of the analysis goals and guidelines, combined with the identification of specific evaluation criteria, will normally dictate the data output requirements for the life-cycle cost analysis (i.e., the type of data desired from the analysis and the pre-

---

[5] The subject of cost estimation is a significant area which should be investigated further. For additional material, see P. F. Ostwald, *Cost Estimating,* 2nd ed., (Englewood Cliffs, N.J.: Prentice-Hall), 1984, R. D. Stewart, *Cost Estimating,* John Wiley & Sons, Inc., (New York: John Wiley), 1982, and G. J. Thuesen and W. J. Fabrycky, *Engineering Economy,* 7th ed., (Englewood Cliffs, N.J.: Prentice-Hall) 1989.

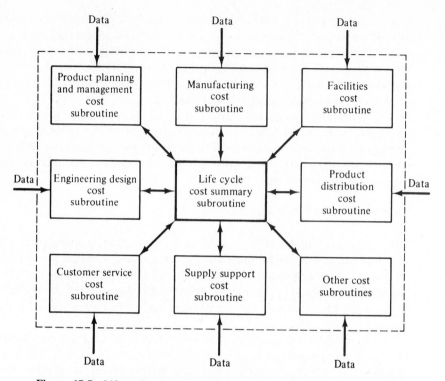

Figure 17.7 Life-cycle model configuration. *Note:* Although not shown here, many of the subroutines actually interact with each other in terms of data input and output.

ferred format in which the data are to be presented). With the analysis output requirements defined, the cost analyst develops the methodologies and relationships necessary to produce the desired results. This is accomplished through the development of the cost breakdown structure and the selection of the cost model, where system parameters, estimating relationships, and cost factors are identified. The completion of these steps leads to the identification of the input data necessary for accomplishing the analysis itself.

The acquisition of the right type of input data in a timely manner, and presented in a manageable format, is one of the most important steps in the overall life-cycle cost analysis process. The requirements for data must be carefully defined since the application of too little data, too much data, the wrong type of data, and so on, can invalidate the overall analysis, resulting in poor decisions which may be quite costly in the long term. Further, every effort should be made to avoid the unnecessary expenditure of valuable resources in generating data that may turn out not to be required at all. Often, there is a tendency to undertake elaborate analytical exercises to develop meaningless precise quantitative factors at times when only top system-level estimates are required to satisfy the immediate need.

In the development of cost data, the requirements may vary considerably, depending on the program phase, the extent of system/product definition, and the type and depth of the analysis being accomplished. During the early planning and conceptual design stages of system development, available data are limited, and the cost analyst must depend primarily on the use of various parametric cost estimating techniques in the development of cost data, as indicated in Figure 17.8. As the system design progresses, more complete design information becomes available and the analyst is able to develop cost estimates by comparing the characteristics of the new system with similar systems where historical cost data are recorded. The generation of cost data is based on analogous estimating methods. Finally, as the system design configuration becomes firm, design data (to include drawings, specifications, parts lists, predictions, etc.) are produced which allow for the development of detail engineering and manufacturing estimates. Such estimates incorporate standard cost factors, fixed overhead rates, and so on.

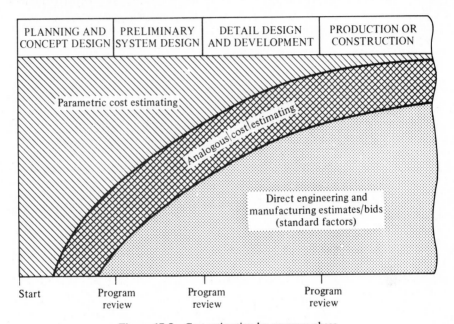

Figure 17.8   Cost estimation by program phase.

The most general cost estimating technique uses parametric methods to determine future costs in areas where there are few or no meaningful data available. Parametric cost estimating relationships are basically "rules of thumb" which relate various categories of cost-to-cost-generating or explanatory variables of one form or another. These explanatory variables usually represent characteristics of system performance, physical features, effectiveness factors, or even other cost elements. Estimating relationships may take different forms (i.e., continuous or discontinuous, mathematical or nonmathematical, linear or nonlinear, etc.). Some of these are presented in the following paragraphs.

Simple linear functions.   Simple linear estimating functions are expressed as

$$y = a + bx \qquad (17.1)$$

where y and x are the dependent and independent variables, and a and b are parameters. Linear functions are useful because many cost relationships are of this form. Sometimes linear relationships are developed employing curve-fitting techniques or normal regression analysis. Such linear functions are utilized for forecasting purposes and may relate one cost parameter to another, a cost parameter to a noncost parameter, and/or a noncost parameter to another noncost parameter. Two illustrations are given in Figure 17.9.

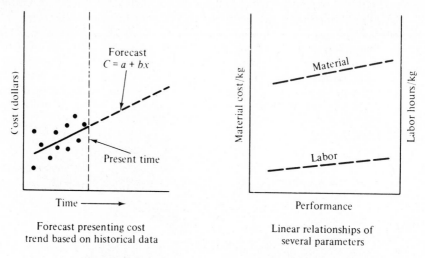

Figure 17.9   Simple linear cost estimating relationships.

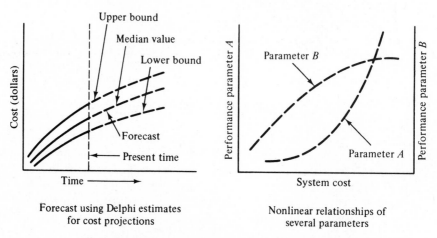

Figure 17.10   Simple nonlinear cost estimating relationships.

Simple nonlinear functions.   Not all cost functions are linear. Some cost relationships may be exponential in nature, hyperbolic, or may fit some other curve. Examples of nonlinear forms involving a single explanatory variable are

$$y = ab^x \qquad \text{(exponential)} \qquad (17.2)$$

$$y = a + bx - cx^2 \qquad \text{(parabolic)} \qquad (17.3)$$

$$y = \frac{1}{a + bx} \qquad \text{(hyperbolic)} \qquad (17.4)$$

Figure 17.10 illustrates two nonlinear applications.

Discontinuous step functions.   The estimating relationships introduced imply a continuous function involving cost and other variables. However, in many instances cost can be constant over a specific range of the explanatory variable, then suddenly increase to a higher level at some point, remain constant again, then increase to another level, and so on. This type of relationship, known as a *step function,* is illustrated in Figure 17.11. These kinds of functions are useful in illustrating the cost behavior of quantity procurements in production, support activities which are represented in small noncontinuous increments, price-quantity relationships, and so on.

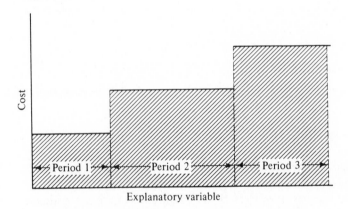

Figure 17.11   Discontinuous step function.

Other estimating forms.   Of a more general nature, the analyst may use analogies as a form of estimating. Historical data from events in the past may be employed in terms of future estimates on the basis of similarity. Such estimates may be used directly or may be factored to some extent to compensate for slight differences. This is often known as *analogous cost estimating.*

Another approach may involve *rank-order cost estimation.* A series of comparable activities are evaluated in terms of cost and then ranked on the basis of the magnitude of the cost (i.e., the highest-cost activity on down to the lowest-cost activity). After the ranking is accomplished, the various activities are viewed in relation to each other, and the initial cost values may be adjusted if the specific relationships appear to be unrealistic. This may require an iterative process.

When analyzing historical cost data, one will find that the actual cost of a given activity, when completed on a number of occasions, will vary. This variance may assume any form of distribution such as illustrated in Figure 17.12. In the prediction of future costs for comparable activities on a new program, the cost analyst may wish to assume a distribution and determine the median or mean value, vari-

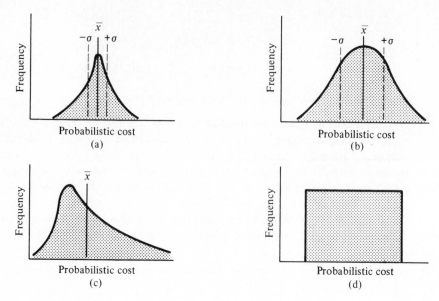

Figure 17.12 Cost distributions.

ance, standard deviation, and so on, in order to assess risk in terms of probabilities of possible cost variations. The distributions associated with historical costs will facilitate this task.

Developing Cost Data

In developing cost data for a life-cycle cost analysis, the cost analyst should initially investigate all possible data sources to determine what is available for direct application in support of analysis objectives. If the required data are not available, the use of parametric cost estimating techniques may be appropriate. However, one should first determine what can be derived from existing data banks, initial system planning data, supplier documentation, reliability and maintainability predictions, logistic support analyses, test data, field data, and so on. Some of these data sources are discussed below.

Existing data banks.   Actual historical information on existing systems, similar in configuration and function to the item(s) being developed, may be used when applicable. Often it is feasible to employ such data and apply adjustment factors as necessary to compensate for any differences in technology, configuration, projected operational environment, and time frame. Included in this category of existing data are standard cost factors which have been derived from historical experience that can be applied to specific functions or activities. Standard cost factors may cover such areas as:

1. The cost of engineering labor—dollars per labor hour for principal engineer, senior engineer, technician, and so on.

2. The cost of manufacturing labor by classification—dollars per labor hour per classification.

3. Overhead rate—dollars per direct labor cost (or percent).

4. Training cost—dollars per student week.

5. Shipping cost—dollars per pound per mile.

6. The cost of fuel—dollars per gallon.

7. The cost of maintaining inventory—percent of the inventory value per year.

8. The cost of facilities—dollars per cubic foot of occupancy.

9. The cost of material x—dollars per pound or per foot.

Figure 17.13 presents an example of some of the standard cost factors that may be used.

 These and comparable factors, where actual quantitative values can be directly applied, are usually established from known rates and costs in the market place, and are a direct input to the analysis. However, care must be exercised to ensure that the necessary inflationary and deflationary adjustments are incorporated on a year-to-year basis.

 Advanced system/product planning data.   Advanced planning data for the system or product being evaluated usually includes market analysis data, definition of system operational requirements and the maintenance concept, the re-

---

A. *Inventory*
- Purchase order ($/order)
  repairable 200
  nonrepairable 100
- New item entry ($/item)
  repairable 375
  nonrepairable 290
- Inventory maintenance ($/$/yr)
  25% of inventory value/yr

B. *Personnel Labor* ($/hr)
- Operator 15.25
- Maintenance
  organization 8.50
  intermediate 10.25
  factory 15.50

C. *Training* ($/student week)
- Operator 300
- Maintenance 450

D. *Transporation* ($/pound)
- Domestic 1.65
- Foreign 5.45

E. *Packing* ($/pound)
- Domestic 2.35
- Foreign 5.55

F. *Data* ($/page)
- Operating instructions 250
- Maintenance instructions 300
- Failure report 100

G. *Facilities*
- Construction–adjust for
  geographical location
- Operation–$50/sq ft/yr
- Maintenance–$100/sq ft/yr

G. *Support Equipment Spares*
- 20% of support equipment material

Figure 17.13   Illustration of standard cost factors.

sults of technical feasibility studies, and program management data. The cost analyst needs information pertaining to the proposed physical configuration and major performance features of the system, the anticipated mission to be performed and associated utilization factors, system effectiveness parameters, the geographical location and environmental aspects of the system, the maintenance concept and logistic support philosophy, and so on. This information serves as the baseline from which all subsequent program activities evolve. If the basic information is not available, the analyst must make some assumptions and proceed accordingly. These assumptions must then be thoroughly documented.

Individual cost estimates, predictions, and analyses. Throughout the early phases of a program, cost estimates are usually generated on a somewhat continuing basis. These estimates may cover research and development activities, production or construction activities, and/or system operating and support activities. Research and development activities, which are basically nonrecurring in nature, are usually covered by initial engineering cost estimates or by cost-to-complete projections. Such projections primarily reflect labor costs and include inflationary factors, cost growth due to design changes, and so on.

Production cost estimates are often presented in terms of both nonrecurring costs and recurring costs. Nonrecurring costs are handled in a manner similar to research and development costs. On the other hand, recurring costs are frequently based on individual manufacturing cost standards, value engineering data, industrial engineering standards, and so on. Quite often, the individual standard cost factors that are used in estimating recurring manufacturing costs are documented separately, and are revised periodically to reflect labor and material inflationary effects, supplier price changes, effects of learning curves, and so on.

System operating and support costs are based on the projected activities throughout the operational use and logistic support phase of the life cycle and are undoubtedly the most difficult to estimate. Operating costs are a function of system or product mission requirements and utilization factors. Support costs are basically a function of the inherent reliability and maintainability characteristics in the system design and the logistics requirements necessary to support all scheduled and unscheduled maintenance actions throughout the programmed life cycle. Logistic support requirements include maintenance personnel and training, supply support (spares, repair parts, and inventories), test and support equipment, transportation and handling, facilities, and certain facets of technical data. Thus, individual operation and support cost estimates are based on the predicted frequency of maintenance or the mean of the maintenance (MTBM) factor, and on the logistic support resources required when maintenance actions occur. These costs are derived from reliability and maintainability prediction data (refer to Chapters 13 and 14), logistic support analysis (LSA) data (refer to Chapter 16), and other supporting information, all of which is based on system/product engineering design data.

Supplier documentation. Proposals, catalogs, design data, and reports covering special studies conducted by suppliers (or potential suppliers) may be used as a data source when appropriate. Quite often, major elements of a system are either procured off the shelf or developed through a subcontracting arrangement of some type. Various potential suppliers will submit proposals for consideration, and these proposals may include not only acquisition cost factors but (in some instances) life-cycle cost projections. If supplier cost data are used, the cost analyst must become completely knowledgeable as to what is and is not included. Omissions and/or the double counting of costs must not occur.

Engineering test and field data.   During the latter phases of system development and production and when the system or product is being tested or is in operational use, the experience gained represents the best source of data for actual analysis and assessment purposes (refer to Section 6.5). Such data are collected and used as an input to the life-cycle cost analysis. Also, field data are utilized to the extent possible in assessing the life-cycle cost impact that may result from any proposed modifications on prime equipment, software, and/or the elements of logistic support.

These five main sources of data identified above for life-cycle costing purposes are presented in a summary manner to provide an overview as to what the cost analyst should look for. In pursuing the data requirements further, the analyst will find that a great deal of experience has been gained in determining research and development and production/construction costs. However, very few historical cost data are currently available in the operations and support area. Accounting for operation and support costs has been lacking in the past, but this situation should ultimately rectify itself as the emphasis on life-cycle costing continues to increase.

Treatment of Cost over the Life Cycle

With the system/product CBS defined and cost estimating approaches established, it is appropriate to apply the resultant data to the life cycle. In accomplishing this, one needs to understand the steps required in developing cost profiles, aspects of inflation, the effects of learning curves, the time value of money, and so on. These are discussed in the following paragraphs.

Development of cost profile.   In developing a cost profile, there are different procedures that may be followed. The steps noted below are suggested.

1. Identify all activities throughout the life cycle that will generate costs of one type or another. This includes functions associated with planning, research and development, test and evaluation, production, construction, product distribution, system/product operational use, logistic support, and so on (refer to Figure 2.1).

2. Relate each activity identified in item 1 to a specific cost category in the cost breakdown structure (refer to Figure 17.6). All program activities should fall into one or more of the categories in the CBS.

3. Establish the appropriate cost factors in constant dollars for each activity in the CBS where constant dollars reflect the general purchasing power of the dollar

at the time of decision (i.e., today). Relating costs in terms of constant dollars will allow for a direct comparison of activity levels from year to year prior to the introduction of inflationary cost factors, changes in price levels, economic effects of contractual agreements with suppliers, and so on, which often cause some confusion in the evaluation of alternatives. Also, using constant dollars tends to assure consistency in accomplishing comparative studies.

4. Within each cost category in the CBS, the individual cost elements are next projected into the future on a year-to-year basis over the life cycle as applicable. The result should be a cost stream in constant dollars for the activities that are included.

5. For each cost category in the CBS, and for each applicable year in the life cycle, introduce next the appropriate inflationary factors, economic effects of learning curves, changes in price levels, and so on. The modified values constitute a new cost stream and reflect realistic costs as they are anticipated for each year of the life cycle (i.e., expected 1990 costs in 1990, 1991 costs in 1991, etc.). These costs may be used directly in the preparation of future budget requests, since they reflect the actual dollar needs anticipated for each year in the life cycle.

6. Summarize next the individual cost streams by major categories in the CBS and develop a top-level cost profile.

The results from the foregoing sequence of steps are presented in Figure 17.14. First, it is possible and often beneficial to evaluate the cost stream for individual activities of the life-cycle such as research and development, production, operation and support, and so on. Second, these individual cost streams may be shown in the context of the total cost spectrum. Finally, the total cost profile may be viewed from the standpoint of the logical flow of activities and the proper level and timely expenditure of dollars. The profile in Figure 17.14 represents a budgetary estimate covering future resource needs.

When dealing with two or more alternative system configurations, each will include different levels of activity, different design approaches, different logistic support requirements, and so on. No two alternatives will be identical. Thus, individual profiles will be developed for each alternative and ultimately compared on an equivalent basis utilizing the economic analyses techniques of Chapter 8. Figure 17.15 illustrates several alternative profiles.

Dealing with inflation.   When developing time-phased cost profiles, the reality of inflation should be considered for each future year in the life cycle. During the past several decades, inflation has been a significant factor in the rising costs of products and services and in the reduction of the purchasing power of the dollar. Inflation is a broad term covering the general increase(s) in the unit cost of an item or activity and is related primarily to labor and material costs, as follows:

1. Inflation factors applied to labor costs are due to salary and wage increases, cost of living increases, and increases in overhead rates due to the rising costs of personnel fringe benefits, retirement benefits, insurance, and so on. Inflation

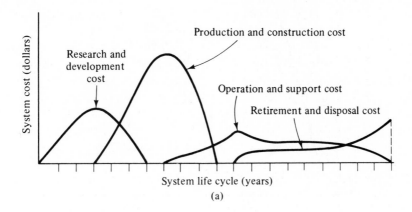

(a)

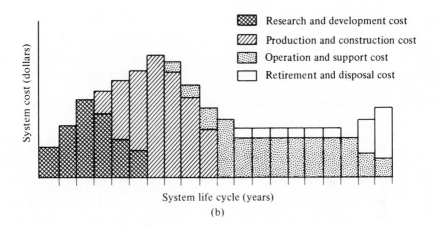

(b)

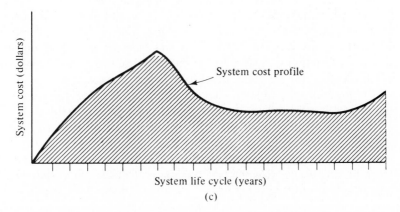

(c)

Figure 17.14   Development of cost profiles.

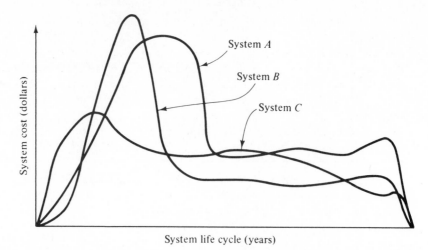

Figure 17.15 Life-cycle cost profiles of alternatives.

factors should be determined for different categories of labor (i.e., engineering labor, technician labor, manufacturing labor, construction labor, customer service personnel labor, management labor, etc.) and should be estimated for each year in the life cycle.

2. Inflation factors applied to material costs are due to material availability (or unavailability), supply and demand characteristics, the increased costs of material processing, and increases in material handling and transportation costs. Inflation factors will often vary with each type of material and should be estimated for each year in the life cycle.

Increasing costs of an inflationary nature often occur as a result of new contract provisions with suppliers, new labor agreements and union contracts, revisions in procurement policies, shifts in sources of supply, the introduction of engineering changes, program schedule shifts, changes in productivity levels, changes in item quantities, and for other comparable reasons. Also, inflation factors are influenced to some extent by geographical location and competition. When reviewing the various causes of inflation, one must be extremely careful to avoid overestimating and double counting for the effects of inflation. For instance, a supplier's proposal may include provisions for inflation, and unless this fact is noted, there is a chance that an additional factor for inflation will be included for the same reason.

Inflation factors should be estimated on a year-to-year basis if at all possible. Since inflation estimates may change considerably with general economic conditions at the national level, cost estimates far out in the future (i.e., five years and more) should be reviewed at least annually and adjusted as required. Inflation factors may be established by using price indices or by the application of a uniform escalation rate.

Application of learning curves. When performing a repetitive activity or process, learning takes place and the experience gained often results in reduced cost.

Although learning and the associated cost variations occur at different activity points throughout the life cycle, the greatest impact of learning on cost is realized in the production of large quantities of a given item. In such instances, the cost of the first unit produced is generally higher than the cost of the 25th unit, which may be higher than the cost of the 50th unit, and so on. This is primarily due to job familiarization by the workers in the production facility, development of more efficient methods for item fabrication and assembly, the use of more efficient tools, and improvement in overall management. The effects of learning generally result in the largest portion of any cost savings taking place relatively early in a uniform production run, with a leveling off taking place later.

Learning curves are commonly derived on the basis of assuming a constant percentage cost savings for each doubling of the quantity of production units. For example, 80% *unit* learning curve implies that the second unit costs 80% of the first, the fourth unit costs 80% of the second, the eighth units costs 80% of the fourth, and so on. This learning curve is then applied to find the production cost profile.

Unit learning curves may vary considerably, depending on the expected magnitude of the cost savings estimated for the second unit, the 10th unit, the 20th unit, etc. Because of product complexity, an 80% unit cost reduction may not be realized until the production of the 10th unit. In this case, an 80% learning curve will be based on the 10th, 20th, 40th, and 80th units as being the major milestones for cost measurement. Thus, we still may utilize an 80% learning curve, but the cost factors will be different. The analyst should evaluate the complexity of tasks in the production process and attempt to determine the type of unit learning curve that is most appropriate for the situation at hand. A variety of unit learning curves are presented in Figure 17.16.

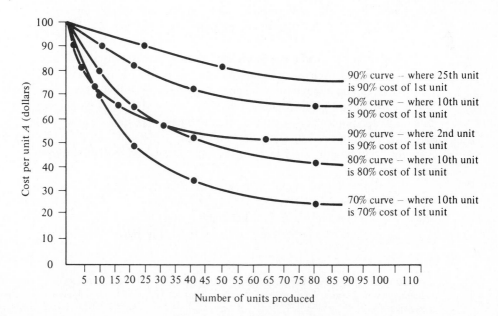

Figure 17.16   Unit learning curves.

Sometimes in the application of learning curves it may be more appropriate to use a cumulative average learning curve. If it turns out that the projected average cost of producing the first 20 units is 80% of the average cost of producing the first 10 units, the process follows an 80% *cumulative average* learning curve.

Learning curves can also cover material cost, although the percentage(s) may be different. While labor costs relate to both the required personnel skill levels and labor-hours to accomplish a given function, material costs may vary as a function of factory tooling, material scrapage rates (or percent of raw material utilized after the fabrication process), procurement methods, and inventory policies. Again, there are many factors involved, and different learning curves may be applied depending on the specific situation.

In the application of learning curves the cost analyst must ensure that the production process is indeed continuous and relatively void of design changes, manufacturing changes in producing the product, and/or organization changes that will ultimately cancel out the effects of a learning curve altogether, if not create a negative learning curve where producing the tenth unit is more costly than the first. The proper application of learning curves is considered significant in life-cycle costing. Although the effects of learning curves may be minimal for research and development activities, they may be significant in determining production costs and follow-on system or product operation and maintenance support costs.[6]

Time value of money. The development of a given system or product requires many decisions to be made. Such decisions evolve from the evaluation of alternative proposals of one type or another. Each proposal considered in the evaluation process represents a potential investment and should be viewed from the standpoint of anticipated revenues (i.e., benefits) and costs that will occur over the designated life cycle. Alternatives such as the investment in configuration *A*, configuration *B*, or the investment of money in a bank, are evaluated in a like manner.

Since revenues and costs are related to different activities at different points in time over the life cycle, a common point of reference must be assumed so that all alternatives can be compared on an equivalent basis. The flow of revenues and costs, having time value, for each alternative being considered must be equated to a common reference point. This point is generally the present time or now, when decisions that have a significant impact on the future are made; thus, all future revenues and costs for each year in the life cycle may be discounted to their present equivalent amounts.

The time value of money and the calculation of equivalence is presented in detail in Chapter 8. In performing life-cycle cost analyses for decision-making purposes, the principles discussed in that chapter must be incorporated. In other words, in the evaluation of alternatives, such as those illustrated by the cost profiles in

---

[6] Chapter 11 in W. J. Fabrycky, P. M. Ghare, and P. E. Torgersen, *Applied Operations Research and Management Science,* (Englewood Cliffs, N.J.: Prentice-Hall, 1984) gives a quantitative treatment of learning curves.

Figure 17.15, all costs must be converted to a common point in time in order to view these alternatives on an equivalent basis.[7]

Summarization of costs.   Table 17.1 reflects a summary listing of costs and a breakdown showing the percent contribution of each major category to the total. The categories listed are those indicated in the cost breakdown structure in Figure 17.6. These costs constitute the summation of individual costs for each category, for each year in the life cycle, and are discounted to the present value. The application of discounting here assumes that the configuration reflected by the cost summary is being compared with other configurations. If the purpose is to view system cost from a budgetary standpoint (representing only this configuration), then undiscounted costs would be presented.

TABLE 17.1 LIFE-CYCLE COST SUMMARY

COST CATEGORY (REFER TO COST BREAKDOWN STRUCTURE)	COST($)	PERCENT CONTRIBUTION(%)
1. Research and development (C_R)	$ 130,579	10.3
a. System/product management (C_{RM})	19,016	1.5
b. Product planning (C_{RP})	2,536	0.2
c. Product research (C_{RR})	6,339	0.5
d. Engineering design (C_{RE})	68,459	5.4
e. Design documentation (C_{RD})	10,142	0.8
f. System/product software (C_{RS})	8,874	0.7
g. System/product test and evaluation (C_{RT})	15,213	1.2
2. Production and construction cost (C_P)	574,296	45.3
a. Production/construction management (C_{PA})	10,021	0.8
b. Industrial engineering and operations analysis (C_{PI})	13,945	1.1
c. Manufacturing (C_{PM})	448,908	35.4
d. Construction (C_{PC})	67,191	5.3
e. Quality control (C_{PQ})	11,411	0.9
f. Initial logistics support (C_{PL})	22,820	1.8
3. Operation and maintenance support cost (C_O)	505,836	39.9
a. System/product life-cycle management (C_{OA})	19,016	1.5
b. System/product operations (C_{OO})	40,568	3.2
c. System/product distribution (C_{OD})	111,563	8.8
d. System/product maintenance (C_{OM})	192,699	15.2
e. Inventory—spares and material support (C_{OI})	59,585	4.7
f. Operator and maintenance training (C_{OT})	70,995	5.6
g. Technical data (C_{OP})	11,410	0.9
h. System/product modifications (C_{OR})	—	—
4. Retirement and disposal cost (C_D)	57,049	4.5
Grand total (C)	$1,267,760	100.0%

[7] In addition to Chapter 8, the concept of economic equivalence through use of the time value of money is presented in G. J. Thuesen and W. J. Fabrycky, *Engineering Economy,* 7th ed. (Englewood Cliffs, N.J.: Prentice-Hall, 1989).

Referring to Table 17.1, categories where the percent contribution is relatively high should be broken down into the different subcategories included therein, and the high-cost areas should be investigated further in order to determine the cause(s). The breakout of costs in this fashion not only allows for a comparison of different activities for a given system or product configuration, but also facilitates the direct comparison with other systems where costs are presented in a like manner.

17.4. EVALUATING ECONOMIC ALTERNATIVES

The completion of a life-cycle cost analysis as outlined in Section 17.3 serves as an aid in the decision-making process pursuant to the evaluation of alternatives. It also provides a basis for estimating budgetary requirements for a defined system configuration over its life cycle.

Design for economic feasibility includes the definition of a need (or a problem statement), the identification of feasible alternatives in response to the defined need, the selection of a preferred approach, and the refinement of the configuration selected for the purposes of attaining economic improvements. Life-cycle cost is the evaluation criterion adopted. Three design examples are presented below with the objective of illustrating some of the steps involved in the evaluation of alternatives on the basis of life-cycle cost.[8]

Evaluating Two Design Alternatives

Suppose that there is a need for a new complex system to perform a certain function six years from now. Each of two potential suppliers has submitted a proposal providing anticipated system life-cycle cost and activities (i.e., research and development,

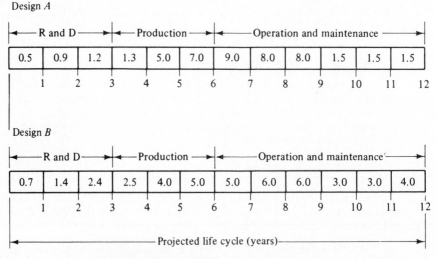

Figure 17.17 Life-cycle costs for two designs.

[8] A more extensive illustration of a life-cycle cost analysis is presented in Chapter 19.

production, and system operation and maintenance). The alternative system costs, represented by design *A* and design *B*, are shown in Figure 17.17. The cost streams are illustrated with the cost factors for each year presented in millions of dollars. Costs, although incurred throughout the year, are accumulated at the end of each year for purposes of simplicity. This is consistent with the money flow model presented in Chapter 8. The objective is to select either design *A* or design *B* on the basis of the given life-cycle cost information, assuming a nominal annual interest rate of 8%.

The two life-cycle cost streams can be compared on an equivalent basis by calculating the present cost from Equation 8.2 using the interest factors in Appendix D, Table D.4. This is shown in Table 17.2, with design *B* having the lowest present equivalent cost.

TABLE 17.2 PRESENT EQUIVALENT COST COMPARISON

Year, n	UNDISCOUNTED COST (MILLION $) Design A	Design B	$P/F, 8, n$ ( )	PRESENT COST (MILLION $) Design A	Design B
1	0.5	0.7	0.9259	0.4629	0.6481
2	0.9	1.4	0.8573	0.7716	1.2002
3	1.2	2.4	0.7938	0.9526	1.9051
4	1.3	2.5	0.7350	0.9555	1.8375
5	5.0	4.0	0.6806	3.4030	2.7224
6	7.0	5.0	0.6302	4.4114	3.1510
7	9.0	5.0	0.5835	5.2515	2.9175
8	8.0	6.0	0.5403	4.3224	3.2418
9	8.0	6.0	0.5003	4.0024	3.0018
10	1.5	3.0	0.4632	0.6948	1.3896
11	1.5	3.0	0.4289	0.6434	1.2867
12	1.5	4.0	0.3971	0.5957	1.5884
			Total	$26.4672	$24.8901

Consideration should be given next to the point when the preferred alternative actually assumes a favorable position. In some instances, a given alternative may be preferable on the basis of the quantitative life-cycle cost figure of merit; but the time preference of the cost profile may not be favorable when compared to other opportunities, particularly if the perceived advantage is not realized until a point far out in the life cycle. Therefore, a break-even analysis should be done prior to arriving at a final decision. Such an analysis may assume a variety of forms as presented in Section 8.6. Figure 17.18 illustrates the application of the break-even concept to the two alternatives identified in Figure 17.17. In this instance, the cumulative life-cycle costs of each alternative are calculated, and the least costly approach assumes a favorable position in approximately six years and three months.

One must ask the question: Is the break-even point (or crossover point) reasonable in terms of possible system obsolescence, competition, and business risk and uncertainty? If the break-even point is too far out in time, it may be more feasible to select design A in lieu of design B. Also, if the magnitude of the benefits derived are

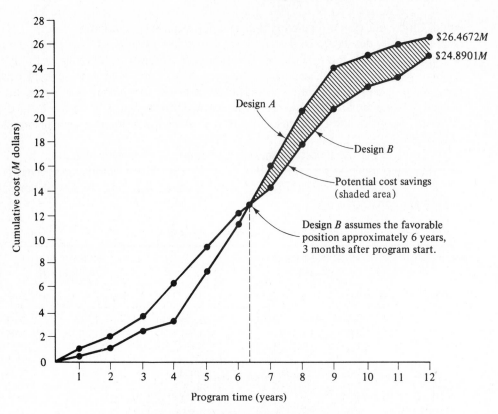

Figure 17.18   Comparison of cost alternatives.

small by accepting *B* (i.e., potential cost savings), then *A* may be preferable when considering other factors. When evaluating all factors in this example, design *B* remains as the preferred alternative. It is felt that the break-even point is close enough in time to be reasonable with little risk involved, and the potential $1,577,100 in present equivalent cost savings is significant enough to justify this choice.

Capital Equipment Evaluation

Suppose that a manufacturing firm is considering the possibility of introducing a new product which is expected to meet sales expectations for at least ten years. The firm must invest in capital equipment in order to manufacture the product. Based on a survey of potential sources, there are two equipment alternatives considered to be feasible.

   Table 17.3 gives the projected cash flow for the two equipment alternatives (both revenues and costs are given). A six-year life cycle and an interest rate of 10% are assumed. There is an initial investment of $15,000 for equipment *A* and $20,000 for equipment *B*. Benefits (in terms of anticipated net revenues) and costs are shown in the table along with the interest factors from Appendix D, Table D.6. The net present equivalent value (NPV) for equipment *A* is $30,486, and the NPV for equipment *B* is $7,960. Thus, equipment *A* is preferred.

TABLE 17.3 NET PRESENT-VALUE COMPARISON

EQUIPMENT A

	Cash Flow		P/F, 10, n	Net Present Value	
Year, n	Benefits	Costs	( )	Benefits	Costs
0	—	$15,000	1.0000	—	$15,000.00
1	—	6,000	0.9091	—	5,454.60
2	$ 5,000	3,000	0.8264	$ 4,132.00	2,479.20
3	$12,000	—	0.7513	9,015.60	—
4	16,500	—	0.6830	11,269.50	—
5	25,800	—	0.6209	16,019.22	—
6	23,000	—	0.5645	12,983.50	—
Total	$82,300	$24,000		$53,419.82	$22,933.80

EQUIPMENT B

	Cash Flow		P/F, 10, n	Net Present Value	
Year, n	Benefits	Costs	( )	Benefits	Costs
0	—	$20,000	1.0000	—	$20,000.00
1	—	12,000	0.9091	—	10,909.20
2	$ 4,000	6,000	0.8264	$ 3,305.60	4,958.40
3	13,000	5,000	0.7513	9,766.90	3,756.50
4	17,000	3,000	0.6830	11,611.00	2,049.00
5	22,000	—	0.6209	13,659.80	—
6	20,000	—	0.5645	11,290.00	—
Total	$76,000	$46,000		$49,633.30	$41,673.10

Prior to making a final choice, the analyst should perform a break-even analysis and establish the payback points for each project. Although other factors may affect the ultimate decision, the project exhibiting an early payback is usually desirable when considering risk and uncertainty. Figure 17.19 illustrates cash flows and payback points, and project *A* retains its preferred status.

Evaluating System Design Alternatives

A ground vehicle currently under development requires the incorporation of radio communication equipment. A decision is needed as to the type of equipment deemed most feasible from the standpoint of performance, reliability, and life-cycle cost. Budget limitations require that the equipment unit cost (based on life-cycle cost) not exceed $20,000. Review of all possible supplier sources indicates that there are two design configurations that appear (based on preliminary design data) to meet the specified requirements. Each is to be evaluated on an equivalent basis in terms of reliability and total life-cycle cost.

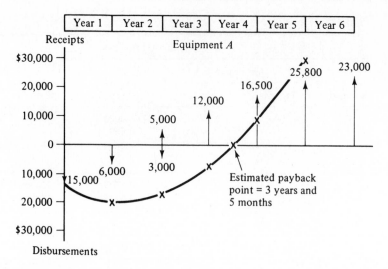

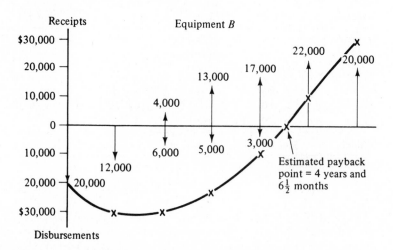

Figure 17.19   Cash flow and payback comparisons.

The communication equipment is to be installed in a light vehicle. The equipment shall enable communication with other vehicles at a range of 200 miles, overhead aircraft at an altitude of 10,000 ft or less, and a centralized area communication facility. The system must have a reliability MTBF of 450 hours, a $\overline{M}ct$ of 30 minutes, and a MLH/OH requirement of 0.2. The operational and maintenance concepts and program time frame are illustrated in Figure 17.20.

The decision maker needs to evaluate each of the two proposed design configurations on the basis of total cost. Research and development costs and investment (or production) costs are determined from engineering and manufacturing cost estimates prepared by the respective suppliers. Operation and maintenance cost is

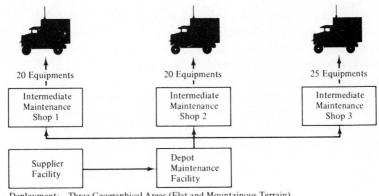

20 Equipments   20 Equipments   25 Equipments

| Intermediate Maintenance Shop 1 | Intermediate Maintenance Shop 2 | Intermediate Maintenance Shop 3 |

| Supplier Facility | Depot Maintenance Facility |

Deployment: Three Geographical Areas (Flat and Mountainous Terrain)
Utilization: Four (4) hr/day Throughout Year (Average)

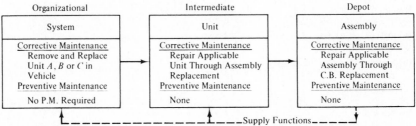

Organizational	Intermediate	Depot
System	Unit	Assembly
Corrective Maintenance Remove and Replace Unit A, B or C in Vehicle Preventive Maintenance No P.M. Required	Corrective Maintenance Repair Applicable Unit Through Assembly Replacement Preventive Maintenance None	Corrective Maintenance Repair Applicable Assembly Through C.B. Replacement Preventive Maintenance None

Supply Functions

The Illustrated Maintenance Concept should be expanded to include such factors as
$\overline{M}_{ct}$, TAT, MMH/OH, Pipeline, etc., for each level. Refer to Chapter 9, Figure 3.5.

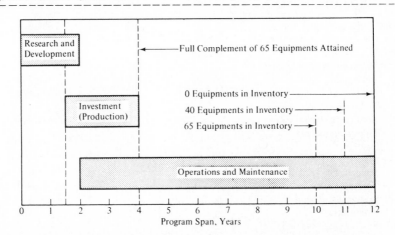

Research and Development

Full Complement of 65 Equipments Attained

Investment (Production)

0 Equipments in Inventory
40 Equipments in Inventory
65 Equipments in Inventory

Operations and Maintenance

Program Span, Years

Figure 17.20   System concepts for radio communication equipment.

also provided based on reliability prediction reports, maintainability prediction reports, and the results of the logistic support analysis. These costs are related to the cost breakdown structure, and the analysis is accomplished following the steps described earlier.

A breakdown of life-cycle cost is presented in Table 17.4. The CBS used here is different than that illustrated in Figure 17.6 in that it is tailored to the evaluation being made. Note that the acquisition cost (Research and Development and Investment) is higher for configuration *A* ($478,033 versus $384,131). This is partially due to a better design using more reliable components. Although the initial cost is higher, the overall life-cycle cost is lower, owing to a reduction in anticipated maintenance actions which result in lower operation and maintenance costs. Thus, configuration *A* is preferred.

TABLE 17.4 LIFE-CYCLE COST BREAKDOWN

	CONFIGURATION *A*		CONFIGURATION *B*	
COST CATEGORY	Present Cost	% of Total	Present Cost	% of Total
1. Research and development	$ 70,219	7.8	$ 53,246	4.2
(a) Program management	9,374	1.1	9,252	0.8
(b) Advanced *R* and *D*	4,152	0.5	4,150	0.4
(c) Engineering design	41,400	4.5	24,581	1.9
(d) Equipment development				
and test	12,176	1.4	12,153	0.9
(e) Engineering data	3,117	0.3	3,110	0.2
2. Investment	407,814	45.3	330,885	26.1
(a) Manufacturing	333,994	37.1	262,504	20.8
(b) Construction	45,553	5.1	43,227	3.4
(c) Initial logistic support	28,267	3.1	25,154	1.9
3. Operations and maintenance	422,217	46.9	883,629	69.7
(a) Operations	37,811	4.2	39,301	3.1
(b) Maintenance	384,406	42.7	844,328	66.6
• Maintenance personnel				
and support	210,659	23.4	407,219	32.2
• Spare/repair parts	103,520	11.5	228,926	18.1
• Test and support				
equipment maintenance	47,713	5.3	131,747	10.4
• Transportation and				
handling	14,404	1.6	51,838	4.1
• Maintenance training	1,808	0.2	2,125	Neg.
• Maintenance facilities	900	0.1	1,021	Neg.
• Technical data	5,402	0.6	21,452	1.7
(c) System/equipment				
modifications	. . . .	. . . .	. . . .	. . . .
(d) System phase-out and				
disposal	. . . .	. . . .	. . . .	. . . .
GRAND TOTAL	$900,250	100%	$1,267,760	100%

Figure 17.21 shows the relationship of the two configurations in terms of the specified reliability requirement and the design to unit life-cycle cost goal. Both configurations meet the specified requirements. However, configuration A is clearly preferred, owing to a higher reliability and a lower unit life-cycle cost.

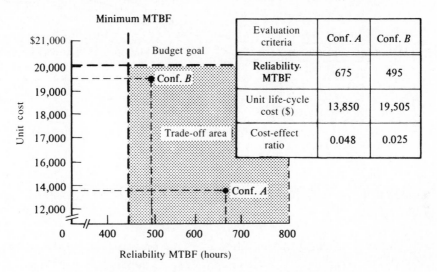

Evaluation criteria	Conf. *A*	Conf. *B*
Reliability, MTBF	675	495
Unit life-cycle cost ($)	13,850	19,505
Cost-effect ratio	0.048	0.025

Figure 17.21   Reliability versus unit life-cycle cost.

As stated, configuration A is preferred on the basis of total life-cycle cost. Prior to a final decision, however, the analyst should perform a break-even analysis to determine the point in time that configuration A becomes more desirable than configuration B. Figure 17.22 illustrates a payback point of six years and five months, or a little more than two years after the equipment is introduced. This point is early enough in the life cycle to support the decision.

By referring to Table 17.4, the analyst can readily pick out the high-cost contributors (those which contribute more than 10% of the total cost). These are the areas where a more refined analysis is required and greater emphasis is needed in providing valid input data. For instance, maintenance personnel and support cost (C_{OMM}) and spare/repair parts cost (C_{OMX}) contribute 23.4% and 11.5%, respectively, to the total cost for configuration *A*. This leads the analyst to reevaluate the design in terms of impact on personnel support and spares; the prediction methods used in determining maintenance frequencies and inventory requirements; the analytical model to ensure that the proper parameter relationships are established; and cost factors such as personnel labor cost, spares material costs, inventory holding cost; and so on. If the analyst wishes to determine the sensitivity of these areas to input variations, he or she may perform a sensitivity analysis. In this instance, it is appropriate to vary MTBF as a function of maintenance personnel and support cost (C_{OMM}) and spare/repair parts cost (C_{OMX}). Figure 17.23 presents the results of their sensitivity analysis.

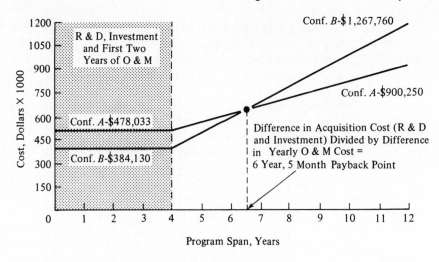

Figure 17.22   Investment payback (breakeven analysis).

MTBF Multiplier	P.V. Cost, Dollars
0.67	323,140
**1.00	210,659
1.33	162,325
2.00	112,565

**Baseline Configuration *A*

MTBF Multiplier	P.V. Cost, Dollars
0.67	199,576
**1.00	103,520
1.33	92,235
2.00	80,130

**Baseline Configuration *A*

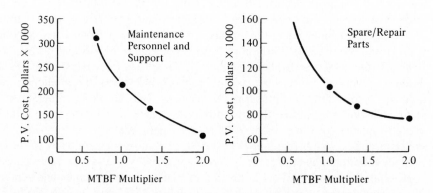

Figure 17.23   Sensitivity analysis.

The analyst (or decision maker) should review the break-even analysis in Figure 17.22 and determine how far out in time he or she is willing to go and remain with configuration A. Assuming that the selected maximum payback point is seven years, the difference in alternatives is equivalent to approximately $65,000 (the present-cost difference between the two configurations at the seven-year point). This indicates the range of input variations allowed. For instance, if design configuration A changes, or if the reliability prediction is in error resulting in a MTBF as low as 450 hours (the specified system requirement), the maintenance personnel and support cost (C_{OMM}) will increase to approximately $324,000. This is an increase of about $113,340 above the baseline value. Thus, although the system reliability is within the specified requirements, the cost increase due to the input MTBF variation causes a decision shift in favor of configuration B. The analyst should always assess the sensitivity of significant input parameters and determine their impact on the ultimate decision.

17.5. LIFE-CYCLE COSTING IN PROGRAM EVALUATION

In performing any life-cycle analysis, there are many factors to be considered, assumptions to be made, data to collect and evaluate, parameter relationships to establish, and so on. The process as such allows for the introduction of a great deal of error if one is not careful. Although it is obviously impossible to eliminate all error (and risk in terms of using the analysis in decision making), one should take every precaution to validate the analysis to the extent possible. Thus, in making a final recommendation as to a proposed course of action (based on analysis results), the analyst should review the overall analysis process in terms of problem definition, validity of stated assumptions, model parameter relationships, inclusions or exclusions, adequacy of data input, and stated conclusions. The main question is Does the preferred configuration *clearly* have the advantage over other alternative considerations? In response, the analyst may wish to pose a number of specific questions in the form of a checklist as an aid in assessing the final output results. Figure 17.24 presents a sample checklist, including some basic but significant questions that may pertain to any type of analysis effort.

Once the system design has been selected and program requirements established, it is essential that an ongoing review, evaluation, and control function be initiated. Management must conduct a periodic assessment to compare current life-cycle cost estimates with the specified LCC/DTC requirements. The selected review points should be a part of, or tied directly to, the normal scheduled program and/or design reviews. Usually, major reviews occur after a significant stage or level of activity in the program prior to entering the next phase. Figure 17.25 identified the major reviews for one type of program.

In preparation for each individual program review, a life-cycle cost analysis is accomplished for the system/product configuration as it exists at that time. A comparison is made between the target life-cycle cost and the estimated value based on

A. *Assumptions*
   1. Are all assumptions adequately identified?
   2. Do any of the specified assumptions treat quantitative uncertainties as facts?
   3. Do any of the specified assumptions treat qualitative uncertainties as facts?
   4. Are major assumptions reasonable?
B. *Alternatives*
   1. Are current capabilities adequately considered among alternatives?
   2. Are mixtures of system components considered among the alternatives?
   3. Are any feasible and significant alternatives omitted?
C. *Documentation*
   1. Is the study adequately documented?
   2. Are the facts stated correctly?
   3. Are the facts stated with proper qualification?
   4. Are the applicable reference sources listed?
D. *Model relationships*
   1. Does the model adequately address the problem?
   2. Are cost and effectiveness parameters linked logically?
   3. Does the model allow for a timely response?
   4. Does the model provide valid (comprehensive) and reliable (repeatable) results?
E. *Effectiveness parameters*
   1. Are the measures of effectiveness identified?
   2. Is the effectiveness measure appropriate to the mission function? Are operational and maintenance concepts adequately defined?
   3. Do the effectiveness measures employed ignore some objectives and concentrate on others?
   4. Are performance measures mistaken for effectiveness measures?
   5. Does the effectiveness of a future system take into account the time dimension?
   6. Are expected and average values used correctly to measure effectiveness?
   7. If quantitative measures of effectiveness are unattainable, is a qualitative comparison feasible?
   8. Is the effectiveness measure sensitive to changes in assumptions?
   9. In the event that two or more effectiveness measures are appropriate, are the measures properly weighted (the relative weighting in terms of significance or level of importance of each applicable criterion factor employed)?
F. *Cost*
   1. Is the cost model employed adequately described?
   2. Are cost categories adequately defined?
   3. Are cost estimates relevant?
   4. Are incremental and marginal costs considered?
   5. Are variable and fixed costs separately identifiable?
   6. Are escalation factors specified and employed?
   7. Is the discount rate specified and employed?

Figure 17.24   Sample Checklist

8. Are all costs elements considered?
   (a) Feasibility studies
   (b) Design and development
   (c) Production and test
   (d) Installation and checkout
   (e) Personnel and training
   (f) Technical data
   (g) Facility construction and maintenance
   (h) Spare/repair parts
   (i) Support equipment/tools
   (j) Inventory maintenance
   (k) Customer support (field service)
   (l) Program management
9. Are the cost aspects of all alternatives treated in a consistent and comparable manner?
10. Are the cost estimates (cost estimating relationships) reasonably accurate? Are areas of risk and uncertainty identified?
11. Is cost amortization employed? If so, how?
12. Has the sensitivity of cost estimates been properly addressed through a sensitivity analysis?

G. *Conclusions and recommendations*
1. Are the conclusions and recommendations logically derived from the material contained in the study?
2. Have all the significant ramifications been considered in arriving at the conclusions and recommendations presented?
3. Are the conclusions and recommendations really feasible in light of political, cultural, policy or other considerations?
4. Do the conclusions and recommendations indicate bias?
5. Are the conclusions and recommendations based on external considerations?
6. Are the conclusions and recommendations based on insignificant differences?

Figure 17.24   (continued)

the analysis. Figure 17.25 illustrates the results at the third formal review scheduled one year after program start. The projections convey expected cumulative expenditures. Note that the illustration indicates that actual research and development cost exceeds the initial projection (i.e., the costs in the first year); the expected cost at the end of production will exceed the anticipated value; and that the life cycle cost at the end of the ten-year life period will be less than the target due to a reduction in operation and support cost.

If the projected life-cycle cost at any given program review exceeds the target value, one should evaluate the different categories of cost in the CBS and identify the high-cost contributors. Also, one should determine the relative relationships of these costs with each other, and with the overall life-cycle cost figure, to see if any

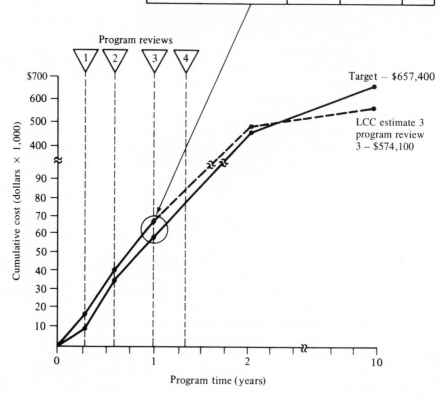

Figure 17.25 Program review based on life-cycle cost estimates.

significant changes have occurred since the previous review. Areas of concern should be investigated in terms of the possible cause(s) for the high cost, and recommendations for corrective action should be initiated where appropriate. Such recommendations may take the form of design changes, production or process changes, logistic support policy revisions, and/or changes involving the management of resources. Recommendations where significant life-cycle cost reduction can be realized should be documented and submitted for management action.

The program review and control function is an iterative process of life-cycle cost assessment. Inherent in this process is not only the accomplishment of life-cycle cost analyses, but the feedback and corrective action that is required when problems occur and requirements are not being met. Management emphasis is necessary in both areas. It is not sufficient to merely review and assess a system or product in terms of life-cycle cost unless corrective measures can and will be taken when required.

SELECTED REFERENCES

(1) Blanchard, B. S., *Design and Manage to Life Cycle Cost*. Forest Grove, Ore.: M/A Press, 1978.

(2) Blanchard, B. S., *Logistics Engineering and Management* (3rd ed.). Englewood Cliffs, N.J.: Prentice-Hall, 1986.

(3) DARCOM P700–6 (Army), NAVMAT P5242 (Navy), AFLCP/AFSCP 800–19 (Air Force), *Joint-Design-to-Cost Guide, Life Cycle Cost as a Design Parameter*. Department of Defense, 1977.

(4) DOD Directive 4245.3, *Design to Cost*. Department of Defense.

(5) Earles, M. E., *Factors, Formulas and Structures for Life Cycle Costing*. Concord, Mass.: Eddins-Earles, 1978.

(6) English, J. M., ed., *Economics of Engineering and Social Systems*. New York: John Wiley, 1972.

(7) Fabrycky, W. J., and B. S. Blanchard, *Life-Cycle Cost and Economic Analysis*. Englewood Cliffs, N.J.: Prentice-Hall, 1991.

(8) Fabrycky, W. J., and G. J. Thuesen, *Economic Decision Analysis* (2nd ed.). Englewood Cliffs, N.J.: Prentice-Hall, 1980.

(9) Fisher, G. H., *Cost Considerations in System Analysis*. New York: American Elsevier, 1971.

(10) Grant, E. L., W. G. Ireson, and R. S. Leavenworth, *Principles of Engineering Economy*, (7th ed.). John Wiley, 1982.

(11) Humphreys, K. K., and S. Katell, *Basic Cost Engineering*. New York: Marcel Dekker, 1981.

(12) Jelen, F. C., and J. H. Black, *Cost and Optimization Engineering*. New York: McGraw-Hill, 1983.

(13) Kendall, M. G., ed., *Cost-Benefit Analyses*. New York: American Elsevier, 1971.

(14) MIL-STD-1390B, "Level Of Repair", Naval Publications Center, 5801 Tabor Avenue, Philadelphia, Pennsylvania, 19120.

(15) Ostwald, P. F., *Cost Estimating* (2nd ed.). Englewood Cliffs, N.J.: Prentice-Hall, 1984.

(16) Riggs, J. L., *Engineering Economics* (2nd ed.). New York: McGraw-Hill, 1982.

(17) Stewart, R. D., *Cost Estimating*. New York: John Wiley, 1982.

(18) Stewart, R. D., and R. M. Wyskida, *Cost Estimator's Reference Manual*. New York: John Wiley, 1987.

(19) Thuesen, G. J., and W. J. Fabrycky, *Engineering Economy* (7th ed.). Englewood Cliffs, N.J.: Prentice-Hall, 1989.

QUESTIONS AND PROBLEMS

1. Select a system or product of your choice and define the life cycle. Illustrate the life cycle in a manner similar to the approach conveyed in Figure 2.1.

2. Describe life-cycle cost in your own words and discuss what is included (excluded). Describe DTC, cost effectiveness, life-cycle cost analysis, and economic feasibility.

3. How would you create cost emphasis in the planning and development of a new system or product?

4. What is the purpose of cost allocation? (Refer to Figure 17.5.)

5. Why do we need to consider life-cycle cost? Name some of the benefits associated with life-cycle costing.

6. Describe the following in your own words: total cost, unit cost, indirect cost, variable cost, fixed cost, recurring cost, nonrecurring cost, incremental cost, functional cost, and marginal cost.

7. What is the cost breakdown structure? What purpose does the CBS serve? What characteristics should be incorporated in the CBS? How does the CBS relate to the work breakdown structure?

8. Describe some of the more commonly used cost estimating methods. Under what conditions should they be applied? Provide some examples.

9. Discuss how you would treat inflation in determining life-cycle cost.

10. How would you apply learning curves in the system/product life-cycle? How do learning curves affect LCC? What factors influence learning curves?

11. How can life-cycle cost analysis be successfully employed as a management tool in system/product planning, development, production, operational use, and logistic support?

12. Assume that the average system maintenance cost is $5,000 per year for the next 10 years. What is the present value of the maintenance cost stream? The interest rate is 10%.

13. What is the present value of the cost stream illustrated using an interest factor of 8%?

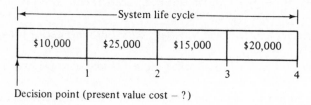

14. What is a break-even analysis? Why is it important?

15. Two different projects are identified by the cost streams illustrated. Using a 10% discount rate, determine which project is preferred. At what point in time does the preferred project assume a favorable position? Illustrate by accomplishing a break-even analysis.

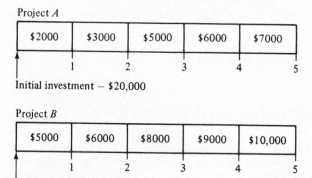

16. Calculate the anticipated life-cycle cost for your personal automobile.

17. Refer to example 3 (Figure 17.20 and Table 17.4). What would be the likely impact on LCC if

   (a) The system MTBF is decreased?

   (b) The $\overline{M}$ct is increased?

   (c) The MLH/OH is increased?

   (d) System utilization is increased?

   (e) The fault-isolation capability in the system were inadequate?

   (f) The transportation time between the organizational and intermediate levels of maintenance is increased?

   (g) The turnaround time at the intermediate maintenance shop is decreased?

   (h) The reliability of the test and support equipment at the intermediate maintenance shop is decreased?

18. The ABC Corporation is considering the possibility of introducing product X into the market. A market analysis indicates that the corporation could sell over 500 of these products per year (if available) at a price of $50 each for at least ten years in the future. The product is not repairable and is discarded at failure. To manufacture the product, the corporation needs to invest in some capital equipment. Based on a survey of potential sources, there are three alternatives considered feasible to meet the need.

   (1) Configuration A includes a machine that is automatic, will produce up to 350 products per year, and can be purchased at a price of $11,050. The expected reliability (MTBF) of this machine is 210 hours, and the anticipated cost per corrective maintenance action is $75. Preventive maintenance is required every six months, and the cost per maintenance action is $50. The average machine operation cost is $0.50 per hour, and the estimated salvage value after ten years of operation is $300.

   (2) Configuration B includes a machine that is also automatic, will produce 440 products per year, and can be purchased at a price of $9,725. The expected reliability (MTBF) of this machine is 385 hours, and the anticipated cost per corrective maintenance action is $100. Preventive maintenance is required every six months, and the cost per maintenance action is $60. The average machine operating cost is $0.65 per hour, and the estimated salvage value after ten years of operation is $550.

   (3) Configuration C includes a machine that is semiautomatic (requiring a part-time machine operator), will produce 300 products per year, and can be purchased at a price of $5,075. The expected reliability (MTBF) of this machine is 150 hours, and the anticipated cost per corrective maintenance action is $50. Preventive maintenance is required every six months, and the cost per maintenance action is $80. The average machine operating cost is $4.20 per hour, and the estimated salvage value after ten years of operation is $100.

   The expected machine utilization will be eight hours per day and 270 days per year. Machine output is expected to be one-half (0.5) during the first year of operation, at full capacity during year two, and on. The manufacturing cost (materials and labor associated with material procurement, material handling, quality control, inspection and test, and packaging) for product X is $20 using machine A, $30 using machine B, and $40 using machine C. The expected allocated distribution cost (transportation, warehousing, etc.) is $5 per item.

   Evaluate each of the three configurations and select one on the basis of life-cycle cost. An 8% interest rate is assumed for evaluation purposes. Construct break-even curves showing the preferences in terms of time.

19. Based on the information provided below, compute the life-cycle cost in terms of present value using a 10% discount factor for System *XYZ*. Indicate the total value at the start of the program (decision point) and plot the cost stream or profile.

System *XYZ* is installed in an aircraft that will be deployed at five operational bases. Each base will have a maximum force level of 12 aircraft, with the bases being activated in series (e.g., base 1 at the end of year three, base two in year five, etc.). The total number of System *XYZ*s in operation are

YEAR NUMBER									
1	2	3	4	5	6	7	8	9	10
0	0	0	10	20	40	60	55	35	25

System *XYZ* is a newly designed configuration packaged in three units (unit *A*, unit *B*, and unit *C*) with the following specified requirements for each of the units.

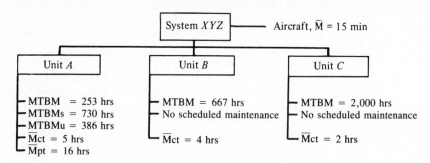

The average System *XYZ* use is four hours per day and units *A*, *B*, and *C* are operating 100% of the time when system *XYZ* is on. One of the aircraft crew members will be assigned to operate several different systems throughout flight and it is assumed that 10% of his time is allocated to System *XYZ*.

Relative to the maintenance concept, System *XYZ* incorporates a built-in self-test that enables rapid system checkout and fault isolation to the unit level. No external support equipment is required at the aircraft. In the event of a *no-go* condition, fault isolation is accomplished to the unit and the applicable unit is removed, replaced with a spare, and sent to an intermediate-level maintenance shop (located at the operational base) for corrective maintenance. Unit repair is accomplished through module replacement, with the modules being discarded at failure. Scheduled (preventive) maintenance is accomplished on unit A in the intermediate shop every 6 months. No depot maintenance is required; however, the depot does provide backup supply and support functions as required.

The requirements for System *XYZ* dictate the program profile shown. Assume that life-cycle costs are broken down into three categories, represented by the blocks in the program profile.

In an attempt to simplify the problem, the following additional factors are provided.
(a) RDT & E costs for System *XYZ* (to include labor and material) are $100,000 for year one, $200,000 for year two, and $250,000 for year three.
(b) RDT & E costs for special support equipment are $50,000 in year two and $10,000 in year three.

Year Number									
1	2	3	4	5	6	7	8	9	10

Research, Design, Test, Evaluation.

Production

Operations and Maintenance

(c) System *XYZ* operational models are produced, delivered, and purchased in the year prior to the operational deployment need. System unit costs are unit A = $6,000, unit B = $3,000, and unit C = $1,000. Recurring and nonrecurring manufacturing costs are amortized on a unit basis.

(d) Support equipment is required at each intermediate maintenance shop at the start of the year when System *XYZ* operational models are deployed. In addition, a backup support equipment set is required at the depot when the first operational base is activated. The cost per set of support equipment is $20,000. Support equipment maintenance is based on a burden rate of $0.50 per direct maintenance labor hour for the prime equipment.

(e) Spares units are required at each intermediate maintenance shop at the time of base activation. Assume that two unit As, one unit B, and one unit C are provided at each shop as safety stock. Also, assume that one spare System *XYZ* is stocked at the depot for backup support.

Additional spares constitute modules. Assume that material costs are $100 per corrective maintenance action and $50 per preventive maintenance action. This includes inventory maintenance costs.

In the interests of simplicity, the effects of the total logistics pipeline and shop turnaround time on spares are ignored in this problem.

(f) For each maintenance action at the system level, one low-skilled technician at $9.00 per direct maintenance labor-hour is required on a full-time basis. $\overline{M}$ is 15 minutes.

For each corrective maintenance action involving unit A, unit B, or unit C, two technicians are required on a full-time basis. One technician is low-skilled at $9.00 per hour and one technician is high-skilled at $11.00 per hour. Direct and indirect costs are included in these rates. For each preventive maintenance action, one high-skilled technician at $11.00 per hour is required on a full-time basis.

(g) System operator personnel costs are $12.00 per hour.

(h) Facility costs are based on a burden rate of $0.20 per direct maintenance labor hour associated with the prime equipment.

(i) Maintenance data costs are assumed to be $20 per maintenance action.

Assume that the design-to-unit-acquisition cost (i.e., unit flyaway cost) requirement is $15,000. Has this requirement been met? Assume that the design-to-unit-O&M cost requirement is $20,000. Has this requirement been met?

In solving this problem, be sure to state all assumptions in a clear and concise manner.

PART V: SYSTEMS ENGINEERING MANAGEMENT

18

Systems Engineering Management

Systems engineering management involves planning, organizing, staffing, monitoring, and controlling the process of designing, developing, and producing a system that will meet a stated need in an effective and efficient manner. It provides the necessary overview function(s) to ensure that all needed engineering disciplines and related specialties are properly integrated. It ensures that the system being developed contains the proper mix of hardware, software, facilities, personnel, data, and so on. The objective of systems engineering management is to provide the right item, at the right location, at the right time, with a minimum expenditure of human and physical resources.

Systems engineering management, as presented in this chapter, relates primarily to those functions required for the successful implementation of the process illustrated in Figure 2.4. This includes the necessary system planning activities in the conceptual design phase, the appropriate integration of preliminary system design and detail design functions, and the monitoring and control functions to ensure that the desired output will be attained. It also includes a presentation of important interfaces between the consumer, producer, and supplier which must be managed.

18.1. SYSTEMS ENGINEERING PLANNING

Referring to Figure 2.4, early system planning starts with the identification of a need and extends through conceptual design into preliminary design. The results of such planning may be classified in terms of *technical* requirements and *management* requirements. The relationship between technical requirements (presented in the form of specifications) and management requirements (included in various program plans) are identified in Figure 3.8. Of particular interest for system engineering objectives are the System Specification (Type A Specification) and the System Engineering Management Plan (SEMP).

The relationships between specifications and plans (introduced in Figure 3.8) has been expanded as presented in Figure 18.1. The System Specification and the SEMP must "talk" to each other. The management approach must support technical objectives, and technical objectives must be specified in a realistic manner, consistent with organizational and scheduling goals. While both documents are important, the emphasis in this chapter will be on the SEMP (Section 3.3 covers the basic contents of the System Specification).

The System Engineering Management Plan (SEMP) is the key management document, covering those activities and milestones necessary to accomplish the system engineering objectives discussed throughout this text. The SEMP directly supports the Program Management Plan (PMP) which, for discussion purposes, is considered to be the top management document for a given program or project. The objectives of the SEMP are to provide the structure, policies, and procedures to foster the integration of the various engineering-related activities needed for system design and development. Referring to Figure 18.1, the SEMP is developed to provide the integration of a number of individual program plans (reliability program plan, maintainability program plan, etc.) and to promote the communication with other top-level planning documents (configuration management plan, test and evaluation master plan, etc.). A proposed outline for the SEMP is illustrated in Figure 18.2.[1]

As indicated in Figure 18.2, the proposed format contains three basic parts. Part I includes a statement of work, an organization structure, a work breakdown structure (WBS), a listing of tasks, a program schedule, cost projections, and all other management functions necessary in carrying out a successful program in response to system engineering objectives. Part II describes the system engineering process, or those functions illustrated in Figure 2.4 and described in Chapters 3, 4, 5, and 6. Part III covers the integration of key individual program plans that address engineering design-related functions. A few of these plans are discussed in Chapter 13 (Reliability Program Plan), Chapter 14 (Maintainability Program Plan), Chapter 15 (Human Factors Program Plan), and Chapter 16 (Integrated Logistic Support Plan).

As indicated, the SEMP is developed early in conceptual design, and is usually completed in time to provide the necessary top-down guidance as the program pro-

---

[1] The proposed SEMP outline is consistent with the outline format presented in the Defense Systems Management College publication, *Systems Engineering Management Guide,* Fort Belvoir, Virginia 22060, December 1986.

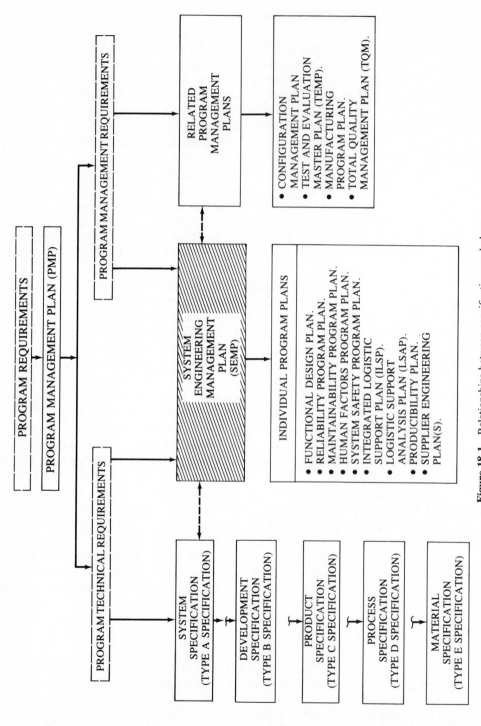

Figure 18.1 Relationships between specifications and plans

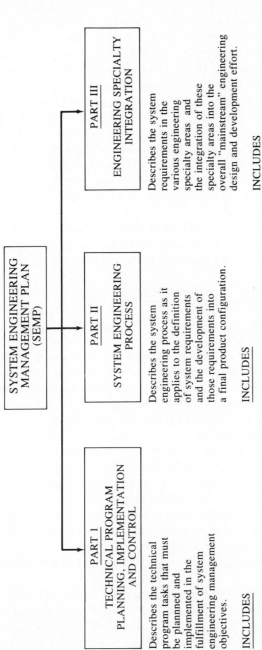

SYSTEM ENGINEERING MANAGEMENT PLAN (SEMP)

PART 1
TECHNICAL PROGRAM PLANNING, IMPLEMENTATION AND CONTROL

Describes the technical program tasks that must be plannned and implemented in the fulfillment of system engineering management objectives.

INCLUDES

- Program requirements (statement of work).
- Organization.
- Supplier or subcontractor requirements (contracting).
- Program interfaces.
- Work breakdown structure (WBS).
- Scheduling.
- Technical performance measurement (TPM).
- Cost estimating/reporting.
- Program monitoring/control.
- Risk management.

PART II
SYSTEM ENGINEERING PROCESS

Describes the system engineering process as it applies to the definition of system requirements and the development of those requirements into a final product configuration.

INCLUDES

- Identification of need.
- Feasibility analysis.
- Operational requirements.
- System maintenance concept.
- Functional analysis.
- Requirements allocation.
- System analysis and trade-offs (optimization).
- System design.
- System test and evaluation.

PART III
ENGINEERING SPECIALTY INTEGRATION

Describes the system requirements in the various engineering specialty areas and the integration of these specialty areas into the overall "mainstream" engineering design and development effort.

INCLUDES

- Reliability engineering.
- Maintainability engineering.
- Human factors engineering.
- Safety engineering.
- Value/cost engineering.
- Logistics engineering.
- Components engineering.
- Producibility.
- Quality engineering.
- Other engineering and design-related disciplines.

Figure 18.2   The system engineering management plan (SEMP)

gresses through the steps described in Part Two of this text. Figure 3.9 presents some major program milestones (to include the SEMP) in the context of the overall system acquisition process. Figure 18.3 extends this process by identifying some major system engineering activities in the form of a sample milestone chart for a typical program. Basically, tasks A1 through A7 reflect those activities that normally occur during conceptual design, tasks B1 through B7 constitute the major activities accomplished during preliminary design, and so on.

Although a milestone chart is sufficient in many instances, its use does not always force the task integration that must occur to meet systems engineering requirements. Therefore, one needs to employ a scheduling technique that identifies the input-output needs for individual task accomplishment and shows task interrelationships. Further, the employment of a technique suitable to activities early in the life cycle (where precise times are not readily available) is desirable. The networking approach or PERT-CPM (program evaluation and review technique, critical path method) is the appropriate scheduling technique for this purpose. PERT effectively shows the interrelations of combined activities and introduces the aspects of probability by which better management decisions can be made. Figure 18.4 presents a partial example of a PERT network as related to conceptual and preliminary design activities.[2]

In applying PERT to a project, one must identify all interdependent activities and events for each applicable phase of the project. Events are related to the program milestone dates based on management objectives. Managers and programmers work with engineering organizations to define these objectives and identify specific tasks and subtasks. When this is accomplished to the necessary level of detail, networks are developed which start with a summary network and work down to detail networks covering specific segments of a program. The partial network illustrated in Figure 18.4 represents a summary network covering the key system engineering activities reflected in Figure 18.3. A detail network, for example, may constitute a reliability program evolving from event B6.

When actually constructing networks, one starts with an end objective (e.g., system delivered to the consumer) and works backward until the beginning or starting event is identified. Each event is labeled, coded, and checked in terms of program time frame. Activities are then identified and checked to ensure that they are properly sequenced. Activity times are estimated and these times are stated in terms of their probability of occurrence. Some activities can be performed on a concurrent basis, but others must be accomplished in series. For each completed network, there is one beginning event and one ending event. All activities must lead to the ending event.

The application of PERT scheduling is appropriate for both small- and large-scale projects and is of particular value for one-of-a-kind system or for those program phases where repetitive tasks are not predominant. PERT is readily adaptable to advance planning and forces the precise definition of tasks, task sequences, and task interrelationships. The technique enables management and engineering to pre-

---

[2] As this chapter presents only a *survey* of key system engineering management techniques, the student is advised to become familiar with networking techniques through further review of available texts on project management, industrial engineering, and the like. See Section 12.4.

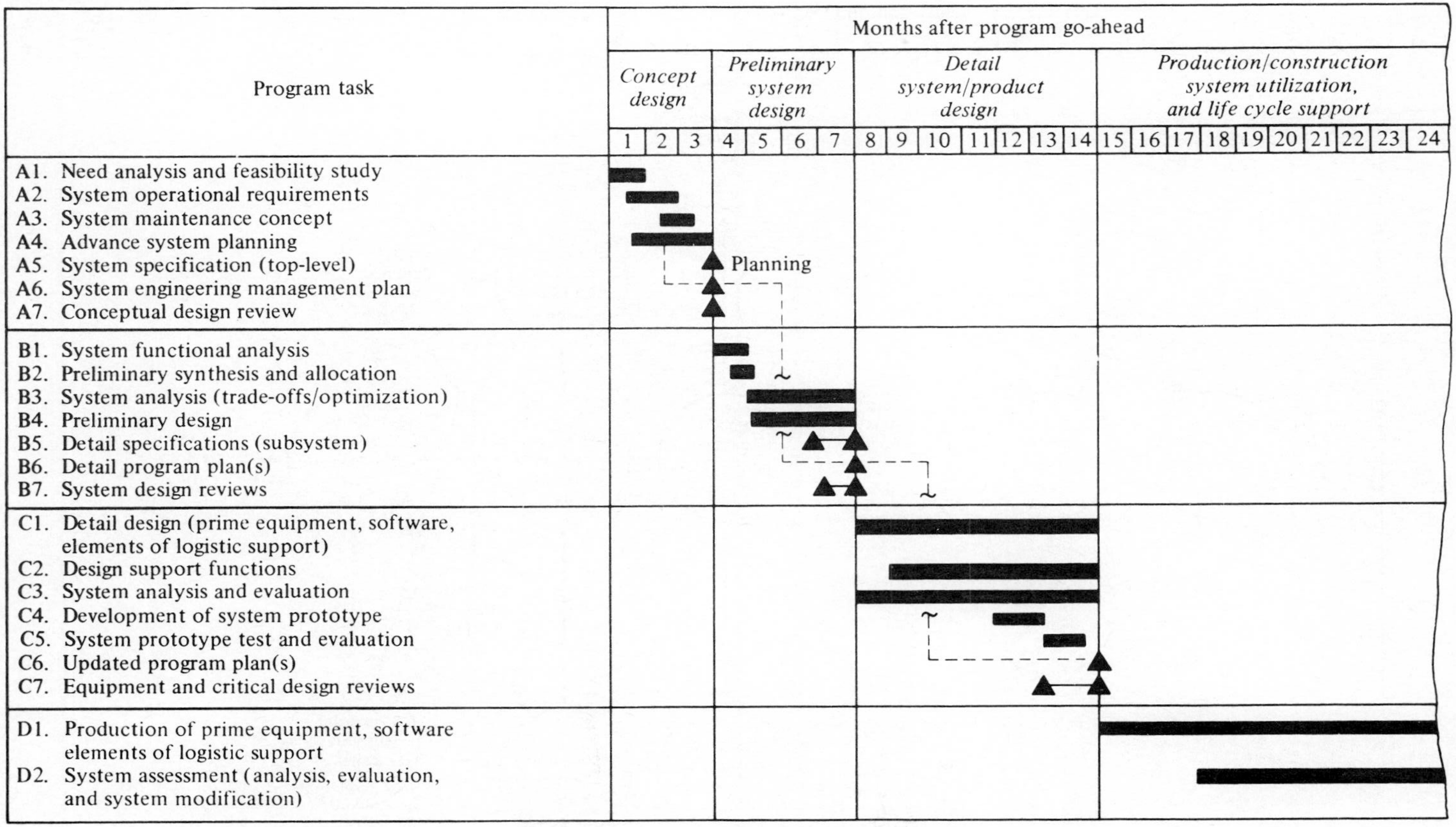

	Months after program go-ahead																							
Program task	Concept design			Preliminary system design				Detail system/product design								Production/construction system utilization, and life cycle support								
	1	2	3	4	5	6	7	8	9	10	11	12	13	14	15	16	17	18	19	20	21	22	23	24

Planning

A1. Need analysis and feasibility study
A2. System operational requirements
A3. System maintenance concept
A4. Advance system planning
A5. System specification (top-level)
A6. System engineering management plan
A7. Conceptual design review

B1. System functional analysis
B2. Preliminary synthesis and allocation
B3. System analysis (trade-offs/optimization)
B4. Preliminary design
B5. Detail specifications (subsystem)
B6. Detail program plan(s)
B7. System design reviews

C1. Detail design (prime equipment, software, elements of logistic support)
C2. Design support functions
C3. System analysis and evaluation
C4. Development of system prototype
C5. System prototype test and evaluation
C6. Updated program plan(s)
C7. Equipment and critical design reviews

D1. Production of prime equipment, software elements of logistic support
D2. System assessment (analysis, evaluation, and system modification)

Figure 18.3 Basic milestone chart.

553

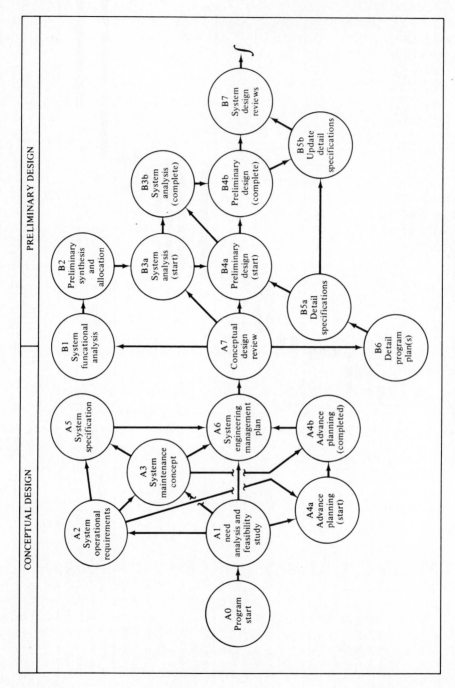

Figure 18.4 Partial summary program network.

dict with some degree of certainty the probable time that it will take to achieve an objective. It also enables the rapid assessment of progress and the detection of problems and delays. Thus, PERT is a valuable scheduling technique and is particularly adaptable to computer methods.

Other network scheduling techniques include CPM (critical path method) and PERT-COST. CPM is analogous to PERT except that emphasis is placed on critical activities (those activities requiring the greatest amount of time for completion). The critical path represents the longest path through a network and identifies those activities which may pose problems of both a technical and administrative nature if schedule slippage occurs. These are the activities that must be closely monitored and controlled throughout the program.

PERT-COST, an extension of PERT and CPM, deals not only with the element of time but includes cost. A cost network can be superimposed upon a PERT network by estimating the total cost and cost function for each activity line. Figure 18.5 shows a sample activity cost function. Such functions can be generated for the entire network.

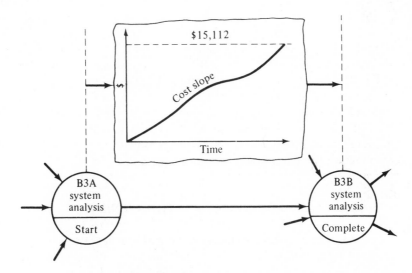

Figure 18.5   Activity-cost function.

When using the PERT-COST technique, there is a time-cost option which enables management to evaluate alternatives relative to the allocation of resources for activity accomplishment. In many instances, time can be saved by applying more resources or, conversely, cost may be reduced by extending the time to complete an activity. Time and cost alternatives are evaluated with the objective of selecting the lowest-cost approach within the maximum allowable time limit.

The events and activities discussed constitute tasks that may be organized into work packages. The identified work packages are then evaluated from the standpoint of task type, complexity, and the required completion schedule. In addition, estimated costs are established for each activity and are related to work packages. Figure 18.6 illustrates the work package concept as derived from an overall program

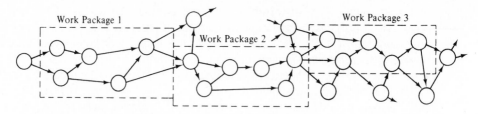

Figure 18.6 Work package identification.

network. The individual work packages are ultimately combined and integrated with the program work breakdown structure (WBS). The WBS links objectives and tasks with resources and is an excellent management tool for program planning and budgeting. A partial sample WBS is presented in Figure 18.6[3]

The WBS is a logical separation of work-related units. Work packages are identified against each block (see Block 3A1100, Figure 18.7, for an example of work package identification by block). These packages and WBS blocks are then related to organizational groups, departments, small project sections, suppliers, and so on. Cost estimating is accomplished for each work package and identified by WBS block. The WBS is structured and coded in such a manner that program costs may be initially targeted and then collected against each block.[4] Costs may be accumulated both vertically and horizontally to provide summary amounts for various categories of work. This can be accomplished expeditiously by developing an effective WBS project coding or numbering system and by employing computerized methods for input-output data manipulations. These cost data combined with the program milestone charts and networks provide management with the necessary tools for program planning, evaluation, and control.

18.2. ORGANIZATION FOR SYSTEMS ENGINEERING

A second major aspect of systems engineering management is that of organizing and staffing. Organization is necessary in response to the planning activity discussed above and consists of (1) determining what activities need to be accomplished; (2) grouping the identified activities in terms of a functionally-oriented structure of some type (unit, group, department, division, section); and (3) staffing the structure with the appropriate personnel skills to perform the designated activities in a coordinated manner.

---

[3] In developing a WBS, *all* program tasks must be covered. In addition, *all* elements of the system (like prime mission equipment, software, elements of logistic support) must be identified in the various categories of the WBS.

[4] In terms of cost collection, the WBS and cost breakdown structure used in life-cycle cost analyses (refer to Chapter 17) are quite similar. The WBS serves as a program management tool and is often used for contracting purposes. The cost breakdown structure is a tool used in performing system or product analysis and evaluation. Although their purposes are quite different, many of the cost input factors are the same. The two efforts should be coordinated to the maximum extent possible to ensure compatibility in cost estimating.

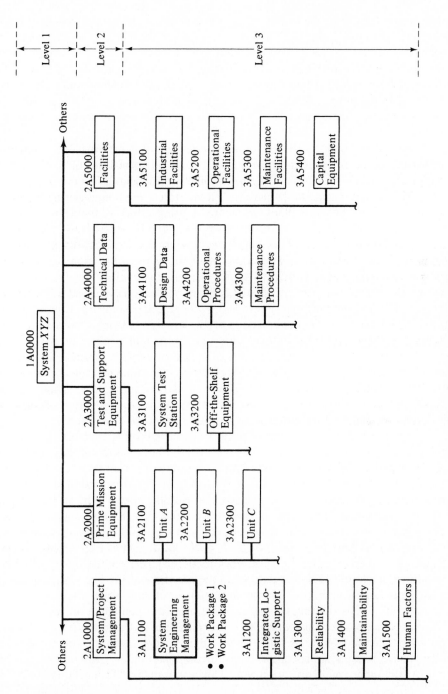

Figure 18.7 Partial work breakdown structure.

The establishment of an organization structure for a system project generally depends on the project objectives, the type of system and extent of new development required, the tasks to be accomplished, and the project schedule requirements. The manager responsible for the organization may develop a completely new structure, restructure an existing organization, or take over a current capability as is. Subsequently, the initially defined organization may change as the system moves from conceptual design to preliminary design, detail design, and so on. Organization design is dynamic, and organization structures must be flexible in adapting to changing situations.

A primary building block for most organizational patterns is the functional approach, which involves the grouping of functional specialities or disciplines into separately identifiable entities. The intent is to perform similar work within one organizational component. In the pure functional structure, all engineering work is the responsibility of one executive, all manufacturing effort is the responsibility of another executive, and so on. In this case, the same organizational group will accomplish the same type of work for *all* ongoing projects on a concurrent basis.

A partial functional organization, including the identification of major work packages, is illustrated in Figure 18.8. The systems engineering function comes within the overall engineering organization and includes the tasks described in Section 18.1. The organization is responsible for accomplishing these tasks as they apply to all projects. This approach is often desirable for small firms or agencies, since it is easier to manage a homogenous group of similar functions and personnel with comparable backgrounds. Also, the duplication of effort is minimized. On the other hand, for large operations this is not necessarily preferable, owing to impractical centralization of responsibility. Thus, for large, multiproduct firms, the pure functional approach is modified somewhat and organized in terms of individual projects, with some cross-disciplinary activity introduced as appropriate.

Figure 18.9 illustrates a pure project organization that is solely responsive to the planning, design and development, production, operational use, and support of a single system or product. It is time-limited and directly oriented to the life cycle of the system. The commitment of the varied skills and resources required is purely for the purpose of accomplishing tasks associated with the system. The project will have its own engineering functions, its own testing function, its own support function, and so on. Thus, there will be a systems engineering group for project A, a different systems engineering group for project B, and a third systems engineering group for project C, as shown in Figure 18.9. Each group may perform the same overall tasks identified in Section 18.1, but tailored to the specific project. The pure project organization is usually appropriate in the development of large-scale systems.

Figure 18.10 illustrates a project-functional staff configuration, with a mix of the pure functional organization illustrated in Figure 18.8 and the project organization illustrated in Figure 18.9. For large single-system operations, where the scope of work and funding are adequate, the configuration in Figure 18.9 may be appropriate. At the other end of the spectrum, if there are many different in-house product lines where each individually can not justify a separate project organization, the concept in Figure 18.8 may be preferred. Often, however, the actual situation encoun-

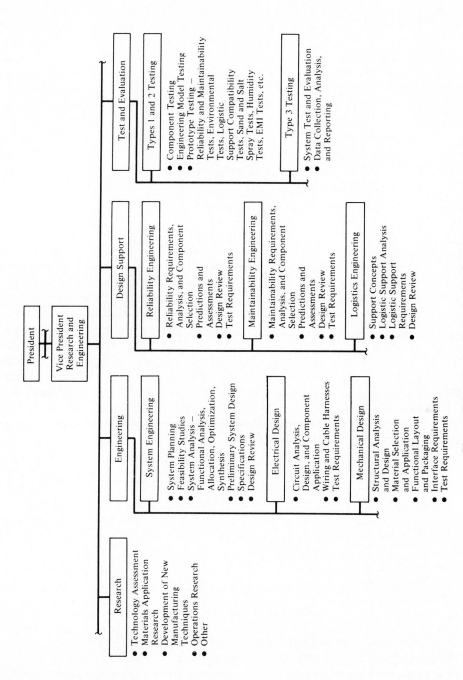

Figure 18.8 Functional organization (partial).

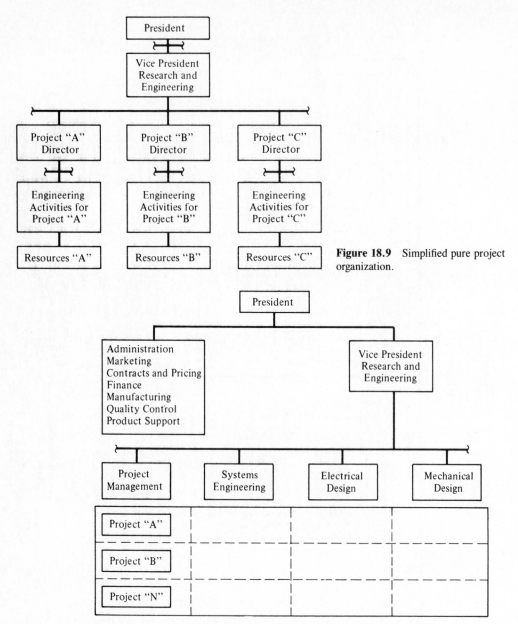

Figure 18.9 Simplified pure project organization.

Figure 18.10 A matrix organization.

tered requires an approach somewhere in between; that is, a firm or agency may be dealing with a variety of products, involving activities differing in magnitude, levels of funding, duration of the life cycle, and so on. When combining the various individual requirements, it may be appropriate to perform some functions through a project structure while others are better accomplished by specialized staff groups.

An activity matrix for a mixed project-functional staff configuration is illustrated in Figure 18.10. In this case the systems engineering group provides staff support for project *A*, project *B*, and so on, as required. Each project does not have its own systems engineering activity. In contrast, Figure 18.11 presents a project-staff organization which varies slightly in format, although the operational concept is the same. In this case, each project incorporates a systems engineering activity, while engineering support activities are provided through staff functions.

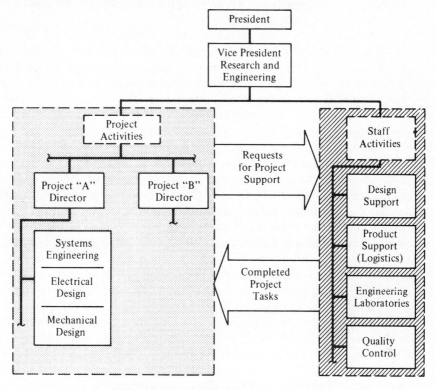

Figure 18.11   Typical project-staff organization.

18.3. DIRECTION AND CONTROL OF SYSTEMS ENGINEERING

Program implementation consists of day-to-day managerial functions associated with influencing, guiding, and supervising subordinates in the accomplishment of the organizational activities identified through planning. This requires a continuing awareness, on the part of the manager, of the specifics of activities performed and whether the results are fulfilling the program objectives in an effective manner. Controlling involves (1) the monitoring of program activities to ensure that the end objectives are being met (e.g., system design is in accordance with the specified requirements, program costs are within the established bounds, tasks are completed within the prescribed scheduled times), and (2) the initiation of corrective action as required to overcome deficiencies if the objectives are not being met effectively.

Project direction deals with the identification of responsibilities and the enforcement of procedures to ensure that project objectives are met. Project objectives have been introduced in the form of (1) technical requirements presented in the system specification, and (2) planning requirements identified through schedules and cost projections. These objectives are included in the project management plan and systems engineering management plan (SEMP), along with the organization responsibilities and procedures for the accomplishment of the project tasks necessary for the attainment of such objectives.

Project control is the sustaining ongoing management activity that will guide, monitor, and evaluate project accomplishments in terms of the stated objectives. With the requirements established, it is essential that a management information system be developed to provide data that enable the measurement and control of certain selected factors as the project progresses.[5] This is necessary to give assurance that the ongoing system development process is producing the desired results.

Project review dates are established and data are collected and incorporated into the management information system. The output results are compared with the initially established requirements for compatibility assessment. Input data are taken from accounting records, predictions, analyses, and so on. The comparison process should answer such questions as Is the program on schedule? Are program costs within the established budget limitations? Will the currently envisioned system design configuration meet all performance requirements? Is the predicted system availability compatible with the established operational requirements? Will the reliability, maintainability, and life-cycle cost goals be met? These and other questions will have to be answered on a number of occasions, particularly in the early phases of system development when there are many unknowns and the risks are high.

In establishing objectives during the early stages of a project, one must determine the parameters to be measured and then select a management information system to provide needed data at the right time. Descriptive information should contain input data needs and the methods for data acquisition, the format(s) for data reporting, and the corrective action process to be followed in the event that problem areas are detected. Data reporting will likely vary from project to project. Typical examples are presented in Figure 18.12 through 18.16. Figures 18.12 and 18.13 indicate task progress (in terms of work completed) against program activities and milestones. Figure 18.14 shows a typical output from PERT time-cost reporting. Figure 18.15 illustrates a cumulative summary of life-cycle cost projections and reliability predictions and compares the results with initial target objectives. Figure 18.16 illustrates the results of an evaluation of three specific system parameters.

The data from a management information system should readily point out existing problem areas. Also, potential areas where problems are likely to occur if program operations continue as originally planned should become visible. To deal with these situations, planning should establish a corrective-action procedure that will include the following:

---

[5] The factors selected will depend on system requirements and will likely vary from project to project. In addition, the management reporting requirements for a given project will differ for each phase in the life cycle.

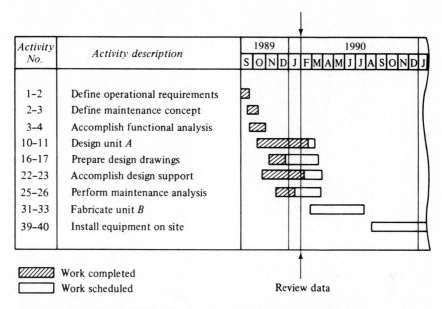

Figure 18.12   Bar chart showing progress against plan.

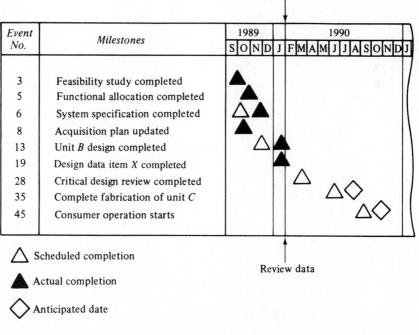

Figure 18.13   A partial milestone chart.

PERT TIME-COST STATUS REPORT

PROJECT: System XYZ					CONTRACT NO.: 4B-3015				REPORT DATE: 4/10/90			
Item/identification					Time status				Cost status			
WBS no.	Project no.	Beginning event no.	Ending event no.	Exp. elap. time (t_e) (weeks)	Earliest completion date (D_E)	Latest completion date (D_L)	Slack $D_L - D_E$ (weeks)	Actual date completed	Cost est. ($)	Actual cost to date ($)	Latest revised est. ($)	Overrun (underrun) ($)
3A1100	814763	36	37	5.2	3/4/86	4/11/90	7.8	4/5/90	2,000	1,950	1,950	(50)
2A3100	824823	239	240	3.0	5/15/86	4/30/90	−3.2		3,500	3,650	4,000	500
3A5400	833110	312	313	4.4	6/15/86	8/2/90	9.6		5,660	3,550	5,660	0

COST AND SCHEDULE OVERVIEW

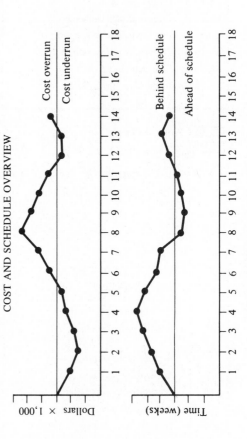

Figure 18.14 Project status reporting.

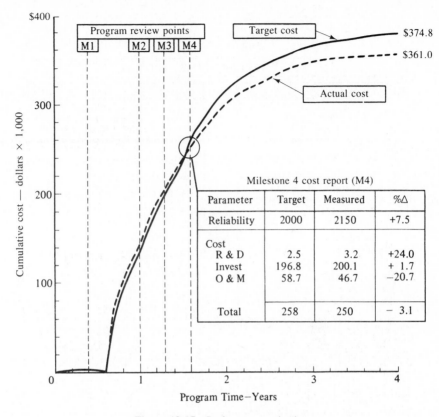

Figure 18.15   Project cost projection.

1. Problems are identified and ranked in order of importance.
2. Each problem is evaluated on the basis of the ranking, addressing the most critical ones first. Alternative possibilities for corrective action are considered in terms of (a) effects on program schedule and cost, (b) effects on system performance and effectiveness, and (c) the risks associated with the decision as to whether to take corrective action.
3. When the decision to take corrective action is reached, planning is initiated to take steps to resolve the problem. This may be in the form of a change in management policy, a contractual change, an organizational change, or a system/equipment configuration change.
4. After corrective action has been implemented, some follow-up activity is required to (a) ensure that the incorporated change(s) actually has resolved the problem, and (b) assess other aspects of the program to ensure that additional problems have not been created as a result of the change.

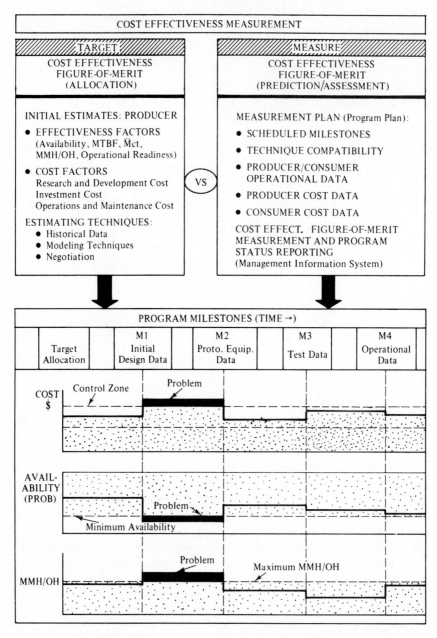

Figure 18.16 Parameter measurement and evaluation.

18.4. CONSUMER, PRODUCER, AND SUPPLIER INTERFACES

Previous sections of this chapter highlighted the planning, organization, and control aspects of various projects with an emphasis on systems engineering management. The material presented primarily addresses the general relationships between the consumer and the producer of a system. For many small systems, much project activity will be conducted within the producer's organization and the aspects of planning, organization, and control are directed mainly toward producer-consumer interfaces. Supplier activities may be limited to providing various types of off-the-shelf components already in inventory for use in the system. Extensive design and development are not required in this instance. Supplier management includes the traditional functions of supplier selection, purchasing and contracting, quality control and inspection, and so on. In many instances, systems engineering management activities are accomplished within the producer's organization, not within the supplier's organization, although the impact will probably affect supplier procurement requirements.

On the other hand, for large-scale systems where major portions of the system are subcontracted by the producer to one or more suppliers, the same philosophy that exists between the consumer and producer will exist between the producer and supplier. Suppliers of major system elements will need to establish a systems engineering capability, conduct the necessary systems engineering planning, and accomplish the functions described in Sections 18.1 through and 18.3. In such instances, the systems engineering management plan is prepared and implemented by the producer, and those functions applicable at the subsystem level are accomplished by the supplier. For example, the accomplishment of system analyses and trade-off studies at the subsystem level are necessary for subsystem design definition. Also, the requirement to ensure the proper integration of various engineering disciplines, such as electrical engineering, reliability engineering, maintainability engineering, and so on, exists at this level. Thus, there are specific project requirements at the system level that must be specified and closely coordinated at the supplier level.

When dealing with large systems, there is a great deal of communication required between the consumer, the producer, and the supplier(s). The consumer and producer organization structures should be compatible to the extent that comparable functions can be readily identified. This is so that the proper coordination at all appropriate levels of the organization can be established. In some instances, the producer will strive to set up a parallel organization structure with that of the consumer so as to simplify communications by establishing a one-to-one relationship for the functions being accomplished. Although the organizations may not have the same name or perform identical functions, there should be a commonality of tasks established between the producer and the supplier organizations. This point is illustrated to a certain extent in Figure 18.17. The goal is to identify comparable functions between the consumer and the producer and between the producer and its suppliers, and to establish vertical project direction and control within each organization while allowing some technical (horizontal) coordination across organization lines. This technical communication on a day-to-day basis is particularly significant in the per-

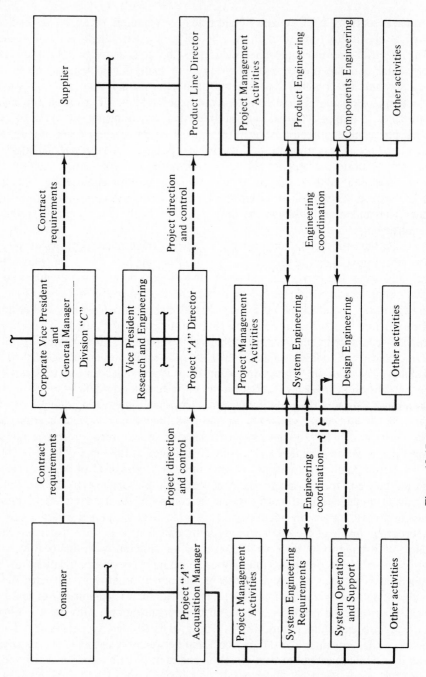

Figure 18.17 Consumer–producer–supplier organization relationships.

formance of systems engineering functions, as a complete initial understanding of all system requirements and interfaces is necessary.

18.5. SUMMARY

Systems engineering management covers a wide spectrum of activities within the overall management domain of planning, organizing, directing, and controlling. This chapter addresses some of the basic concerns of systems engineering management in a rather cursory manner. Because of the comprehensiveness of the subject area, an effective summary might well include a review of the following questions:

1. Have system qualitative and quantitative requirements been identified? Is the system justified? Have operational requirements been defined? Has the maintenance concept been defined?
2. Has a system specification been developed?
3. Have systems engineering functions and tasks been identified? Have task schedules and cost projections been prepared?
4. Has the responsibility for systems engineering functions been established within the consumer organization, producer organization, and/or supplier organizations as applicable?
5. Has a system engineering management plan (SEMP) been developed? Does the plan contain the information specified in Section 18.1?
6. Have detail program plans been developed for reliability engineering? Maintainability engineering? Human factors? Integrated logistic support? Integrated test and evaluation? Quality control? Other engineering disciplines (as applicable)? Are these plans compatible to and integrated with the systems engineering management plan?
7. Are all engineering functions properly integrated through the systems engineering activity?
8. Have the results of system analyses and trade-off studies been adequately documented?
9. Has a method for system measurement and evaluation been established?
10. Has a corrective action procedure been established to handle proposed system modifications or changes?
11. Have conceptual, system, equipment, and critical design reviews been scheduled?

SELECTED REFERENCES

(1) Blanchard, B. S., *Engineering Organization and Management*. Englewood Cliffs, N.J.: Prentice-Hall, 1976.
(2) Blanchard, B. S., *System Engineering Management*. New York: John Wiley, 1991.

(3) Chase, W. P., *Management of System Engineering*. New York: John Wiley, 1974.

(4) Chestnut, H., *Systems Engineering Methods*. New York: John Wiley, 1967.

(5) Cleland, D. I., and W. R. King, *Project Management Handbook,* 2nd Edition. New York: Van Nostrand Reinhold, Inc., 1989.

(6) Defense Systems Management College (DSMC), *Systems Engineering Management Guide*. DSMC, Fort Belvoir, Virginia 22060, December 1986.

(7) Jenkins, G. M., and P. V. Youle, *Systems Engineering: A Unifying Approach In Industry And Society*. Oxford: Alden & Mowbray, 1971.

(8) Karger, D., and R. Murdick, *Managing Engineering and Research*. New York: Industrial Press, 1969.

(9) Kerzner, H., *Project Management: A Systems Approach to Planning, Scheduling, and Controlling*, 3rd Edition. New York: Van Nostrand Reinhold, 1989.

(10) Wymore, A. W., *Systems Engineering Methodology for Interdisciplianry Teams*. New York: John Wiley, 1976.

QUESTIONS AND PROBLEMS

1. What is the purpose of the system specification? The project management plan? What are the basic differences between the two?

2. What is the purpose of the systems engineering management plan? What information should be included? How does it relate to the reliability program plan? Maintainability program plan? Integrated logistic support plan?

3. What project functions/tasks should be included in systems engineering? Why?

4. List some of the techniques generally used in the scheduling of tasks. Discuss the applications and indicate some of the advantages and disadvantages of each.

5. What is PERT? PERT-COST? CPM?

6. When employing PERT-COST, the time-cost option applies. What is meant by the time-cost option? How can it affect the critical path in a network?

7. What is a WBS (discuss its purpose and application)? How can the activities and events in a PERT network be related to the WBS? How can the cost information used in PERT-COST be related to the WBS? How does the WBS relate to the cost breakdown structure used in life-cycle costing?

8. Why is the establishment of a good management information system so important?

9. Refer to Figure 18.14. There is a trend toward a program cost overrun. How would you determine the cause of the overrun (where the problem exists)? What would you do about it?

10. Why is the initiation of a corrective-action procedure so important in the management of a project?

11. What is the purpose of the systems engineering organization? Why is the establishment of organization goals and objectives important? What are they based on?

12. What are the basic differences between line and staff organization functions?

13. What are the advantages and disadvantages of a functional organization? Project organi-

zation? Project-staff organization? Under which type of organization structure will systems engineering provide the best results?

14. How does the WBS tie in with the organization structure?

15. Assume that you are just assigned the responsibility of establishing a systems engineering group for a given project. What steps would you take in acquiring a fully operating functional organization?

16. What type of personnel (in terms of background and experience) would you attempt to hire in your systems engineering group (assuming that you are the manager of such a group)?

PART VI: SYSTEM DESIGN APPLICATIONS

19

Communication System Design[1]

A large metropolitan area wishes to upgrade its overall communications capability by acquiring a new communication system for installation in patrol aircraft, helicopters, ground vehicles, certain designated ground facilities, and in the central communications control facility. Two alternative system configurations are to be considered and evaluated. The objective is to define system operational requirements, the maintenance concept, and required program planning information so that the system configuration with the minimum life-cycle cost can be selected.[2]

19.1. INTRODUCTION

The new communication system should be adaptable for use in all designated applications. The only configuration difference should be in the interface connections, mounting fixtures, and so on. In addition, the communication system must meet the following objectives:

---

[1] This system design application was adapted from B. S. Blanchard, *Design and Manage to Life Cycle Cost* (Forest Grove, Ore.: M/A Press, 1978).

[2] The system in this example is the communication unit which will be installed in aircraft, ground vehicles, and designated facilities.

1. *Objective 1*—The system is to be installed in three low-flying light aircraft (10,000 ft or less) in quantities of one per aircraft. The system will enable communication with loitering helicopters dispersed within a 200-mile radius and with the central communication control facility. It is anticipated that each aircraft will fly 15 times per month, with an average flight duration of 3 hours. The system utilization requirement is 1.1 hours of system operation for every hour of aircraft operation, which includes air time plus some ground time. It is assumed that *all* functions of the system are fully operational throughout this time period. The system must exhibit a MTBM of at least 500 hours and a $\overline{M}ct$ not to exceed 15 minutes.

2. *Objective 2*—The system is to be installed in each of five helicopters and will enable communication with patrol aircraft through a 200-mile range, other helicopters with a 50-mile radius, and with the central communications control facility. It is anticipated that each helicopter will fly 25 times per month, with an average flight duration of 2 hours. The utilization requirement is 0.9 hours of system operation for every flight hour of helicopter operation. The system must meet a 500-hours MTBM requirement and a $\overline{M}ct$ of 15 minutes or less.

3. *Objective 3*—The system is to be installed in each of 50 police patrol vehicles (one per vehicle) and will enable communication with other vehicles within a 10-mile radius, with any one of the five fixed communications facilities at a range of 50 miles, and with the central communications control facility at a range of 50 miles or less. Each vehicle will be in operation on the average of 5 hours per day, 5 days per week, and will be utilized 100% during that time. The required MTBM is 400 hrs, and the $\overline{M}ct$ should be 30 minutes or less.

4. *Objective 4*—Two systems are to be installed in each of five fixed communications facilities optimally located throughout the metropolitan geographical area. Each system will enable communication with patrol vehicles at a range of 25 miles and with the central communications control facility at a range of 50 miles. System utilization requirements are 120 hours per month, the required MTBM is 200 hours, and the $\overline{M}ct$ should not exceed 60 minutes.

5. *Objective 5*—Ten systems are to be installed in the central communications control facility and will enable communication with patrol aircraft and loitering helicopters at a range of 400 miles, the five fixed communication facilities at a range of 50 miles, and with patrol vehicles within a 10-mile radius. In addition, each system will be able to communicate with the intermediate maintenance facilities. The average utilization requirement for each system is 3 hours per day for 360 days per year. The MTBM requirement is 200 hours, and the $\overline{M}ct$ should not exceed 45 minutes.

6. *Objective 6*—Two mobile vans will be used to support intermediate-level maintenance at the five fixed communications facilities. Each van will incorporate one communication system which will be used on the average of 2 hours per day for a 360-day year. The required MTBM is 400 hours, and the $\overline{M}ct$ is 15 minutes.

These deployment objectives are illustrated in the communications network shown in Figure 19.1. Note that the requirements for 80 communication systems are

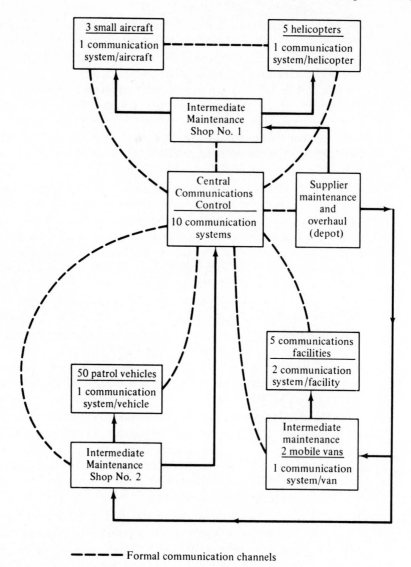

—————— Formal communication channels

Figure 19.1   Communications network.

shown along with the proposed maintenance support facilities. This information serves as the baseline required to commence with the analysis.

In line with the stated operational objectives, there is a need to acquire a new communication system that will meet the performance and effectiveness requirements (e.g., voice transmission range, clarity of message, MTBM, $\overline{M}$ct, etc.) Further, budget limitations suggest that the system life-cycle cost should not exceed $35,000.[3] Advanced program planning indicates that a full complement of systems must be in operation five years after the start of the program, and that this capability

---

[3] System life-cycle cost in this instance is determined by dividing the total life-cycle cost figure of merit by the number of operating systems (i.e., 80).

must be maintained through the eleventh year of the program. The significant program milestones and projected number of systems in operational use are presented in Figure 19.2. This forms the basis for defining the system life cycle, the major life-cycle functions, and the life-cycle cost.

Based on a review of the available sources of supply, there is no known existing system that will completely fulfill the need; but there are two new candidate design configurations that should suffice assuming that all design goals are met. The objective is to evaluate each configuration in terms of its life-cycle cost and to recommend a preferred system configuration.

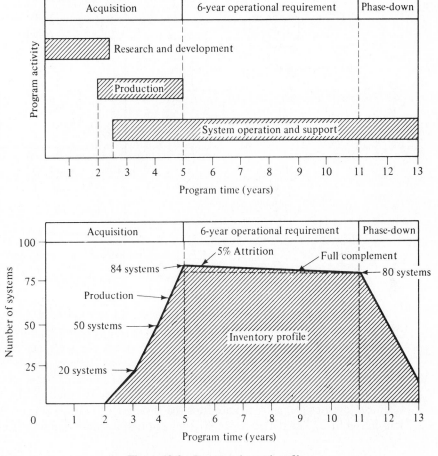

Figure 19.2   Program plan and profile.

19.2. THE ANALYSIS APPROACH

The analysis approach generally follows the steps illustrated in Figure 3.7 and is described further in Section 17.3. However, prior to the identification of a cost breakdown structure (CBS) and the development of cost factors, the analyst needs to further expand the description of the baseline system configuration and its maintenance concept. As a start, an assumed packaging scheme is developed and shown in Figure 19.3. This configuration (including units, assemblies, and modules) is developed from conceptual design information and is fairly representative of each of the candidates being considered.[4]

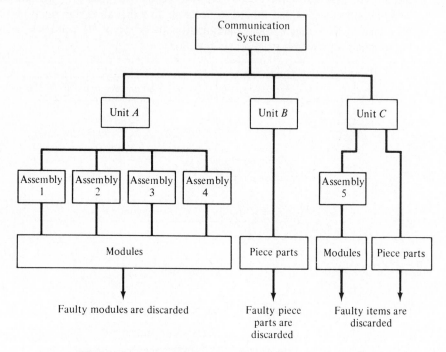

Figure 19.3 System packaging scheme (baseline configuration).

As indicated in Chapter 3, the maintenance concept can be defined as a series of statements and/or illustrations that include criteria covering maintenance levels, support policies, effectiveness factors (e.g., maintenance time constraints, turnaround times, transportation times), and basic logistic support requirements. The maintenance concept is a prerequisite to the system or product design, whereas a detailed maintenance plan reflects the results of design and is used for the acquisition of those logistics elements required for the sustaining life-cycle support of the system in the field. The maintenance concept can best be illustrated by Figure 19.4.[5]

[4] The configuration reflected in Figure 19.3 obviously does not represent *final* design but is close enough for life-cycle cost purposes (particularly for analyses accomplished in the early stages of a program). Further design definition will occur as the program progresses.

[5] The quantitative support factors presented in Figure 19.4 are considered as minimum design requirements. For each alternative configuration being evaluated in a life-cycle cost analysis, the support factors may vary somewhat as a function of the specific design characteristics. However, the minimum requirements still must be met.

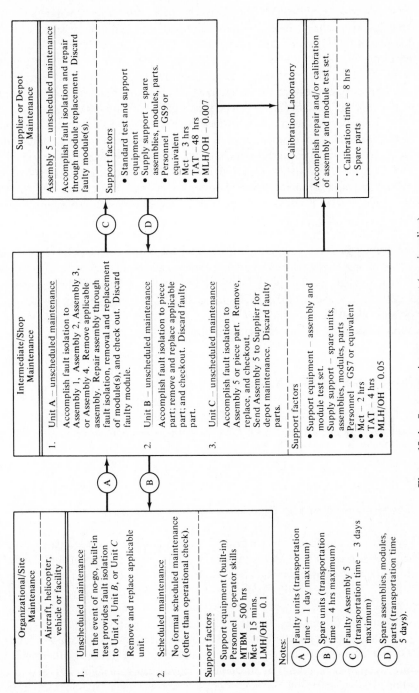

Figure 19.4 System maintenance concept (repair policy).

Referring to Figure 19.4, there are three levels of maintenance to consider. *Organizational maintenance* is performed in the aircraft or helicopters, in the patrol vehicles, or in the various communication facilities (as applicable) by user or operator personnel. *Intermediate maintenance,* or the second level of maintenance, is accomplished in a remote shop facility by trained personnel possessing the skills necessary to perform the assigned functions. *Depot maintenance,* the highest level of maintenance, constitutes the specialized repair or overhaul of complex components at the supplier's facility by highly skilled supplier maintenance personnel.

The specific functions scheduled to be accomplished on the communication system at each level of maintenance are noted. In the event of a malfunction, fault isolation is performed to the applicable unit by using the built-in test capability (i.e., unit *A*, *B*, or *C*). Units are removed and replaced at the organizational level and sent to the intermediate maintenance shop for corrective maintenance. At the intermediate shop, units are repaired through assembly and/or part replacement, assemblies are repaired through module replacement, and so on.

The maintenance concept illustrated in Figure 19.4 shows the functions that are anticipated for each level of maintenance, the effectiveness requirements in terms of maintenance frequency and times, and the major elements of logistic support to include personnel skill levels, test and support equipment, supply support requirements, and facilities. This information is not only required as an input to the system design process, but serves as the basis for determining operation and support costs in the total life-cycle cost spectrum.

19.3. LIFE-CYCLE COST ANALYSIS

With the problem (or need) defined, and with a description of the system operation requirements and maintenance concept, it is now appropriate to proceed with the specific steps involved in the life-cycle cost analysis of the proposed alternative configurations. These steps, leading to the generation of cost data, are given in this section.

Development of the Cost Breakdown Structure

The CBS assumed for the purpose of this evaluation is presented in Figure 19.5, which is similar to the CBS in Figure 17.6 (also refer to Appendix C). Although not all of the cost categories may be relevant or significant in terms of the magnitude of cost as a function of total life-cycle cost, this CBS does serve as a good starting point. Initially, *all* costs must be considered, with the subsequent objective of concentrating on those cost categories reflecting the high contributors.

The evaluation of two alternative communication system configurations and the selection of a preferred approach are required. In both instances, there will be activities involving planning, management, engineering design, test and evaluation, production, distribution, system operation, maintenance and logistic support, and ultimate equipment disposal. In an attempt to be more specific for the life-cycle cost

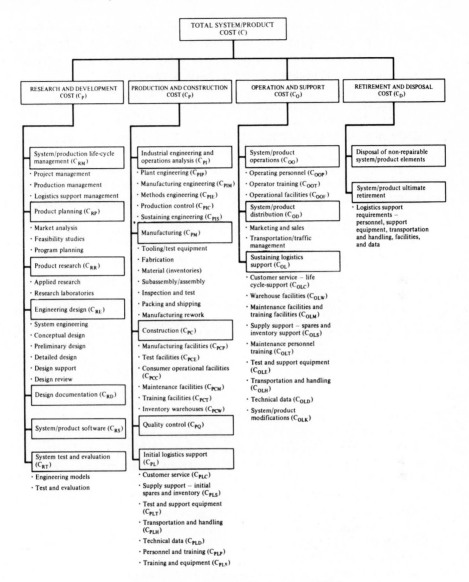

Figure 19.5   Cost breakdown structure (CBS).

analysis, the analyst may wish to consider the following steps:

1. Identify *all* anticipated program activities that will generate costs of one type or another in the life cycle for *each* of the two alternatives.

2. Relate each identified activity to a specific cost category in the CBS. Each activity should fall into one or more of the categories in the CBS. If it does not, the CBS should be expanded or revised as appropriate to cover the required effort.

3. Develop a matrix-type worksheet for the purposes of recording costs for each applicable category by year in the life cycle.[6] Table 19.1 illustrates an example of data format presentation. The assumed life cycle and major activities are based on the program plan presented in Figure 19.2.

4. Generate cost input data for each applicable activity listed in the matrix, and record the results in Table 19.1.

Generation of Input Cost Data[7]

Referring to Figure 19.5, the major cost categories are C_R, C_P, and C_O. It is assumed that C_D will be negligible in this instance. The following paragraphs present an overview of the major sources supporting these relevant areas of cost.

Research and development cost (C_R). Research and development costs include those early life-cycle costs that will be incurred by the metropolitan agency responsible for the acquisition of the communication system (i.e., the customer or consumer), and those costs incurred by the supplier in the development of the system (i.e., the contractor). There will be some common costs to the customer associated with both alternatives and relating to initial program planning, the accomplishment of feasibility studies, the development of operational requirements and the maintenance concept, the preparation of top-level system specifications, and general management activities. Also, there will be supplier costs that are peculiar to each alternative and that are included in the supplier's proposal as submitted to the customer.

Although the analyst may ultimately wish to evaluate only delta or incremental costs (i.e., those costs peculiar to one alternative or the other), the approach here is to address total life-cycle costs.[8] In determining such costs, the analyst uses a combination of customer cost projections and the proposals submitted by each of two potential suppliers as source data for the life-cycle cost analysis. Table 19.2 presents a summary of research and development costs.

It should be noted that the costs presented are primarily nonrecurring, constituting management and engineering labor with the proper inflationary factors included. Labor costs are developed from personnel projections indicating the class of labor (i.e., manager, supervisor, senior engineer, engineer, technician, etc.) and the labor hours of effort required by class by month for each functional activity. The labor hours per month are then converted to dollars by applying standard cost factors. Both indirect and direct costs are included, along with a 10% per year inflation factor. Material costs are included in test and evaluation (category C_{RT}), since engineering prototype models must be produced in order to verify the system performance, effectiveness, and supportability characteristics.

---

[6] The matrix desired could be a direct printout from the life-cycle cost model. Model design must consider the various data output requirements in terms of both content and format.

[7] The material covered here is presented in enough detail to convey the *overall approach* used in the evaluation of the two alternative communication system configurations. An in-depth discussion of *each* cost input factor would be rather extensive.

[8] It is important to look at total life-cycle cost in order to properly assess major cost drivers. The aspect of delta cost can be reviewed later.

TABLE 19.1 COST COLLECTION WORKSHEET

PROGRAM ACTIVITY	COST CATEGORY DESIGNATION	COST BY PROGRAM YEAR ($)													TOTAL COST (CONSTANT $)	TOTAL COST (ACTUAL $)	PERCENT CONTRIBUTION
		1	2	3	4	5	6	7	8	9	10	11	12	13			
Alternative *A*																	
1. Research and development	C_R																
a. Life-cycle management	C_{RM}																
b. Product planning	C_{RP}																
(1) Feasibility studies	—																
(2) Program planning																	
2.																	
3.																	
Others																	
Alternative *B*																	
1. Research and development	C_R																
2.																	

TABLE 19.2 RESEARCH AND DEVELOPMENT COST SUMMARY

PROGRAM ACTIVITY*	COST CATEGORY DESIGNATOR	COST BY PROGRAM YEAR (INFLATED DOLLARS)			TOTAL ACTUAL COST ($)
		Year 1	Year 2	Year 3	
A. Customer costs—alternative *A*					
1. System/product management	C_{RM}	40,248	44,273	48,700	133,221
2. Production planning	C_{RP}	15,960	—	—	15,960
B. Supplier costs—alternative *A*					
1. System/product management	C_{RM}	35,604	38,165	40,081	113,850
2. Product planning	C_{RP}	10,062	11,068	—	21,130
3. Engineering design	C_{RE}	50,728	82,736	103,552	237,016
4. Design documentation	C_{RD}	10,110	18,115	20,200	48,425
5. System test and evaluation	C_{RT}	—	—	67,648	67,648
Total research and development cost for alternative *A*	C_R	162,712	194,357	280,181	637,250
A. Customer costs—alternative *B*					
1. System/product management	C_{RM}	40,248	44,273	48,700	133,21
2. Product planning	C_{RP}	15,960	—	—	15,960
B. Supplier costs—alternative *B*					
1. System/product management	C_{RM}	32,508	35,759	39,335	107,602
2. Product planning	C_{RP}	—	—	—	—
3. Engineering design	C_{RE}	65,194	81,175	91,750	238,119
4. Design documentation	C_{RD}	12,214	12,110	15,300	39,624
5. System test and evaluation	C_{RT}	—	—	60,531	60,531
Total research and development cost for alternative *B*	C_R	166,124	173,317	255,616	595,057

*Only applicable cost categories are listed. There are no costs associated with product research (C_{RR}) and system software (C_{RS}).

The supplier costs for alternative *A* and *B* in Table 19.2 are derived directly from two individual supplier proposals. Direct costs, indirect costs, inflation factors for both labor and material, general and administrative expenses, and projected supplier profits are included.

Production and construction cost (C_P).   This category includes the recurring and nonrecurring costs associated with the production of the required 80 operational systems plus the four additional systems intended to compensate for possible attrition. Referring to Figure 19.2, the leading edge of the inventory profile represents the production requirements. Obviously, the analyst must convert the projected population growth to a specific production profile, as shown in Figure 19.6. This profile becomes the basis for determining production costs.

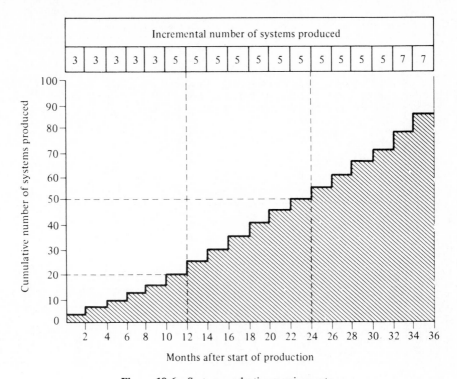

Figure 19.6   System production requirements.

If permitted the option, each of the two potential suppliers may propose an entirely different production scheme while still meeting the inventory requirements illustrated in Figure 19.2. This, of course, would in all probability create a significant variation in the input planning factors and cost. However, in this instance the specified production scheme in Figure 19.6 is assumed for both alternatives.

Production and construction cost considers the appropriate factors indicated in Figure 19.5, and the results are tabulated in Table 19.3. A description of these costs follows:

1. *System or product management cost* (C_{RM})—includes the ongoing management activity required throughout the production phase. Costs in this category are a continuation of the system or product management cost stream reflected in the R&D cost summary in Table 19.2.

2. *Industrial engineering and operations analysis cost* (C_{PI})—addresses the functions of production planning, manufacturing engineering, methods engineering, and so on. A minimal level of effort is shown in Table 19.3.

3. *Manufacturing recurring cost* (C_{PMR})—includes those activities related directly to the fabrication, assembly, inspection, and test of the 84 systems being produced. Each of the two potential suppliers submitted a proposal covering functions compatible with and in direct support of the production scheme illus-

TABLE 19.3 PRODUCTION/CONSTRUCTION COST SUMMARY

PROGRAM ACTIVITY	COST CATEGORY DESIGNATOR	COST BY PROGRAM YEAR (INFLATED DOLLARS)				TOTAL ACTUAL COST ($)
		Year 2	Year 3	Year 4	Year 5	
A. Customer costs—alternative A						
1. System/Product management	C_{RM}	—	—	53,570	58,927	112,497
B. Supplier costs—alternative A						
1. Industrial engineering and operations analysis	C_{PI}	10,000	18,000	20,000	20,200	68,200
2. Manufacturing						
a. Recurring cost	C_{PMR}	—	174,000	226,500	234,600	635,100
b. Nonrecurring cost	C_{PMN}	48,900	—	—	—	48,900
3. Quality control	C_{PQ}	—	25,400	26,000	27,102	78,502
4. Initial logistics support						
a. Supply support	C_{PLS}	—	9,600	19,200	19,200	48,000
b. Test and support equipment	C_{PLT}	5,000	65,000	—	—	70,000
c. Technical data	C_{PLD}	5,100	—	—	—	5,100
d. Personnel training	C_{PLP}	16,400	15,000	15,000	—	46,400
Total production and construction cost for alternative A	C_{P}	85,400	307,000	360,270	360,029	1,112,699
A. Customer costs—alternative B						
1. System/product management	C_{RM}	—	—	53,570	58,927	112,497
B. Supplier costs—alternative B						
1. Industrial engineering and operations analysis	C_{PI}	15,000	20,000	16,000	15,000	66,000
2. Manufacturing						
a. Recurring cost	C_{PMR}	—	167,000	235,500	255,000	657,500
b. Nonrecurring cost	C_{PMN}	55,100	—	—	—	55,100
3. Quality control	C_{PQ}	—	19,800	22,100	31,090	72,990
4. Initial logistics support						
a. Supply support	C_{PLS}	—	18,340	19,340	19,340	57,020
b. Test and support equipment	C_{PLT}	5,000	65,000	—	—	70,000
c. Technical data	C_{PLD}	5,350	—	—	—	5,350
d. Personnel training	C_{PLP}	16,400	15,000	15,000	—	46,400
Total production and construction cost for alternative B	C_{P}	96,850	305,140	361,510	379,357	1,142,857

trated in Figure 19.6. Since the majority of such activities are repetitive in nature, each supplier estimated the cost of the first system and then projected a learning curve to reflect the cost of subsequent systems. The proposed learning curves for alternatives *A* and *B* are illustrated in Figure 19.7. These curves support the recurring manufacturing costs included in Table 19.3.

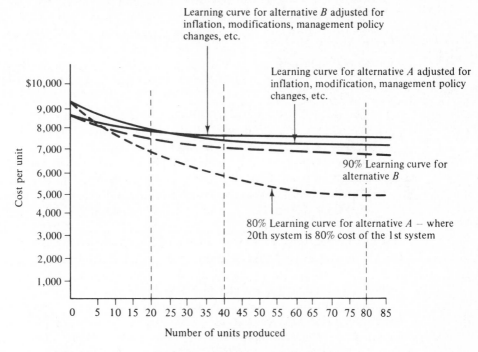

Figure 19.7   Production learning curves.

4. *Manufacturing nonrecurring cost* (C_{PMN})—covers all costs associated with the acquisition and installation of special tooling, fixtures and jigs, and factory test equipment. These costs, included in the supplier's proposal, are basically a one-time expenditure during year 2 in anticipation of the production requirements commencing in year 3.

5. *Quality control cost* (C_{PQ})—includes the category of quality assurance, which is a sustaining level of activity required to ensure that good overall product quality exists throughout the production process, and the category of qualification testing, which constitutes the testing of a representative sample of items to verify that the level of quality inherent in the items being produced is adequate. The costs for each supplier's proposal are included.

6. *Initial spares and inventory cost* (C_{PLS})—covers the acquisition of major units to support organizational maintenance, and a few assemblies plus parts to provide support at the intermediate level of maintenance. These items represent the inventory safety stock factor and are located at the intermediate maintenance shops and the supplier facility identified in Figure 19.1. Replenishment spares for the sustaining support of the system in operational use

throughout the life cycle are covered in category C_{OLS}. This category covers an initial limited procurement, whereas the replenishment spares are based on realistic consumption and demand factors. The assumptions used in determining costs are as follows:

    a. *Alternative A*—one complete set of units (unit A, unit B, and unit C) are required for shop 1, shop 2, each of the mobile vans, and for the supplier facility as a backup. The acquisition price for unit A is $3,800, unit B is $2,150, and unit C is $2,650. Five sets are equivalent to $43,000, with $8,600 in year 3, $17,200 in year 4, and $17,200 in year 5. The cost of assemblies and parts is $5,000 ($1,000 in year 3, $2,000 in year 4, and $2,000 in year 5).

    b. *Alternative B*—six sets of units (unit A, unit B, and unit C) are required. The additional unit (above and beyond the requirements for alternative A) is to be located at shop 2 to cover the anticipated increased number of maintenance actions. The acquisition price for unit A is $3,530, unit B is $2,060, and unit C is $2,580. Six sets are equivalent to $49,020, with $16,340 in each of years 3, 4, and 5. The cost of assemblies and parts is $8,000 ($2,000 in year 3, $3,000 in year 4, and $3,000 in year 5).

7. *Test and support equipment acquisition cost* (C_{PLT})—covers the assembly and module test set located in each intermediate maintenance facility, and several items of commercial and standard equipment located at the supplier facility to support depot maintenance (refer to Figures 19.1 and 19.4). The design cost associated with the test set is $5,000 expended in year 2. The production acquisition price of each test set is $15,000 and there are four test sets required. Referring to Table 19.4, there are systems being introduced into operational use in year 3 at all locations; thus, the four test sets must be available in year 3. The commercial and standard equipment needed for depot maintenance requires no additional design effort, and can be acquired for $5,000. Thus, the costs for year 3 are $65,000. The test and support equipment requirements for each alternative are considered to be comparable. The sustaining annual maintenance and logistics requirements for test equipment are included in cost category C_{OLE}.

8. *Technical data cost* (C_{PLD})—relates to the preparation and publication of system installation and test instructions, operating procedures, and maintenance procedures. These data are required to operate and maintain the system in the field throughout its programmed life cycle. The acquisition cost of this data is $5,100 for alternative *A* and $5,350 for alternative B. Data costs are applicable to year 2.

9. *Personnel training cost* (C_{PLP})—covers the initial cost of training system operators and maintenance technicians. For operator training, it is assumed that 20 operators are trained in year 2, 30 operators in year 3, and 30 operators in year 4. The cost of training is $500 per student (for each alternative configuration). In the maintenance area, formal training will be given to two technicians assigned to each of the four intermediate maintenance facilities. These techni-

TABLE 19.4 UNITS IN USE AND OPERATING TIMES

CATEGORIES OF SYSTEMS	PROGRAM YEAR												
	1	2	3	4	5	6	7	8	9	10	11	12	13
Communication systems in use	—	—	20	50	80	80	80	80	80	80	80	29	10
1. Aircraft application (3)													
a. Quantity of systems	—	—	1	3	3	3	3	3	3	3	3	3	1
b. System operating time (hr)	—	—	594	1,782	1,782	1,782	1,782	1,782	1,782	1,782	1,782	1,782	594
2. Helicopter application (5)													
a. Quantity of systems	—	—	2	5	5	5	5	5	5	5	5	5	3
b. System operating time (hr)	—	—	1,080	2,700	2,700	2,700	2,700	2,700	2,700	2,700	2,700	2,700	1,620
3. Patrol vehicles (50)													
a. Quantity of systems	—	—	7	22	50	50	50	50	50	50	50	10	2
b. System operating time (hr)	—	—	1,100	28,600	65,000	65,000	65,000	65,000	65,000	65,000	65,000	13,000	2,600
4. Communication facilities (5)													
a. Quantity of systems	—	—	5	10	10	10	10	10	10	10	10	5	—
b. System operating time (hr)	—	—	7,200	14,400	14,400	14,400	14,400	14,400	14,400	14,400	14,400	7,200	—
5. Central communications control (1)													
a. Quantity of systems	—	—	4	8	10	10	10	10	10	10	10	4	3
b. System operating time (hr)	—	—	4,320	8,640	10,800	10,800	10,800	10,800	10,800	10,800	10,800	5,400	3,240
6. Mobile vans (2)													
a. Quantity of systems	—	—	1	2	2	2	2	2	2	2	2	2	1
b. System operating time (hr)	—	—	720	1,440	1,440	1,440	1,440	1,440	1,440	1,440	1,440	1,440	720

cians will accomplish on-the-job training for the additional personnel in the shops and vans. The cost of maintenance training is $800 per student week, or $6,400 in year 2.

Operation and support cost (C_O).   This category includes the cost of operating and supporting the system throughout its programmed life cycle. These costs primarily constitute user costs and are based on the program planning information in Figure 19.2. The inventory profile in the figure is expanded as shown in Table 19.4 to indicate the specific quantity of systems in use and the total operating time (in hours) for all systems in each applicable year of the life cycle.

System use is based on the individual operating times stated in the six objectives described as part of the problem definition, and is determined from Equations 19.1 through 19.6 as follows:

Operating time for aircraft application (hours)

$$= \text{(quantity of system in aircraft)(15 flights/month)(12)} \qquad (19.1)$$
$$\times \text{ (3 hrs/flight)(1.1)}$$

Operating time for helicopter application (hours)

$$= \text{(quantity of systems in helicopters)(25 flights/month)} \qquad (19.2)$$
$$\times \text{ (12 months)(2 hrs/flight)(0.9)}$$

Operating time for patrol vehicle application (hours)

$$= \text{(quantity of systems in patrol vehicles)(5 hrs/day)} \qquad (19.3)$$
$$\times \text{ (5 days/weeks)(52 weeks/year)(1.0)}$$

Operating time for communication facility application (hours)

$$= \text{(quantity of systems in facilities)(120 hrs/month)(12)} \qquad (19.4)$$

Operating time for central communication control application (hours)

$$= \text{(quantity of systems)(3 hrs/day)(360 days/year)} \qquad (19.5)$$

Operating time for mobile van application (hours)

$$= \text{(quantity of systems in vans)(2 hrs/day)(360 days/year)} \qquad (19.6)$$

Although the *actual* utilization of the system will vary from operator to operator, from organization to organization, from one geographical area to the next, and so on, the factors included in Table 19.4 are average values and are employed in the baseline example. Also, it is assumed that each alternative configuration being evaluated will be operated in the same manner.

Operation and support cost includes those individual costs associated with system operations, distribution, and sustaining logistic support. Only those significant costs that are applicable to the communication system are discussed below.

1. *Operating personnel cost* (C_{OOP}) covers the total costs of operating the communication system for the various applications. Since the operator is charged with a number of different duties, only that allocated portion of time associated with the direct operation of the communication system is counted. Operating personnel cost is determined from Equation 19.7 using the data in Table 19.4 as a base.

C_{OOP} – (cost of operator labor)(quantity of operators)

$\quad$ = (quantity of systems)(hours of system operation) $\times$ (% allocation)$\qquad$(19.7)

In determining operator costs, different hourly rates are applied for the various applications (e.g., \$14.50 per hour for the aircraft and helicopter application, \$10.50 per hour for the facility application, etc.) and varied allocation factors are applied because of a personnel workload different from one application to the next. The resulting costs are included in Table 19.5.

$\quad$ 2. *System distribution and transportation cost* (C_{ODT}) covers the initial transportation and installation cost (i.e., the packing and shipping of systems from the supplier's manufacturing facility to the point of application, and installation of the system for operational use). Equation 19.8 is used to determine total cost in this category.

C_{ODT} = (cost of transportation) + (cost of packing) + (cost of system installation)

$$\text{(19.8)}$$

where transportation and packing costs are based on dollars per cwt (i.e., \$30 per cwt for transportation and \$40 per cwt for packing), and installation costs are a function of labor cost in dollars per labor-hour and the number of labor hours required.

$\quad$ Costs in this category are based on the quantity of systems indicated in Table 19.4, and on the appropriate transportation rate structures. The analyst should review the latest Interstate Commerce Commission (ICC) documentation on rates in order to determine the proper transportation costs. The figures used in this life-cycle cost analysis are presented in Table 19.5.

$\quad$ 3. *Unscheduled maintenance cost* (C_{OLA}) covers the personnel activity costs associated with the accomplishment of *unscheduled* or corrective maintenance on the communication system. Specifically, this includes the direct and indirect cost in the performance of maintenance actions (a function of maintenance labor-hours and the cost per labor-hour), the material handling cost associated with given maintenance actions, and the cost of documentation for each maintenance action. These costs for the two alternative communication system configurations being evaluated are summarized in Table 19.6. Determining unscheduled maintenance cost is dependent on predicting the number of anticipated maintenance actions that are likely to occur throughout the life cycle (i.e., the expected frequency of unscheduled maintenance, or the reciprocal of the MTBM). Since there is no *scheduled* maintenance permitted in this instance, the MTBM factor assumed here is directly equated with unscheduled maintenance actions. The frequency of maintenance is usually based on the reliability failure rates for individual components of the system and is derived from reliability prediction data, maintainability prediction data, logistic support analysis data, or a combination thereof.

$\quad$ Review of the objectives in the problem definition indicates that the need relative to MTBM requirements differs from one application to the next. Since it is the goal to design a single system configuration for use in all applications (to the maximum extent practicable), the most stringent conditions must be met. Thus, the new

TABLE 19.5 OPERATION AND SUPPORT COST SUMMARY

PROGRAM ACTIVITY	COST CATEGORY DESIGNATION	COST BY PROGRAM YEAR													TOTAL ACTUAL COST ($)
		1	2	3	4	5	6	7	8	9	10	11	12	13	
A. Alternative A															
1. Operating personnel	C_{OOP}	—	—	861	2,584	2,584	2,584	2,584	2,584	2,584	2,584	2,584	2,584	861	24,978
2. Transportation	C_{ODT}	—	—	8,000	20,000	32,000	32,000	32,000	32,000	32,000	32,000	32,000	12,000	4,000	268,000
3. Unscheduled maintenance	C_{OLA}	—	—	3,715	8,886	14,871	14,871	14,871	14,871	14,871	14,871	14,871	4,896	1,536	123,130
4. Maintenance facilities	C_{OLM}	—	—	139	334	559	559	559	559	559	559	559	184	56	4,626
5. Supply support	C_{OLS}	—	—	8,880	21,360	35,760	35,760	35,760	35,760	35,760	35,760	35,760	11,760	3,600	295,920
6. Maintenance personnel training	C_{OLT}	—	—	—	—	1,300	1,300	1,300	1,300	1,300	1,300	1,300	—	—	9,100
7. Test and support equipment	C_{OLE}	—	—	1,625	3,250	3,250	3,250	3,250	3,250	3,250	3,250	3,250	3,250	1,625	32,500
8. Transportation and handling	C_{OLH}	—	—	168	378	630	630	630	630	630	630	630	210	84	5,250
Total operation and support cost for alternative A	C_O	—	—	23,388	56,792	90,954	90,954	90,954	90,954	90,954	90,954	90,954	34,884	11,762	763,504
B. Alternative B															
1. Operating personnel	C_{OOP}	—	—	861	2,584	2,584	2,584	2,584	2,584	2,584	2,584	2,584	2,584	861	24,978
2. Transportation	C_{ODT}	—	—	8,000	20,000	32,000	32,000	32,000	32,000	32,000	32,000	32,000	12,000	4,000	268,000
3. Unscheduled maintenance	C_{OLA}	—	—	4,529	10,881	18,322	18,322	18,322	18,322	18,322	18,322	18,322	6,077	1,812	151,553
4. Maintenance facilities	C_{OLM}	—	—	169	409	690	690	690	690	690	690	690	229	68	5,705
5. Supply support	C_{OLS}	—	—	10,800	26,160	44,160	44,160	44,160	44,160	44,160	44,160	44,160	14,640	4,320	365,040
6. Maintenance personnel training	C_{OLT}	—	—	—	—	1,300	1,300	1,300	1,300	1,300	1,300	1,300	—	—	9,100
7. Test and support equipment	C_{OLE}	—	—	2,113	4,225	4,225	4,225	4,225	4,225	4,225	4,225	4,225	4,225	2,113	42,251
8. Transportation and handling	C_{OLH}	—	—	210	462	756	756	756	756	756	756	756	252	84	6,300
Total operation and support cost for alternative B	C_O	—	—	26,682	64,721	104,037	104,037	104,037	104,037	104,037	104,037	104,037	40,007	13,258	872,927

TABLE 19.6 UNSCHEDULED MAINTENANCE COST

PROGRAM ACTIVITY		1	2	3	4	5	6	7	8	9	10	11	12	13
								PROGRAM YEAR						
A. Alternative A														
1. Organizational maintenance (system level)	Total labor-hours	—	—	9.25	22.25	37.25	37.25	37.25	37.25	37.25	37.25	37.25	12.25	3.75
	Personnel cost	—	—	102	245	410	410	410	410	410	410	410	135	41
	Material handling cost	—	—	370	890	1,490	1,490	1,490	1,490	1,490	1,490	1,490	490	150
	Documentation cost	—	—	370	890	1,490	1,490	1,490	1,490	1,490	1,490	1,490	490	150
2. Intermediate maintenance	Total labor-hours	—	—	111	267	447	447	447	447	447	447	447	147	45
	Personnel cost	—	—	1,443	3,741	5,811	5,811	5,811	5,811	5,811	5,811	5,811	1,911	585
	Material handling cost	—	—	370	890	1,490	1,490	1,490	1,490	1,490	1,490	1,490	490	150
	Documentation cost	—	—	740	1,780	2,980	2,980	2,980	2,980	2,980	2,980	2,980	980	300
3. Depot maintenance	Total labor-hours	—	—	16	36	60	60	60	60	60	60	60	20	8
	Personnel cost	—	—	240	540	900	900	900	900	900	900	900	300	120
	Material handling cost	—	—	40	90	150	150	150	150	150	150	150	50	20
	Documentation cost	—	—	40	90	150	150	150	150	150	150	150	50	20
Total unscheduled maintenance cost		—	—	3,715	8,886	14,871	14,871	14,871	14,871	14,871	14,871	14,871	4,896	1,536
B. Alternative B														
1. Organizational maintenance	Total labor-hours	—	—	11.25	27.25	46.00	46.00	46.00	46.00	46.00	46.00	46.00	15.25	4.50
	Personnel cost	—	—	124	300	506	506	506	506	506	506	506	168	50
	Material handling cost	—	—	450	1,090	1,840	1,840	1,840	1,840	1,840	1,840	1,840	610	180
	Documentation cost	—	—	450	1,090	1,840	1,840	1,840	1,840	1,840	1,840	1,840	610	180
2. Intermediate maintenance	Total labor-hours	—	—	135	327	552	552	552	552	552	552	552	183	54
	Personnel cost	—	—	1,755	4,251	7,176	7,176	7,176	7,176	7,176	7,176	7,176	2,379	702
	Material handling cost	—	—	450	1,090	1,840	1,840	1,840	1,840	1,840	1,840	1,840	610	180
	Documentation cost	—	—	900	2,180	3,680	3,680	3,680	3,680	3,680	3,680	3,680	1,220	360
3. Depot maintenance	Total labor-hours	—	—	20	44	72	72	72	72	72	72	72	24	8
	Personnel cost	—	—	300	660	1,080	1,080	1,080	1,080	1,080	1,080	1,080	360	120
	Material handling cost	—	—	50	110	180	180	180	180	180	180	180	60	20
	Documentation cost	—	—	50	110	180	180	180	180	180	180	180	60	20
Total unscheduled maintenance cost		—	—	4,529	10,881	18,322	18,322	18,322	18,322	18,322	18,322	18,322	6,077	1,812

system is required to exhibit a MTBM of 500 hrs or greater. Response to this requirement by the two suppliers is illustrated in Figure 19.8. Note that the predicted MTBM for alternative *A* is 650 hours and for alternative *B* is 525 hours. These values are further broken down to unit levels compatible with the system packaging scheme in Figure 19.3 and illustrated maintenance concept in Figure 19.4. Although failures are randomly distributed in general, these values are used to determine an average factor for the frequency of maintenance.

Using the information in Figure 19.8, the next step is to calculate the average number of maintenance actions for each system (and unit-level) configuration by year through the life cycle. This is accomplished by dividing the system operating time in Table 19.4 by the MTBM. The results (rounded off to the nearest whole number), are presented in Table 19.7.[9]

Personnel labor cost covering the accomplishment of unscheduled maintenance is a function of maintenance labor hours and the cost per labor hour. Maintenance labor hour and labor cost factors in this instance are based on the number of technicians (with specific skill levels) assigned to a given maintenance action and the length of time that the technicians are assigned. The assumed factors are

> Organizational maintenance—0.25 MLH per maintenance action at a cost of $11.00 per MLH.
>
> Intermediate maintenance—3 MLH per maintenance action at a cost of $13.00 per MLH.
>
> Supplier (or depot) maintenance—4 MLH per maintenance action at a cost of $15.00 per MLH.

The labor-hour factors used are related to personnel requirements for direct maintenance where the elapsed times are indicated by the $\overline{M}ct$ values. The hourly rates include direct dollars plus a burden or overhead rate. Labor-hour calculations, personnel cost, material handling cost (i.e., $10 per maintenance action), and documentation cost (i.e., $30 per maintenance action, or $20 at the intermediate level and $10 at the organizational level) are noted in Table 19.6 for each of the configurations being evaluated. The total cost factors are also included in Table 19.5.

4. *Maintenance facilities cost* (C_{OLM}) is based on the occupancy, utilities, and facility maintenance costs as prorated to the communication system. Facilities cost in this instance is primarily related to the intermediate level of maintenance.

5. *Supply support cost* (C_{OLS}) includes the cost of spare parts required as a result of system failures; spare parts required to fill the logistics pipeline to compensate for delays due to active repair times, turnaround times, and supplier lead times; spare parts required to replace repairable items which are condemned or phased out of the inventory for one reason or another (e.g., those items that are damaged to the

---

[9] A more precise method is to use Monte Carlo analysis to determine the number of maintenance actions and then to assess each individual anticipated maintenance action in terms of expected logistic support resource requirements. The analyst should adapt to the needs of the analysis, but it is felt that the approach used herein is adequate for the purposes intended.

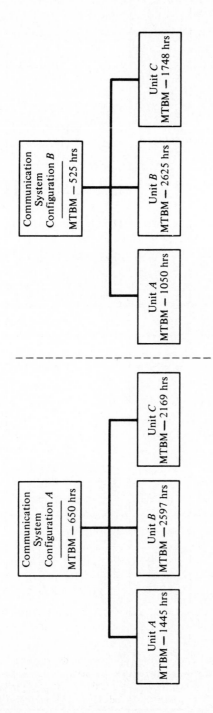

Note: The frequency of maintenance in this instance is MTBM. It may be preferable to work with frequency factors rather than with MBTM values.

Figure 19.8 System maintenance factors.

TABLE 19.7 NUMBER OF UNSCHEDULED MAINTENANCE ACTIONS

ALTERNATIVE	APPLICATION	SYSTEM OR UNIT	1	2	3	4	5	6	7	8	9	10	11	12	13
	Aircraft application	System	—	—	1	3	3	3	3	3	3	3	3	3	1
		Unit A	—	—	1	1	1	1	1	1	1	1	1	1	1
		Unit B	—	—	—	1	1	1	1	1	1	1	1	1	—
		Unit C	—	—	—	1	1	1	1	1	1	1	1	1	—
	Helicopter application	System	—	—	2	4	4	4	4	4	4	4	4	4	3
		Unit A	—	—	1	2	2	2	2	2	2	2	2	2	1
		Unit B	—	—	—	1	1	1	1	1	1	1	1	1	1
		Unit C	—	—	1	1	1	1	1	1	1	1	1	1	1
Alternative configuration A	Patrol vehicle application	System	—	—	14	44	100	100	100	100	100	100	100	20	4
		Unit A	—	—	6	20	45	45	45	45	45	45	45	9	2
		Unit B	—	—	4	11	25	25	25	25	25	25	25	5	1
		Unit C	—	—	4	13	30	30	30	30	30	30	30	6	1
	Communications facilities	System	—	—	11	22	22	22	22	22	22	22	22	11	—
		Unit A	—	—	5	10	10	10	10	10	10	10	10	5	—
		Unit B	—	—	2	5	5	5	5	5	5	5	5	2	—
		Unit C	—	—	4	7	7	7	7	7	7	7	7	4	—
	Central communications control	System	—	—	7	13	17	17	17	17	17	17	17	8	5
		Unit A	—	—	3	6	8	8	8	8	8	8	8	4	2
		Unit B	—	—	2	3	4	4	4	4	4	4	4	2	1
		Unit C	—	—	2	4	5	5	5	5	5	5	5	2	2
	Mobile vans	System	—	—	2	3	3	3	3	3	3	3	3	3	2
		Unit A	—	—	1	1	1	1	1	1	1	1	1	1	1
		Unit B	—	—	—	1	1	1	1	1	1	1	1	1	—
		Unit C	—	—	1	1	1	1	1	1	1	1	1	1	1

TABLE 19.7 CONTINUED

ALTERNATIVE	APPLICATION	SYSTEM OR UNIT	PROGRAM YEAR												
			1	2	3	4	5	6	7	8	9	10	11	12	13
	Aircraft application	System	—	—	2	4	4	4	4	4	4	4	4	4	2
		Unit A	—	—	1	2	2	2	2	2	2	2	2	2	1
		Unit B	—	—	—	1	1	1	1	1	1	1	1	1	—
		Unit C	—	—	1	1	1	1	1	1	1	1	1	1	1
	Helicopter application	System	—	—	2	5	5	5	5	5	5	5	5	5	3
		Unit A	—	—	1	2	2	2	2	2	2	2	2	2	1
		Unit B	—	—	—	1	1	1	1	1	1	1	1	1	1
		Unit C	—	—	1	2	2	2	2	2	2	2	2	2	1
Alternative configuration B	Patrol vehicle application	System	—	—	17	54	124	124	124	124	124	124	124	25	5
		Unit A	—	—	9	27	62	62	62	62	62	62	62	12	2
		Unit B	—	—	3	11	25	25	25	25	25	25	25	5	1
		Unit C	—	—	5	16	37	37	37	37	37	37	37	8	2
	Communications facilities	System	—	—	14	27	27	27	27	27	27	27	27	14	—
		Unit A	—	—	7	14	14	14	14	14	14	14	14	7	—
		Unit B	—	—	3	5	5	5	5	5	5	5	5	3	—
		Unit C	—	—	4	8	8	8	8	8	8	8	8	4	—
	Central communications control	System	—	—	8	16	21	21	21	21	21	21	21	10	6
		Unit A	—	—	4	8	11	11	11	11	11	11	11	5	3
		Unit B	—	—	2	3	4	4	4	4	4	4	4	2	1
		Unit C	—	—	2	5	6	6	6	6	6	6	6	3	2
	Mobile vans	System	—	—	2	3	3	3	3	3	3	3	3	3	2
		Unit A	—	—	1	1	1	1	1	1	1	1	1	1	1
		Unit B	—	—	—	1	1	1	1	1	1	1	1	1	—
		Unit C	—	—	1	1	1	1	1	1	1	1	1	1	1

extent beyond which it is economically feasible to accomplish repair); and the cost of maintaining the inventory throughout the designated period of support.

Referring to the illustrated maintenance concept in Figure 19.4, spare units are required in the intermediate maintenance shops to support unit replacements at the organizational level. Spare assemblies, modules, and certain designated piece parts are required to support intermediate maintenance actions, and some assemblies and parts are required to support supplier repair activities or depot maintenance. In other words, the maintenance concept indicates the type of spares required at each level of maintenance, and the system network in Figure 19.1 (with specific geographical locations defined) conveys the quantity of maintenance facilities providing overall system support. A logistic support analysis is accomplished to provide additional maintenance data as required.

The next step is to determine the quantity of spare parts required at each location. Too many spares, or a large inventory, could be extremely costly in terms of investment and inventory maintenance. On the other hand, not enough spares could result in a stock-out condition which will in turn cause systems to be inoperative, and the defined objectives of the communication network will not be met. This condition may also be quite costly. The goal is to analyze the inventory requirements in terms of a profile similar to the illustration in Figure 16.4 and obtain a balance between the cost of acquiring spares and the cost of maintaining inventory as described in Section 16.3. Referring to Figure 16.4, the critical factor constitutes consumption or demand and the probability of having the right type of spare part available when required. Relative to this analysis, demand rates are a function of unit, assembly, module, or part reliability and are based on the Poisson distribution. Intuitive in the model employed herein is the expression in Equation 16.7, which is used in spare-part-quantity determination.

Another consideration is the turnaround time (TAT) and the transportation time between facilities for repairable items (i.e., the total time from the point of failure until the item is repaired and recycled back into the inventory and ready for use, or the time that it takes to acquire an item from the source of supply). These time factors are identified in Figure 19.4, have a significant impact on spare-part requirements, and are considered in the procurement lead time calculation shown in Figure 9.8.

The process employed for determining the costs associated with supply support is fairly comprehensive, and all of the specific detailed steps used in this analysis are not included here. However, the concepts used are discussed. In essence, inventory requirements are covered in two categories. An initial procurement of spare units and assemblies, basically representing safety stock, is reflected under initial logistics support cost (i.e., category C_{PLS}). The spares required for sustaining support, to include both repairable and nonrepairable items, are covered in this category. These items are directly related to consumption and the quantity of unscheduled maintenance actions identified in Table 19.7. The associated costs include both material costs and the cost of maintaining the inventory. Annual inventory holding cost is assumed to be 20% of the inventory value.

6. *Maintenance personnel training cost* (C_{OLT}) is covered in two categories. Initially, when the system is first introduced, there is a requirement to train operators and maintenance technicians. The cost for this initial training is included in category C_{PLP}. Subsequently, the cost of training relates to personnel attrition and the addition of new operators and/or maintenance technicians. A figure of $1,300 per year is assumed for formal sustaining training until that point in time when system phase-out commences.

7. *Test and support equipment cost* (C_{OLE}) is presented in two categories. Category C_{PLT} includes the design and acquisition of the test and support equipment required for the intermediate and depot levels of maintenance. This category includes the sustaining support of these items on a year-to-year basis (i.e., the unscheduled and scheduled maintenance actions associated with the test equipment). Unscheduled maintenance is a function of the use of the test equipment, which in turn directly relates to the unscheduled maintenance actions noted in Table 19.7. In addition, the reliability and maintainability characteristics of the test equipment itself will significantly influence the cost of supporting that test equipment. Scheduled maintenance constitutes the periodic 180-day calibration of certain elements of the test equipment in the calibration laboratory (refer to Figure 19.4). Calibration is required to maintain the proper test traceability to primary and secondary standards. The costs associated with test equipment maintenance and logistic support can be derived through an in-depth logistic support analysis (described in Chapter 16). However, the magnitude of the test equipment required in this case is relatively small when compared to other systems. Based on past experience with comparable items, a factor of 5% of the acquisition cost ($3,250) is considered to be appropriate for the annual maintenance and logistics cost for the test equipment associated with alternative *A*. A factor of 6.5% ($4,225) is assumed for alternative *B*, since the test equipment use requirements will be greater.

8. *Transportation and handling costs* (C_{OLH}) includes the annual costs associated with the movement of materials among the organizational, intermediate, and depot levels of maintenance. This is in addition to the costs of initial distribution and system installation covered in category C_{ODT}.

For the system being analyzed, the movement of materials between the organizational and intermediate levels of maintenance is not considered to be significant in terms of relative cost. However, the shipment of materials between the intermediate maintenance facilities and the supplier for depot maintenance is considered to be significant, and can be determined from Equation 19.9.

$$C_{OLH} = \text{(cost of transportation)} + \text{(cost of packing)}$$
$$\times \text{(quantity of one-way shipments)} \qquad (19.9)$$

where transportation and packing costs are $30 per cwt and $40 per cwt, respectively. The material being moved between the intermediate maintenance facilities and the supplier includes assembly 5 of Unit *C*, which is supported at the depot level of maintenance (refer to Figure 19.4). The estimated quantity of one-way trips are given in Table 19.8.

TABLE 19.8 TRIPS BETWEEN MAINTENANCE
FACILITIES AND SUPPLIER

	ALTERNATIVE A	ALTERNATIVE B
Year 3	8	10
Year 4	18	22
Year 5–11	30 per year	36 per year
Year 12	10	12
Year 13	4	4

19.4. EVALUATION OF ALTERNATIVES

The problem is to select the best of the two alternatives on the basis of total life-cycle cost. A comparison of alternatives A and B using this criterion is presented in Table 19.9. The costs are listed for those major categories of the cost breakdown structure that are relevant to this analysis. The other categories of cost that are included in Figure A.5, and not here, are considered as being not applicable to this case study.

Referring to Figure 19.9., the initial concern is to determine whether the candidates meet the specified requirements (i.e., performance, the unit life-cycle cost goal of $35,000, the MTBM of 500 hours etc.). In this instance, both alternatives meet the requirements and fall within the trade-off area identified in Figure 19.9.

When evaluating two or more alternatives on a relative basis, the future cost projections for each alternative must be reduced to their present amounts. The present costs for alternatives A and B are presented in Table 19.10, and the cost profiles are illustrated in Figure 19.10 based on an interest rate of 10%. The present equivalent cost for alternative A is

$$P = \$162,712(\overset{P/F,10,1}{0.9091}) + \$279,757(\overset{P/F,10,2}{0.8265}) + \cdots + \$11,762(\overset{P/F,10,13}{0.2897})$$

$$= \$1,663,200.$$

For alternative B the present equivalent cost is

$$P = \$166,124(\overset{P/F,10,1}{0.9091}) + \$270,167(\overset{P/F,10,1}{0.8265}) + \cdots + \$13,258(\overset{P/F,10,13}{0.2897})$$

$$= \$1,704,799.$$

The results of this analysis support alternative A as the preferred configuration on the basis of present life-cycle cost. Note that the research and development (R&D) cost is higher for alternative A; however, the overall life-cycle cost is lower, owing to a significantly lower operation and support (O&S) cost. This would tend to indicate that the equipment design for reliability pertaining to alternative A is somewhat better. Although this increased reliability results in higher R&D cost, the anticipated quantity of maintenance actions is lower resulting in lower O&S costs.

TABLE 19.9 LIFE-CYCLE COST BREAKDOWN

COST CATEGORY	ALTERNATIVE A Cost ($)	ALTERNATIVE A % of Total	ALTERNATIVE B Cost ($)	ALTERNATIVE B % of Total
1. Research and development cost (C_R)				
(a) System/product management (C_{RM})	247,071	9.8	240,823	9.2
(b) Product planning (C_{RP})	37,090	1.5	15,960	0.7
(c) Engineering design (C_{RE})	237,016	9.4	238,119	9.1
(d) Design data (C_{RD})	48,425	1.9	39,624	1.5
(e) System test and evaluation (C_{RT})	67,648	2.7	60,531	2.3
Subtotal	637,250	25.3	595,057	22.8
2. Production and construction cost (C_P)				
(a) System/product management (C_{RM})	112,497	4.5	112,497	4.3
(b) Industrial engineering and operations analysis (C_{PI})	68,200	2.8	66,000	2.5
(c) Manufacturing—recurring (C_{PMR})	635,100	25.3	657,500	25.2
(d) Manufacturing—nonrecurring (C_{PMN})	48,900	1.9	55,100	2.1
(e) Quality control (C_{PQ})	78,502	3.1	72,990	2.8
(f) Initial logistics support (C_{PL})				
(1) Supply support—initial (C_{PLS})	48,000	1.9	57,020	2.2
(2) Test and support equipment (C_{PLT})	70,000	2.8	70,000	2.7
(3) Technical data (C_{PLD})	5,100	0.2	5,350	0.2
(4) Personnel training (C_{PLP})	46,400	1.8	46,400	1.8
Subtotal	1,112,699	44.3	1,142,857	43.8
3. Operation and support cost (C_0)				
(a) Operating personnel (C_{OOP})	24,978	0.9	24,978	0.9
(b) Distribution—transportation (C_{ODT})	268,000	10.7	268,000	10.4
(c) Unscheduled maintenance (C_{OLA})	123,130	4.9	151,553	5.8
(d) Maintenance facilities (C_{OLM})	4,626	0.2	5,705	0.2
(e) Supply support (C_{OLS})	295,920	11.8	365,040	14.0
(f) Maintenance personnel training (C_{OLT})	9,100	0.4	9,100	0.3
(g) Test and support equipment (C_{OLE})	32,500	1.3	42,251	1.6
(h) Transportation and handling (C_{OLH})	5,250	0.2	6,300	0.2
Subtotal	763,504	30.4	872,927	33.4
Grand total	2,513,453	100%	2,610,841	100%

Prior to a final decision on which alternative to select, the analyst should accomplish a break-even analysis to determine the point in time when alternative A becomes more economical than alternative B. Figure 19.11 indicates that the break-even point, or the point in time when alternative A becomes less costly, is approximately four years and ten months after the program start. This point is early enough in the life cycle to support the decision. If this crossover point were much further out in time, the decision might be questioned.

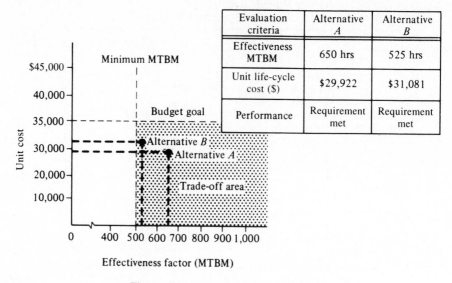

Evaluation criteria	Alternative A	Alternative B
Effectiveness MTBM	650 hrs	525 hrs
Unit life-cycle cost ($)	$29,922	$31,081
Performance	Requirement met	Requirement met

Figure 19.9   Effectiveness versus unit cost.

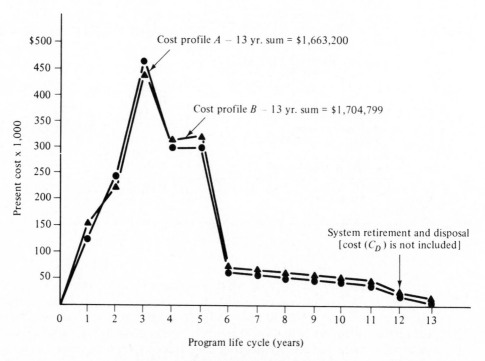

Figure 19.10   Alternative cost profiles.

TABLE 19.10 COST ALLOCATION BY PROGRAM YEAR

PROGRAM ACTIVITY	COST CATEGORY DESIGNATION	COST BY PROGRAM YEAR													TOTAL ACTUAL COST ($)
		1	2	3	4	5	6	7	8	9	10	11	12	13	
A. Alternative A															
1. Research and development	C_R	162,712	194,357	280,181	—	—	—	—	—	—	—	—	—	—	637,250
2. Production/construction	C_P		85,400	307,000	360,270	360,270	—	—	—	—	—	—	—	—	1,112,699
3. Operation and support	C_O		—	23,388	56,792	90,954	90,954	90,954	90,954	90,954	90,954	90,954	34,884	11,762	763,504
Total actual cost	C	162,712	279,757	610,569	417,062	450,983	90,954	90,954	90,954	90,954	90,954	90,954	34,884	11,762	$2,513,453
Total present cost	$C_{10\%}$	147,921	231,191	458,720	284,853	280,015	51,344	46,659	42,430	38,574	35,063	31,879	11,114	3,407	$1,663,200
B. Alternative B															
1. Research and development	C_R	166,124	173,317	255,616	—	—	—	—	—	—	—	—	—	—	595,057
2. Production/construction	C_P		96,850	305,140	361,510	379,357	—	—	—	—	—	—	—	—	1,142,857
3. Operation and support	C_O		—	26,682	64,721	104,037	104,037	104,037	104,037	104,037	104,037	104,037	40,007	13,258	872,927
Total actual cost	C	166,124	270,167	587,438	426,231	483,394	104,037	104,037	104,037	104,037	104,037	104,037	40,007	13,258	$2,610,841
Total present cost	$C_{10\%}$	151,023	223,266	441,342	291,116	300,139	58,729	53,371	48,533	44,122	40,106	36,465	12,746	3,841	$1,704,799

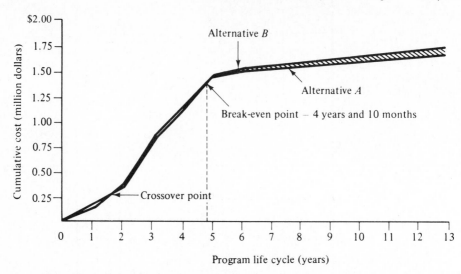

Figure 19.11 Break-even analysis.

19.5. *SENSITIVITY ANALYSIS*

When considering the analysis results, it is noted that the delta cost between the two alternatives is only $41,599 (refer to Table 19.10), and that the cost profiles are relatively close to each other (refer to Figure 19.10). These factors do not overwhelmingly support a clear-cut decision in favor of alternative *A* without introducing some risk. In view of the possible inaccuracies associated with the input data, the analyst may wish to perform a sensitivity analysis to determine the effects of input parameter variations on the present life-cycle cost. The analyst should determine how much variation can be tolerated before the decision shifts in favor of alternative *B*.

Referring to Table 19.9, the analyst should select the *high contributors* (those which contribute more than 10% of the total cost), determine the cause-and-effect relationships, and identify the various input data factors that directly impact cost. In instances where such factors are based on highly questionable prediction techniques, the analyst should vary these factors over a probable range of values and assess the results. For instance, key input parameters in this analysis include the system operating time (in hours) and the MTBM factor. Using the life-cycle cost model, the analyst will apply a multiple factor to the operating time values and the MTBM and determine the delta cost associated with each variation. From this, trend curves are projected, as illustrated in Figure 19.12.

Through analysis of the information presented, it is quite evident that a very *small* variation in operating time and/or MTBM may cause the decision to shift in favor of alternative B. This provides an indication that there is a high degree of risk associated with making a wrong decision. Thus, the analyst should make every effort to reduce this risk by improving the input data to the greatest extent possible. Also, where the results are particularly close, the magnitude of risk associated with the decision must be determined.

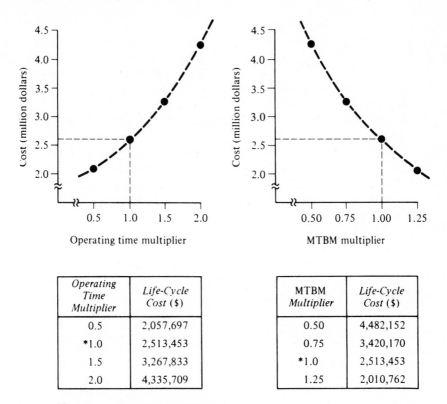

Operating Time Multiplier	Life-Cycle Cost ($)
0.5	2,057,697
*1.0	2,513,453
1.5	3,267,833
2.0	4,335,709

MTBM Multiplier	Life-Cycle Cost ($)
0.50	4,482,152
0.75	3,420,170
*1.0	2,513,453
1.25	2,010,762

*Baseline configuration for Alternative *A*

Figure 19.12   Sensitivity analysis.

19.6. SUMMARY

This project illustrates the application of a life-cycle cost analysis to support a design decision where two alternative configurations are being evaluated in response to a stated need. The objective is to select the configuration that will fulfill all system performance and effectiveness requirements, fall within a specified budget constraint, and reflect the lowest life-cycle cost.

Inherent in the process of evaluating the two design alternatives are the initial steps of the system design process illustrated in Figure 2.4. A need is defined through a needs analysis, system operational requirements are identified, and a maintenance concept is described in accordance with the concepts presented in Chapter 3. Based on this information, design requirements are described in terms of a functional packaging approach (refer to Figure 19.3), and quantitative effectiveness factors are established in terms of MTBM, $\overline{M}ct$, MLH/OH, and cost.

The evaluation itself addresses two candidate systems, each of which meets the specified performance and effectiveness requirements and falls within the allocated budget. However, each configuration exhibits different performance and effectiveness characteristics, and the objective is to select the best in terms of life-cycle cost. The evaluation follows the systems analysis process illustrated in Figure 3.7.

Accomplishing the life-cycle cost analysis incorporates the steps described in Chapter 17. A cost breakdown structure is developed (refer to Figure 19.5), cost-generating variables and factors are identified, and costs are summarized year by year. Finally, cost profiles are developed, a break-even analysis is accomplished, and the best configuration is selected using the economic analysis techniques of Chapter 8.

In developing data for the life-cycle cost analysis, many different factors are employed, particularly with regard to the determination of costs for system operation and support. Reliability factors are used in the determination of maintenance frequencies (i.e., the number of unscheduled maintenance actions listed in Table 19.7). These are derived from reliability prediction data described in Chapter 13. Maintainability factors ($\overline{M}ct$ and MLH/OH) are employed to determine personnel costs. These factors are developed from maintainability prediction data described in Chapter 14. Logistic support requirements include supply support factors, test and support equipment requirements, transportation requirements, and so on, developed through the LSA described in Chapter 16. Much of the data developed in this case study utilizes the concepts described in Part Four, supported by some of the analysis techniques described in Part Three.

A final step in the life-cycle cost analysis involves the application of sensitivity analysis to assess the risk(s) associated with a given decision. Further, the accomplishment of a sensitivity analysis aids the analyst to evaluate cause-and-effect relationships and the interactions that occur with system parameter variation (refer to Figure 19.12).

It must be realized that the system considered in this case is actually one unit in a finite population of similar units. The population is deployed to meet the demand for an improved overall communication capability in a metropolitan area. An optimal population design requires an approach presented in the next system design application.

20

Repairable Equipment Population System Design[1]

The system under consideration in this system design application may be described as follows: a finite population of repairable equipment is procured and maintained in operation to meet a demand. As repairable equipment units fail or become unserviceable, they are repaired and returned to service. As they age, the older units are removed from the system and replaced with new units. The general equipment design problem is to determine the population size, the replacement age of units, the number of repair channels, the design mean time between failures, and the design mean time to repair, so that the sum of all costs associated with the system will be minimized.

20.1. INTRODUCTION

Two problems are treated in this system design application. The first is to determine the population size, the replacement age of units, and the number of repair channels so that the sum of all costs associated with the system will be minimized. The second design problem is to generate and evaluate candidate systems by specifying the

[1] This system design application was adapted from W. J. Fabrycky and J. T. Hart, "Economic Optimization of a Finite Population System Deployed to Meet a Demand," Proceedings of the Joint Conference, American Association of Cost Engineers/American Institute of Industrial Engineers, Houston, Texas, 1975 and from J. Banks and W. J. Fabrycky, *Procurement and Inventory Systems Analysis* (Englewood Cliffs, N.J.: Prentice-Hall, 1987).

unit mean time between failures, MTBF, and the unit mean time to repair, MTTR, as a function of unit cost, as well as the population size, the replacement age of units, and the number of repair channels.

The Operational System

The repairable equipment population system (REPS) illustrated in Figure 20.1 is designed and deployed to meet a demand, D. Units within the system can be separated into two groups; those in operation and available to meet demand, and those out of operation and hence unavailable to meet demand. It is assumed that units are not discarded upon failure, but are repaired and returned to operation.

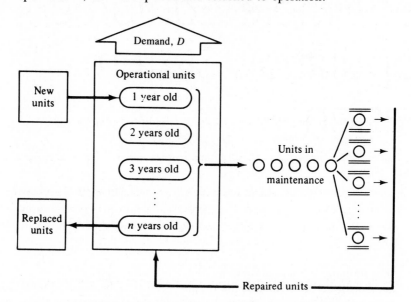

Figure 20.1 The repairable equipment population system.

As units age, they become less reliable and their maintenance costs increase. Accordingly, it is important to determine the optimum replacement age. It is assumed that the number of new units procured each year is constant and that the number of units in each age group is equal to the ratio of the total number required in the population and the desired number of age groups. Although the analysis deals with the life cycle of the units, the objective is to optimize the total system of which the units are a part.

In Design Problem I, the design process consists of specifying a population of units, a number of maintenance channels, and a replacement schedule for bringing new units into the system. For Design Problem II, the design process is extended to include establishing the unit's reliability and maintainability characteristics. In either case the system is to be designed to meet the demand for equipment optimally.

The foregoing paragraphs have described, in general terms, the finite population queuing system under investigation. This model describes the operation of numerous systems. For example, both the airlines and the military operate and maintain aircraft with these system characteristics. In ground transit, vehicles such as rental automobiles, taxis, and commercial trucks constitute repairable equipment populations. Production equipment types such as weaving looms, drill presses, and autoclaves are populations of equipment which fit the repairable classification. But the repairable unit may also be an inventory of components for the larger entities mentioned. For example, aircraft hydraulic pumps, automobile starters and alternators, truck engines, electric motors, and electronic controllers for equipment also constitute repairable equipment population systems.

Scope and Assumptions

Repairable equipment population systems normally come into being over a non-steady-state build-up phase. They then operate over a steady-state interval of years, after which a phase-out period is entered. Only the steady-state mode of operation will be considered here.

The following assumptions are adopted in the development of the mathematical model and algorithm for REPS.

1. The interarrival times are exponentially distributed.
2. The repair times are exponentially distributed.
3. The number of units in the population is small such that finite population queuing formulations must be used.
4. The interarrival times are statistically independent of the repair times.
5. The repair channels are parallel and each is capable of similar performance.
6. The population size will always be larger than or at least equal to the number of service channels.
7. Each channel performs service on one unit at a time.
8. MTBF and MTTR values vary for each age group and represent the expected value for these variables for that age group.
9. Units completing repair return to operation with the same operational characteristics of their age group.

System Design Evaluation

Design Problem I may be evaluated by using the decision evaluation function of Equation 7.1. Here the objective is to find optimal values for controllable design variables in the face of uncontrollable system parameters. In Design Problem II, the decision evaluation function of Equation 7.2 is used to seek the best candidate system. This is accomplished by establishing values for controllable design dependent parameters in the face of uncontrollable design independent parameters and optimal values for design variables.

Design variables.   Three design variables are identified in the repairable equipment population system. These controllables are the number of units to deploy, the replacement age of units, and the number of repair channels. Optimal values are sought for these variables so that the sum of all costs associated with the repairable equipment population system will be minimized.

In Design Problem I the focus is entirely on optimizing design variables as the only controllable factors. This situation arises when the system is in existence and the objective is to optimize its operation in the face of uncontrollable system parameters. The focus shifts to seeking the best candidate system in Design Problem II. In this activity, optimal values for design variables are secondary. They are needed as a means for comparing candidate systems equivalently. Then, when the best system is identified, they are used to specify its optimal operation.

System parameters.   Demand is the primary stimulus on the repairable equipment population system and the justification for its existence. This uncontrollable system parameter is assumed to be constant over time. Other uncontrollable system parameters are economic in nature. They include the shortage penalty cost which arises when there are insufficient units operational to meet demand, the cost of providing repair capability, and the time value of money on invested capital.

Some system parameters are uncontrollable in Design Problem I, but controllable in Design Problem II. These are the design MTBF and MTTR, the energy efficiency of equipment units, the design life of units, and the first cost and salvage value of these units. It is through these design dependent system parameters that the best candidate system may be identified.

20.2. MATHEMATICAL MODEL FORMULATION

A mathematical model for system design evaluation can be formulated using the structure outlined in Section 7.3. The model utilizes annual equivalent life-cycle cost as the evaluation measure expressed as

$$AELCC = PC + OC + RC + SC \tag{20.1}$$

where

$AELCC$ = annual equivalent life-cycle cost
PC = annual equivalent population cost
OC = annual operating cost
RC = annual repair facility cost
SC = annual shortage penalty cost

Annual Equivalent Population Cost

The annual equivalent cost of a deployed population of N equipment units is

$$PC = C_i N$$

with[2]

$$C_i = P \left(\overset{A|P,i,n}{}\right) - B \left(\overset{A|F,i,n}{}\right) \tag{20.2}$$

[2] Alternatively, $(P - F)\left(\overset{A|P,i,n}{}\right) + Fi$ could be used as given by Equation 8.7.

Book value, B, in Equation 20.2 is used to represent the original value of a unit minus its accumulated depreciation at any point in time. The depreciation of a unit over its lifetime by the straight-line method gives an expression for book value as

$$B = P - n\frac{P - F}{L} \qquad (20.3)$$

where

C_i = annual equivalent cost per unit
P = first or acquisition cost of a unit
F = estimated salvage value of a unit
B = book value of a unit at the end of year n
L = estimated life of the unit
N = number of units in the population
n = retirement age of units $n > 1$
i = annual interest rate

Annual Operating Cost

The annual cost of operating a population of N deployed equipment units is

$$OC = (EC + LC + PMC + \text{Other})N \qquad (20.4)$$

where

EC = annual cost of energy consumed
LC = annual cost of operating labor
PMC = annual cost of preventative maintenance

Other annual operating costs may be incurred. These include all recurring annual costs of keeping the population of equipment units in service such as storage cost, insurance premiums, and taxes.

Annual Repair Facility Cost

The annual cost of providing a repair facility to repair failed equipment units is

$$RC = C_r M \qquad (20.5)$$

where

C_r = annual fixed and variable repair cost per repair channel
M = number of repair channels

If there are a number of repair channel components with different estimated lives, then C_r is the sum of their annual costs. Some of the repair facility cost items that could be included are the cost of the building, maintenance supplies, test equipment, and so on, expressed on a per channel basis. The administrative, maintenance manpower, and other overhead costs would also be computed on a yearly basis and on a per channel basis.

Annual Shortage Penalty Cost

When failed equipment units cause the number in an operational state to fall below the demand, an out-of-operation or shortage cost is incurred. The annual shortage cost is the product of the shortage cost per unit short per year and the expected number of units short expressed as

$$SC = C_s[E(S)]. \tag{20.6}$$

The expected number of units short can be found from the probability distribution of n units short, P_n, as developed from Equations 11.38 and 11.39 in Section 11.6.

Define the quantity $N - D$ as the number of extra items to be held in the population. For $n = 0, 1, 2, \ldots, N - D$ there is no shortage of items. However, when

$$n = N - D + 1, \text{ a shortage of 1 item exists}$$

$$n = N - D + 2, \text{ a shortage of 2 items exists}$$

$$\vdots$$

$$n = N, \qquad \text{a shortage of } D \text{ items exists}$$

We can now compute the expected number of items short, $E(S)$, by multiplying the number of items short by the probability of that occurrence as

$$E(S) = \sum_{j=1}^{D} jP_{(N-D+j)}. \tag{20.7}$$

20.3. DESIGN PROBLEM I

In this REPS design example, the decision maker has no control over system parameters but can only choose the number of equipment units to procure and deploy, the age at which units should be replaced, and the number of channels in the repair facility.

Assume that the demand, D, is for 15 identical equipment units. Table 20.1 lists system parameters for this design example.

Table 20.2. exhibits a set of design variables for the example under consideration. The example computations which follow are based on these variables and the system parameters in Table 20.1.

Annual Equivalent Population Cost

First compute the book value, B, of the units at retirement age after four years using Equation 20.3 as

$$\$52,000 - 4\left(\frac{\$52,000 - \$7,000}{6}\right) = \$22,000.$$

TABLE 20.1 SYSTEM PARAMETERS FOR
DESIGN PROBLEM I EXAMPLE

PARAMETER	VALUE	
Unit acquisition cost	$52,000	
Unit design life	6 years	
Unit salvage value at end of design life	$7,000	
Unit operating cost:		
energy and fuel	$500	
operating labor	450	
preventive maintenance	400	
other operating costs	400	
Annual repair channel cost	$45,000	
Annual shortage cost	$73,000	
Annual interest rate	10%	
Age cohorts	MTBF	MTTR
0–1	0.20	0.03
1–2	0.24	0.04
2–3	0.29	0.05
3–4	0.29	0.05
4–5	0.26	0.06
5–6	0.22	0.07

TABLE 20.2 DESIGN VARIABLES FOR DESIGN PROBLEM 1

POPULATION, N	REPAIR CHANNELS, M	RETIREMENT AGE, n
19	3	4

The annual equivalent unit cost per unit from Equation 20.2 is

$$C_i = \$52,000\left[\frac{0.10\,(1.10)^4}{(1.10)^4 - 1}\right] - \$22,000\left[\frac{0.10}{(1.10)^4 - 1}\right]$$

$$= \$16,404 - \$4,740 = \$11,664$$

from which the annual equivalent population cost is

$$PC = \$11,664(19) = \$221,616.$$

Annual Operating Cost

Annual operating cost for the deployed population is found from Equation 20.4 to be

$$OC = (\$500 + \$450 + \$400 + \$400)(19) = \$33,250.$$

Annual Repair Facility Cost

The annual equivalent repair channel cost for three channels from Equation 20.5 is

$$RC = \$45,000(3) = \$135,000.$$

Annual Shortage Cost

Calculation of the shortage cost is based on the MTBF and MTTR values from Table 20.1 for years 0-4. From these values, the average MTBF and MTTR for the population can be computed as[3]

$$MTBF = (1/4)(0.20 + 0.24 + 0.29 + 0.29) = 0.2550$$

$$MTTR = (1/4)(0.03 + 0.04 + 0.05 + 0.05) = 0.0425$$

The failure rate of an item and the repair rate at a repair channel are given by

$$= (1/MTBF) = (1/0.2550) = 3.9215$$

$$= (1/MTTR) = (1/0.0425) = 23.5294$$

from which $\left(\dfrac{\lambda}{\mu}\right) = 1/6$.

Next compute C_n for $n = 0, 1, \ldots, 3$ from Equation 11.39 as

$$C_0 = \frac{19!(1/6)^0}{19!0!} = 1$$

$$C_1 = \frac{19!(1/6)^1}{18!1!} = 3.1665$$

$$C_2 = \frac{19!(1/6)^2}{17!2!} = 4.7496$$

$$C_3 = \frac{19!(1/6)^3}{16!3!} = 4.4856$$

Computing C_n for $n = 4, \ldots, 19$, also from Equation 11.39,

---

[3] Since the units are homogenous, the aggregate MTBF and MTTR for the population can be found as

$$MTBF = \frac{1}{n} \sum_{j=1}^{n} MTBF_j$$

$$MTTR = \frac{1}{n} \sum_{j=1}^{n} MTTR_j$$

where the subscript j represents age groups and n is the number of these age groups. This follows from the superposition of Poisson processes.

$$C_4 = \frac{19!(1/6)^4}{15!3!3^1} = 3.9813$$

$$C_5 = \frac{19!(1/6)^5}{14!3!3^2} = 3.3224$$

$$C_6 = \frac{19!(1/6)^6}{13!3!3^3} = 2.5840$$

$$\vdots$$

$$C_{18} = \frac{19!(1/6)^{18}}{1!3!5^{15}} = 0.0000$$

$$C_{19} = \frac{19!(1/6)^{19}}{0!3!3^{16}} = 0.0000$$

Now,

$$\sum_{n=0}^{19} C_n = 27.9390$$

and from Equation 11.38

$$P_0 = \frac{1}{\sum_{n=0}^{19} C_n} = \frac{1}{27.9390} = 0.0358.$$

P_n for $n = 1, 2, \ldots, N$ can now be computed from $P_n = P_o C_n = (0.0358)C_n$ as follows:

$$P_0 = 0.0358 \times 1 \qquad = 0.0358$$
$$P_1 = 0.0358 \times 3.1665 = 0.1134$$
$$\vdots$$
$$P_5 = 0.0358 \times 3.3224 = 0.1189$$
$$P_6 = 0.0358 \times 2.5840 = 0.0925$$
$$\vdots$$
$$P_{18} = 0.0358 \times 0.0000 = 0.0000$$
$$P_{19} = 0.0358 \times 0.0000 = 0.0000$$

Now the expected number of units short can be calculated from Equation 20.7 as

$$E(S) = \sum_{j=1}^{D} jP_{(N-D+j)} = \sum_{j=1}^{15} jP_{(4+j)}$$

$$= 1(0.1189) + 2(0.0925) + \ldots + 15(0.0000)$$

$$= 1.00663$$

from which the annual shortage cost is

$$SC = C_S[E(S)]$$

$$= \$73,000(1.00663) = \$73,484.$$

The total system annual equivalent cost may now be summarized as

$$TC = PC + OC + RC + SC$$

$$= \$221,616 + \$33,250 + \$135,000 + \$73,484 = \$463,350$$

As can be seen from Table 20.3, this solution is actually the optimum with neighbouring points given for N, M, and n.

TABLE 20.3 POINTS IN THE OPTIMUM REGION

RETIREMENT AGE, n	NUMBER OF UNITS, N	NUMBER OF REPAIR CHANNELS, M		
		2	3	4
3	19	$598,395	$465,985	$469,130
4	18	592,920	464,770	465,755
	19	600,720	463,350*	464,295
	20	610,775	466,610	468,755
5	19	643,050	480,375	467,735

*Optimum

The shortage distribution can be calculated from the P_n values and plotted as a histogram of $Pr(S = s) = N - D + s$. In this example, $Pr(S = s) = P_{4+s}$

$$Pr(S = 0) = 0.622$$

$$Pr(S = 1) = 0.119$$

$$Pr(S = 2) = 0.093$$

$$Pr(S = 3) = 0.067$$

$$Pr(S = 4) = 0.046$$

$$Pr(S = 5) = 0.027$$

$$Pr(S = 6) = 0.015$$

$$Pr(S = 7) = 0.008$$

$$Pr(S = 8) = 0.003$$

The shortage probability histogram for this example is exhibited in Figure 20.2.

In the next section the adequacy of this "baseline" design will be considered in the face of design requirements. Additionally, challengers in the form of alternative candidate systems will be presented and evaluated.

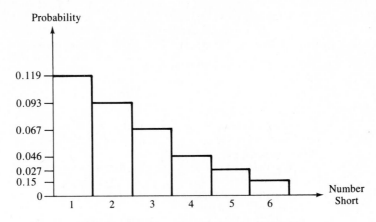

Figure 20.2   Shortage probability histogram.

20.4. DESIGN PROBLEM II

In Design Problem II, the decision maker has control over a set of design dependent parameters in addition to the set of design variables. Accordingly, the decision evaluation function of Equation 7.2 applies. In this section, design dependent parameters will be altered in the baseline design of the previous section to seek a REPS candidate more acceptable than the baseline design.

Considering Design Requirements

In the absence of design requirements, the baseline design might be acceptable. However, assume that it is not acceptable due to the existence of the following design requirements:

1. The cost per deployed unit must not exceed $50,000.
2. The probability of no units short of demand must be at least 0.70.
3. The mean MTBF for the unit over its life must be at least 0.20 years.
4. A unit must not be kept in service more than 4 years.

Note that the baseline design does not meet requirement 1 and 2, but it does meet requirements 3 and 4. Also its annual equivalent life-cycle cost is noted to be $463,350. A search must now be made to develop one or more candidate systems that will meet all design requirements at an acceptable life-cycle cost.

Generating Candidate Systems

Suppose that two candidate systems are generated in the face of the demand for 15 equipment units and the design independent parameters of Table 20.1. Design dependent parameters for these candidates are given in Table 20.4.

TABLE 20.4 DESIGN DEPENDENT PARAMETERS FOR CANDIDATE SYSTEMS

PARAMETER	CANDIDATE SYSTEM 1		CANDIDATE SYSTEM 2	
Unit acquisition cost	$45,000		$43,000	
Unit design life	6 years		6 years	
Unit salvage value at end of design life	$6,000		$5,000	
Unit operating cost:				
energy and fuel	$600		$800	
operating labor	500		700	
preventative maintenance	400		400	
other operating costs	400		400	
Age cohorts	MTBF	MTTR	MTBF	MTTR
0–1	0.16	0.04	0.18	0.04
1–2	0.21	0.04	0.21	0.04
2–3	0.26	0.05	0.25	0.05
3–4	0.26	0.06	0.25	0.05
4–5	0.26	0.06	0.23	0.06
5–6	0.24	0.06	0.20	0.06

Optimization over the design variables for each set of design dependent parameters gives the results summarized in Table 20.5 for the candidate systems. Also included are the optimized results for the baseline design. The following observations are made:

1. Only Candidate System 2 meets all design requirements.
2. Candidate System 2 has the lowest cost per unit and, therefore, the lowest investment cost for the deployed population.
3. Candidate System 2 has the highest probability of meeting the demand for 15 servicable equipment units.
4. Candidate System 2 consumes the most energy and requires the highest expenditure for operating labor.

TABLE 20.5 EVALUATION SUMMARY FOR REPS CANDIDATES

	BASELINE DESIGN	CANDIDATE SYSTEM 1	CANDIDATE SYSTEM 2
Number deployed, N	19	20	20
Repair channels, M	3	4	4
Retirement age, n	4	5	4
Unit cost, P	$52,000	$45,000	$43,000
Mean MTBF	0.26	0.23	0.22
Probability of no units short	0.622	0.663	0.730
Annual equivalent life-cycle cost	$463,350	$478,470	$468,825

5. Candidate System 2 has an annual equivalent life-cycle cost penalty of $468,825 less $463,350 or $5,475 over the baseline design but, has a lower investment cost than Candidate System 1.

There may be other candidate systems that meet all design requirements and which have life-cycle costs equal to or less than the LCC for Candidate System 2. The objective of Design Problem II is to formalize the evaluation of candidates, with each candidate being characterized by its design dependent parameter values. Comparison of the candidates is made only after the optimized (minimized) values are found for life-cycle cost. This is in keeping with the idea of equivalence set forth in Equations 7.2 and 8.7.

20.5. SUMMARY

This system design application models and optimizes a repairable equipment population system composed of operational units and repair facilities used to keep the units operational. In Design Problem I it is assumed that the units are procured with a predetermined MTBF and MTTR, with these values varying as a function of the age of units. The annual equivalent life-cycle cost is minimized by finding the population size, the replacement age, and the number of repair channels.

In Design Problem II, an extension is made by considering units with MTBF and MTTR characteristics which are a function of the first cost of the units. Here it is assumed that improvements in MTBF and MTTR can be obtained at a price. This leads to the multiple application of the optimization routine for Design Problem I; an application to each possible MTBF and MTTR case. The result is the specification of MTBF and MTTR values for the units to be deployed.

This system design application is an illustration of several tools of system analysis. Queuing theory for the finite population is applied from Chapter 11. Concepts of reliability and maintainability are illustrated from Chapters 13 and 14. Finally, economic feasibility with emphasis on annual equivalent life-cycle cost is illustrated from Chapters 8 and 17.

21

Energy Storage
System Design[1]

Conventional systems for the production of electrical energy require a generating plant with sufficient capacity to meet peak demand. Within the limits of present technology, fossil fuels, hydro energy, and nuclear fuels are converted through a series of processes to useful electric power which is transmitted to and used by the consumer in essentially the same instant of time. This requires a high capital investment with low utilization for most of the demand cycle. The feasibility of providing energy storage as an alternative to production capacity to meet peak demand is considered in this system design application.

21.1. INTRODUCTION

There are many energy sources available to humankind, most of which are derived from the radiant energy of the sun. But, energy in a useful form is the result of a basic three-step process involving the operations of conversion and storage. In the first step, solar energy is *converted* to a form that can be readily stored. In the second, the actual *storage* takes place. Finally, the stored energy is converted to a useful form.

---

[1] This system design application was adapted from A. Bruckner, W. J. Fabrycky, and J. E. Shamblin, "Economic Optimization of Energy Conversion with Storage," *IEEE Spectrum*, vol. 5, no. 4 (1969).

618

In the case of fossil fuels, nature has performed the first two steps. The energy of the sun was converted into complex and marvelous hydrocarbon molecules and stored in this form through the ages. In the case of hydro energy, solar energy is converted into potential energy in water stored behind a dam. Here again, nature has provided the first steps. Since much of the power generated makes use of fossil fuels or hydro power, nature provides the initial conversion and subsequent storage. Only the step of conversion to usable form is provided by man.

If an economical energy storage system were developed, sources of energy now considered to be irrelevant could be utilized. Many attempts have been made to harness the sun, the tides, and the wind, but all have been of limited value because human beings have not been able to duplicate the operations of conversion and storage performed by nature. Therefore, the fundamental problem in making use of energy sources other than fossil fuels and hydro power is a result of our inability to gather energy at exactly the rate and quantity in which we desire to use it. Humankind must, therefore, seek a means to make input and output energy rates independent of each other. Figure 21.1 illustrates the schematic concept of energy conversion with storage. In an experimental energy storage system under consideration, energy sources are called upon to produce electric power by the use of a conventional steam plant, hydro turbines, solar cells, or wind-driven generators. The resulting current is used to electrolyze water, which, because of the closed nature of the process, produces hydrogen and oxygen gas at high pressure. The hydrogen may then be stored in steel tanks or in underground caverns. The oxygen may be similarly stored or used as a by-product.

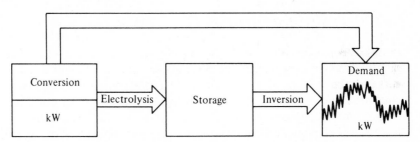

Figure 21.1   Energy conversion and storage schematic.

An automatic control system is used to accelerate or decelerate the electrolysis process, based on the pressure in the hydrogen storage tank. The hydrogen in the pressure tanks is released through a pressure reducer to a fuel-cell system. The amount of hydrogen released to the fuel cell is controlled by the load demand. Filtered air or pure oxygen must also be pumped into the fuel system, and the resulting water by-product must be recirculated into the pressure electrolysis system. Assuming that it is desirable to have reasonably high-voltage alternating currents, the fuel cell system output can be fed to an inverter and be made available to meet demand.

There is a distinct advantage of a system such as the one just described and illustrated in Figure 21.1. During periods of low demand, excess power from the conversion block is used to store energy in the storage block. In such a system, the storage block supplies demand loads greater than the capacity of the conversion block. Thus, an energy conversion-storage system provides a means for making the demand for energy independent of energy supply.

21.2. DEMAND FUNCTION

Demand for electrical energy is the primary justification for the design and construction of an energy conversion and storage system. A hypothetical demand function is shown in Figure 21.2. It spans a one-year period and exhibits a peak kilowatt demand that is approximately four times the minimum kilowatt demand (when including daily variation, the peak-to-low demand ratio is approximately 8:1). This ratio is typical of a demand environment incorporating variations from daily, weekly, and seasonal consumer and industrial energy requirements.

The demand function shown in Figure 21.2 consists of 2,190 points, one point for every four hours. If demand data were plotted less frequently than every four hours, an error of significant magnitude would be introduced because of the lack of resolution between true kilowatt-hours and peak capacity. One-hour intervals would be more desirable. These 2,190 points provide the basic data for the determination of system physical characteristics and the subsequent cost analysis. Conclusions concerning design of the system are based on this unique demand function with its individual capacity peak, total kilowatt-hours, and particular pattern.

21.3. THE OPTIMIZATION ALGORITHM

In this design application, total system equivalent annual cost is expressed as a function of conversion capacity as the controllable decision (design) variable. Two major cost components are involved. The first reflects the decreasing equivalent annual cost of energy conversion (steam plant, solar, hydro, nuclear, etc.) as the required conversion capacity or scale of the plant decreases. The second reflects the increasing equivalent annual cost of energy storage capability needed to compensate for the reduction in conversion capacity for a specific storage efficiency.

These major cost components, together with their subordinate cost components, are shown in Figure 21.3. The figure also shows the total system cost of a conversion-storage installation as a function of the energy conversion capacity. The ordinate intercept point gives the total cost of the conversion-storage system if no storage is provided and is, therefore, the cost of a conversion plant without energy storage.

If a minimum point exists on the total cost function, a system incorporating energy storage is economically feasible. The conversion capacity at the minimum point is the least-cost scale of the plant. This is a design value useful in developing an optimal conversion-storage system, since this minimum point also establishes the energy storage component capacities required for the system. If no minimum point exists on the total cost function, it may be concluded that energy storage is not feasible for the specific application under study.

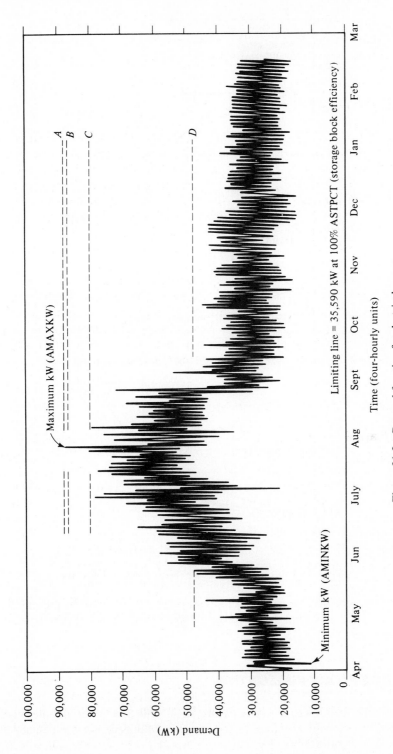

Figure 21.2 Demand function for electrical energy.

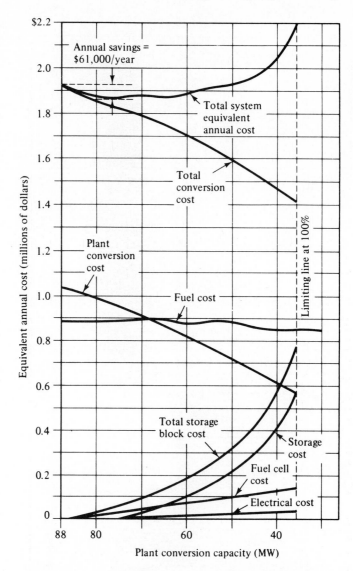

Figure 21.3 Cost components for energy conversion with storage.

In the sections that follow, a representative situation will be presented to illustrate the foregoing model. Much of the computational work was performed on a digital computer.

21.4. STORAGE BLOCK EVALUATION

Consider the line designated *A* in Figure 21.2. This line represents the required capacity of the conversion block if no storage is provided. For the demand function shown, this is 88,000 kW. The minimum capacity is 11,200 kW. Lines *B*, *C*, and *D*

TABLE 21.1 COST AS A FUNCTION OF STORAGE EFFICIENCY

(1) ASTPCT	(2) AKWUP	(3) AONE	(4) ATWO	(5) A2THEO	(6) AKWDWN	(7) AKWFC	(8) AKWEL	(9) IDTUP	(10) IDTDWN	(11) CSTTOT
1.0000	87230	3072	3072	3072	11960	768	768	1	1	$882166
1.0000	86460	6144	6144	6144	12730	1536	1536	1	1	$883227
1.0000	85690	9216	9216	8960	13120	2304	1920	1	2	$883149
1.0000	84920	12288	12288	12032	13500	3072	2304	1	2	$881997
1.0000	84160	15360	15360	15488	13730	3840	2534	1	5	$879006
1.0000	83390	18432	18432	18560	13880	4608	2688	1	5	$878270
.	.	.	.	.	.	.	.	.	.	.
1.0000	35770	47308568	47308568	47234102	35460	52220	24260	808	1377	$846594
1.0000	35590	47906832	47906832	48083474	35620	52400	24420	820	1392	$854178
0.9000	87230	3072	3413	3379	12040	768	845	1	1	$882511
0.9000	86460	6144	6827	7117	12880	1536	1689	1	1	$883560
0.9000	85690	9216	10240	10189	13270	2304	2073	1	2	$883304
.	.	.	.	.	.	.	.	.	.	.

623

represent decremental reductions in the capacity of the conversion block, which necessitate increased capacity in the storage block.

The ordinate of Figure 21.2, from kW minimum (AMINKW) to kW maximum (AMAXKW), is divided into 100 such decrements. The results shown in Table 21.1 provide the basis for developing physical system characteristics. Column 1, AST-PCT, gives the efficiency of the storage block. Each value given in column 2, AKWUP, is the difference between kW maximum and successive kW capacity decrements within a given efficiency. These values are the decremental reduction in conversion capacity.

For each conversion capacity considered, a given area (AONE) in kWh must be met by stored energy. This requires that certain kWh be stored (ATWO). Note that AONE = ATWO when the storage efficiency is 100%. The computations for a given efficiency terminate at a *limiting line*, defined as that minimum conversion capacity below which it would not be possible to satisfy the demand requirements in the kWh for the year.

The tabulated values in Table 21.1 are given as a function of conversion capacity for a specific efficiency. Curves portraying these data may be developed. Their range for a given storage block efficiency indicates the effect of the appropriate limiting line.

Figure 21.4 exhibits the electrolysis capacity in kilowatts (AKWEL) as a function of the conversion capacity for various efficiencies of storage. It is developed from Table 21.1 by reference to columns 1, 2, and 8. Figure 21.4 also shows fuel-cell capacity in kilowatts (AKWFC) as a function of plant conversion capacity. It is developed from Table 21.1 by reference to columns 1, 2, and 7. Figure 21.5 shows the kilowatt-hours of energy (ATWO) needed to be stored as a function of conversion capacity. Figure 21.6 shows the fuel efficiency as a function of running load. Information from Figures 21.4 through 21.6 may now be used with the data in Table 21.1 to develop cost extensions.

21.5. COST EXTENSION COMPUTATIONS

Conversion cost and storage cost make up the total cost function, as illustrated in Figure 21.3. This section is concerned with the economic analysis required in the development of these cost functions. It is based on the system physical characteristics described.

Assumed cost relationships are given in Figures 21.7 through 21.10. These curves are only hypothetical and are used to demonstrate the analysis procedure. The system is experimental, making precise cost information difficult to develop.

Conversion Cost

Suppose that the equivalent annual installed cost of the conversion block is a function of its capacity in kilowatts, as given in Figure 21.7. Note that the total cost increases at a decreasing rate. Although the function shown is not applicable to all

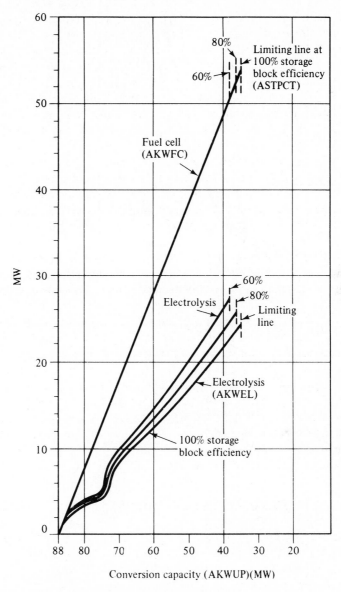

Figure 21.4   Electrolysis and fuel cell capacity vs. conversion capacity.

conversion plants, it represents a typical steam or hydro plant. It is assumed here that the time value of money was considered in deriving the equivalent annual cost.

The second major component of conversion cost is the cost of fuel if a steam plant is being considered. Table 21.1 gives the annual fuel cost as a function of the capacity of the conversion block and the efficiency of the storage block. The results are based on previously computed data of columns 2, 6, 9, and 10 in Table 21.1, as well as the efficiency function of Figure 21.6. When no storage is provided, the fuel cost is computed as follows: the demand function is divided into vertical segments;

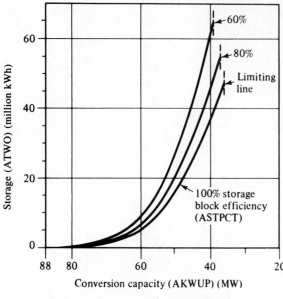

Figure 21.5 Storage requirements as a function of conversion capacity.

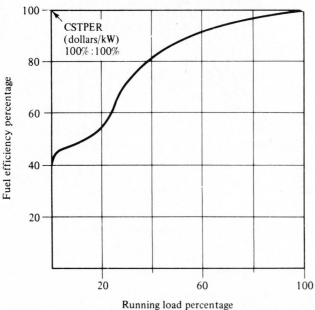

Figure 21.6 Fuel efficiency as a function of running load.

the average ordinate value of each segment determines the integer number of generators (four-generator plant) and the overall running load percentage. As an example, when

   A segment average ordinate intercept = 80,000 kW

   Required integer number of generators = 4

   Generators at 100% running load = 4 − 1 = 3

   Kilowatts per generator = 22, 000

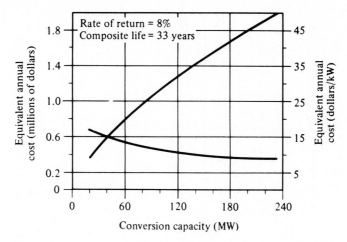

Figure 21.7   Power plant costs as a function of plant conversion capacity.

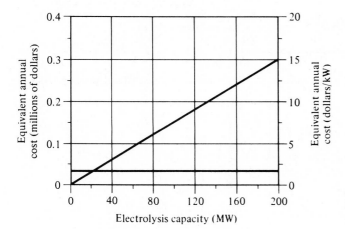

Figure 21.8   Electrolysis cost vs. electrolysis capacity.

the plant overall running load is

$$\frac{(80,000 - 66,000) + 3(22,000)}{1(22,000) + 3(22,000)} 1 = 91\%.$$

By computer look-up reference to Figure 21.7, the overall plant fuel efficiency for the given segment is 99%. For this segment, the fuel cost is

$$\frac{\$ \text{ per kWh at 100\%}}{\text{efficiency (\%)}} \times \text{av. kW} \times \frac{t}{\text{segment}} \times \frac{\text{hours}}{t}.$$

When storage is used, a comparably equivalent adjustment is made for the kW per generator. The mix of generator kW is established as being equal respectively to the two plateaus of uniform running load (AKWUP, AKWDWN) until the limiting line of constant running load for the full year is reached. This represents the best de-

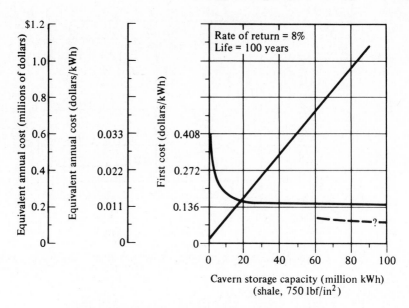

Figure 21.9   Cavern storage cost vs. storage capacity.

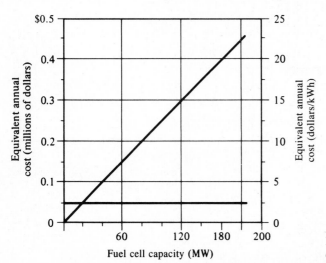

Figure 21.10   Fuel cell cost vs. fuel cell capacity.

sign of plant in that uniform running load portions of the year are run at highest fuel efficiency. Load in kW per generator is established for each plateau such that the generator kW selected is most equivalent to the defined four equal-sized generators. The low plateau number of generators and kW per generator are determined by comparing AKWDWN per uniform kW per generator such that if the decimal portion of the ratio is greater than 0.5, the next highest integer number of generators is chosen. For this plateau, the kW per generator is then AKWDWN per number of generators. A similar method determines the balance of the generator mix and kW for AKWUP, except that the balance of required generation capacity is AKWUP minus AKWDWN.

These segment costs are cumulatively added across that time portion of the conversion function where demand is variable. The total fuel cost for the year requires only the additional cost computation for the two plateau portions of the year that run at 100% efficiency. These expressions are as follows:

$$\text{IDTUP} = \text{number of time periods of upper plateau}$$

$$\text{Fuel cost up} = \text{AKWUP} \times \frac{\text{CSTPER}}{100\%} \times \text{IDTUP} \times \text{ADT}$$

$$\text{IDTDWN} = \text{number of time periods of lower plateau}$$

$$\text{Fuel cost down} = \text{AKWDWN} \times \frac{\text{CSTPER}}{100\%} \times \text{IDTDWN} \times \text{ADT}$$

Therefore,

$$\text{total fuel cost} = \text{cost up} + \text{cost down} + \sum \text{segment fuel costs}$$

where

$$\begin{aligned}
\text{ADT} &= \text{hours per time increment} \\
\text{IDTUP} &= \text{number of upper plateau time increments} \\
\text{IDTDWN} &= \text{number of lower plateau time increments} \\
\text{CSTPER} &= \$ \text{ per kWh only at 100\% efficiency} \\
&\quad (\text{equals \$0.0027 kWh in this study}).
\end{aligned}$$

Note that there is an inverse relationship between storage efficiency and overall fuel efficiency. The latter increases as storage block efficiency decreases since the portion of the year at uniform running load is greater.

Storage Cost

As indicated earlier, the storage block is composed of the electrolysis equipment, storage equipment, and the fuel-cell-inversion equipment. The cost of these components is assumed to be as shown in Figure 21.7 through 21.10.

Total System Cost

Column 5 of Table 21.1 represents the sum of the component costs given in the previous columns; it is the total equivalent annual cost of the three-block system of Figure 21.1 as a function of the conversion capacity of the system. Note that the total equivalent annual cost is $1,922,000 for no storage and that the conversion capacity required is 88,000 kW. As the conversion capacity decreases, the total equivalent annual cost decreases to a minimum value of $1,861,000. Beyond this point, the total equivalent annual cost begins increasing beyond bound. Thus, it may be concluded that the optimum conversion capacity for this situation is 74,940 kW if the efficiency of storage is 100%.

21.6. DESIGN SPECIFICATION

The entire system may now be designed from the parameters given at the minimum point. Specifically, it is required that the conversion block (plant) be 74,940 kW (with a generator mix level of 74,940 and 15,570 for the two plateaus in this case), that the capacity of the storage block be 52,224 kWh, that the capacity of the electrolysis equipment be 4,377 kW, and that the capacity of the fuel-cell-inversion equipment be 13,050 kW. These design specifications, which are a direct result of the computations leading up to Table 21.2, appear in Table 21.1.

If the total equivalent annual cost function had no minimum point, a system without energy storage is the least-cost system. Such a condition could be forced by a storage block efficiency that is very low.

TABLE 21.2 TOTAL SYSTEM EQUIVALENT ANNUAL COST (STORAGE BLOCK EFFICIENCY = 100%)

PLANT CONVERSION CAPACITY (kW)	ANNUAL COST OF CONVERSION BLOCK (000s)	ANNUAL FUEL COST (000s)	ANNUAL COST OF TOTAL STORAGE BLOCK (000s)	TOTAL SYSTEM EQUIVALENT ANNUAL COST (000s)
88,000	$1,040	$882	$ 0	$1,922
87,230	1,030	882	3	1,915
84,160	1,010	881	14	1,905
81,080	994	875	24	1,893
78,010	963	880	32	1,875
74,940	944	875	42	1,861
71,870	920	903	81	1,904
68,800	890	894	112	1,896
65,720	860	884	131	1,875
62,650	835	877	161	1,873
59,580	804	869	194	1,867
56,510	780	908	232	1,920
53,440	743	889	279	1,911
50,360	719	872	329	1,920
47,290	680	861	385	1,926
44,220	660	853	460	1,973
41,150	620	848	549	2,017
38,080	585	847	650	2,082
35,590	563	854	745	2,162
Limiting line				

21.7. SUMMARY

The optimization algorithm presented provides a means for determining design parameters for an energy conversion-storage system. The required background computations were performed on a digital computer. This type of generalized modeling is useful in predicting the effects of parameter changes on the total system while using sufficient empirical data to describe an actual system. This algorithm can analyze any actual demand curve with individual annual cost parameters for conversion capacity and other equipment. Fuel cost can be treated individually on a regional basis.

More important, however, is the capability of this algorithm to assist design for both system physical requirements and economic values. Some typical questions might be

1. What is the economic limit to use of storage versus the physical limit (limiting line) for a particular demand curve?
2. What is the effect on the minimum-cost point of the total system by an increase in fuel-cell efficiency?
3. If one component of the system can achieve unusually high performance by a high-priced part, will the system gain warrant the marginal part cost?
4. How much fuel and fuel cost can be saved by a uniform running load for an individual demand curve?
5. What are the capacity specifications for an electrolysis unit in a plant design for a given demand curve and an expected efficiency level?
6. For which component or subcomponent will additional research or operating control efforts offer the best potential yield in savings?
7. What will be the effects on a plant design and cost of operation if the shape of the demand curve is affected by a new industrial consumer?

Even though the present state of the art is not yet sufficiently advanced for commercial application, energy storage systems do offer some potential advantages. Generally inherent is the ability of such systems to come up to load level rapidly and independently in case of power blackouts. Storage systems combined with nuclear plants could offer the joint advantages of a relatively constant running load and a much smaller plant for the same demand requirements.

APPENDICES

A
Functional Analysis

Functional analysis includes the process of translating top-level system requirements into specific qualitative and quantitative design requirements. Given the identified need for a system, supported by the definition of operational requirements and the maintenance concept for that system (described in Chapter 3), it is necessary to translate this information into meaningful design criteria. This translation task constitutes an iterative process of breaking down system-level requirements into successive levels of detail, and a convenient mechanism for communicating this information is the *functional flow diagram*:

Functional flow diagrams (or functional block diagrams) are developed to describe the system in functional terms. These are developed to reflect both *operational* and *support* activities as they occur throughout the system life cycle, and they are structured in a manner that illustrates the hierarchal aspects of a system (see Figure 4.2, Chapter 4). Some key features of the overall functional flow diagram process are noted as follows:

1. The functional block diagram approach should include coverage of *all* activities throughout the system life cycle, and the method of presentation should reflect proper activity sequences and interface interrelationships.
2. The information included within the functional blocks should be concerned with *what* is required before looking at *how* it should be accomplished.
3. The process should be flexible to allow for expansion if additional definition is required or reduction if too much detail is presented. The objective is to pro-

gressively and systematically work down to the level where resources can be identified with *how* a task should be accomplished.

In the development of functional flow diagrams, some degree of standardization is necessary for communications in defining the system. Thus, certain basic practices and symbols should be used, whenever possible, in the physical layout of functional diagrams. The paragraphs below provide some guidance in this direction.

1. *Function block*. Each separate function on a functional diagram should be presented in a single box enclosed by a solid line. Blocks used for reference to other flows should be indicated as partially enclosed boxes labeled "Ref." Each function may be as gross or detailed as required by the level of functional diagram on which it appears, but it should stand for a definite, finite, discrete action to be accomplished by equipment, personnel, facilities, software, or any combination thereof. Questionable or tentative functions should be enclosed in dotted blocks.

2. *Function numbering*. Functions identified on the functional flow diagrams at each level should be numbered in a manner which preserves the continuity of functions and provides information with respect to function origin throughout the system. Functions on the top-level functional diagram should be numbered 1.0, 2.0, 3.0, and so on. Functions which further indenture these top functions should contain the same parent identifier and should be coded at the next decimal level for each indenture. For example, the first identure of function 3.0 would be 3.1, the second 3.1.1, the third 3.1.1.1, and so on. For expansion of a higher-level function within a particular level of indenture, a numerical sequence should be used to preserve the continuity of function. For example, if more than one function is required to amplify function 3.0 at the first level of indenture, the sequence should be 3.1, 3.2, 3.3, . . . , 3.*n*. For expansion of function 3.3 at the second level, the numbering shall be 3.3.1, 3.3.2, . . . , 3.3.*n*. Where several levels of indentures appear on a single functional diagram, the same pattern should be maintained. While the basic ground rule should be to maintain a minimum level of indentures on any one particular flow, it may become necessary to include several levels to preserve the continuity of functions and to minimize the number of flows required to functionally depict the system.

3. *Functional reference*. Each functional diagram should contain a reference to its next higher functional diagram through the use of a reference block. For example, function 4.3 should be shown as a reference block in the case where the functions 4.3.1, 4.3.2, . . . , 4.3.*n*, and so on, are being used to expand function 4.3. Reference blocks shall also be used to indicate interfacing functions as appropriate.

4. *Flow connection*. Lines connecting functions should indicate only the functional flow and should not represent either a lapse in time or any intermediate activity. Vertical and horizontal lines between blocks should indicate that all functions so interrelated must be performed in either a parallel or series se-

quence. Diagonal lines may be used to indicate alternative sequences (cases where alternative paths lead to the next function in the sequence).

5. *Flow directions*. Functional diagrams should be laid out so that the functional flow is generally from left to right and the reverse flow, in the case of a feedback functional loop, from right to left. Primary input lines should enter the function block from the left side; the primary output, or *go* line, should exit from the right, and the *no-go* line should exit from the bottom of the box.

6. *Summing gates*. A circle should be used to depict a summing gate. As in the case of functional blocks, lines should enter and/or exit the summing gate as appropriate. The summing gate is used to indicate the convergence, or divergence, or parallel or alternative functional paths and is annotated with the term AND or OR. The term AND is used to indicate that parallel functions leading into the gate must be accomplished before proceeding to the next function, or that paths emerging from the AND gate must be accomplished after the preceding functions. The term OR is used to indicate that any of several alternative paths (alternative functions) converge to, or diverge from, the OR gate. The OR gate thus indicates that alternative paths may lead or follow a particular function.

7. *Go and no-go paths*. The symbols G and $\overline{G}$ are used to indicate go and no-go paths, respectively. The symbols are entered adjacent to the lines leaving a particular function to indicate alternative functional paths.

8. *Numbering procedure for changes to functional diagrams*. Additions of functions to existing data should be accomplished by locating the new function in its correct position without regard to sequence of numbering. The new function should be numbered using the first unused number at the level of indenture appropriate for the new function.

With the objective of illustrating how some of these general guidelines are employed, Figures A.1 through A.6 are included to present a few simple applications.

1. Figure A.1 provides an example of the basic format used in the development of functional block diagrams.

2. Figure A.2 includes an illustration of a functional block diagram approach as applied to automobile transportation. Three levels of operational functional flows and one level covering maintenance are noted.

3. Figure A.3 includes a simplified illustration of functional block diagrams applied to a home lawn mowing system. Two levels of operational flows and two levels of maintenance flows are noted.

4. Figure A.4 includes a functional block diagram for a segment of a radar system.

5. Figures A.5 and A.6 present a top-level overview of a space system application, showing operational functions and maintenance functions, respectively.

Although these sample block diagrams do not cover the selected system in entirety, it is hoped that this material is presented in enough detail to provide the appropriate level of guidance in the development of functional block diagrams.

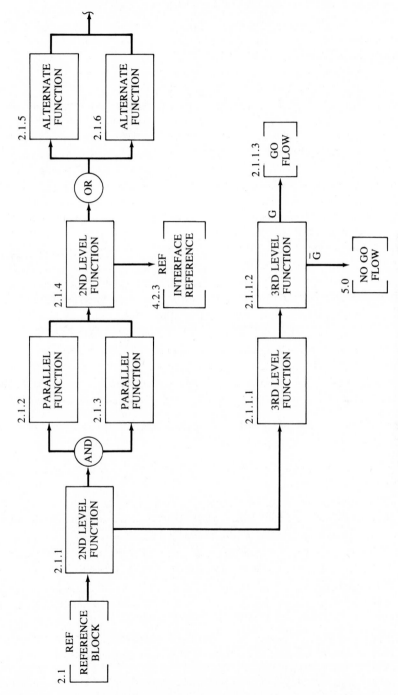

Figure A.1 General format of functional block diagrams.

OPERATIONAL FUNCTIONAL FLOWS – THREE LEVELS

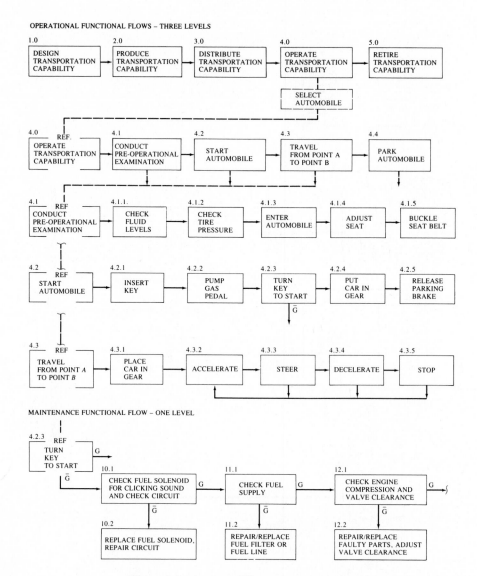

Figure A.2   An automobile system (example).

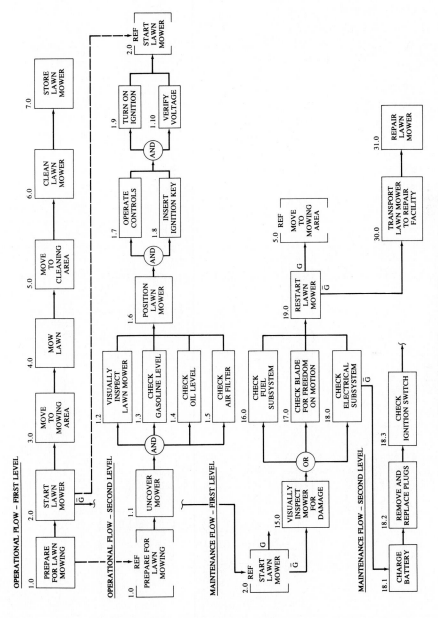

Figure A.3 A lawn mowing system (example).

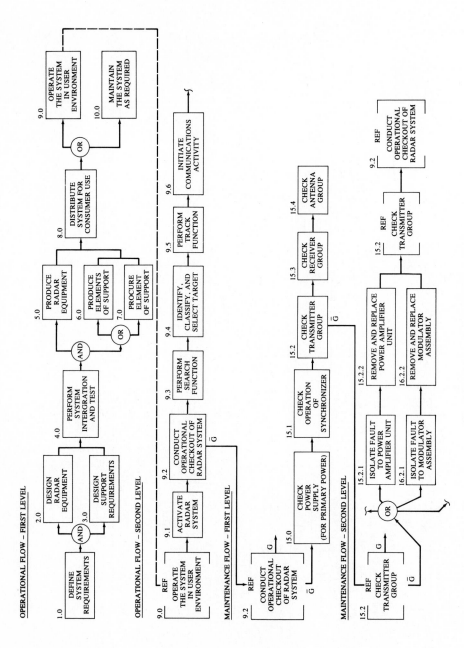

Figure A.4 Radar system flow (example).

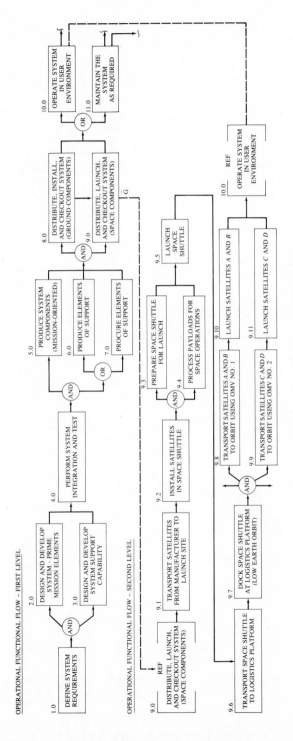

OPERATIONAL FUNCTIONAL FLOW – FIRST LEVEL

OPERATIONAL FUNCTIONAL FLOW – SECOND LEVEL

Figure A.5 Space system operational flow (example).

639

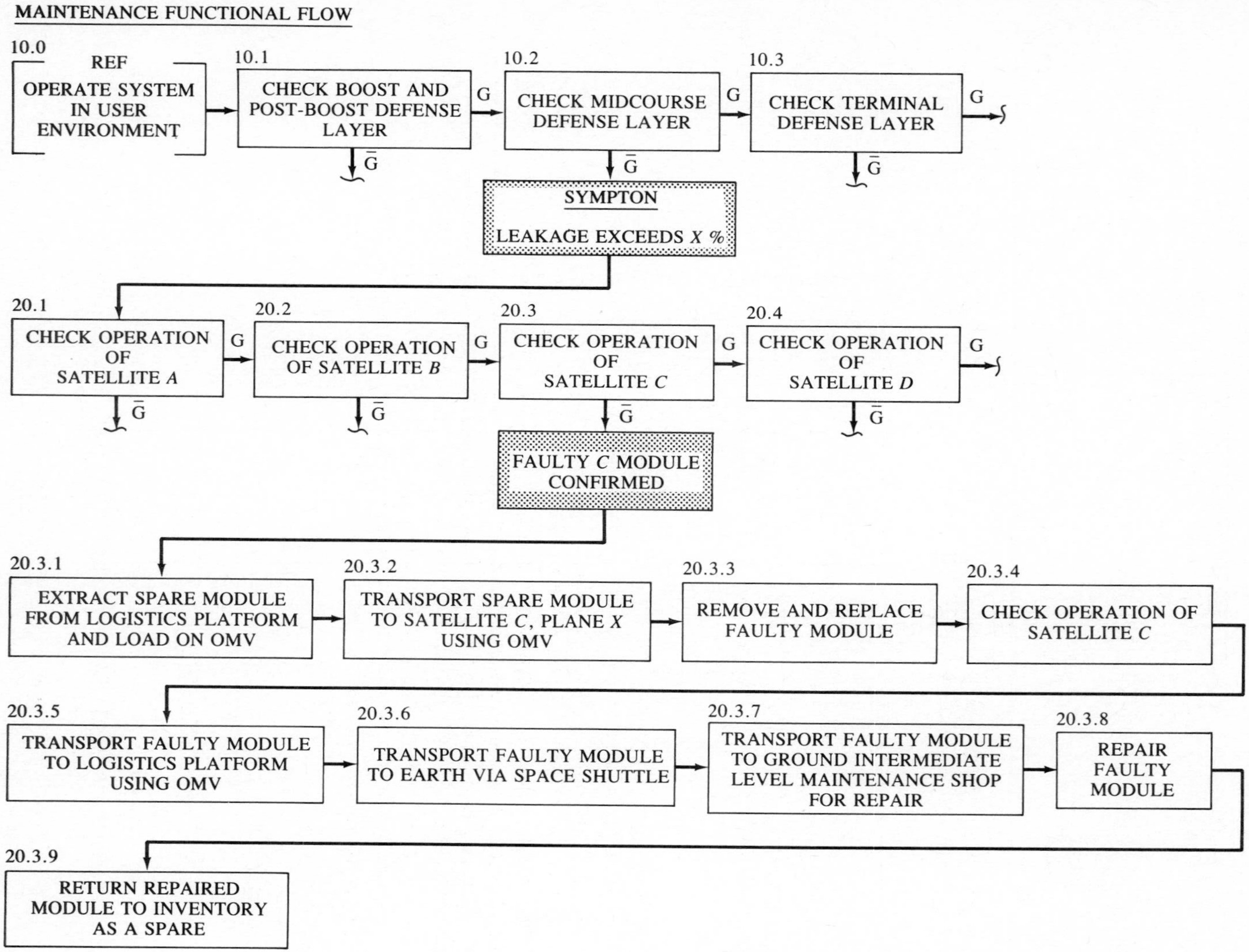

MAINTENANCE FUNCTIONAL FLOW

Figure A.6   Space system maintenance flow (example).

B
Design Review Criteria

On a periodic basis, throughout the design and development process, it is beneficial to do an informal review to assess (1) whether the appropriate characteristics have been considered and incorporated into the system or product design and (2) whether the necessary tasks have been accomplished on programs to ensure optimum overall results. Questions, presented in a checklist format, have been developed to reflect certain features. Reliability, maintainability, human factors, supportability, and economic considerations are included. Not all questions are applicable in all reviews; however, the answer to those questions that are applicable should be yes to reflect desirable results. For many questions, a more in-depth study of appropriate subject-oriented reference material will be required prior to arriving at a decision. These questions directly support the abbreviated checklist presented in Figure 5.6, Chapter 5.

GENERAL

1.0   System Operational Requirements

1.1 Has the need for the system or product been established and justified?

1.2 Has the overall system technical design approach been justified through a feasibility analysis?

1.3 Has the mission for the system been defined through mission scenarios or profiles?

1.4 Have all basic system performance parameters been defined (technical performance measures)?

1.5 Has the system or product life cycle been defined (design, development, test and evaluation, production and/or construction, distribution, operational use, sustaining support, retirement and disposal)?

1.6 Has the planned operational deployment and distribution been defined (customer requirements, quantity of items per user location, distribution schedule)?

1.7 Have system use requirements been defined? This includes projected hours of system operation or the quantity of operational cycles in a given time period.

1.8 Has the operational environment been defined in terms of temperature extremes, humidity, vibration and shock, storage, transportation and handling? A dynamic scenario is desired.

2.0 Effectiveness Factors

2.1 Have the appropriate system effectiveness and cost effectiveness factors been defined for the system or product (availability, dependability, capability, readiness, life cycle cost, or equivalent)?

2.2 Have applicable quantitative factors for reliability, maintainability, usability, and supportability been specified? This may include MTBM, MTBF, λ, MDT, $\overline{M}$, ADT, LDT, $\overline{M}$ct, $\overline{M}$pt, MMH/OH, Cost/OH, Cost/MA, TAT, and so on.

2.3 Are the effectiveness factors that have been identified directly traceable to the system operational requirements (mission scenario)?

2.4 Can each of the effectiveness factors identified for the system be measured? Have test and evaluation provisions been incorporated for the purposes of verification?

2.5 In the event that two or more effectiveness measures are appropriate, are the measures properly weighted? The relative weighting is in terms of significance or level of importance.

3.0 System Maintenance Concept

3.1 Have the levels of maintenance been identified and defined?

3.2 Have basic maintenance functions been identified for each level?

3.3 Have level-of-repair policies been established? Repair versus discard? Repair at intermediate level or at depot level?

3.4 Have the criteria for level-of-repair decisions been adequately defined?

3.5 Have the appropriate effectiveness factors been established for design of the system support capability? This may include quantitative factors covering spare part demand rates, inventory locations and levels, test equipment reliability and process time, facility utilization, and so on.

3.6 Have criteria been established for test and support equipment at each level of maintenance? Built-in versus external test equipment? Diagnostic requirements?

3.7 Have software requirements for system or component testing been established? Test language requirements?

3.8 Have criteria been established for personnel quantities and/or skills at each level of maintenance?

3.9 Have criteria been established for transportation and handling requirements?

3.10 Have the maintenance and support environments been defined in terms of temperature extremes, humidity, vibration and shock, transportation and handling, and storage?

3.11 Have the organizational responsibilities for maintenance and system or product support been identified (user support, contractor support, third-party maintenance)?

4.0 Functional Analysis and Allocation

4.1 Has the system been defined in *functional* terms using the functional block diagram approach?

4.2 Have all major system operational functions and maintenance functions been defined?

4.3 Does the functional analysis directly evolve from the system operational requirements and the maintenance concept? Are the functions directly traceable to top system-level requirements (such as the mission scenario)?

4.4 Is the functional analysis presented in enough detail for development of the system specification?

4.5 Have the appropriate system-level requirements been allocated to the depth necessary for adequate design definition (subsystem level and below)? This may include the allocation of quantitative reliability requirements, maintainability requirements, supportability factors, and cost parameters.

5.0 System Trade-off Studies

5.1 Have trade-off evaluations and analyses been accomplished to support major design decisions?

5.2 Have all feasible alternatives been considered in trade-off studies?

5.3 Have such analyses been accomplished with life-cycle considerations in mind (decisions based on life cycle impacts)?

5.4 Have system trade-off studies been adequately documented?

6.0 System and Supporting Specifications

6.1 Has a system specification been prepared, that is, a Type *A* specification?

6.2 Have the appropriate development, procurement, process, and material specifications been prepared (Types B, C, D, and E)?

6.3 Has a specification tree been developed to note governing specifications presented in a hierarchical manner?

6.4 Does the system specification include operational requirements, the maintenance concept, and a functional definition of the system?

6.5 Does the system specification include effectiveness requirements (reliability, maintainability, human factors, supportability)?

6.6 Are the various specifications that are applicable compatible with each other? Have conflicting specification requirements been eliminated?

7.0 System Engineering Management Plan (SEMP)

7.1 Has a system Engineering Management Plan (SEMP) been developed?

7.2 Does the plan address the overall system life cycle and its phases and activities?

7.3 Does the plan describe the system engineering process?

7.4 Does the plan properly integrate the different engineering specialties involved in the system or product design process?

7.5 Does the System Engineering Management Plan properly integrate other plans such as the Reliability Program Plan, Maintainability Program Plan, Safety Engineering Plan, Integrated Logistics Support Plan, and the like?

7.6 Does the plan adequately support the system specification?

7.7 Are program tasks, organizational responsibilities, WBS and schedule requirements, cost estimates, and program monitoring and control functions included?

7.8 Have formal design reviews been covered?

8.0 Design Documentation

8.1 Has a design documentation tree been developed?

8.2 Has the system or product design been adequately defined in design documentation?

8.3 Does the design documentation truly reflect the system or product configuration as it currently exists? Are all changes incorporated?

8.4 Is the design documentation adequate for procurement and contracting purposes?

8.5 Can the identical system or product be reproduced in multiple quantities from the available design documentation?

8.6 Are drawings, design data bases, software logic flows, technical publications, and the like compatible?

8.7 Has a formal configuration management program been established?

9.0 Logistic Support Requirements

9.1 Has an Integrated Logistic Support Plan (ILSP) or a Logistic Support Analysis Plan (LSAP) been developed?

9.2 Has a logistic support strategy been defined; that is, has a description of the environment in which the system is to be supported been set? Does the maintenance concept support this strategy?

9.3 Have the major elements of support been defined through the Logistic Support Analysis (test and support equipment, supply support, personnel and training, transportation and handling, computer resources facilities, and technical data requirements)?

9.4 Have supportability design criteria been defined for the prime mission-oriented elements of the system?

9.5 Have supportability design criteria been defined for the system support capability (design input requirements for the elements of logistic support)?

9.6 Do the elements of the Logistic Support Analysis (LSA) back up the system maintenance concept?

10.0 Ecological Requirements

10.1 Has an environmental impact study been completed (to determine if the system will have an adverse impact on the environment)?

10.2 Are the required standards associated with air quality, water quality, noise levels, solid waste processing, and so on, being maintained in spite of the introduction, operation, and sustaining support of the system or product?

10.3 Have potentially degrading ecological effects been identified? Is corrective action being taken to eliminate problems in this area?

11.0 Societal Requirements

11.1 Does the system or product satisfy societal needs?

11.2 Have the societal effects from introducing the system or product into the inventory been evaluated to determine the impact of system operation and support on community life?

11.3 Have all adverse societal effects caused by the introduction, operation, and support of the system or product been minimized, if not eliminated?

12.0 Economic Factors

12.1 Has the system or product been justified in terms of total life-cycle revenues and costs?

12.2 Are all cost elements considered?

12.3 Are all cost categories adequately defined?

12.4 Are cost estimates relevant?

12.5 Are variable and fixed costs separately identifiable?

12.6 Are escalation factors specified and employed where applicable?

12.7 Are learning curves specified and employed where applicable?

12.8 Is the discount rate specified and employed in the economic evaluation of alternatives?

12.9 Are the cost aspects employed in the evaluation of alternatives treated in a consistent manner?

12.10 Are the cost-estimating relationships used in economic analyses reasonably accurate?

12.11 Are areas of risk and uncertainty identified? Have the high-cost contributors been identified?

12.12 Has the sensitivity of cost estimates been properly addressed through a contingency or sensitivity analysis?

DESIGN FEATURES (Representative Sample)

1.0 Accessibility

1.1 Are key system components directly accessible for the performance of both operator and maintenance tasks?

1.2 Is access easily attained?

1.3 Are access requirements compatible with the frequency of maintenance or the importance of the need? Accessibility for items requiring frequent maintenance should be greater than for items requiring infrequent maintenance.

1.4 Are access doors provided where appropriate? Are hinged doors used? Can access doors that are hinged be supported in the open position?

1.5 Are access openings adequate in size and optimally located for the access required?

1.6 Are access door fasteners minimized?

1.7 Are access door fasteners of the quick-release variety?

1.8 Can access be attained without the use of tools?

1.9 If tools are required to gain access, are the number of tools held to a minimum? Are the tools of the standard variety?

1.10 Are access provisions between modules and components adequate?

1.11 Are access doors and openings labeled in terms of items that are accessible from within?

2.0 Adjustments and Alignments

2.1 Have adjustment and alignment requirements been minimized, if not eliminated?

2.2 Are adjustment requirements and frequencies known where applicable?

2.3 Are adjustment points accessible?

2.4 Are adjustment-point locations compatible with the maintenance level at which the adjustment is made?

2.5 Have adjustment and alignment interaction effects been eliminated?

2.6 Are factory adjustments specified?

2.7 Are adjustment points adequately labeled?

2.8 Can adjustments and alignments be made without the requirement for special tools?

3.0 Cables and Connectors

3.1 Are cables fabricated in removable sections?

3.2 Are cables routed to avoid sharp bends?

3.3 Are cables routed to avoid pinching?

3.4 Is cable labeling adequate?

3.5 Is cable clamping adequate?

3.6 Are the connectors used of the quick-disconnect variety?

3.7 Are connectors that are mounted on surfaces far enough apart so that they can be firmly grasped for connecting and disconnecting?

3.8 Are connectors and receptacles labeled?

3.9 Are connectors and receptacles keyed?

3.10 Are connectors standardized?

3.11 Do the connectors incorporate provisions for moisture prevention?

4.0 Calibration

4.1 Have calibration requirements been minimized, if not eliminated?

4.2 Are calibration requirements known where applicable?

4.3 Are calibration frequencies and tolerances known?

4.4 Have the facilities for calibration been identified?

4.5 Are the necessary standards available for calibration?

4.6 Have calibration procedures been prepared?

4.7 Is traceability to the National Bureau of Standards possible?

4.8 Are calibration requirements compatible with the maintenance concept and the logistic support analysis (LSA)?

5.0 Packaging and Mounting

5.1 Is the packaging design attractive from the standpoint of consumer appeal (color, shape, size)?

5.2 Is functional packaging incorporated to the maximum extent possible? Interaction effects between packages should be minimized. It should be possible to limit maintenance to the removal of one module (the one containing the failed part) when a failure occurs and not require the removal of two, three, or four modules in order to resolve the problem.

5.3 Is the packaging design compatible with level of repair analysis decisions? Repairable items are designed to include maintenance provisions such as test points, accessibility, and plug-in components. Items to be discarded upon failure should be encapsulated and relatively low in cost. Maintenance provisions within the disposable module are not required.

5.4 Are disposable modules incorporated to the maximum extent practical? It is highly desirable to reduce overall support through a no-maintenance-design concept as long as the items being discarded are relatively high in reliability and low in cost.

5.5 Are plug-in modules and components used to the maximum extent possible (unless the use of plug-in components significantly degrades the equipment reliability)?

5.6 Are accesses between modules adequate to allow for hand grasping?

5.7 Are modules and components mounted such that the removal of any single item for maintenance will not require the removal of other items? Component stacking should be avoided where possible.

5.8 In areas where module stacking is necessary because of limited space, are the modules mounted in such a way that access priority has been assigned in accordance with the predicted removal and replacement frequency? Items that require frequent maintenance should be more accessible.

5.9 Are modules and components, not of a plug-in variety, mounted with four fasteners or less? Modules should be securely mounted, but the number of fasteners should be held to a minimum.

5.10 Are shock-mounting provisions incorporated where shock and vibration requirements are excessive?

5.11 Are provisions incorporated to preclude installation of the wrong module?

5.12 Are plug-in modules and components removable without the use of tools? If tools are required, they should be of the standard variety.

5.13 Are guides (slides or pins) provided to facilitate module installation?

5.14 Are modules and components labeled?

5.15 Are module and component labels located on top or immediately adjacent to the item and in plain sight?

5.16 Are the labels permanently affixed so that they will not come off during a maintenance action or as a result of environment? Is the information on the label adequate? Disposable modules should be so labeled. In equipment racks, are the heavier items mounted at the bottom of the rack? Unit weight should decrease with the increase installation height.

5.17 Are operator panels optimally positioned? For personnel in the standing position, panels should be located between 40 and 70 inches above the floor. Critical or precise controls should be between 48 and 64 inches above the floor. For personnel in the sitting position, panels should be located 30 inches above the floor.

5.18 Are drawers in equipment racks mounted on roll-out slides?

6.0 *Disposability*

6.1 Has the equipment been designed for disposability (selection of materials, packaging)?

6.2 Have procedures been prepared to cover system-equipment-component disposal?

6.3 Can the components or materials used in system or equipment design be recycled for use in other products?

6.4 If component and material recycling is not feasible, can decomposition be accomplished?

6.5 Can recycling and/or decomposition be accomplished using existing logistic support resources?

6.6 Are recycling and/or decomposition methods and results consistent with environmental, ecological, safety, political, and social requirements?

6.7 Are the methods used for recycling and/or decomposition economically feasible?

7.0 *Environment*

7.1 Has system or product design considered all possible phases of activity from an environmental standpoint (environmental requirements during system and handling operation or use, transportation, storage, and maintenance)?

7.2 Has system or product design considered the following: temperature, humidity, vibration, shock, pressure, wind, salt spray, sand, and dust? Have the ranges and extreme conditions been specified and properly addressed in design? Have the proper environmental profiles been addressed?

7.3 Is the design compatible with air and water quality standards?

7.4 Have provisions been made to specify and control noise, illumination, temperature, and humidity in areas where personnel are required to perform operating and maintenance tasks?

8.0 *Fasteners*

8.1 Are quick-release fasteners used on doors and access panels?

8.2 Are the total number of fasteners minimized?

8.3 Are the number of different type of fasteners held to a minimum? This relates to standardization.

8.4 Have fasteners been selected based on the requirement for standard tools rather than special tools?

9.0 *Handling*

9.1 For heavy items, are hoist lugs (lifting eyes) or base-lifting provisions for forklift-truck application incorporated? Hoist lugs should be provided on all items weighing more than 150 pounds.

9.2 Are hoist and base-lifting points identified relative to lifting capacity?

9.3 Are weight labels provided?

9.4 Are packages, units, components, or other items weighing over 10 pounds provided with handles? Are the proper-sized handles used and are they located in the right position? Are the handles optimally located from the weight distribution standpoint? Handles should be located over the center of gravity.

9.5 Are packages, units, or other items weighing more than 40 pounds provided with two handles for two-man carrying?

9.6 Can normal packing materials be used for shipping? If not, are special containers, cases, or covers provided to protect component vulnerable areas from damage during handling?

10.0 *Human Factors*

10.1 Has a system analysis been done to verify optimum human-machine interfaces? Are automated and manual functions adequately identified?

10.2 Are the identified automated and manual functions consistent with the results of the overall system-level functional analysis?

10.3 Have operational sequence diagrams (OSDs) been prepared where appropriate?

10.4 Has a detailed *operator* task analysis been done to verify task sequences, task complexities, personnel skills, and so on?

10.5 Has a detailed *maintenance* task analysis been done to verify maintenance task sequences, task complexities, personnel skills, and so on?

10.6 Is the detailed maintenance task analysis compatible with reliability data, maintainablity data, and logistic support analysis (LSA) data?

10.7 Are the detailed operator and maintenance task analysis compatible with system or product operating and maintenance procedures (task sequences, depth of explanatory material based on task complexity)?

10.8 For human-interface functions, is the system or product design optimum when considering anthropometric factors, human sensory factors, psychological factors, and physiological factors? For manual tasks, does the design reflect ease of operation by low-skilled personnel? Is the design such that potential human error rates are minimized?

10.9 Has a detailed training plan for operator and maintenance personnel been prepared? Have training facility, equipment, material, software, and data requirements been identified?

10.10 Is the human factors effort compatible with safety engineering requirements?

10.11 Has an approach been established for personnel test and evaluation?

11.0 Interchangeability

11.1 Are equipment, modules, and/or components that perform similar operations electrically, functionally, and physically interchangeable?

11.2 Can replacements of like items be made without adjustments and/or alignments?

12.0 Maintainability

12.1 Is the system or product maintainable in terms of troubleshooting and diagnostic provisions, accessibility, ease of replacement, handling capabilities, accuracy of test and verification, and economics in the performance of maintenance (corrective and preventive)? Actually, many of the other items in this checklist may appropriately be included under maintainability depending on the organization involved in the design.

12.2 Have maintainability requirements for the system or equipment been adequately defined? Are they compatible with system performance, reliability, supportability, and effectiveness factors?

12.3 Have maintainability requirements been allocated to the appropriate level (MTBM, MDT, MMH/OH, $\overline{M}_{ct}$, M_{pt}, \$/MA to the unit assembly, subassembly, and/or other appropriate component of the system)?

12.4 Have anticipated system or product corrective and preventive maintenance requirements been identified through a detailed maintenance engineering analysis? Have the proper trade-off studies been conducted to attain the proper balance between corrective and preventive maintenance? Too much preventive maintenance can be costly and can significantly impact correct maintenance requirements. Are the results compatible with logistic support analysis (LSA) data?

12.5 Has a level of repair analysis (LORA) been completed? Are the results consistent with the maintenance concept and the logistic support analysis (LSA)?

12.6 Have maintainability predictions been accomplished to assess the design in terms of the specified requirements? Do the predictions indicate compliance with the requirements?

12.7 Have maintainability demonstrations been conducted? Do the results indicate compliance with the requirements?

13.0 Panel Displays and Controls

13.1 Are controls standardized?

13.2 Are controls sequentially positioned?

13.3 Is control spacing adequate?

13.4 Is control labeling adequate?

13.5 Have the proper control and display relationships been incorporated, based on good human factors criteria?

13.6 Are the proper type of panel switches used?

13.7 Is the control panel lighting adequate?

13.8 Are the controls placed according to frequency and/or criticality of use?

14.0 Producibility

14.1 Does the design lend itself to economic production? Can simplified fabrication and assembly techniques be employed?

14.2 Has the design stabilized (minimum change)? If not, are changes properly controlled through good configuration management methods?

14.3 Is the design such that rework requirements are minimized? Are spoilage factors held to a minimum?

14.4 Has the design been verified through prototype testing, environmental qualification, reliability qualification, maintainability demonstration, and the like?

14.5 Is the design such that many models of the same item can be produced with identical results? Are fabrication steps, manufacturing processes and assembly methods adequately controlled through good quality assurance procedures?

14.6 Has adequate consideration been given to the application of Just In Time (JIT), Taguchi, Material Requirements Planning (MRP), and related methods in the production process?

14.7 Are production drawings, CAD, CAM, CALS data, material lists, and the like, adequate for production needs?

14.8 Can currently available facilities, standard tools, and existing personnel be used for fabrication, assembly, manufacturing and test operations?

14.9 Is the design such that automated manufacturing processes (like CA or numerical control techniques) can be applied for high-volume repetitive functions?

14.10 Is the design definition such that two or more suppliers can produce the system or product from a given set of data with identical results?

15.0 Reliability

15.1 Is the design simple? Has the number of component parts been kept to a minimum?

15.2 Are standard high-reliability parts being used?

15.3 Are item failure rates known? Has the mean life been determined?

15.4 Have parts been selected to meet reliability requirements?

15.5 Have parts with excessive failure rates (unreliable parts) been identified?

15.6 Have adequate derating factors been established and adhered to where appropriate?

15.7 Have the shelf-life and wear-out characteristics of parts been determined?

15.8   Have all critical-useful-life items been eliminated from the design? If not, have they been identified with inspection and replacement requirements specified? Has a critical-useful-life analysis been accomplished?

15.9   Have critical parts which require special procurement methods, testing, and handling provisions been identified?

15.10 Has the need for the selection of matching parts been eliminated?

15.11 Have fail-safe provisions been incorporated where possible (protection against secondary failures resulting from primary failures)?

15.12 Has the use of adjustable components been minimized?

15.13 Have safety factors and safety margins been used in the application of parts?

15.14 Have component failure modes and effects been identified? Has a FMEA, a FMECA, and/or a fault-tree analysis (FTA) been accomplished?

15.15 Has a stress-strength analysis been accomplished?

15.16 Have cooling provisions been incorporated in design "hot spot" areas? Is cooling directed toward the most critical items?

15.17 Has redundancy been incorporated in the design where needed to meet specified reliability requirements?

15.18 Are the best available methods for reducing the adverse effects of operational and maintenance environments on critical components being incorporated?

15.19 Have the risks associated with critical-item failures been identified and accepted? Is corrective action in design being taken?

15.20 Have reliability requirements for spares and repair parts been considered?

15.21 Have reliability predictions been accomplished?

15.22 Have reliability testing requirements been defined? Test requirements in design? Test requirements in production and/or construction?

15.23 Has a reliability failure analysis and corrective action capability been installed?

16.0 Safety

16.1   Has an integrated safety engineering plan been prepared and implemented?

16.2   Has a hazard analysis been accomplished to identify potential hazardous conditions? Is the hazard analysis compatible with the reliability FMECA or FMEA?

16.3   Have system or product hazards from heat, cold, thermal change, barometric change, humidity change, shock, vibration, light, mold, bacteria, corrosion, rodents, fungi, odor, chemicals, oils, greases, handling and tranportation, and so on, been eliminated?

16.4   Have fail-safe provisions been incorporated in the design?

16.5   Have protruding devices been eliminated or are they suitably protected?

16.6   Have provisions been incorporated for protection against high voltages? Are all external metal parts adequately grounded?

16.7 Are sharp metal edges, access openings, and corners protected with rubber, fillets, fiber, or plastic coating?

16.8 Are electrical circuit interlocks employed?

16.9 Are standoffs or handles provided to protect system component from damage during the performance of shop maintenance?

16.10 Are tools that are used near high-voltage areas adequately insulated at the handle or at other parts of the tool which the maintenance man is likely to touch?

16.11 Are the environments such that personnel safety is ensured? Are noise levels within a safe range? Is illumination adequate? Is the air clean? Are the temperatures at a proper level? Are OSHA requirements being maintained?

16.12 Has the proper protective clothing been identified for areas where the environment could be detrimental to human safety? Radiation, intense cold or heat, gas and loud noise are examples.

16.13 Are safety equipment requirements identified in areas where ordnance devices (and the like) are activated?

17.0 Selection of Parts and Materials

17.1 Have appropriate standards been consulted for the selection of components and materials?

17.2 Have all component parts and materials selected for the design been adequately evaluated prior to their procurement and application? Evaluation should consider performance parameters, reliablity, maintainablity, supportablity, human factors, quality, and cost.

17.3 Have supplier sources for component-part and material procurement been established?

17.4 Are the established supplier sources reliable in terms of quality level, ability to deliver on time, and willingness to accept component-warranty provisions? There is an on-going concern for control specifications, process variations, stresses, tolerances, item interchangeability, and the like.

17.5 Have the alternatve supplier sources been identified in the event that the prime source fails to deliver?

18.0 Servicing and Lubrication

18.1 Have servicing requirements been held to a minimum?

18.2 When servicing is indicated, are the specific requirements identified?

18.3 Are procurment sources for servicing materials known?

18.4 Are servicing points accessible?

18.5 Have personnel and equipment requirements for servicing been identified? This includes handling equipment, vehicles, cart, and the like.

18.6 Does the design include servicing indicators?

19.0 Software

19.1 Have all system software requirements for operating and maintenance functions been identified? Have these requirements been developed through the system-level functional analysis to provide traceablity?

19.2 Is the software complete in terms of scope and depth of coverage?

19.3 Is the software compatible relative to the equipment with which it interfaces? Is operating software compatible with maintenance software? With other elements of the system?

19.4 Are the language requirements for operating software and maintenance software compatible?

19.5 Is all software adequately covered through good documentation (logic functional flows, coded programs, and so on)?

19.6 Has the software been adequately tested and verified for accuracy (performance), reliablity, and maintainability?

20.0 Standardization

20.1 Are standard off-the-shelf components and parts incorporated in the design to the maximum extent possible (except for items not compatible with effectiveness factors)? Maximum standardization is desirable.

20.2 Are the same items and/or parts used in similar applications?

20.3 Are identifying equipment labels and nomenclature assignments standardized to the maximum extent possible?

20.4 Are equipment-control-panel positions and layouts (from panel to panel) the same or similar when a number of panels are incorporated and provide comparable functions?

21.0 Supportability

21.1 Have spare and repair part requirements been minimized to the greatest extent possible? Are the number of different part types used throughout the design minimized?

21.2 Are the types and quantity of spare and repair parts compatible with the system maintenance concept, the logistic support analysis (LSA), and level of repair analysis data?

21.3 Are the types and quantity of spare and repair parts designated for a given location appropriate for the estimated demand at that location? Too many or too few spares can be costly.

21.4 Have the distribution channels and inventory points for spare and repair parts been established?

21.5 Are spare and repair part provisioning factors (like replacement frequencies) directly traceable to reliability and maintainability predictions?

21.6 Are the specified logistics pipeline times compatible with effective supply support? Long pipeline times place a tremendous burden on logistic support.

21.7 Have spare and repair parts been identified and provisioned for preoperational support activities (interim producer or supplier support, test programs)?

21.8 Have test and acceptance procedures been developed for spare and repair parts? Spare and repair parts should be processed, produced, and accepted on a similar basis with their equivalent components in the prime equipment.

21.9 Have the consequences (risks) of stock depletion been defined in terms of effect on mission requirements and cost?

21.10 Has an inventory safety stock level been defined?

21.11 Has a provisioning or procurement cycle been defined (procurement or order frequency)? Have EOQ factors been determined?

21.12 Has a supply-availability requirement been established (the probability of having a spare available when required)?

21.13 Have the test and support equipment requirements been defined for each level of maintenance?

21.14 Have standard test and support equipment items been selected? Newly designed equipment should not be necessary unless standard equipment is unavailable.

21.15 Are the selected test and support equipment items compatible with the prime equipment? Does the test equipment do the job?

21.16 Are the test and support equipment requirements compatible with maintenance concept, logistic support analysis (LSA), and level of repair analysis data?

21.17 Have test and support equipment requirements (both in terms of variety and quantity) been minimized to the greatest extent possible?

21.18 Are the reliability and maintainability features in the test and support equipment compatible with those equivalent features in the prime equipment? It is not practical to select an item of support equipment which is not as reliable as the item it supports.

21.19 Have logistic support requirements for the selected test and support equipment been defined? This includes maintenance tasks, calibration equipment, spare and repair parts, personnel and training, data, and facilities.

21.20 Is the test and support equipment selection process based on cost-effectiveness considerations (life-cycle cost)?

21.21 Have test and maintenance software requirements been adequately defined?

21.22 Have operational and maintenance personnel requirements (quantity and skill levels) been defined?

21.23 Are operational and maintenance personnel requirements minimized to the greatest extent possible?

21.24 Are operational and maintenance personnel requirements compatible with logistic support analysis (LSA) and with human factors data? Personnel quantities and skill levels should track both sources.

21.25 Are the planned personnel skill levels at each location compatible with the complexity of the operational and maintenance tasks specified?

21.26 Has maximum consideration been given to the use of existing personnel skills for the new system?

21.27 Have personnel attrition rates been established?

21.28 Have personnel effectiveness factors been determined (actual time that work is accomplished per the total time allowed for work accomplishment)?

21.29 Have operational and maintenance training requirements been specified? This includes consideration of both initial training and replenishment training throughout the life cycle.

21.30 Have specific training programs been planned? The type of training, frequency of training, duration of training, and student entry requirements should be identified.

21.31 Are the planned training programs compatible with the personnel skill level requirements specified for the performance of operational and maintenance tasks?

21.32 Have training equipment requirements been defined? Acquisitioned?

21.33 Have maintenance provisions for training equipment been planned?

21.34 Have training data requirements been met?

21.35 Are the planned operating and maintenance procedures (designated for support of the system throughout its life cycle) used to the maximum extent possible in the training programs?

22.0 Testability

22.1 Have self-test provisions been incorporated where appropriate?

22.2 Is reliability degradation due to the incorporation of built-in test minimized? The BIT capability should not significantly impact the reliability of the overall system.

22.3 Is the extent or depth of self-testing compatible with the level of repair analysis?

22.4 Are self-test provisions automatic?

22.5 Have direct fault indicators been provided (a fault light, an audio signal, or a means of determining that a malfunction positively exists)? Are continuous conditions monitoring provisions incorporated where appropriate?

22.6 Are test points provided to enable checkout and fault isolation beyond the level of self-test? Test point for fault isolation within an assembly should not be incorporated if the assembly is to be discarded at failure. Test point provisions must be compatible with the level of repair analysis.

22.7 Are test points accessible? Accessibility should be compatible with the extent of maintenance performed. Test points on the operator's front panel are not required for a depot maintenance action.

22.8 Are test points functionally and conveniently grouped to allow for sequential testing (following a signal flow), testing of similar functions, or frequency of use when access is limited?

22.9 Are test points provided for a direct test of all replaceable items?

22.10 Are test points adequately labeled? Each test point should be identified with a unique number, and the proper signal or expected measured output should be specified on a label located adjacent to the test point.

22.11 Are test points adequately illuminated to allow the technician to see the test point number and labeled signal value?

22.12 Can the component malfunctions that could possibly occur be detected through a no-go indication at the system level? Are false alarm rates minimized? This is a measure of test thoroughness.

22.13 Will the prescribed maintenance software provide adequate diagnostic information?

The preceding questions are representative (not be considered all-inclusive) of what should be considered in conducting a system or product design review. These questions stem from specific qualitative and quantitative design criteria that serve as an initial input into the design process.

C

Life-Cycle Cost Analysis

Through this text, a *life cycle* approach to system design and development has been assumed and addressed. Consistent with the objective of promoting effective and efficient results is the evaluation of various design alternatives in terms of life-cycle cost (LCC). While life-cycle cost is discussed throughout, the subject is emphasized in Chapter 17. The requirements for life-cycle costing and the process used in accomplishing a life-cycle cost analysis are included.

In accomplishing a life-cycle cost analysis, one needs to develop a cost breakdown structure (CBS), or cost tree, to facilitate the initial allocation of costs (top-down) and the subsequent collection of costs on a functional basis (bottom-up). The cost breakdown structure must include the consideration of *all* costs and is intended to aid in providing overall cost visibility. It is tailored to a specific system requirement, and the cost categories will vary somewhat in terms of depth of coverage depending on the type of system being evaluated.

The cost breakdown structure, in the context of the life-cycle cost analysis, is discussed in detail in Section 17.3 of Chapter 17. As a supplement, this section includes a sample cost breakdown structure in Figure C.1 and a description of the numerous cost categories in Tables C.1 and C.2. Through inspection, one can gain an appreciation for the many variables that must be addressed in a typical life-cycle cost analysis. Also, review of the different quantitative expressions in Table C.1 should lead to better understanding the relationships of the various input parameters. The combination of the material in Chapter 17 and the cost breakdown structure data presented here is intended to provide a "road map" for accomplishing a traditional life-cycle cost analysis.[1]

---

[1] Figure C.1 and Tables C.1 and C.2 were taken with permission from B.S. Blanchard, *Logistics Engineering and Management*, 3rd ed. (Englewood Cliffs, N.J.: Prentice-Hall).

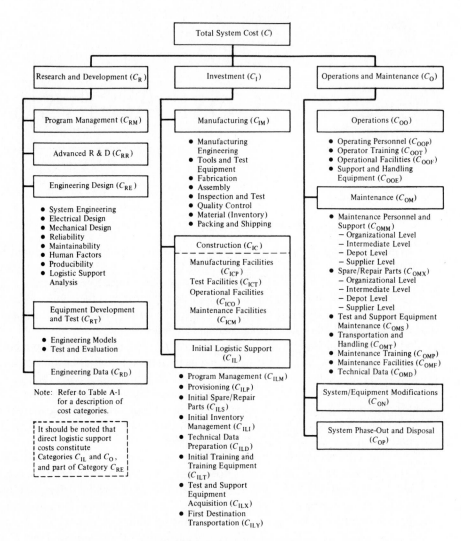

Figure C.1   Cost breakdown structure.

TABLE C-1 DESCRIPTION OF COST CATEGORIES

Cost Category (Reference Figure C-1)	Method of Determination (Quantitative Expression)	Cost Category Description and Justification
Total system cost (C)	$C = [C_R + C_I + C_O]$ $C_R = R$ and D cost $C_I =$ Investment cost $C_O =$ Operations and maintenance cost	Includes all future costs associated with the acquisition, utilization, and subsequent disposition of the system/equipment.
Research and development (C_R)	$C_R = [C_{RM} + C_{RR} + C_{RE} + C_{RT} + C_{RD}]$ $C_{RM} =$ Program management cost $C_{RR} =$ Advanced R and D cost $C_{RE} =$ Engineering design cost $C_{RT} =$ Equipment development and test cost $C_{RD} =$ Engineering data cost	Includes all costs associated with conceptual/feasibility studies, basic research, advanced research and development, engineering design, fabrication and test of engineering prototype models (hardware), and associated documentation. Also covers all related program management functions. These costs are basically nonrecurring.
Program management (C_{RM})	$C_{RM} = \sum_{i=1}^{N} C_{RM_i}$ $C_{RM_i} =$ Cost of specific activity i $N =$ Number of activities	Costs of management oriented activity applicable (across-the-board) to conceptual/feasibility studies, research, engineering design (including logistic support in the design process), equipment development and test, and related data/documentation. Such costs cover the program manager and his administrative staff; marketing; contracts; procurement; configuration management; logistics management; data management; etc. Management functions relate to C_{RR}, C_{RE}, C_{RT}, and C_{RD}.
Advanced research and development (C_{RR})	$C_{RR} = \sum_{i=1}^{N} C_{RR_i}$ $C_{RR_i} =$ Cost of specific activity i	Advanced research includes conceptual/feasibility studies conducted to determine and/or justify a need for a specific requirement. This includes

TABLE C-1 DESCRIPTION OF COST CATEGORIES (Continued)

Cost Category (Reference Figure C-1)	Method of Determination (Quantitative Expression)	Cost Category Description and Justification
	N = Number of activities	effort oriented to defining mission scenarios, system operational requirements (Chapter 3), preliminary maintenance concept (Chapter 4), etc., accomplished early in a program.
Engineering design (C_{RE})	$C_{RE} = \sum_{i=1}^{N} C_{RE_i}$ C_{RE_i} = Cost of specific design activity i N = Number of design activities	Includes all initial design effort associated with system/equipment definition and development. Specific areas include system engineering; design engineering (electrical, mechanical, drafting); reliability and maintainability engineering; human factors; functional analysis and allocation (Chapter 5); logistic support analysis (Chapter 6); components; producibility; standardization; safety; etc. Design associated with modifications is covered in C_{ON}.
Equipment development and test (C_{RT})	$C_{RT} = [C_{RDL} + C_{RDM} + \sum_{i=1}^{N} C_{RDT_i}]$ C_{RDL} = Cost of prototype fabrication and assembly labor C_{RDM} = Cost of prototype material C_{RDT_i} = Cost of test operations and support associated with specific test i N = Number of identifiable tests	The fabrication, assembly, test and evaluation of engineering prototype models (in support of engineering design activity–C_{RE}) is included herein. Specifically, this constitutes fabrication and assembly; instrumentation; quality control and inspection; material procurement and handling; logistic support (personnel, training, spares, facilities, support equipment, etc.); data collection; and evaluation of prototypes. Initial logistic support for operational system/equipment is covered in C_{IL}.

662

TABLE C-1 DESCRIPTION OF COST CATEGORIES (Continued)

Cost Category (Reference Figure C-1)	Method of Determination (Quantitative Expression)	Cost Category Description and Justification
Engineering data (C_{RD})	$C_{RD} = \sum_{i=1}^{N} C_{RD_i}$ C_{RD_i} = Cost of specific data N = Number of data items	This category includes the preparation, printing, publication, and distribution of all data/documentation associated with C_{RM}, C_{RE}, and C_{RT}, C_{RR}. This covers program plans; R and D reports; design data; test plans and reports; analyses; preliminary operational and maintenance procedures; and all effort related to a specific documentation requirement.
Investment (C_I)	$C_I = [C_{IM} + C_{IC} + C_{IL}]$ C_{IM} = System/equipment manufacturing cost C_{IC} = System construction cost C_{IL} = Cost of initial logistic support	Includes all costs associated with the acquisition of systems/equipment (once that design and development has been completed). Specifically this covers manufacturing (recurring and nonrecurring); manufacturing management; system construction; and initial logistic support.
Manufacturing (C_{IM})	$C_{IM} = [C_{IN} + C_{IR}]$ C_{IN} = Nonrecurring manufacturing cost C_{IR} = Recurring manufacturing cost	This covers all recurring and nonrecurring costs associated with the production and test of multiple quantities of prime systems/equipments. Facility construction, capital equipment, and facility maintenance are covered under C_{IC}.
Nonrecurring manufacturing cost (C_{IN})	$C_{IN} = [C_{INM} + C_{INT} + C_{INA} + C_{INP}$ $+ \sum_{i=1}^{N} C_{INQ} + \sum_{j=1}^{N} C_{INS}]$ C_{INM} = Manufacturing engineering cost C_{INT} = Tools and factory test equipment cost (excluding capital equipment) C_{INA} = Quality assurance cost	Includes all fixed *nonrecurring* costs associated with the production and test of operational systems/equipments. Specifically, this covers manufacturing management; manufacturing engineering; initial tooling and factory test equipment; quality assurance; first article qualification test (reliability test, maintainability

TABLE C-1 DESCRIPTION OF COST CATEGORIES (Continued)

Cost Category (Reference Figure C-1)	Method of Determination (Quantitative Expression)	Cost Category Description and Justification
	C_{INP} = Manufacturing management cost C_{INQ} = Cost of qualification test i C_{INS} = Cost of production sampling test j N = Number of individual tests	demonstration, support equipment compatibility, technical data verification, personnel test and evaluation, interchangeability, environmental test) and related support; production sampling tests and related support. Logistic support for each individual qualification and sampling test is included in the cost of the individual test.
Recurring manufacturing cost (C_{IR})	$C_{IR} = [C_{IRE} + C_{IRL} + C_{IRM} + C_{IRI} + C_{IRT}]$ C_{IRE} = Recurring manufacturing engineering support cost C_{IRL} = Production fabrication and assembly labor cost C_{IRM} = Production material and inventory cost C_{IRI} = Inspection and test cost C_{IRT} = Packing and initial transportation cost	This category covers all *recurring* production costs to include fabrication; subassembly and assembly; material and inventory control; inspection and test; packing and shipping to the point of first destination. Sustaining engineering support required on a recurring basis is also included. Costs are associated with the production of prime equipment. Operational test and support equipment, training equipment, and spare/repair parts material costs are included in C_{IL}. Manufacturing management cost is included in C_{IN}.
Construction cost (C_{IC})	$C_{IC} = [C_{ICP} + C_{ICT} + C_{ICO} + C_{ICM}]$ C_{ICP} = Manufacturing facilities cost C_{ICT} = Test facilities cost C_{ICO} = Operational facilities acquisition cost C_{ICM} = Maintenance facilities acquisition cost	Includes all initial acquisition costs associated with manufacturing, test, operational and/or maintenance facilities (real property, plant, and equipment), and utilities (gas, electric power, water, telephone, heat, air conditioning, etc.). Facility costs cover the development of new building projects, the modification of existing

TABLE C-1 DESCRIPTION OF COST CATEGORIES (Continued)

Cost Category (Reference Figure C-1)	Method of Determination (Quantitative Expression)	Cost Category Description and Justification
	For each item, one should consider the following. $C_{IC(\ )} = [C_{ICA} + C_{ICB} + C_{ICU} + C_{ICC}]$ C_{ICA} = Construction labor cost C_{ICB} = Construction material cost C_{ICU} = Cost of utilities C_{ICC} = Capital equipment cost	facilities, and/or the occupancy of existing facilities without modification. Work areas plus family housing are considered. Category costs include preliminary surveys; real estate; building constructions; roads and pavement; railroad sidings; etc. Cost items include construction labor, construction material, capital equipment, and utility installation. (a) Manufacturing facilities support the operations described in C_{IM}, C_{IN}, C_{IR}. Initial and sustaining costs are included. (b) Test facilities cover any peculiar requirements (beyond that covered under existing categories) for evaluation test. (c) Operational facilities are required for system operation throughout its life-cycle. Sustaining costs are covered in C_{OOF}. (d) Maintenance facilities are required to support the maintenance needs of the system throughout its life-cycle. Sustaining costs are covered in C_{OMF}.
Initial logistic support cost (C_{IL})	$C_{IL} = [C_{ILM} + C_{ILP} + C_{ILS} + C_{ILI} + C_{ILD} + C_{ILT} + C_{ILX} + C_{ILY}]$ C_{ILM} = Logistic program management cost C_{ILP} = Cost of provisioning C_{ILS} = Initial spare/repair part material cost	Includes all integrated logistic support planning and control functions associated with the development of system support requirements, and the transition of such requirements from supplier(s) to the applicable operational site. Elements cover (a) Logistic program management cost—man-

TABLE C-1 DESCRIPTION OF COST CATEGORIES (Continued)

Cost Category (Reference Figure C-1)	Method of Determination (Quantitative Expression)	Cost Category Description and Justification
	C_{ILI} = Initial inventory management cost C_{ILD} = Cost of technical data preparation C_{ILT} = Cost of initial training and training equipment C_{ILX} = Acquisition cost of operational test and support equipment C_{ILY} = Initial transportation and handling cost	agement, control, reporting, corrective action system, budgeting, planning, etc. (b) Provisioning cost—preparation of data which is needed for the procurement of spare/repair parts and test and support equipment. (c) Initial spare/repair part material cost—spares material stocked at the various inventory points to support the maintenance needs of prime equipment, test and support equipment, and training equipment. Replenishment spares are covered in C_{OMX}. (d) Initial inventory management cost—cataloging, listing, coding, etc., of spares entering the inventory. (e) Technical data preparation cost—development of operating and maintenance instructions, test procedures, maintenance cards, tapes, etc. Also includes reliability and maintainability data, test data, etc., covering production and test operations. (f) Initial training and training equipment cost—design and development of training equipment, training aids/data, and the training of personnel initially assigned to operate and maintain the prime equipment, test and support equipment, and training equipment. Personnel training costs include instructor time; supervision; student pay and allow-

TABLE C-1 DESCRIPTION OF COST CATEGORIES (Continued)

Cost Category (Reference Figure C-1)	Method of Determination (Quantitative Expression)	Cost Category Description and Justification
		ances; training facilities; and student transportation. Training accomplished on a sustaining basis throughout the system life-cycle (due to personnel attrition) is covered in C_{OOT} and C_{OMP}.
		(g) Test and support equipment acquisition cost —design, development, and acquisition of test and support equipment plus handling equipment needed to operate and maintain prime equipment in the field. The maintenance of test and support equipment throughout the system life-cycle is covered in C_{OOE} and C_{OMS}.
		(h) Initial transportation and handling cost (first destination transportation of logistic support elements from supplier to the applicable operational site).
		Initial facility costs are identified in C_{IC}.
		Specific logistic support requirements are defined in the Logistic Support Analysis (LSA) accomplished during engineering design as discussed in Chapter 6 and Appendix B.
Operations and maintenance cost (C_O)	$C_O = [C_{OO} + C_{OM} + C_{ON} + C_{OP}]$ C_{OO} = Cost of system/equipment life-cycle operations C_{OM} = Cost of system/equipment life-cycle maintenance	Includes all costs associated with the operation and maintenance support of the system throughout its life-cycle subsequent to equipment delivery in the field. Specific categories cover the cost of system operation, maintenance, sustaining

TABLE C-1 DESCRIPTION OF COST CATEGORIES (Continued)

Cost Category (Reference Figure C-1)	Method of Determination (Quantitative Expression)	Cost Category Description and Justification
	C_{ON} = Cost of system/equipment modifications C_{OP} = Cost of system/equipment phase-out and disposal	logistic support, equipment modifications, and system/equipment phase-out and disposal. Costs are generally determined for each year throughout life-cycle.
Operations cost (C_{OO})	$C_{OO} = [C_{OOP} + C_{OOT} + C_{OOF} + C_{OOE}]$ C_{OOP} = Operating personnel cost C_{OOT} = Cost of operator training C_{OOF} = Cost of operational facilities C_{OOE} = Cost of support and handling equipment	Includes all costs associated with the actual operation (not maintenance) of the system throughout its life-cycle. Specific categories cover the costs of system/equipment operational personnel (system operator); the formal training of operators; operational facilities; and support and handling equipment necessary for system operation.
Operating personnel cost (C_{OOP})	$C_{OOP} = [(T_O)(C_{PO})(Q_{PO})(N_{PO}) \times$ (% allocation)] T_O = Hours of system operation C_{PO} = Cost of operator labor Q_{PO} = Quantity of operators/system N_{PO} = Number of operating systems	This category covers the costs of operating personnel as allocated to the system. A single operator may operate more than one system, but costs should be allocated on an individual system basis. Such costs include base pay or salary and allowances; fringe benefits (insurance, medical, retirement); travel; clothing allowances; etc. Both direct and overhead costs are included.
Operator training cost (C_{OOT})	$C_{OOT} = [(Q_{SO})(T_T)(C_{TOP})]$ Q_{SO} = Quantity of student operators T_T = Duration of training program (weeks) C_{TOP} = Cost of operator training ($/student-week)	Initial operator training is included in C_{ILT}. This category covers the *formal* training of personnel assigned to operate the system. Such training is accomplished on a periodic basis throughout the system life-cycle to cover personnel replacements due to attrition. Total costs include instructor

TABLE C-1 DESCRIPTION OF COST CATEGORIES (Continued)

Cost Category (Reference Figure C-1)	Method of Determination (Quantitative Expression)	Cost Category Description and Justification
		time; supervision; student pay and allowances while in school; training facilities (allocation of portion of facility required specifically for formal training); training aids, equipment, and data; and student transportation as applicable.
Operational facilities cost (C_{OOF})	$C_{OOF} = [(C_{PPE} + C_U)(\% \text{ Allocation}) \times (N_{OS})]$ $C_{PPE} = $ Cost of operational facility support ($/site) $C_U = $ Cost of utilities ($/site) $N_{OS} = $ Number of operational sites *Alternate Approach* $C_{OOF} = [(C_{PPF})(N_{OS})(S_O)]$ $C_{PPF} = $ Cost of operational facility space ($/square foot/site). Utility cost allocation is included. $S_O = $ Facility space requirements (square feet)	Initial acquisition cost for operational facilities is included in C_{ICO}. This category covers the annual recurring costs associated with the occupancy and maintenance (repair, paint, etc.) of operational facilities throughout the system life-cycle. Utility costs are also included. Facility and utility costs are proportionately allocated to each system.
Support and handling equipment cost (C_{OOE})	$C_{OOE} = [C_{OOO} + C_{OOU} + C_{OOS}]$ $C_{OOO} = $ Cost of operation $C_{OOU} = $ Cost of equipment corrective maintenance $C_{OOS} = $ Cost of equipment preventive maintenance	Initial acquisition cost for operational support and handling equipment is covered in C_{ILX}. This category includes the annual recurring usage and maintenance costs for those items which are required to support system operation throughout the life-cycle (e.g., launchers, dollies, vehicles, etc.). The costs specifically cover equipment operation (not covered elsewhere); equipment corrective maintenance; and preventive main-

TABLE C-1 DESCRIPTION OF COST CATEGORIES (Continued)

Cost Category (Reference Figure C-1)	Method of Determination (Quantitative Expression)	Cost Category Description and Justification
	$C_{OOU} = [(Q_{CA})(M_{MHC})(C_{OCP}) + (Q_{CA})(C_{MHC}) + (Q_{CA})(C_{DC})](N_{OS})$ Q_{CA} = Quantity of corrective maintenance actions (M_A). Q_{CA} is a function of (T_O) (λ). M_{MHC} = Corrective maintenance manhours/M_A. C_{OCP} = Corrective maintenance labor cost ($\$/M_{MHC}$) C_{MHC} = Cost of material handling/ corrective M_A C_{DC} = Cost of corrective maintenance documentation/M_A N_{OS} = Number of operational sites $C_{OOS} = [(Q_{PA})(M_{MHP})(C_{OPP}) + (Q_{PA})(C_{MPH}) + (Q_{PA})(C_{DP})](N_{OS})$ Q_{PA} = Quantity of preventive maintenance actions (M_A). Q_{PA} relates to fpt M_{MHP} = Preventive maintenance manhours/M_A C_{OPP} = Preventive maintenance labor cost ($\$/M_{MHP}$) C_{MHP} = Cost of material. handling/ preventive M_A C_{DP} = Cost of preventive maintenance documentation/M_A N_{OS} = Number of operational sites	tenance. Spares and consumables are included in C_{OMX}. Corrective and preventive maintenance requirements are derived from the Logistic Support Analysis (LSA) discussed in Chapter 6 and Appendix B.

TABLE C-1 DESCRIPTION OF COST CATEGORIES (Continued)

Cost Category (Reference Figure C-1)	Method of Determination (Quantitative Expression)	Cost Category Description and Justification
Maintenance cost (C_{OM})	$C_{OM} = [C_{OMM} + C_{OMX} + C_{OMS} + C_{OMT} + C_{OMP} + C_{OMF} + C_{OMD}]$ C_{OMM} = Maintenance personnel and support cost C_{OMX} = Cost of spare/repair parts C_{OMS} = Test and support equipment maintenance cost C_{OMT} = Transportation and handling cost C_{OMF} = Cost of maintenance facilities C_{OMD} = Cost of technical data	Includes all sustaining maintenance labor, spare/repair parts, test and support equipment, transportation and handling, replenishment training, support data, and facilities necessary to meet the maintenance needs of the prime equipment throughout its life-cycle. Such needs include both corrective and preventive maintenance requirements at all echelons—organizational, intermediate, depot, and factory.
Maintenance personnel and support cost (C_{OMM})	$C_{OMM} = [C_{OOU} + C_{OOS}]$ C_{OOU} = Cost of equipment corrective maintenance C_{OOS} = Cost of equipment preventive maintenance Total cost is the sum of the C_{OMM} values for each echelon of maintenance.	Includes corrective and preventive maintenance labor, associated material handling, and support-ing documentation. When a system/equipment malfunction occurs or when a scheduled maintenance action is performed, personnel manhours are expended, the handling of spares and related material takes place, and maintenance action reports are completed. This category includes all directly related costs.
Corrective maintenance cost (C_{OOU})	$C_{OOU} = [(Q_{CA})(M_{MHC})(C_{OCP}) + (Q_{CA})(C_{MHC}) + (Q_{CA})(C_{DC})](N_{MS})$ Q_{CA} = Quantity of corrective maintenance actions (M_A). $Q_{CA} = (T_O)(\lambda)$ M_{MHC} = Corrective maintenance manhours/M_A C_{OCP} = Corrective maintenance labor cost ($\$/M_{MHC}$)	This category includes the personnel activity costs associated with the accomplishment of corrective maintenance. Related spares, test and support equipment, transportation, training, and facility costs are covered in C_{OMX}, C_{OMS}, C_{OMT}, and C_{OMF}, respectively. Total cost includes the sum of individual costs for each maintenance action multiplied by the quantity of

TABLE C-1 DESCRIPTION OF COST CATEGORIES (Continued)

Cost Category (Reference Figure C-1)	Method of Determination (Quantitative Expression)	Cost Category Description and Justification
	C_{MHC} = Cost of material handling/ corrective M_A C_{DC} = Cost of documentation/ corrective M_A N_{MS} = Number of maintenance sites Determine C_{OOU} for each appropriate echelon of maintenance.	maintenance actions anticipated over the entire system life-cycle. A maintenance action includes any requirement for corrective maintenance resulting from catastrophic failures, dependent failures, operator/maintenance induced faults, manufacturing defects, etc. The cost per maintenance action considers the personnel labor expended for direct tasks (localization, fault isolation, remove and replace, repair, verification), associated administrative/logistic delay time, material handling, and maintenance documentation (failure reports, spares issue reports). The corrective maintenance labor cost, C_{OCP}, will of course vary with the personnel skill level required for task performance. Both direct labor and overhead costs are included.
Preventive maintenance cost (C_{OOS})	$C_{OOS} = [(Q_{PA})(M_{MHP})(C_{OPP}) + (Q_{PA})(C_{MHP})$ $+ (Q_{PA})(C_{DP})](N_{MS})$ Q_{PA} = Quantity of preventive maintenance actions (M_A). Q_{PA} relates to fpt M_{MHP} = Preventive maintenance manhours/M_A C_{OPP} = Preventive maintenance labor cost ($/$M_{MHP}$) C_{MHP} = Cost of material handling/ preventive M_A	This category includes the personnel activity costs associated with the accomplishment of preventive or scheduled maintenance. Related spares/consumables, test and support equipment, transportation, training, and facility costs are covered in C_{OMX}, C_{OMS}, C_{OMT}, and C_{OMF}, respectively. Total cost includes the sum of individual costs for each preventive maintenance action multiplied by the quantity of maintenance actions anticipated over the system life-cycle. A maintenance action includes servicing, lubrica-

TABLE C-1 DESCRIPTION OF COST CATEGORIES (Continued)

Cost Category (Reference Figure C-1)	Method of Determination (Quantitative Expression)	Cost Category Description and Justification
	C_{DP} = Cost of documentation/preventive M_A N_{MS} = Number of maintenance sites Determine C_{OOS} for each appropriate echelon of maintenance.	tion, inspection, overhaul, calibration, periodic system check-outs, and the accomplishment of scheduled critical item replacements. The cost per maintenance action considers the personnel labor expended for preventive maintenance tasks, associated administrative/logistic delay time, material handling, and maintenance documentation. The preventive maintenance labor cost, C_{OPP}, will of course vary with the personnel skill level required for task performance. Both direct labor and overhead costs are included.
Spare/repair parts cost (C_{OMX})	$C_{OMX} = [C_{SO} + C_{SI} + C_{SD} + C_{SS} + C_{SC}]$ C_{SO} = Cost of organizational spare/repair parts C_{SI} = Cost of intermediate spare/repair parts C_{SD} = Cost of depot spare/repair parts C_{SS} = Cost of supplier spare/repair parts C_{SC} = Cost of consumables $C_{SO} = \sum_{i=1}^{N_{MS}} [(C_A)(Q_A) + \sum_{i=1} (C_{Mi})(Q_{Mi})$ $+ \sum_{i=1} (C_{Hi})(Q_{Hi})]$ C_A = Average cost of material purchase order (\$/order) Q_A = Quantity of purchase orders C_M = Cost of spare item i	Initial spare/repair part costs are covered in C_{ILS}. This category includes all replenishment spare/repair parts and consumable materials (e.g., oil, lubricants, fuel, etc.) that are required to support maintenance activities associated with prime equipment, operational support and handling equipment (C_{OOE}), test and support equipment (C_{OMS}), and training equipment at each echelon (organizational, intermediate, depot, supplier). This category covers the cost of purchasing; the actual cost of the material itself; and the cost of holding or maintaining items in the inventory. Costs are assigned to the applicable level of maintenance. Specific quantitative requirements for spares (Q_M) are derived from the Logistic Support Analysis (LSA) discussed

TABLE C-1 DESCRIPTION OF COST CATEGORIES (Continued)

Cost Category (Reference Figure C-1)	Method of Determination (Quantitative Expression)	Cost Category Description and Justification
	Q_M = Quantity of i items required or demand C_H = Cost of maintaining spare item i in the inventory (\$/\$ value of the inventory) Q_H = Quantity of i items in the inventory N_{MS} = Number of maintenance sites C_{SI}, C_{SD}, and C_{SS} are determined in a similar manner.	in Chapter 6 and Appendix B. These requirements are based on the criteria described in Chapter 2, Section 2E3. The optimum quantity of purchase orders (Q_A) is based on the EOQ criteria described in Section 2E3. Support equipment spares are based on the same criteria used in determining spare part requirements for prime equipment.
Test and support equipment cost (C_{OMS})	$C_{OMS} = [C_{SEO} + C_{SEI} + C_{SED}]$ C_{SEO} = Cost of organizational test and support equipment C_{SEI} = Cost of intermediate test and support equipment C_{SED} = Cost of depot test and support equipment $C_{SEO} = [C_{OOU} + C_{OOS}]$ C_{OOU} = Cost of equipment corrective maintenance C_{OOS} = Cost of equipment preventive maintenance $C_{OOU} = [(Q_{CA})(M_{MHC})(C_{OCP}) + (Q_{CA})(C_{MHC}) + (Q_{CA})(C_{DC})](N_{MS})$ Q_{CA} = Quantity of corrective maintenance actions (M_A) or $Q_{CA} = (T_O)(\lambda)$	Initial acquisition cost for test and support equipment is covered in C_{1Lx}. This category includes the annual recurring life-cycle maintenance cost for test and support equipment at each echelon. Support equipment operational costs are actually covered by the tasks performed in C_{OMM}. Maintenance constitutes both corrective and preventive maintenance, and the costs are derived on a similar basis with prime equipment (C_{OOU} and C_{OOS}). Spares and consumables are included in C_{OMX}. In some instances, specific items of test and support equipment are utilized for more than one (1) system, and in such cases, associated costs are allocated proportionately to each system concerned.

TABLE C-1 DESCRIPTION OF COST CATEGORIES (Continued)

Cost Category (Reference Figure C-1)	Method of Determination (Quantitative Expression)	Cost Category Description and Justification
	M_{MHC} = Corrective maintenance manhours/M_A	
	C_{OCP} = Corrective maintenance labor cost ($\$/M_{MHC}$)	
	C_{MHC} = Cost of material handling/ corrective M_A	
	C_{DC} = Cost of documentation/ corrective M_A	
	N_{MS} = Number of maintenance sites (involving organizational maintenance)	
	$C_{OOS} = [(Q_{PA})(M_{MHP})(C_{OPP}) + (Q_{PA})(C_{MHP}) + (Q_{PA})(C_{DP})](N_{MS})$	
	Q_{PA} = Quantity of preventive maintenance actions (M_A). Q_{PA} = fpt	
	M_{MHP} = Preventive maintenance manhours/M_A	
	C_{OPP} = Preventive maintenance labor cost ($\$/M_{MHP}$)	
	C_{MHP} = Cost of material handling/ preventive M_A	
	C_{DP} = Cost of documentation/ preventive M_A	
	N_{MS} = Number of maintenance sites (involving organizational maintenance)	
	C_{SEI} and C_{SED} are determined in a similar manner.	

TABLE C-1 DESCRIPTION OF COST CATEGORIES (Continued)

Cost Category (Reference Figure C-1)	Method of Determination (Quantitative Expression)	Cost Category Description and Justification
Transportation and handling cost (C_{OMT})	$C_{OMT} = [(C_T)(Q_T) + (C_P)(Q_T)]$ C_T = Cost of transportation C_P = Cost of packing Q_T = Quantity of one-way shipments $C_T = [(W)(C_{TS})]$ W = Weight of item (lb) C_{TS} = Shipping cost ($/lb) C_{TS} will of course vary with the distance (in miles) of the one-way shipment. $C_P = [(W)(C_{TP})]$ C_{TP} = Packing cost ($/lbs) Packing cost and weight will vary depending on whether reusable containers are employed.	Initial (first destination) transportation and handling costs are covered in C_{ILY}. This category includes all sustaining transportation and handling (or packing and shipping) between organizational, intermediate, depot, and supplier facilities in support of maintenance operations. This includes the return of faulty material items to a higher echelon; the transportation of items to a higher echelon for preventive maintenance (overhaul, calibration); and the shipment of spare/repair parts, personnel, data, etc., from the supplier to forward echelons.
Maintenance training cost (C_{OMP})	$C_{OMP} = [(Q_{SM})(T_T)(C_{TOM})]$ Q_{SM} = Quantity of maintenance students C_{TOM} = Cost of maintenance training ($/student-week) T_T = Duration of training program (weeks)	Initial maintenance training cost is included in C_{ILT}. This category covers the *formal* training of personnel assigned to maintain the prime equipment, test and support equipment, and training equipment. Such training is accomplished on a periodic basis throughout the system life-cycle to cover personnel replacements due to attrition. Total costs include instructor time; supervision; student pay and allowances while in school; training facilities (allocation of portion of facility required specifically for formal training); training aids and data; and student transportation as applicable.

TABLE C-1 DESCRIPTION OF COST CATEGORIES (Continued)

Cost Category (Reference Figure C-1)	Method of Determination (Quantitative Expression)	Cost Category Description and Justification
Maintenance facilities cost (C_{OMF})	$C_{OMF} = [(C_{PPM} + C_U) \times (\% \text{ allocation})(N_{MS})]$ $C_{PPM} = $ Cost of maintenance facility support (\$/site) $C_U = $ Cost of utilities (\$/site) $N_{MS} = $ Number of maintenance sites *Alternate approach* $C_{OMF} = [(C_{PPO})(N_{MS})(S_O)]$ $C_{PPO} = $ Cost of maintenance facility space (\$/square foot/site). Utility cost allocation is included. $S_O = $ Facility space requirements (square feet) Determine C_{OMF} for each appropriate echelon of maintenance.	Initial acquisition (construction) cost for maintenance facilities is included in C_{ICM}. This category covers the annual recurring costs associated with the occupancy and support (repair, modification, paint, etc.) of maintenance shops at all echelons throughout the system life-cycle. On some occasions, a given maintenance shop will support more than one (1) system, and in such cases, associated costs are allocated proportionately to each system concerned.
Technical data cost (C_{OMD})	$C_{OMD} = \sum_{i=1}^{N} C_{OMDi}$ $C_{OMDi} = $ Cost of specific data item i $N = $ Number of data items	Initial technical data preparation costs are covered in C_{ILD}. Individual data reports covering specific maintenance actions are included in C_{OOE}, C_{OMM}, and C_{OMS}. This category includes any other data (developed on a sustaining basis) necessary to support the operation and maintenance of the system throughout its life-cycle.
System/equipment modification cost (C_{ON})	$C_{ON} = \sum_{i=1}^{N} C_{ONi}$ $C_{ONi} = $ Cost of specific modification i	Throughout the system life-cycle after equipment has been delivered in the field, modifications are often proposed and initiated to improve system performance, effectiveness, or a combination of

677

TABLE C-1 DESCRIPTION OF COST CATEGORIES (Continued)

Cost Category (Reference Figure C-1)	Method of Determination (Quantitative Expression)	Cost Category Description and Justification
	N = Number of system/equipment modifications	both. This category includes modification kit design (R and D); material; installation and test instructions; personnel and supporting resources for incorporating the modification kit; technical data change documentation; formal training (as required) to cover the new configuration; spares; etc. The modification may affect all elements of logistics.
System phase-out and disposal cost (C_{OP})	$C_{OP} = [(F_C)(Q_{CA})(C_{DIS} - C_{REC})]$ F_C = Condemnation factor Q_{CA} = Quantity of corrective maintenance actions C_{DIS} = Cost of system/equipment disposal C_{REC} = Reclamation value	This category covers the liability or assets incurred when an item is condemned or disposed. This factor is applicable throughout the system/equipment life-cycle when phase-out occurs. This category represents the only element of cost that may turn out to have a negative value—resulting when the reclamation value of the end item is larger than the disposal cost.

TABLE C-2 SUMMARY OF TERMS

C	Total system life-cycle cost
C_A	Average cost of material purchase order (\$/order)
C_{DC}	Cost of maintenance documentation/data for each corrective maintenance action (\$/M_A)
C_{DIS}	Cost of system/equipment disposal
C_{DP}	Cost of maintenance documentation/data for each preventive maintenance action (\$/M_A)
C_H	Cost of maintaining spare item i in the inventory or inventory holding cost (\$/dollar value of the inventory)
C_I	Total investment cost
C_{IC}	Construction cost
C_{ICA}	Construction fabrication labor cost
C_{ICB}	Construction material cost
C_{ICC}	Capital equipment cost
C_{ICM}	Maintenance facilities acquisition cost
C_{ICO}	Operational facilities acquisition cost
C_{ICP}	Manufacturing facilities cost (acquisition and substaining)
C_{ICT}	Test facilities cost (acquisition and sustaining)
C_{ICU}	Cost of utilities
C_{IL}	Initial logistic support cost
C_{ILD}	Cost of technical data preparation
C_{ILI}	Initial inventory management cost
C_{ILM}	Logistics program management cost
C_{ILP}	Cost of provisioning (preparation of procurement data covering spares, test and support equipment, etc.)
C_{ILS}	Initial spare/repair part material cost
C_{ILT}	Cost of initial training and training equipment
C_{ILX}	Acquisition cost of operational test and support equipment
C_{ILY}	Initial transportation and handling cost
C_{IM}	Manufacturing cost
C_{IN}	Nonrecurring manufacturing/production cost
C_{INA}	Quality assurance cost
C_{INM}	Manufacturing engineering cost
C_{INP}	Manufacturing management cost
C_{INQ}	Cost of qualification test (first article)
C_{INS}	Cost of production sampling test
C_{INT}	Tools and factory equipment cost (excluding capital equipment)
C_{IR}	Recurring manufacturing/production cost
C_{IRE}	Recurring manufacturing engineering support cost
C_{IRI}	Inspection and test cost
C_{IRL}	Production fabrication and assembly labor cost
C_{IRM}	Production material and inventory cost
C_{IRT}	Packing and initial transportation cost
C_M	Cost of spares item i
C_{MHC}	Cost of material handling for each corrective maintenance action (\$/M_A)
C_{MHP}	Cost of material handling for each preventive maintenance action (\$/M_A)
C_O	Operations and maintenance cost

TABLE C-2 SUMMARY OF TERMS (Continued)

C_{OCP}	Corrective maintenance labor cost ($/$M_{MHC}$)
C_{OM}	Cost of system/equipment life-cycle maintenance
C_{OMD}	Cost of technical data
C_{OMF}	Cost of maintenance facilities
C_{OMM}	Maintenance personnel cost
C_{OMP}	Cost of replenishment maintenance training
C_{OMS}	Test and support equipment maintenance cost
C_{OMT}	Transportation and handling cost
C_{OMX}	Spare/repair parts cost (replenishment spares)
C_{ON}	Cost of system/equipment modifications
C_{OO}	Cost of system/equipment life-cycle operations
C_{OOE}	Cost of support and handling equipment
C_{OOF}	Cost of operational facilities
C_{OOO}	Cost of operation for support and handling equipment
C_{OOP}	Operating personnel cost
C_{OOS}	Cost of equipment preventive (scheduled) maintenance
C_{OOT}	Cost of replenishment training
C_{OOU}	Cost of equipment corrective (unscheduled) maintenance
C_{OP}	Cost of system/equipment phase-out and disposal
C_{OPP}	Preventive maintenance labor cost ($/$M_{MHP}$)
C_P	Cost of packing
C_{PO}	Cost of operators labor ($/hour)
C_{PPE}	Cost of operational facility support ($/operational site)
C_{PPF}	Cost of operational facility space ($/square foot/site)
C_{PPM}	Cost of maintenance facility support ($/maintenance site)
C_{PPO}	Cost of maintenance facility space ($/square foot/site)
C_R	Total research and development cost
C_{RD}	Engineering data cost
C_{RDL}	Prototype fabrication and assembly labor cost
C_{RDM}	Prototype material cost
C_{RDT}	Prototype test and evaluation cost
C_{RE}	Engineering design cost
C_{REC}	Reclamation value
C_{RM}	Program management cost
C_{RR}	Advanced research and development cost
C_{RT}	Equipment development and test cost
C_{SC}	Cost of consumables
C_{SD}	Cost of depot spare/repair parts
C_{SED}	Cost of depot test and support equipment
C_{SEI}	Cost of intermediate test and support equipment
C_{SEO}	Cost of organizational test and support equipment
C_{SI}	Cost of intermediate spare/repair parts
C_{SO}	Cost of organizational spare/repair parts
C_{SS}	Cost of supplier spare/repair parts
C_T	Cost of transportation
C_{TOM}	Cost of maintenance training ($/student-week)
C_{TOT}	Cost of operator training ($/student-week)
C_{TP}	Packing cost ($/pound)
C_{TS}	Shipping cost ($/pound)

TABLE C-2 SUMMARY OF TERMS (Continued)

C_U	Cost of utilities (\$/operational site)
fc	Condemnation factor (attrition)
fpt	Frequency of preventive maintenance (actions/hour of equipment operation)
λ	System/equipment failure rate (failure/hour of equipment operation)
M_{MHC}	Corrective maintenance manhours/maintenance action (MA)
M_{MHP}	Preventive maintenance manhours/maintenance action (MA)
N_{MS}	Number of maintenance sites
N_{OS}	Number of operational sites
N_{PO}	Number of operating systems
Q_A	Quantity of purchase orders
Q_{CA}	Quantity of corrective maintenance actions (MA)
Q_H	Quantity of i items in the inventory
Q_M	Quantity of i items required or demanded
Q_{PA}	Quantity of preventive maintenance actions (MA)
Q_{PO}	Quantity of operators/system
Q_{SM}	Quantity of maintenance students
Q_{SO}	Quantity of student operators
Q_T	Quantity of one-way shipments
S_O	Facility space requirements (square feet)
T_O	Hours of system operation
T_T	Duration of training program (weeks)
W	Weight of item (pounds)

D

Interest Factor Tables

TABLES D.1–D.14. INTEREST FACTORS FOR ANNUAL COMPOUNDING

Values for each of the six interest formulas derived in Chapter 8 are given for interest rates from 5 to 30%. [*These tables are reproduced from W. J. Fabrycky and G. J. Thuesen*, Economic Decision Analysis. (*Englewood Cliffs, N. J.: Prentice-Hall, Inc., 1980*).]

TABLE D.1 5% INTEREST FACTORS FOR ANNUAL COMPOUNDING

	Single Payment		Equal-Payment Series			
	Compound-Amount Factor	Present-Worth Factor	Compound-Amount Factor	Sinking-Fund Factor	Present-Worth Factor	Capital-Recovery Factor
n	To Find F Given P $F/P, i, n$	To Find P Given F $P/F, i, n$	To Find F Given A $F/A, i, n$	To Find A Given F $A/F, i, n$	To Find P Given A $P/A, i, n$	To Find A Given P $A/P, i, n$
1	1.050	0.9524	1.000	1.0000	0.9524	1.0500
2	1.103	0.9070	2.050	0.4878	1.8594	0.5378
3	1.158	0.8638	3.153	0.3172	2.7233	0.3672
4	1.216	0.8227	4.310	0.2320	3.5460	0.2820
5	1.276	0.7835	5.526	0.1810	4.3295	0.2310
6	1.340	0.7462	6.802	0.1470	5.0757	0.1970
7	1.407	0.7107	8.142	0.1228	5.7864	0.1728
8	1.477	0.6768	9.549	0.1047	6.4632	0.1547
9	1.551	0.6446	11.027	0.0907	7.1078	0.1407
10	1.629	0.6139	12.587	0.0795	7.7217	0.1295
11	1.710	0.5847	14.207	0.0704	8.3064	0.1204
12	1.796	0.5568	15.917	0.0628	8.8633	0.1128
13	1.866	0.5303	17.713	0.0565	9.3936	0.1065
14	1.980	0.5051	19.599	0.0510	9.8987	0.1010
15	2.079	0.4810	21.579	0.0464	10.3797	0.0964
16	2.183	0.4581	23.658	0.0423	10.8378	0.0923
17	2.292	0.4363	25.840	0.0387	11.2741	0.0887
18	2.407	0.4155	28.132	0.0356	11.6896	0.0856
19	2.527	0.3957	30.539	0.0328	12.0853	0.0828
20	2.653	0.3769	33.066	0.0303	12.4622	0.0803
21	2.786	0.3590	35.719	0.0280	12.8212	0.0780
22	2.925	0.3419	38.505	0.0260	13.1630	0.0760
23	3.072	0.3256	41.430	0.0241	13.4886	0.0741
24	3.225	0.3101	44.502	0.0225	13.7987	0.0725
25	3.386	0.2953	47.727	0.0210	14.0940	0.0710
26	3.556	0.2813	51.113	0.0196	14.3752	0.0696
27	3.733	0.2679	54.669	0.0183	14.6430	0.0683
28	3.920	0.2551	58.403	0.0171	14.8981	0.0671
29	4.116	0.2430	62.323	0.0161	15.1411	0.0661
30	4.322	0.2314	66.439	0.0151	15.3725	0.0651
31	4.538	0.2204	70.761	0.0141	15.5928	0.0641
32	4.765	0.2099	75.299	0.0133	15.8027	0.0633
33	5.003	0.1999	80.064	0.0125	16.0026	0.0625
34	5.253	0.1904	85.067	0.0118	16.1929	0.0618
35	5.516	0.1813	90.320	0.0111	16.3742	0.0611
40	7.040	0.1421	120.800	0.0083	17.1591	0.0583
45	8.985	0.1113	159.700	0.0063	17.7741	0.0563
50	11.467	0.0872	209.348	0.0048	18.2559	0.0548
55	14.636	0.0683	272.713	0.0037	18.6335	0.0537
60	18.679	0.0535	353.584	0.0028	18.9293	0.0528
65	23.840	0.0420	456.798	0.0022	19.1611	0.0522
70	30.426	0.0329	588.529	0.0017	19.3427	0.0517
75	38.833	0.0258	756.654	0.0013	19.4850	0.0513
80	49.561	0.0202	971.229	0.0010	19.5965	0.0510
85	63.254	0.0158	1245.087	0.0008	19.6838	0.0508
90	80.730	0.0124	1594.607	0.0006	19.7523	0.0506
95	103.035	0.0097	2040.694	0.0005	19.8059	0.0505
100	131.501	0.0076	2610.025	0.0004	19.8479	0.0504

TABLE D.2   6% INTEREST FACTORS FOR ANNUAL COMPOUNDING

	Single Payment		Equal-Payment Series			
	Compound-Amount Factor	Present-Worth Factor	Compound-Amount Factor	Sinking-Fund Factor	Present-Worth Factor	Capital-Recovery Factor
n	To Find F Given P $F/P, i, n$	To Find P Given F $P/F, i, n$	To Find F Given A $F/A, i, n$	To Find A Given F $A/F, i, n$	To Find P Given A $P/A, i, n$	To Find A Given P $A/P, i, n$
1	1.060	0.9434	1.000	1.0000	0.9434	1.0600
2	1.124	0.8900	2.060	0.4854	1.8334	0.5454
3	1.191	0.8396	3.184	0.3141	2.6730	0.3741
4	1.262	0.7921	4.375	0.2286	3.4651	0.2886
5	1.338	0.7473	5.637	0.1774	4.2124	0.2374
6	1.419	0.7050	6.975	0.1434	4.9173	0.2034
7	1.504	0.6651	8.394	0.1191	5.5824	0.1791
8	1.594	0.6274	9.897	0.1010	6.2098	0.1610
9	1.689	0.5919	11.491	0.0870	6.8017	0.1470
10	1.791	0.5584	13.181	0.0759	7.3601	0.1359
11	1.898	0.5268	14.972	0.0668	7.8869	0.1268
12	2.012	0.4970	16.870	0.0593	8.3839	0.1193
13	2.133	0.4688	18.882	0.0530	8.8527	0.1130
14	2.261	0.4423	21.015	0.0476	9.2950	0.1076
15	2.397	0.4173	23.276	0.0430	9.7123	0.1030
16	2.540	0.3937	25.673	0.0390	10.1059	0.0990
17	2.693	0.3714	28.213	0.0355	10.4773	0.0955
18	2.854	0.3504	30.906	0.0324	10.8276	0.0924
19	3.026	0.3305	33.760	0.0296	11.1581	0.0896
20	3.207	0.3118	36.786	0.0272	11.4699	0.0872
21	3.400	0.2942	39.993	0.0250	11.7641	0.0850
22	3.604	0.2775	43.392	0.0231	12.0416	0.0831
23	3.820	0.2618	46.996	0.0213	12.3034	0.0813
24	4.049	0.2470	50.816	0.0197	12.5504	0.0797
25	4.292	0.2330	54.865	0.0182	12.7834	0.0782
26	4.549	0.2198	59.156	0.0169	13.0032	0.0769
27	4.822	0.2074	63.706	0.0157	13.2105	0.0757
28	5.112	0.1956	68.528	0.0146	13.4062	0.0746
29	5.418	0.1846	73.640	0.0136	13.5907	0.0736
30	5.744	0.1741	79.058	0.0127	13.7648	0.0727
31	6.088	0.1643	84.802	0.0118	13.9291	0.0718
32	6.453	0.1550	90.890	0.0110	14.0841	0.0710
33	6.841	0.1462	97.343	0.0103	14.2302	0.0703
34	7.251	0.1379	104.184	0.0096	14.3682	0.0696
35	7.686	0.1301	111.435	0.0090	14.4983	0.0690
40	10.286	0.0972	154.762	0.0065	15.0463	0.0665
45	13.765	0.0727	212.744	0.0047	15.4558	0.0647
50	18.420	0.0543	290.336	0.0035	15.7619	0.0635
55	24.650	0.0406	394.172	0.0025	15.9906	0.0625
60	32.988	0.0303	533.128	0.0019	16.1614	0.0619
65	44.145	0.0227	719.083	0.0014	16.2891	0.0614
70	59.076	0.0169	967.932	0.0010	16.3846	0.0610
75	79.057	0.0127	1300.949	0.0008	16.4559	0.0608
80	105.796	0.0095	1746.600	0.0006	16.5091	0.0606
85	141.579	0.0071	2342.982	0.0004	16.5490	0.0604
90	189.465	0.0053	3141.075	0.0003	16.5787	0.0603
95	253.546	0.0040	4209.104	0.0002	16.6009	0.0602
100	339.302	0.0030	5638.368	0.0002	16.6176	0.0602

TABLE D.3 7% INTEREST FACTORS FOR ANNUAL COMPOUNDING

	Single Payment		Equal-Payment Series			
	Compound-Amount Factor	Present-Worth Factor	Compound-Amount Factor	Sinking-Fund Factor	Present-Worth Factor	Capital-Recovery Factor
n	To Find F Given P $F/P, i, n$	To Find P Given F $P/F, i, n$	To Find F Given A $F/A, i, n$	To Find A Given F $A/F, i, n$	To Find P Given A $P/A, i, n$	To Find A Given P $A/P, i, n$
1	1.070	0.9346	1.000	1.0000	0.9346	1.0700
2	1.145	0.8734	2.070	0.4831	1.8080	0.5531
3	1.225	0.8163	3.215	0.3111	2.6243	0.3811
4	1.311	0.7629	4.440	0.2252	3.3872	0.2952
5	1.403	0.7130	5.751	0.1739	4.1002	0.2439
6	1.501	0.6664	7.153	0.1398	4.7665	0.2098
7	1.606	0.6228	8.654	0.1156	5.3893	0.1856
8	1.718	0.5820	10.260	0.0975	5.9713	0.1675
9	1.838	0.5439	11.978	0.0835	6.5152	0.1535
10	1.967	0.5084	13.816	0.0724	7.0236	0.1424
11	2.105	0.4751	15.784	0.0634	7.4987	0.1334
12	2.252	0.4440	17.888	0.0559	7.9427	0.1259
13	2.410	0.4150	20.141	0.0497	8.3577	0.1197
14	2.579	0.3878	22.550	0.0444	8.7455	0.1144
15	2.759	0.3625	25.129	0.0398	9.1079	0.1098
16	2.952	0.3387	27.888	0.0359	9.4467	0.1059
17	3.159	0.3166	30.840	0.0324	9.7632	0.1024
18	3.380	0.2959	33.999	0.0294	10.0591	0.0994
19	3.617	0.2765	37.379	0.0268	10.3356	0.0968
20	3.870	0.2584	40.996	0.0244	10.5940	0.0944
21	4.141	0.2415	44.865	0.0223	10.8355	0.0923
22	4.430	0.2257	49.006	0.0204	11.0613	0.0904
23	4.741	0.2110	53.436	0.0187	11.2722	0.0887
24	5.072	0.1972	58.177	0.0172	11.4693	0.0872
25	5.427	0.1843	63.249	0.0158	11.6536	0.0858
26	5.807	0.1722	68.676	0.0146	11.8258	0.0846
27	6.214	0.1609	74.484	0.0134	11.9867	0.0834
28	6.649	0.1504	80.698	0.0124	12.1371	0.0824
29	7.114	0.1406	87.347	0.0115	12.2777	0.0815
30	7.612	0.1314	94.461	0.0106	12.4091	0.0806
31	8.145	0.1228	102.073	0.0098	12.5318	0.0798
32	8.715	0.1148	110.218	0.0091	12.6466	0.0791
33	9.325	0.1072	118.933	0.0084	12.7538	0.0784
34	9.978	0.1002	128.259	0.0078	12.8540	0.0778
35	10.677	0.0937	138.237	0.0072	12.9477	0.0772
40	14.974	0.0668	199.635	0.0050	13.3317	0.0750
45	21.002	0.0476	285.749	0.0035	13.6055	0.0735
50	29.457	0.0340	406.529	0.0025	13.8008	0.0725
55	41.315	0.0242	575.929	0.0017	13.9399	0.0717
60	57.946	0.0173	813.520	0.0012	14.0392	0.0712
65	81.273	0.0123	1146.755	0.0009	14.1099	0.0709
70	113.989	0.0088	1614.134	0.0006	14.1604	0.0706
75	159.876	0.0063	2269.657	0.0005	14.1964	0.0705
80	224.234	0.0045	3189.063	0.0003	14.2220	0.0703
85	314.500	0.0032	4478.576	0.0002	14.2403	0.0702
90	441.103	0.0023	6287.185	0.0002	14.2533	0.0702
95	618.670	0.0016	8823.854	0.0001	14.2626	0.0701
100	867.716	0.0012	12381.662	0.0001	14.2693	0.0701

TABLE D.4 8% INTEREST FACTORS FOR ANNUAL COMPOUNDING

	Single Payment		Equal-Payment Series			
n	Compound-Amount Factor	Present-Worth Factor	Compound-Amount Factor	Sinking-Fund Factor	Present-Worth Factor	Capital-Recovery Factor
	To Find F Given P $F/P, i, n$	To Find P Given F $P/F, i, n$	To Find F Given A $F/A, i, n$	To Find A Given F $A/F, i, n$	To Find P Given A $P/A, i, n$	To Find A Given P $A/P, i, n$
1	1.080	0.9259	1.000	1.0000	0.9259	1.0800
2	1.166	0.8573	2.080	0.4808	1.7833	0.5608
3	1.260	0.7938	3.246	0.3080	2.5771	0.3880
4	1.360	0.7350	4.506	0.2219	3.3121	0.3019
5	1.469	0.6806	5.867	0.1705	3.9927	0.2505
6	1.587	0.6302	7.336	0.1363	4.6229	0.2163
7	1.714	0.5835	8.923	0.1121	5.2064	0.1921
8	1.851	0.5403	10.637	0.0940	5.7466	0.1740
9	1.999	0.5003	12.488	0.0801	6.2469	0.1601
10	2.159	0.4632	14.487	0.0690	6.7101	0.1490
11	2.332	0.4289	16.645	0.0601	7.1390	0.1401
12	2.518	0.3971	18.977	0.0527	7.5361	0.1327
13	2.720	0.3677	21.495	0.0465	7.9038	0.1265
14	2.937	0.3405	24.215	0.0413	8.2442	0.1213
15	3.172	0.3153	27.152	0.0368	8.5595	0.1168
16	3.426	0.2919	30.324	0.0330	8.8514	0.1130
17	3.700	0.2703	33.750	0.0296	9.1216	0.1096
18	3.996	0.2503	37.450	0.0267	9.3719	0.1067
19	4.316	0.2317	41.446	0.0241	9.6036	0.1041
20	4.661	0.2146	45.762	0.0219	9.8182	0.1019
21	5.034	0.1987	50.423	0.0198	10.0168	0.0998
22	5.437	0.1840	55.457	0.0180	10.2008	0.0980
23	5.871	0.1703	60.893	0.0164	10.3711	0.0964
24	6.341	0.1577	66.765	0.0150	10.5288	0.0950
25	6.848	0.1460	73.106	0.0137	10.6748	0.0937
26	7.396	0.1352	79.954	0.0125	10.8100	0.0925
27	7.988	0.1252	87.351	0.0115	10.9352	0.0915
28	8.627	0.1159	95.339	0.0105	11.0511	0.0905
29	9.317	0.1073	103.966	0.0096	11.1584	0.0896
30	10.063	0.0994	113.283	0.0088	11.2578	0.0888
31	10.868	0.0920	123.346	0.0081	11.3498	0.0881
32	11.737	0.0852	134.214	0.0075	11.4350	0.0875
33	12.676	0.0789	145.951	0.0069	11.5139	0.0869
34	13.690	0.0731	158.627	0.0063	11.5869	0.0863
35	14.785	0.0676	172.317	0.0058	11.6546	0.0858
40	21.725	0.0460	259.057	0.0039	11.9246	0.0839
45	31.920	0.0313	386.506	0.0026	12.1084	0.0826
50	46.902	0.0213	573.770	0.0018	12.2335	0.0818
55	68.914	0.0145	848.923	0.0012	12.3186	0.0812
60	101.257	0.0099	1253.213	0.0008	12.3766	0.0808
65	148.780	0.0067	1847.248	0.0006	12.4160	0.0806
70	218.606	0.0046	2720.080	0.0004	12.4428	0.0804
75	321.205	0.0031	4002.557	0.0003	12.4611	0.0803
80	471.955	0.0021	5886.935	0.0002	12.4735	0.0802
85	693.456	0.0015	8655.706	0.0001	12.4820	0.0801
90	1018.915	0.0010	12723.939	0.0001	12.4877	0.0801
95	1497.121	0.0007	18701.507	0.0001	12.4917	0.0801
100	2199.761	0.0005	27484.516	0.0001	12.4943	0.0800

TABLE D.5 9% INTEREST FACTORS FOR ANNUAL COMPOUNDING

	Single Payment		Equal-Payment Series			
	Compound-Amount Factor	Present-Worth Factor	Compound-Amount Factor	Sinking-Fund Factor	Present-Worth Factor	Capital-Recovery Factor
n	To Find F Given P F/P, i, n	To Find P Given F P/F, i, n	To Find F Given A F/A, i, n	To Find A Given F A/F, i, n	To Find P Given A P/A, i, n	To Find A Given P A/P, i, n
1	1.090	0.9174	1.000	1.0000	0.9174	1.0900
2	1.188	0.8417	2.090	0.4785	1.7591	0.5685
3	1.295	0.7722	3.278	0.3051	2.5313	0.3951
4	1.412	0.7084	4.573	0.2187	3.2397	0.3087
5	1.539	0.6499	5.985	0.1671	3.8897	0.2571
6	1.677	0.5963	7.523	0.1329	4.4859	0.2229
7	1.828	0.5470	9.200	0.1087	5.0330	0.1987
8	1.993	0.5019	11.028	0.0907	5.5348	0.1807
9	2.172	0.4604	13.021	0.0768	5.9953	0.1668
10	2.367	0.4224	15.193	0.0658	6.4177	0.1558
11	2.580	0.3875	17.560	0.0570	6.8052	0.1470
12	2.813	0.3555	20.141	0.0497	7.1607	0.1397
13	3.066	0.3262	22.953	0.0436	7.4869	0.1336
14	3.342	0.2993	26.019	0.0384	7.7862	0.1284
15	3.642	0.2745	29.361	0.0341	8.0607	0.1241
16	3.970	0.2519	33.003	0.0303	8.3126	0.1203
17	4.328	0.2311	36.974	0.0271	8.5436	0.1171
18	4.717	0.2120	41.301	0.0242	8.7556	0.1142
19	5.142	0.1945	46.018	0.0217	8.9501	0.1117
20	5.604	0.1784	51.160	0.0196	9.1286	0.1096
21	6.109	0.1637	56.765	0.0176	9.2923	0.1076
22	6.659	0.1502	62.873	0.0159	9.4424	0.1059
23	7.258	0.1378	69.532	0.0144	9.5802	0.1044
24	7.911	0.1264	76.790	0.0130	9.7066	0.1030
25	8.623	0.1160	84.701	0.0118	9.8226	0.1018
26	9.399	0.1064	93.324	0.0107	9.9290	0.1007
27	10.245	0.0976	102.723	0.0097	10.0266	0.0997
28	11.167	0.0896	112.968	0.0089	10.1161	0.0989
29	12.172	0.0822	124.135	0.0081	10.1983	0.0981
30	13.268	0.0754	136.308	0.0073	10.2737	0.0973
31	14.462	0.0692	149.575	0.0067	10.3428	0.0967
32	15.763	0.0634	164.037	0.0061	10.4063	0.0961
33	17.182	0.0582	179.800	0.0056	10.4645	0.0956
34	18.728	0.0534	196.982	0.0051	10.5178	0.0951
35	20.414	0.0490	215.711	0.0046	10.5668	0.0946
40	31.409	0.0318	337.882	0.0030	10.7574	0.0930
45	48.327	0.0207	525.859	0.0019	10.8812	0.0919
50	74.358	0.0135	815.084	0.0012	10.9617	0.0912
55	114.408	0.0088	1260.092	0.0008	11.0140	0.0908
60	176.031	0.0057	1944.792	0.0005	11.0480	0.0905
65	270.846	0.0037	2998.288	0.0003	11.0701	0.0903
70	416.730	0.0024	4619.223	0.0002	11.0845	0.0902
75	641.191	0.0016	7113.232	0.0002	11.0938	0.0902
80	986.552	0.0010	10950.574	0.0001	11.0999	0.0901
85	1517.932	0.0007	16854.800	0.0001	11.1038	0.0901
90	2335.527	0.0004	25939.184	0.0001	11.1064	0.0900
95	3593.497	0.0003	39916.635	0.0000	11.1080	0.0900
100	5529.041	0.0002	61422.675	0.0000	11.1091	0.0900

TABLE D.6　10% INTEREST FACTORS FOR ANNUAL COMPOUNDING

	Single Payment		Equal-Payment Series			
	Compound-Amount Factor	Present-Worth Factor	Compound-Amount Factor	Sinking-Fund Factor	Present-Worth Factor	Capital-Recovery Factor
n	To Find F Given P $F/P, i, n$	To Find P Given F $P/F, i, n$	To Find F Given A $F/A, i, n$	To Find A Given F $A/F, i, n$	To Find P Given A $P/A, i, n$	To Find A Given P $A/P, i, n$
1	1.100	0.9091	1.000	1.0000	0.9091	1.1000
2	1.210	0.8265	2.100	0.4762	1.7355	0.5762
3	1.331	0.7513	3.310	0.3021	2.4869	0.4021
4	1.464	0.6830	4.641	0.2155	3.1699	0.3155
5	1.611	0.6209	6.105	0.1638	3.7908	0.2638
6	1.772	0.5645	7.716	0.1296	4.3553	0.2296
7	1.949	0.5132	9.487	0.1054	4.8684	0.2054
8	2.144	0.4665	11.436	0.0875	5.3349	0.1875
9	2.358	0.4241	13.579	0.0737	5.7950	0.1737
10	2.594	0.3856	15.937	0.0628	6.1446	0.1628
11	2.853	0.3505	18.531	0.0540	6.4951	0.1540
12	3.138	0.3186	21.384	0.0468	6.8137	0.1468
13	3.452	0.2897	24.523	0.0408	7.1034	0.1408
14	3.798	0.2633	27.975	0.0358	7.3667	0.1358
15	4.177	0.2394	31.772	0.0315	7.6061	0.1315
16	4.595	0.2176	35.950	0.0278	7.8237	0.1278
17	5.054	0.1979	40.545	0.0247	8.0216	0.1247
18	5.560	0.1799	45.599	0.0219	8.2014	0.1219
19	6.116	0.1635	51.159	0.0196	8.3649	0.1196
20	6.728	0.1487	57.275	0.0175	8.5136	0.1175
21	7.400	0.1351	64.003	0.0156	8.6487	0.1156
22	8.140	0.1229	71.403	0.0140	8.7716	0.1140
23	8.953	0.1117	79.543	0.0126	8.8832	0.1126
24	9.850	0.1015	88.497	0.0113	8.9848	0.1113
25	10.835	0.0923	98.347	0.0102	9.0771	0.1102
26	11.918	0.0839	109.182	0.0092	9.1610	0.1092
27	13.110	0.0763	121.100	0.0083	9.2372	0.1083
28	14.421	0.0694	134.210	0.0075	9.3066	0.1075
29	15.863	0.0630	148.631	0.0067	9.3696	0.1067
30	17.449	0.0573	164.494	0.0061	9.4269	0.1061
31	19.194	0.0521	181.943	0.0055	9.4790	0.1055
32	21.114	0.0474	201.138	0.0050	9.5264	0.1050
33	23.225	0.0431	222.252	0.0045	9.5694	0.1045
34	25.548	0.0392	245.477	0.0041	9.6086	0.1041
35	28.102	0.0356	271.024	0.0037	9.6442	0.1037
40	45.259	0.0221	442.593	0.0023	9.7791	0.1023
45	72.890	0.0137	718.905	0.0014	9.8628	0.1014
50	117.391	0.0085	1163.909	0.0009	9.9148	0.1009
55	189.059	0.0053	1880.591	0.0005	9.9471	0.1005
60	304.482	0.0033	3034.816	0.0003	9.9672	0.1003
65	490.371	0.0020	4893.707	0.0002	9.9796	0.1002
70	789.747	0.0013	7887.470	0.0001	9.9873	0.1001
75	1271.895	0.0008	12708.954	0.0001	9.9921	0.1001
80	2048.400	0.0005	20474.002	0.0001	9.9951	0.1001
85	3298.969	0.0003	32979.690	0.0000	9.9970	0.1000
90	5313.023	0.0002	53120.226	0.0000	9.9981	0.1000
95	8556.676	0.0001	85556.760	0.0000	9.9988	0.1000
100	13780.612	0.0001	137796.123	0.0000	9.9993	0.1000

TABLE D.7 11% INTEREST FACTORS FOR ANNUAL COMPOUNDING

	Single Payment		Equal-Payment Series			
n	Compound-Amount Factor	Present-Worth Factor	Compound-Amount Factor	Sinking-Fund Factor	Present-Worth Factor	Capital-Recovery Factor
	To Find F Given P F/P, i,n	*To Find P Given F P/F, i,n*	*To Find F Given A F/A, i,n*	*To Find A Given F A/F, i,n*	*To Find P Given A P/A, i,n*	*To Find A Given P A/P, i,n*
1	1.110	0.9009	1.000	1.0000	0.9009	1.1100
2	1.232	0.8116	2.110	0.4739	1.7125	0.5839
3	1.368	0.7312	3.342	0.2992	2.4437	0.4092
4	1.518	0.6587	4.710	0.2123	3.1024	0.3223
5	1.685	0.5935	6.228	0.1606	3.6959	0.2706
6	1.870	0.5346	7.913	0.1264	4.2305	0.2364
7	2.076	0.4817	9.783	0.1022	4.7121	0.2122
8	2.305	0.4339	11.859	0.0843	5.1462	0.1943
9	2.558	0.3909	14.164	0.0706	5.5371	0.1806
10	2.839	0.3522	16.722	0.0598	5.8893	0.1698
11	3.152	0.3173	19.561	0.0511	6.2066	0.1611
12	3.498	0.2858	22.713	0.0440	6.4922	0.1540
13	3.883	0.2575	26.212	0.0382	6.7499	0.1482
14	4.310	0.2320	30.095	0.0332	6.9818	0.1432
15	4.785	0.2090	34.405	0.0291	7.1906	0.1391
16	5.311	0.1883	39.190	0.0255	7.3790	0.1355
17	5.895	0.1696	44.501	0.0225	7.5489	0.1325
18	6.544	0.1528	50.396	0.0198	7.7018	0.1298
19	7.263	0.1377	56.939	0.0176	7.8394	0.1276
20	8.062	0.1240	64.203	0.0156	7.9631	0.1256
21	8.949	0.1117	72.265	0.0138	8.0749	0.1238
22	9.934	0.1007	81.214	0.0123	8.1759	0.1223
23	11.026	0.0907	91.148	0.0110	8.2665	0.1210
24	12.239	0.0817	102.174	0.0098	8.3479	0.1198
25	13.586	0.0736	114.413	0.0087	8.4218	0.1187
26	15.080	0.0663	127.999	0.0078	8.4882	0.1178
27	16.739	0.0597	143.079	0.0070	8.5477	0.1170
28	18.580	0.0538	159.817	0.0063	8.6014	0.1163
29	20.624	0.0485	178.397	0.0056	8.6498	0.1156
30	22.892	0.0437	199.021	0.0050	8.6941	0.1150
31	25.410	0.0394	221.913	0.0045	8.7329	0.1145
32	28.206	0.0355	247.324	0.0040	8.7689	0.1140
33	31.308	0.0319	275.529	0.0036	8.8005	0.1136
34	34.752	0.0288	306.837	0.0033	8.8292	0.1133
35	38.575	0.0259	341.590	0.0029	8.8550	0.1129
40	65.001	0.0154	581.826	0.0017	8.9509	0.1117
45	109.530	0.0091	986.639	0.0010	9.0082	0.1110
50	184.565	0.0054	1688.771	0.0006	9.0416	0.1106

TABLE D.8   12% INTEREST FACTORS FOR ANNUAL COMPOUNDING

	Single Payment		Equal-Payment Series			
n	Compound-Amount Factor	Present-Worth Factor	Compound-Amount Factor	Sinking-Fund Factor	Present-Worth Factor	Capital-Recovery Factor
	To Find F Given P $F/P, i, n$	To Find P Given F $P/F, i, n$	To Find F Given A $F/A, i, n$	To Find A Given F $A/F, i, n$	To Find P Given A $P/A, i, n$	To Find A Given P $A/P, i, n$
1	1.120	0.8929	1.000	1.0000	0.8929	1.1200
2	1.254	0.7972	2.120	0.4717	1.6901	0.5917
3	1.405	0.7118	3.374	0.2964	2.4018	0.4164
4	1.574	0.6355	4.779	0.2092	3.0374	0.3292
5	1.762	0.5674	6.353	0.1574	3.6048	0.2774
6	1.974	0.5066	8.115	0.1232	4.1114	0.2432
7	2.211	0.4524	10.089	0.0991	4.5638	0.2191
8	2.476	0.4039	12.300	0.0813	4.9676	0.2013
9	2.773	0.3606	14.776	0.0677	5.3283	0.1877
10	3.106	0.3220	17.549	0.0570	5.6502	0.1770
11	3.479	0.2875	20.655	0.0484	5.9377	0.1684
12	3.896	0.2567	24.133	0.0414	6.1944	0.1614
13	4.364	0.2292	28.029	0.0357	6.4236	0.1557
14	4.887	0.2046	32.393	0.0309	6.6282	0.1509
15	5.474	0.1827	37.280	0.0268	6.8109	0.1468
16	6.130	0.1631	42.753	0.0234	6.9740	0.1434
17	6.866	0.1457	48.884	0.0205	7.1196	0.1405
18	7.690	0.1300	55.750	0.0179	7.2497	0.1379
19	8.613	0.1161	63.440	0.0158	7.3658	0.1358
20	9.646	0.1037	72.052	0.0139	7.4695	0.1339
21	10.804	0.0926	81.699	0.0123	7.5620	0.1323
22	12.100	0.0827	92.503	0.0108	7.6447	0.1308
23	13.552	0.0738	104.603	0.0096	7.7184	0.1296
24	15.179	0.0659	118.155	0.0085	7.7843	0.1285
25	17.000	0.0588	133.334	0.0075	7.8431	0.1275
26	19.040	0.0525	150.334	0.0067	7.8957	0.1267
27	21.325	0.0469	169.374	0.0059	7.9426	0.1259
28	23.884	0.0419	190.699	0.0053	7.9844	0.1253
29	26.750	0.0374	214.583	0.0047	8.0218	0.1247
30	29.960	0.0334	241.333	0.0042	8.0552	0.1242
31	33.555	0.0298	271.293	0.0037	8.0850	0.1237
32	37.582	0.0266	304.848	0.0033	8.1116	0.1233
33	42.092	0.0238	342.429	0.0029	8.1354	0.1229
34	47.143	0.0212	384.521	0.0026	8.1566	0.1226
35	52.800	0.0189	431.664	0.0023	8.1755	0.1223
40	93.051	0.0108	767.091	0.0013	8.2438	0.1213
45	163.988	0.0061	1358.230	0.0007	8.2825	0.1207
50	289.002	0.0035	2400.018	0.0004	8.3045	0.1204

TABLE D.9 13% INTEREST FACTORS FOR ANNUAL COMPOUNDING

	Single Payment		Equal-Payment Series			
n	Compound-Amount Factor	Present-Worth Factor	Compound-Amount Factor	Sinking-Fund Factor	Present-Worth Factor	Capital-Recovery Factor
	To Find F Given P F/P, i,n	To Find P Given F P/F, i,n	To Find F Given A F/A, i,n	To Find A Given F A/F, i,n	To Find P Given A P/A, i,n	To Find A Given P A/P, i,n
1	1.130	0.8850	1.000	1.0000	0.8850	1.1300
2	1.277	0.7831	2.130	0.4695	1.6681	0.5995
3	1.443	0.6931	3.407	0.2935	2.3612	0.4235
4	1.631	0.6133	4.850	0.2062	2.9745	0.3362
5	1.842	0.5428	6.480	0.1543	3.5173	0.2843
6	2.082	0.4803	8.323	0.1202	3.9976	0.2502
7	2.353	0.4251	10.405	0.0961	4.4226	0.2261
8	2.658	0.3762	12.757	0.0784	4.7987	0.2084
9	3.004	0.3329	15.416	0.0649	5.1316	0.1949
10	3.395	0.2946	18.420	0.0543	5.4262	0.1843
11	3.836	0.2607	21.814	0.0458	5.6870	0.1758
12	4.335	0.2307	25.650	0.0390	5.9175	0.1690
13	4.898	0.2042	29.985	0.0334	6.1218	0.1634
14	5.535	0.1807	34.883	0.0287	6.3024	0.1587
15	6.254	0.1599	40.417	0.0247	6.4625	0.1547
16	7.067	0.1415	46.672	0.0214	6.6037	0.1514
17	7.986	0.1252	53.739	0.0186	6.7290	0.1486
18	9.024	0.1108	61.725	0.0162	6.8399	0.1462
19	10.197	0.0981	70.749	0.0141	6.9382	0.1441
20	11.523	0.0868	80.947	0.0124	7.0249	0.1424
21	13.021	0.0768	92.470	0.0108	7.1018	0.1408
22	14.714	0.0680	105.491	0.0095	7.1695	0.1395
23	16.627	0.0601	120.205	0.0083	7.2296	0.1383
24	18.788	0.0532	136.831	0.0073	7.2828	0.1373
25	21.231	0.0471	155.620	0.0064	7.3298	0.1364
26	23.991	0.0417	176.850	0.0057	7.3719	0.1357
27	27.109	0.0369	200.841	0.0050	7.4085	0.1350
28	30.634	0.0326	227.950	0.0044	7.4410	0.1344
29	34.616	0.0289	258.583	0.0039	7.4699	0.1339
30	39.116	0.0256	293.199	0.0034	7.4957	0.1334
31	44.201	0.0226	332.315	0.0030	7.5182	0.1330
32	49.947	0.0200	376.516	0.0027	7.5381	0.1327
33	56.440	0.0177	426.463	0.0023	7.5563	0.1323
34	63.777	0.0157	482.903	0.0021	7.5717	0.1321
35	72.069	0.0139	546.681	0.0018	7.5855	0.1318
40	132.782	0.0075	1013.704	0.0010	7.6342	0.1310
45	244.641	0.0041	1874.165	0.0005	7.6611	0.1305
50	450.736	0.0022	3459.507	0.0003	7.6752	0.1303

TABLE D.10 14% INTEREST FACTORS FOR ANNUAL COMPOUNDING

	Single Payment		Equal-Payment Series			
	Compound-Amount Factor	Present-Worth Factor	Compound-Amount Factor	Sinking-Fund Factor	Present-Worth Factor	Capital-Recovery Factor
n	To Find F Given P $F/P, i,n$	To Find P Given F $P/F, i,n$	To Find F Given A $F/A, i,n$	To Find A Given F $A/F, i,n$	To Find P Given A $P/A, i,n$	To Find A Given P $A/P, i,n$
1	1.140	0.8772	1.000	1.0000	0.8772	1.1400
2	1.300	0.7695	2.140	0.4673	1.6467	0.6073
3	1.482	0.6750	3.440	0.2907	2.3216	0.4307
4	1.689	0.5921	4.921	0.2032	2.9138	0.3432
5	1.925	0.5194	6.610	0.1513	3.4331	0.2913
6	2.195	0.4556	8.536	0.1172	3.8886	0.2572
7	2.502	0.3996	10.730	0.0932	4.2883	0.2332
8	2.853	0.3506	13.233	0.0756	4.6389	0.2156
9	3.252	0.3075	16.085	0.0622	4.9463	0.2022
10	3.707	0.2697	19.337	0.0517	5.2162	0.1917
11	4.226	0.2366	23.045	0.0434	5.4529	0.1834
12	4.818	0.2076	27.271	0.0367	5.6603	0.1767
13	5.492	0.1821	32.089	0.0312	5.8425	0.1712
14	6.261	0.1597	37.581	0.0266	6.0020	0.1666
15	7.138	0.1401	43.842	0.0228	6.1421	0.1628
16	8.137	0.1229	50.980	0.0196	6.2649	0.1596
17	9.277	0.1078	59.118	0.0169	6.3727	0.1569
18	10.575	0.0946	68.394	0.0146	6.4675	0.1546
19	12.056	0.0829	78.969	0.0127	6.5505	0.1527
20	13.744	0.0728	91.025	0.0110	6.6230	0.1510
21	15.668	0.0638	104.768	0.0095	6.6872	0.1495
22	17.861	0.0560	120.436	0.0083	6.7431	0.1483
23	20.362	0.0491	138.297	0.0072	6.7921	0.1472
24	23.212	0.0431	158.659	0.0063	6.8353	0.1463
25	26.462	0.0378	181.871	0.0055	6.8729	0.1455
26	30.167	0.0331	208.333	0.0048	6.9061	0.1448
27	34.390	0.0291	238.499	0.0042	6.9353	0.1442
28	39.205	0.0255	272.889	0.0037	6.9609	0.1437
29	44.693	0.0224	312.094	0.0032	6.9832	0.1432
30	50.950	0.0196	356.787	0.0028	7.0028	0.1428
31	58.083	0.0172	407.737	0.0025	7.0200	0.1425
32	66.215	0.0151	465.820	0.0022	7.0348	0.1422
33	75.485	0.0132	532.035	0.0019	7.0482	0.1419
34	86.053	0.0116	607.520	0.0017	7.0597	0.1417
35	98.100	0.0102	693.573	0.0014	7.0701	0.1414
40	188.884	0.0053	1342.025	0.0008	7.1048	0.1408
45	363.679	0.0027	2590.565	0.0004	7.1230	0.1404
50	700.233	0.0014	4994.521	0.0002	7.1327	0.1402

TABLE D.11   15% INTEREST FACTORS FOR ANNUAL COMPOUNDING

	Single Payment		Equal-Payment Series			
n	Compound-Amount Factor	Present-Worth Factor	Compound-Amount Factor	Sinking-Fund Factor	Present-Worth Factor	Capital-Recovery Factor
	To Find F Given P $F/P, i, n$	*To Find P Given F* $P/F, i, n$	*To Find F Given A* $F/A, i, n$	*To Find A Given F* $A/F, i, n$	*To Find P Given A* $P/A, i, n$	*To Find A Given P* $A/P, i, n$
1	1.150	0.8696	1.000	1.0000	0.8696	1.1500
2	1.323	0.7562	2.150	0.4651	1.6257	0.6151
3	1.521	0.6575	3.473	0.2880	2.2832	0.4380
4	1.749	0.5718	4.993	0.2003	2.8550	0.3503
5	2.011	0.4972	6.742	0.1483	3.3522	0.2983
6	2.313	0.4323	8.754	0.1142	3.7845	0.2642
7	2.660	0.3759	11.067	0.0904	4.1604	0.2404
8	3.059	0.3269	13.727	0.0729	4.4873	0.2229
9	3.518	0.2843	16.786	0.0596	4.7716	0.2096
10	4.046	0.2472	20.304	0.0493	5.0188	0.1993
11	4.652	0.2150	24.349	0.0411	5.2337	0.1911
12	5.350	0.1869	29.002	0.0345	5.4206	0.1845
13	6.153	0.1625	34.352	0.0291	5.5832	0.1791
14	7.076	0.1413	40.505	0.0247	5.7245	0.1747
15	8.137	0.1229	47.580	0.0210	5.8474	0.1710
16	9.358	0.1069	55.717	0.0180	5.9542	0.1680
17	10.761	0.0929	65.075	0.0154	6.0472	0.1654
18	12.375	0.0808	75.836	0.0132	6.1280	0.1632
19	14.232	0.0703	88.212	0.0113	6.1982	0.1613
20	16.367	0.0611	102.444	0.0098	6.2593	0.1598
21	18.822	0.0531	118.810	0.0084	6.3125	0.1584
22	21.645	0.0462	137.632	0.0073	6.3587	0.1573
23	24.891	0.0402	159.276	0.0063	6.3988	0.1563
24	28.625	0.0349	184.168	0.0054	6.4338	0.1554
25	32.919	0.0304	212.793	0.0047	6.4642	0.1547
26	37.857	0.0264	245.712	0.0041	6.4906	0.1541
27	43.535	0.0230	283.569	0.0035	6.5135	0.1535
28	50.066	0.0200	327.104	0.0031	6.5335	0.1531
29	57.575	0.0174	377.170	0.0027	6.5509	0.1527
30	66.212	0.0151	434.745	0.0023	6.5660	0.1523
31	76.144	0.0131	500.957	0.0020	6.5791	0.1520
32	87.565	0.0114	577.100	0.0017	6.5905	0.1517
33	100.700	0.0099	664.666	0.0015	6.6005	0.1515
34	115.805	0.0086	765.365	0.0013	6.6091	0.1513
35	133.176	0.0075	881.170	0.0011	6.6166	0.1511
40	267.864	0.0037	1779.090	0.0006	6.6418	0.1506
45	538.769	0.0019	3585.128	0.0003	6.6543	0.1503
50	1083.657	0.0009	7217.716	0.0002	6.6605	0.1501

TABLE D.12 20% INTEREST FACTORS FOR ANNUAL COMPOUNDING

	Single Payment		Equal-Payment Series			
	Compound-Amount Factor	Present-Worth Factor	Compound-Amount Factor	Sinking-Fund Factor	Present-Worth Factor	Capital-Recovery Factor
n	To Find F Given P $F/P, i, n$	To Find P Given F $P/F, i, n$	To Find F Given A $F/A, i, n$	To Find A Given F $A/F, i, n$	To Find P Given A $P/A, i, n$	To Find A Given P $A/P, i, n$
1	1.200	0.8333	1.000	1.0000	0.8333	1.2000
2	1.440	0.6945	2.200	0.4546	1.5278	0.6546
3	1.728	0.5787	3.640	0.2747	2.1065	0.4747
4	2.074	0.4823	5.368	0.1863	2.5887	0.3863
5	2.488	0.4019	7.442	0.1344	2.9906	0.3344
6	2.986	0.3349	9.930	0.1007	3.3255	0.3007
7	3.583	0.2791	12.916	0.0774	3.6046	0.2774
8	4.300	0.2326	16.499	0.0606	3.8372	0.2606
9	5.160	0.1938	20.799	0.0481	4.0310	0.2481
10	6.192	0.1615	25.959	0.0385	4.1925	0.2385
11	7.430	0.1346	32.150	0.0311	4.3271	0.2311
12	8.916	0.1122	39.581	0.0253	4.4392	0.2253
13	10.699	0.0935	48.497	0.0206	4.5327	0.2206
14	12.839	0.0779	59.196	0.0169	4.6106	0.2169
15	15.407	0.0649	72.035	0.0139	4.6755	0.2139
16	18.488	0.0541	87.442	0.0114	4.7296	0.2114
17	22.186	0.0451	105.931	0.0095	4.7746	0.2095
18	26.623	0.0376	128.117	0.0078	4.8122	0.2078
19	31.948	0.0313	154.740	0.0065	4.8435	0.2065
20	38.338	0.0261	186.688	0.0054	4.8696	0.2054
21	46.005	0.0217	225.026	0.0045	4.8913	0.2045
22	55.206	0.0181	271.031	0.0037	4.9094	0.2037
23	66.247	0.0151	326.237	0.0031	4.9245	0.2031
24	79.497	0.0126	392.484	0.0026	4.9371	0.2026
25	95.396	0.0105	471.981	0.0021	4.9476	0.2021
26	114.475	0.0087	567.377	0.0018	4.9563	0.2018
27	137.371	0.0073	681.853	0.0015	4.9636	0.2015
28	164.845	0.0061	819.223	0.0012	4.9697	0.2012
29	197.814	0.0051	984.068	0.0010	4.9747	0.2010
30	237.376	0.0042	1181.882	0.0009	4.9789	0.2009
31	284.852	0.0035	1419.258	0.0007	4.9825	0.2007
32	341.822	0.0029	1704.109	0.0006	4.9854	0.2006
33	410.186	0.0024	2045.931	0.0005	4.9878	0.2005
34	492.224	0.0020	2456.118	0.0004	4.9899	0.2004
35	590.668	0.0017	2948.341	0.0003	4.9915	0.2003
40	1469.772	0.0007	7343.858	0.0002	4.9966	0.2001
45	3657.262	0.0003	18281.310	0.0001	4.9986	0.2001
50	9100.438	0.0001	45497.191	0.0000	4.9995	0.2000

TABLE D.13 25% INTEREST FACTORS FOR ANNUAL COMPOUNDING

	Single Payment		Equal-Payment Series			
	Compound-Amount Factor	Present-Worth Factor	Compound-Amount Factor	Sinking-Fund Factor	Present-Worth Factor	Capital-Recovery Factor
n	To Find F Given P $F/P, i, n$	To Find P Given F $P/F, i, n$	To Find F Given A $F/A, i, n$	To Find A Given F $A/F, i, n$	To Find P Given A $P/A, i, n$	To Find A Given P $A/P, i, n$
1	1.250	0.8000	1.000	1.0000	0.8000	1.2500
2	1.563	0.6400	2.250	0.4445	1.4400	0.6945
3	1.953	0.5120	3.813	0.2623	1.9520	0.5123
4	2.441	0.4096	5.766	0.1735	2.3616	0.4235
5	3.052	0.3277	8.207	0.1219	2.6893	0.3719
6	3.815	0.2622	11.259	0.0888	2.9514	0.3388
7	4.768	0.2097	15.073	0.0664	3.1611	0.3164
8	5.960	0.1678	19.842	0.0504	3.3289	0.3004
9	7.451	0.1342	25.802	0.0388	3.4631	0.2888
10	9.313	0.1074	33.253	0.0301	3.5705	0.2801
11	11.642	0.0859	42.566	0.0235	3.6564	0.2735
12	14.552	0.0687	54.208	0.0185	3.7251	0.2685
13	18.190	0.0550	68.760	0.0146	3.7801	0.2646
14	22.737	0.0440	86.949	0.0115	3.8241	0.2615
15	28.422	0.0352	109.687	0.0091	3.8593	0.2591
16	35.527	0.0282	138.109	0.0073	3.8874	0.2573
17	44.409	0.0225	173.636	0.0058	3.9099	0.2558
18	55.511	0.0180	218.045	0.0046	3.9280	0.2546
19	69.389	0.0144	273.556	0.0037	3.9424	0.2537
20	86.736	0.0115	342.945	0.0029	3.9539	0.2529
21	108.420	0.0092	429.681	0.0023	3.9631	0.2523
22	135.525	0.0074	538.101	0.0019	3.9705	0.2519
23	169.407	0.0059	673.626	0.0015	3.9764	0.2515
24	211.758	0.0047	843.033	0.0012	3.9811	0.2512
25	264.698	0.0038	1054.791	0.0010	3.9849	0.2510
26	330.872	0.0030	1319.489	0.0008	3.9879	0.2508
27	413.590	0.0024	1650.361	0.0006	3.9903	0.2506
28	516.988	0.0019	2063.952	0.0005	3.9923	0.2505
29	646.235	0.0016	2580.939	0.0004	3.9938	0.2504
30	807.794	0.0012	3227.174	0.0003	3.9951	0.2503
31	1009.742	0.0010	4034.968	0.0003	3.9960	0.2503
32	1262.177	0.0008	5044.710	0.0002	3.9968	0.2502
33	1577.722	0.0006	6306.887	0.0002	3.9975	0.2502
34	1972.152	0.0005	7884.609	0.0001	3.9980	0.2501
35	2465.190	0.0004	9856.761	0.0001	3.9984	0.2501

TABLE D.14 30% INTEREST FACTORS FOR ANNUAL COMPOUNDING

	Single Payment		Equal-Payment Series			
	Compound-Amount Factor	Present-Worth Factor	Compound-Amount Factor	Sinking-Fund Factor	Present-Worth Factor	Capital-Recovery Factor
n	To Find F Given P $F/P, i, n$	To Find P Given F $P/F, i, n$	To Find F Given A $F/A, i, n$	To Find A Given F $A/F, i, n$	To Find P Given A $P/A, i, n$	To Find A Given P $A/P, i, n$
1	1.300	0.7692	1.000	1.0000	0.7692	1.3000
2	1.690	0.5917	2.300	0.4348	1.3610	0.7348
3	2.197	0.4552	3.990	0.2506	1.8161	0.5506
4	2.856	0.3501	6.187	0.1616	2.1663	0.4616
5	3.713	0.2693	9.043	0.1106	2.4356	0.4106
6	4.827	0.2072	12.756	0.0784	2.6428	0.3784
7	6.275	0.1594	17.583	0.0569	2.8021	0.3569
8	8.157	0.1226	23.858	0.0419	2.9247	0.3419
9	10.605	0.0943	32.015	0.0312	3.0190	0.3312
10	13.786	0.0725	42.620	0.0235	3.0915	0.3235
11	17.922	0.0558	56.405	0.0177	3.1473	0.3177
12	23.298	0.0429	74.327	0.0135	3.1903	0.3135
13	30.288	0.0330	97.625	0.0103	3.2233	0.3103
14	39.374	0.0254	127.913	0.0078	3.2487	0.3078
15	51.186	0.0195	167.286	0.0060	3.2682	0.3060
16	66.542	0.0150	218.472	0.0046	3.2832	0.3046
17	86.504	0.0116	285.014	0.0035	3.2948	0.3035
18	112.455	0.0089	371.518	0.0027	3.3037	0.3027
19	146.192	0.0069	483.973	0.0021	3.3105	0.3021
20	190.050	0.0053	630.165	0.0016	3.3158	0.3016
21	247.065	0.0041	820.215	0.0012	3.3199	0.3012
22	321.184	0.0031	1067.280	0.0009	3.3230	0.3009
23	417.539	0.0024	1388.464	0.0007	3.3254	0.3007
24	542.801	0.0019	1806.003	0.0006	3.3272	0.3006
25	705.641	0.0014	2348.803	0.0004	3.3286	0.3004
26	917.333	0.0011	3054.444	0.0003	3.3297	0.3003
27	1192.533	0.0008	3971.778	0.0003	3.3305	0.3003
28	1550.293	0.0007	5164.311	0.0002	3.3312	0.3002
29	2015.381	0.0005	6714.604	0.0002	3.3317	0.3002
30	2619.996	0.0004	8729.985	0.0001	3.3321	0.3001
31	3405.994	0.0003	11349.981	0.0001	3.3324	0.3001
32	4427.793	0.0002	14755.975	0.0001	3.3326	0.3001
33	5756.130	0.0002	19183.768	0.0001	3.3328	0.3001
34	7482.970	0.0001	24939.899	0.0001	3.3329	0.3001
35	9727.860	0.0001	32422.868	0.0000	3.3330	0.3000

E
Statistical Tables

TABLE E.1. CUMULATIVE POISSON PROBABILITIES

Cumulative probabilities $\times$ 1,000 for the Poisson distribution are given for μ up to 24. The tabular values were computed from $\sum (\mu^x e^{-\mu}/x!)$. [*These tables are reproduced from W. J. Fabrycky, P. M. Ghare, and P. E. Torgersen,* Applied Operations Research and Management Science. (*Englewood Cliffs, N. J.: Prentice-Hall, 1984.*)]

TABLE E.2. RANDOM RECTANGULAR VARIATES

Random variates from the rectangular distribution, $f(x) = \frac{1}{10}$, *are presented. [These tables are reproduced with permission from the RAND Corporation,* A Million Random Digits with 100,000 Normal Deviates (*New York: The Free Press, 1955*), *pp. 130–31.*]

TABLE E. 3. CUMULATIVE NORMAL PROBABILITIES

Cumulative probabilities are given from $-\infty$ to $Z = (x - \mu)/\sigma$ for the standard normal distribution. [*Tabular values are adapted with permission from E. L. Grant, and R. S. Leavenworth,* Statistical Quality Control, *4th ed. (New York: McGraw-Hill, 1972).*]

TABLE E.1 CUMULATIVE POISSON PROBABILITIES × 1,000

μ \ x	0	1	2	3	4	5	6	7	8	9	10	11	12	13	14
0.1	905	995	1,000												
0.2	819	982	999	1,000											
0.3	741	963	996	1,000											
0.4	670	938	992	999	1,000										
0.5	607	910	986	998	1,000										
0.6	549	878	977	997	1,000										
0.7	497	844	966	994	999	1,000									
0.8	449	809	953	991	999	1,000									
0.9	407	772	937	987	998	1,000									
1.0	368	736	920	981	996	999	1,000								
1.1	333	699	900	974	995	999	1,000								
1.2	301	663	879	966	992	998	1,000								
1.3	273	627	857	957	989	998	-1,000								
1.4	247	592	833	946	986	997	999	1,000							
1.5	223	558	809	934	981	996	999	1,000							
1.6	202	525	783	921	976	994	999	1,000							
1.7	183	493	757	907	970	992	998	1,000							
1.8	165	463	731	891	964	990	997	999	1,000						
1.9	150	434	704	875	956	987	997	999	1,000						
2.0	135	406	677	857	947	983	995	999	1,000						
2.2	111	355	623	819	928	975	993	998	1,000						
2.4	091	308	570	779	904	964	988	997	999	1,000					
2.6	074	267	518	736	877	951	983	995	999	1,000					

TABLE E.1 CUMULATIVE POISSON PROBABILITIES × 1,000 (Continued)

μ \ x	0	1	2	3	4	5	6	7	8	9	10	11	12	13	14
2.8	061	231	469	692	848	935	976	992	998	999	1,000				
3.0	050	199	423	647	815	916	966	988	996	999	1,000				
3.2	041	171	380	603	781	895	955	983	994	998	1,000				
3.4	033	147	340	558	744	871	942	977	992	997	999	1,000			
3.6	027	126	303	515	706	844	927	969	988	996	999	1,000			
3.8	022	107	269	473	668	816	909	960	984	994	998	999	1,000		
4.0	018	092	238	433	629	785	889	949	979	992	997	999	1,000		
4.2	015	078	210	395	590	753	867	936	972	989	996	999	1,000		
4.4	012	066	185	359	551	720	844	921	964	985	994	998	999	1,000	
4.6	010	056	163	326	513	686	818	905	955	980	992	997	999	1,000	
4.8	008	048	143	294	476	651	791	887	944	975	990	996	999	1,000	
5	007	040	125	265	440	616	762	867	932	968	986	995	998	999	1,000
6	002	017	062	151	285	446	606	744	847	916	957	980	991	996	1,000
7	001	007	030	082	173	301	450	599	729	830	901	947	973	987	994
8	000	003	014	042	100	191	313	453	593	717	816	888	936	966	983
9	000	001	006	021	055	116	207	324	456	587	706	803	876	926	959
10		000	003	010	029	067	130	220	333	458	583	697	792	864	917
11		000	001	005	015	038	079	143	232	341	460	579	689	781	854
12		000	001	002	008	020	046	090	155	242	347	462	576	682	772
13			000	001	004	011	026	054	100	166	252	353	463	573	675
14				000	002	006	014	032	062	109	176	260	358	464	570
15				000	001	003	008	018	037	070	118	185	268	363	466

TABLE E.1 CUMULATIVE POISSON PROBABILITIES × 1,000 (Continued)

x \ μ	15	16	17	18	19	20	21	22	23	24	25	26	27	28	29
7	998	999	1,000												
8	992	996	998	999	1,000										
9	978	989	995	998	999	1,000									
10	951	973	986	993	997	998	999	1,000							
11	907	944	968	982	991	995	998	999	1,000						
12	844	899	937	963	979	988	994	997	999	999	1,000				
13	764	835	890	930	957	975	986	992	996	998	999	1,000			
14	669	756	827	883	923	952	971	983	991	995	997	999	999	1,000	
15	568	664	749	819	875	917	947	967	981	989	994	997	998	999	1,000

x \ μ	0	1	2	3	4	5	6	7	8	9	10	11	12	13	14
16					000	001	004	010	022	043	077	127	193	275	368
17					000	001	002	005	013	026	049	085	135	201	281
18						000	001	003	007	015	030	055	092	143	208
19						000	001	002	004	009	018	035	061	098	150
20							000	001	002	005	011	021	039	066	105
21								000	001	003	006	013	025	043	072
22								000	001	002	004	008	015	028	048
23									000	001	002	004	009	017	031
24										000	001	003	005	011	020

TABLE E.1 CUMULATIVE POISSON PROBABILITIES × 1,000 (Continued)

x → μ ↓	15	16	17	18	19	20	21	22	23	24	25	26	27	28	29
16	467	566	659	742	812	868	911	942	963	978	987	993	996	998	999
17	371	468	564	655	736	805	861	905	937	959	975	985	991	995	997
18	287	375	469	562	651	731	799	855	899	932	955	972	983	990	994
19	215	292	378	469	561	647	725	793	849	893	927	951	969	980	988
20	157	221	297	381	470	559	644	721	787	843	888	922	948	966	978
21	111	163	227	302	384	471	558	640	716	782	838	883	917	944	963
22	077	117	169	232	306	387	472	556	637	712	777	832	877	913	940
23	052	082	123	175	238	310	389	472	555	635	708	772	827	873	908
24	034	056	087	128	180	243	314	392	473	554	632	704	768	823	868

x → μ ↓	30	31	32	33	34	35	36	37	38	39	40	41	42	43	44
16	999	1,000													
17	999	999	1,000												
18	997	998	999	1,000											
19	993	996	998	999	999	1,000									
20	987	992	995	997	998	999	1,000								
21	976	985	991	994	997	998	999	999							
22	959	973	983	989	994	996	998	999	1,000						
23	936	956	971	981	988	993	996	997	999	1,000					
24	904	932	953	969	979	987	992	995	997	998	999	999	1,000		

TABLE E.2   RANDOM RECTANGULAR VARIATES

14541	36678	54343	94932	25238	84928	30668	34992	69955	06633
88626	98899	01337	48085	83315	33563	78656	99440	55584	54178
31466	87268	62975	19310	28192	06654	06720	64938	67111	55091
52738	52893	51373	43430	95885	93795	20129	54847	68674	21040
17444	35560	35348	75467	26026	89118	51810	06389	02391	96061
62596	56854	76099	38469	26285	86175	65468	32354	02675	24070
38338	83917	50232	29164	07461	25385	84838	07405	38303	55635
29163	61006	98106	47538	99122	36242	90365	15581	89597	03327
59049	95306	31227	75288	10122	92687	99971	97105	37597	91673
67447	52922	58657	67601	96148	97263	39110	95111	04682	64873
57082	55108	26992	19196	08044	57300	75095	84330	92314	11370
00179	04358	95645	91751	56618	73782	38575	17401	38686	98435
65420	87257	44374	54312	94692	81776	24422	99198	51432	63943
52450	75445	40002	69727	29775	32572	79980	67902	97260	21050
82767	26273	02192	88536	08191	91750	46993	02245	38659	28026
17066	64286	35972	32550	82167	53177	32396	34014	20993	03031
86168	32643	23668	92038	03096	51029	09693	45454	89854	70103
33632	69631	70537	06464	83543	48297	67693	63137	62675	56572
77915	56481	43065	24231	43011	40505	90386	13870	84603	73101
90000	92887	92668	93521	44072	01785	27003	01851	40232	25842
55809	70237	10368	58664	39521	11137	20461	53081	07150	11832
50948	64026	03350	03153	75913	72651	28651	94299	67706	92507
27138	59012	27872	90522	69791	85482	80337	12252	83388	48909
03534	58643	75913	63557	25527	47131	72295	55801	44847	48019
48895	34733	58057	00195	79496	93453	07813	66038	55245	43168
57585	23710	77321	70662	82884	80132	42281	17032	96737	93284
95913	24669	42050	92757	68677	75567	99777	49246	93049	79863
12981	37145	95773	92475	43700	85253	33214	87656	13295	09721
62349	64163	57369	65773	86217	00135	33762	72398	16343	02263
68193	37564	56257	50030	53951	84887	34590	22038	40629	29562
56203	82226	83294	60361	29924	09353	87021	08149	11167	81744
31945	23224	08211	02562	20299	85836	94714	50278	99818	62489
68726	52274	59535	80873	35423	05166	06911	25916	90728	20431
79557	25747	55585	93461	44360	18359	20493	54287	43693	88568
05764	29803	01819	51972	91641	03524	18381	65427	11394	37447
30187	66931	01972	48438	90716	21847	35114	91839	26913	68893
30858	43646	96984	80412	91973	81339	05548	49812	40775	14263
85117	38268	18921	29519	33359	80642	95362	22133	40322	37826
59422	12752	56798	31954	19859	32451	04433	62116	14899	38825
73479	91833	91122	45524	73871	77931	67822	95602	23325	37718
83648	66882	15327	89748	76685	76282	98624	71547	49089	33105
19454	91265	09051	94410	06418	34484	37929	61070	62346	79970
49327	97807	61390	08005	71795	49290	52285	82119	59348	55986
54482	51025	12382	35719	66721	84890	38106	44136	95164	92935
30487	19459	25693	09427	10967	36164	33893	07087	16141	12734
42998	68627	66295	59360	44041	76909	56321	12978	31304	97444
03668	61096	26292	79688	05625	52198	74844	69815	76591	35398
45074	91457	28311	56499	60403	13658	81838	54729	12365	24082
58444	99255	14960	02275	37925	03852	81235	91628	72136	53070
82912	91185	89612	02362	93360	20158	24796	38284	55328	96041

TABLE E.2   RANDOM RECTANGULAR VARIATES (Continued)

44553	29642	20317	69470	57789	27631	68040	73201	51302	66497
01914	36106	71351	69176	53353	57353	42430	68050	47862	61922
00768	37958	69915	17709	31629	49587	07136	42959	56207	03625
29742	67676	62608	54215	97167	07008	77130	15806	53081	14297
07721	20143	56131	56112	23451	48773	38121	74419	11696	42614
99158	07133	04325	43936	83619	77182	55459	28808	38034	01054
97168	13859	78155	55361	04871	78433	58538	78437	14058	79510
07508	63835	83056	74942	70117	91928	10383	93793	31015	60839
68400	66460	67212	28690	66913	90798	71714	07698	31581	31086
88512	62908	65455	64015	00821	23970	58118	93174	02201	16771
94549	31145	62897	91582	94064	14687	47570	83714	45928	32685
02307	86181	44897	60884	68072	77693	83413	61680	55872	12111
28922	89390	66771	39185	04266	55216	91537	36500	48154	04517
73898	85742	97914	74170	10383	16366	37404	73282	20524	85004
66220	81596	18533	84825	43509	16009	00830	13177	54961	31140
64452	91627	21897	31830	62051	00760	43702	22305	79009	15065
26748	19441	87908	06086	62879	99865	50739	98540	54002	98337
61328	52330	17850	53204	29955	48425	84694	11280	70661	27303
89134	85791	73207	93578	62563	37205	97667	61453	01067	31982
91365	23327	81658	56441	01480	09677	86053	11505	30898	82143
54576	02572	60501	98257	40475	81401	31624	27951	60172	21382
39870	60476	02934	39857	06430	59325	84345	62302	98616	13452
82288	29758	35692	21268	35101	77554	35201	22795	84532	29927
57404	93848	87288	30246	34990	50575	49485	60474	17377	46550
22043	17104	49653	79082	45099	24889	04829	49097	58065	23492
61981	00340	43594	22386	41782	94104	08867	68590	61716	36120
96056	16227	74598	28155	23304	66923	07918	15303	44988	79076
64013	74715	31525	62676	75435	93055	37086	52737	89455	83016
59515	37354	55422	79471	23150	79170	74043	49340	61320	50390
38534	33169	40448	21683	82153	23411	53057	26069	86906	49708
41422	50502	40570	59748	59499	70322	62416	71408	06429	70123
38633	80107	10241	30880	13914	09228	68929	06438	17749	81149
48214	75994	31689	25257	28641	14854	72571	78189	35508	26381
54799	37862	06714	55885	07481	16966	04797	57846	69080	49631
25848	27142	63477	33416	60961	19781	65457	23981	90348	24499
27576	47298	47163	69614	29372	24859	62090	81667	50635	08295
52970	93916	81350	81057	16962	56039	27739	59574	79617	45698
69516	87573	13313	69388	32020	66294	99126	50474	04258	03084
94504	41733	55936	77595	55959	90727	61367	83645	80997	62103
67935	14568	27992	09784	81917	79303	08616	83509	64932	34764
63345	095.	40232	51061	09455	36491	04810	06040	78959	41435
87119	21605	86917	97715	91250	79587	80967	39872	52512	78444
02612	97319	10487	68923	58607	38261	67119	36351	48521	69965
69860	16526	41420	01514	46902	03399	12286	52467	80387	10561
27669	67730	53932	38578	25746	00025	98917	18790	51091	24920
59705	91472	01302	33123	35274	88433	55491	27609	02824	05245
36508	74042	44014	36243	12724	06092	23742	90436	33419	12301
13612	24554	73326	61445	77198	43360	62006	31038	54756	88137
82893	11961	19656	71181	63201	44946	14169	72755	47883	24119
97914	61228	42903	71187	54964	14945	20809	33937	13257	66387

TABLE E.3 CUMULATIVE NORMAL PROBABILITIES

Z	0.09	0.08	0.07	0.06	0.05	0.04	0.03	0.02	0.01	0.00
−3.5	0.00017	0.00017	0.00018	0.00019	0.00019	0.00020	0.00021	0.00022	0.00022	0.00023
−3.4	0.00024	0.00025	0.00026	0.00027	0.00028	0.00029	0.00030	0.00031	0.00033	0.00034
−3.3	0.00035	0.00036	0.00038	0.00039	0.00040	0.00042	0.00043	0.00045	0.00047	0.00048
−3.2	0.00050	0.00052	0.00054	0.00056	0.00058	0.00060	0.00062	0.00064	0.00066	0.00069
−3.1	0.00071	0.00074	0.00076	0.00079	0.00082	0.00085	0.00087	0.00090	0.00094	0.00097
−3.0	0.00100	0.00104	0.00107	0.00111	0.00114	0.00118	0.00122	0.00126	0.00131	0.00135
−2.9	0.0014	0.0014	0.0015	0.0015	0.0016	0.0016	0.0017	0.0017	0.0018	0.0019
−2.8	0.0019	0.0020	0.0021	0.0021	0.0022	0.0023	0.0023	0.0024	0.0025	0.0026
−2.7	0.0026	0.0027	0.0028	0.0029	0.0030	0.0031	0.0032	0.0033	0.0034	0.0035
−2.6	0.0036	0.0037	0.0038	0.0039	0.0040	0.0041	0.0043	0.0044	0.0045	0.0047
−2.5	0.0048	0.0049	0.0051	0.0052	0.0054	0.0055	0.0057	0.0059	0.0060	0.0062
−2.4	0.0064	0.0066	0.0068	0.0069	0.0071	0.0073	0.0075	0.0078	0.0080	0.0082
−2.3	0.0084	0.0087	0.0089	0.0091	0.0094	0.0096	0.0099	0.0102	0.0104	0.0107
−2.2	0.0110	0.0113	0.0116	0.0119	0.0122	0.0125	0.0129	0.0132	0.0136	0.0139
−2.1	0.0143	0.0146	0.0150	0.0154	0.0158	0.0162	0.0166	0.0170	0.0174	0.0179
−2.0	0.0183	0.0188	0.0192	0.0197	0.0202	0.0207	0.0212	0.0217	0.0222	0.0228
−1.9	0.0233	0.0239	0.0244	0.0250	0.0256	0.0262	0.0268	0.0274	0.0281	0.0287
−1.8	0.0294	0.0301	0.0307	0.0314	0.0322	0.0329	0.0336	0.0344	0.0351	0.0359
−1.7	0.0367	0.0375	0.0384	0.0392	0.0401	0.0409	0.0418	0.0427	0.0436	0.0446
−1.6	0.0455	0.0465	0.0475	0.0485	0.0495	0.0505	0.0516	0.0526	0.0537	0.0548
−1.5	0.0559	0.0571	0.0582	0.0594	0.0606	0.0618	0.0630	0.0643	0.0655	0.0668
−1.4	0.0681	0.0694	0.0708	0.0721	0.0735	0.0749	0.0764	0.07̄8	0.0793	0.0808
−1.3	0.0823	0.0838	0.0853	0.0869	0.0885	0.0901	0.0918	0.0934	0.0951	0.0968
−1.2	0.0985	0.1003	0.1020	0.1038	0.1057	0.1075	0.1093	0.1112	0.1131	0.1151
−1.1	0.1170	0.1190	0.1210	0.1230	0.1251	0.1271	0.1292	0.1314	0.1335	0.1357
−1.0	0.1379	0.1401	0.1423	0.1446	0.1469	0.1492	0.1515	0.1539	0.1562	0.1587
−0.9	0.1611	0.1635	0.1660	0.1685	0.1711	0.1736	0.1762	0.1788	0.1814	0.1841
−0.8	0.1867	0.1894	0.1922	0.1949	0.1977	0.2005	0.2033	0.2061	0.2090	0.2119
−0.7	0.2148	0.2177	0.2207	0.2236	0.2266	0.2297	0.2327	0.2358	0.2389	0.2420
−0.6	0.2451	0.2483	0.2514	0.2546	0.2578	0.2611	0.2643	0.2676	0.2709	0.2743
−0.5	0.2776	0.2810	0.2843	0.2877	0.2912	0.2946	0.2981	0.3015	0.3050	0.3085
−0.4	0.3121	0.3156	0.3192	0.3228	0.3264	0.3300	0.3336	0.3372	0.3409	0.3446
−0.3	0.3483	0.3520	0.3557	0.3594	0.3632	0.3669	0.3707	0.3745	0.3783	0.3821
−0.2	0.3859	0.3897	0.3936	0.3974	0.4013	0.4052	0.4090	0.4129	0.4168	0.4207
−0.1	0.4247	0.4286	0.4325	0.4364	0.4404	0.4443	0.4483	0.4522	0.4562	0.4602
−0.0	0.4641	0.4681	0.4721	0.4761	0.4801	0.4840	0.4880	0.4920	0.4960	0.5000

TABLE E.3 CUMULATIVE NORMAL PROBABILITIES (Continued)

Z	0.00	0.01	0.02	0.03	0.04	0.05	0.06	0.07	0.08	0.09
+0.0	0.5000	0.5040	0.5080	0.5120	0.5160	0.5199	0.5239	0.5279	0.5319	0.5359
+0.1	0.5398	0.5438	0.5478	0.5517	0.5557	0.5596	0.5636	0.5675	0.5714	0.5753
+0.2	0.5793	0.5832	0.5871	0.5910	0.5948	0.5987	0.6026	0.6064	0.6103	0.6141
+0.3	0.6179	0.6217	0.6255	0.6293	0.6331	0.6368	0.6406	0.6443	0.6480	0.6517
+0.4	0.6554	0.6591	0.6628	0.6664	0.6700	0.6736	0.6772	0.6808	0.6844	0.6879
+0.5	0.6915	0.6950	0.6985	0.7019	0.7054	0.7088	0.7123	0.7157	0.7190	0.7224
+0.6	0.7257	0.7291	0.7324	0.7357	0.7389	0.7422	0.7454	0.7486	0.7517	0.7549
+0.7	0.7580	0.7611	0.7642	0.7673	0.7704	0.7734	0.7764	0.7794	0.7823	0.7852
+0.8	0.7881	0.7910	0.7939	0.7967	0.7995	0.8023	0.8051	0.8079	0.8106	0.8133
+0.9	0.8159	0.8186	0.8212	0.8238	0.8264	0.8289	0.8315	0.8340	0.8365	0.8389
+1.0	0.8413	0.8438	0.8461	0.8485	0.8508	0.8531	0.8554	0.8577	0.8599	0.8621
+1.1	0.8643	0.8665	0.8686	0.8708	0.8729	0.8749	0.8770	0.8790	0.8810	0.8830
+1.2	0.8849	0.8869	0.8888	0.8907	0.8925	0.8944	0.8962	0.8980	0.8997	0.9015
+1.3	0.9032	0.9049	0.9066	0.9082	0.9099	0.9115	0.9131	0.9147	0.9162	0.9177
+1.4	0.9192	0.9207	0.9222	0.9236	0.9251	0.9265	0.9279	0.9292	0.9306	0.9319
+1.5	0.9332	0.9345	0.9357	0.9370	0.9382	0.9394	0.9406	0.9418	0.9429	0.9441
+1.6	0.9452	0.9463	0.9474	0.9484	0.9495	0.9505	0.9515	0.9525	0.9535	0.9545
+1.7	0.9554	0.9564	0.9573	0.9582	0.9591	0.9599	0.9608	0.9616	0.9625	0.9633
+1.8	0.9641	0.9649	0.9656	0.9664	0.9671	0.9678	0.9686	0.9693	0.9699	0.9706
+1.9	0.9713	0.9719	0.9726	0.9732	0.9738	0.9744	0.9750	0.9756	0.9761	0.9767
+2.0	0.9773	0.9778	0.9783	0.9788	0.9793	0.9798	0.9803	0.9808	0.9812	0.9817
+2.1	0.9821	0.9826	0.9830	0.9834	0.9838	0.9842	0.9846	0.9850	0.9854	0.9857
+2.2	0.9861	0.9864	0.9868	0.9871	0.9875	0.9878	0.9881	0.9884	0.9887	0.9890
+2.3	0.9893	0.9896	0.9898	0.9901	0.9904	0.9906	0.9909	0.9911	0.9913	0.9916
+2.4	0.9918	0.9920	0.9922	0.9925	0.9927	0.9929	0.9931	0.9932	0.9934	0.9936
+2.5	0.9938	0.9940	0.9941	0.9943	0.9945	0.9946	0.9948	0.9949	0.9951	0.9952
+2.6	0.9953	0.9955	0.9956	0.9957	0.9959	0.9960	0.9961	0.9962	0.9963	0.9964
+2.7	0.9965	0.9966	0.9967	0.9968	0.9969	0.9970	0.9971	0.9972	0.9973	0.9974
+2.8	0.9974	0.9975	0.9976	0.9977	0.9977	0.9978	0.9979	0.9979	0.9980	0.9981
+2.9	0.9981	0.9982	0.9983	0.9983	0.9984	0.9984	0.9985	0.9985	0.9986	0.9986
+3.0	0.99865	0.99869	0.99874	0.99878	0.99882	0.99886	0.99889	0.99893	0.99896	0.99900
+3.1	0.99903	0.99906	0.99910	0.99913	0.99915	0.99918	0.99921	0.99924	0.99926	0.99929
+3.2	0.99931	0.99934	0.99936	0.99938	0.99940	0.99942	0.99944	0.99946	0.99948	0.99950
+3.3	0.99952	0.99953	0.99955	0.99957	0.99958	0.99960	0.99961	0.99962	0.99964	0.99965
+3.4	0.99966	0.99967	0.99969	0.99970	0.99971	0.99972	0.99973	0.99974	0.99975	0.99976
+3.5	0.99977	0.99978	0.99978	0.99979	0.99980	0.99981	0.99981	0.99982	0.99983	0.99983

F
Finite Queuing Tables

TABLE F.1–F.3. FINITE QUEUING FACTORS

The probability of a delay, D, and the efficiency factor, F, are given for populations of 10, 20, and 30 units. Each set of values is keyed to the service factor, X, and the number of channels, M. [*These tabular values are adapted with permission from L. G. Peck and R. N. Hazelwood,* Finite Queuing Tables *(New York: John Wiley, 1958).*]

TABLE F.1 FINITE QUEUING FACTORS—POPULATION 10

X	M	D	F	X	M	D	F	X	M	D	F
0.008	1	0.072	0.999		2	0.177	0.990		3	0.182	0.986
0.013	1	0.117	0.998		1	0.660	0.899		2	0.528	0.921
0.016	1	0.144	0.997	0.085	3	0.037	0.999		1	0.954	0.610
0.019	1	0.170	0.996		2	0.196	0.988	0.165	4	0.049	0.997
0.021	1	0.188	0.995		1	0.692	0.883		3	0.195	0.984
0.023	1	0.206	0.994	0.090	3	0.043	0.998		2	0.550	0.914
0.025	1	0.224	0.993		2	0.216	0.986		1	0.961	0.594
0.026	1	0.232	0.992		1	0.722	0.867	0.170	4	0.054	0.997
0.028	1	0.250	0.991	0.095	3	0.049	0.998		3	0.209	0.982
0.030	1	0.268	0.990		2	0.237	0.984		2	0.571	0.906
0.032	2	0.033	0.999		1	0.750	0.850		1	0.966	0.579
	1	0.285	0.988	0.100	3	0.056	0.998	0.180	5	0.013	0.999
0.034	2	0.037	0.999		2	0.258	0.981		4	0.066	0.996
	1	0.302	0.986		1	0.776	0.832		3	0.238	0.978
0.036	2	0.041	0.999	0.105	3	0.064	0.997		2	0.614	0.890
	1	0.320	0.984		2	0.279	0.978		1	0.975	0.549
0.038	2	0.046	0.999		1	0.800	0.814	0.190	5	0.016	0.999
	1	0.337	0.982	0.110	3	0.072	0.997		4	0.078	0.995
0.040	2	0.050	0.999		2	0.301	0.974		3	0.269	0.973
	1	0.354	0.980		1	0.822	0.795		2	0.654	0.873
0.042	2	0.055	0.999	0.115	3	0.081	0.996		1	0.982	0.522
	1	0.371	0.978		2	0.324	0.971	0.200	5	0.020	0.999
0.044	2	0.060	0.998		1	0.843	0.776		4	0.092	0.994
	1	0.388	0.975	0.120	4	0.016	0.999		3	0.300	0.968
0.046	2	0.065	0.998		3	0.090	0.995		2	0.692	0.854
	1	0.404	0.973		2	0.346	0.967		1	0.987	0.497
0.048	2	0.071	0.998		1	0.861	0.756	0.210	5	0.025	0.999
	1	0.421	0.970	0.125	4	0.019	0.999		4	0.108	0.992
0.050	2	0.076	0.998		3	0.100	0.994		3	0.333	0.961
	1	0.437	0.967		2	0.369	0.962		2	0.728	0.835
0.052	2	0.082	0.997		1	0.878	0.737		1	0.990	0.474
	1	0.454	0.963	0.130	4	0.022	0.999	0.220	5	0.030	0.998
0.054	2	0.088	0.997		3	0.110	0.994		4	0.124	0.990
	1	0.470	0.960		2	0.392	0.958		3	0.366	0.954
0.056	2	0.094	0.997		1	0.893	0.718		2	0.761	0.815
	1	0.486	0.956	0.135	4	0.025	0.999		1	0.993	0.453
0.058	2	0.100	0.996		3	0.121	0.993	0.230	5	0.037	0.998
	1	0.501	0.953		2	0.415	0.952		4	0.142	0.988
0.060	2	0.106	0.996		1	0.907	0.699		3	0.400	0.947
	1	0.517	0.949	0.140	4	0.028	0.999		2	0.791	0.794
0.062	2	0.113	0.996		3	0.132	0.991		1	0.995	0.434
	1	0.532	0.945		2	0.437	0.947	0.240	5	0.044	0.997
0.064	2	0.119	0.995		1	0.919	0.680		4	0.162	0.986
	1	0.547	0.940	0.145	4	0.032	0.999		3	0.434	0.938
0.066	2	0.126	0.995		3	0.144	0.990		2	0.819	0.774
	1	0.562	0.936		2	0.460	0.941		1	0.996	0.416
0.068	3	0.020	0.999		1	0.929	0.662	0.250	6	0.010	0.999
	2	0.133	0.994	0.150	4	0.036	0.998		5	0.052	0.997
	1	0.577	0.931		3	0.156	0.989		4	0.183	0.983
0.070	3	0.022	0.999		2	0.483	0.935		3	0.469	0.929
	2	0.140	0.994		1	0.939	0.644		2	0.844	0.753
	1	0.591	0.926	0.155	4	0.040	0.998		1	0.997	0.400
0.075	3	0.026	0.999		3	0.169	0.987	0.260	6	0.013	0.999
	2	0.158	0.992		2	0.505	0.928		5	0.060	0.996
	1	0.627	0.913		1	0.947	0.627		4	0.205	0.980
0.080	3	0.031	0.999	0.160	4	0.044	0.998		3	0.503	0.919

TABLE F.1   FINITE QUEUING FACTORS—POPULATION 10 (Continued)

X	M	D	F	X	M	D	F	X	M	D	F
	2	0.866	0.732		4	0.533	0.906		7	0.171	0.982
	1	0.998	0.384		3	0.840	0.758		6	0.413	0.939
0.270	6	0.015	0.999		2	0.986	0.525		5	0.707	0.848
	5	0.070	0.995	0.400	7	0.026	0.998		4	0.917	0.706
	4	0.228	0.976		6	0.105	0.991		3	0.991	0.535
	3	0.537	0.908		5	0.292	0.963	0.580	8	0.057	0.995
	2	0.886	0.712		4	0.591	0.887		7	0.204	0.977
	1	0.999	0.370		3	0.875	0.728		6	0.465	0.927
0.280	6	0.018	0.999		2	0.991	0.499		5	0.753	0.829
	5	0.081	0.994	0.420	7	0.034	0.998		4	0.937	0.684
	4	0.252	0.972		6	0.130	0.987		3	0.994	0.517
	3	0.571	0.896		5	0.341	0.954	0.600	9	0.010	0.999
	2	0.903	0.692		4	0.646	0.866		8	0.072	0.994
	1	0.999	0.357		3	0.905	0.700		7	0.242	0.972
0.290	6	0.022	0.999		2	0.994	0.476		6	0.518	0.915
	5	0.093	0.993	0.440	7	0.045	0.997		5	0.795	0.809
	4	0.278	0.968		6	0.160	0.984		4	0.953	0.663
	3	0.603	0.884		5	0.392	0.943		3	0.996	0.500
	2	0.918	0.672		4	0.698	0.845	0.650	9	0.021	0.999
	1	0.999	0.345		3	0.928	0.672		8	0.123	0.988
0.300	6	0.026	0.998		2	0.996	0.454		7	0.353	0.954
	5	0.106	0.991	0.460	8	0.011	0.999		6	0.651	0.878
	4	0.304	0.963		7	0.058	0.995		5	0.882	0.759
	3	0.635	0.872		6	0.193	0.979		4	0.980	0.614
	2	0.932	0.653		5	0.445	0.930		3	0.999	0.461
	1	0.999	0.333		4	0.747	0.822	0.700	9	0.040	0.997
0.310	6	0.031	0.998		3	0.947	0.646		8	0.200	0.979
	5	0.120	0.990		2	0.998	0.435		7	0.484	0.929
	4	0.331	0.957	0.480	8	0.015	0.999		6	0.772	0.836
	3	0.666	0.858		7	0.074	0.994		5	0.940	0.711
	2	0.943	0.635		6	0.230	0.973		4	0.992	0.571
0.320	6	0.036	0.998		5	0.499	0.916	0.750	9	0.075	0.994
	5	0.135	0.988		4	0.791	0.799		8	0.307	0.965
	4	0.359	0.952		3	0.961	0.621		7	0.626	0.897
	3	0.695	0.845		2	0.998	0.417		6	0.870	0.792
	2	0.952	0.617	0.500	8	0.020	0.999		5	0.975	0.666
0.330	6	0.042	0.997		7	0.093	0.992		4	0.998	0.533
	5	0.151	0.986		6	0.271	0.966	0.800	9	0.134	0.988
	4	0.387	0.945		5	0.553	0.901		8	0.446	0.944
	3	0.723	0.831		4	0.830	0.775		7	0.763	0.859
	2	0.961	0.600		3	0.972	0.598		6	0.939	0.747
0.340	7	0.010	0.999		2	0.999	0.400		5	0.991	0.625
	6	0.049	0.997	0.520	8	0.026	0.998		4	0.999	0.500
	5	0.168	0.983		7	0.115	0.989	0.850	9	0.232	0.979
	4	0.416	0.938		6	0.316	0.958		8	0.611	0.916
	3	0.750	0.816		5	0.606	0.884		7	0.879	0.818
	2	0.968	0.584		4	0.864	0.752		6	0.978	0.705
0.360	7	0.014	0.999		3	0.980	0.575		5	0.998	0.588
	6	0.064	0.995		2	0.999	0.385	0.900	9	0.387	0.963
	5	0.205	0.978	0.540	8	0.034	0.997		8	0.785	0.881
	4	0.474	0.923		7	0.141	0.986		7	0.957	0.777
	3	0.798	0.787		6	0.363	0.949		6	0.995	0.667
	2	0.978	0.553		5	0.658	0.867	0.950	9	0.630	0.938
0.380	7	0.019	0.999		4	0.893	0.729		8	0.934	0.841
	6	0.083	0.993		3	0.986	0.555		7	0.994	0.737
	5	0.247	0.971	0.560	8	0.044	0.996				

TABLE F.2   FINITE QUEUING FACTORS—POPULATION 20

X	M	D	F	X	M	D	F	X	M	D	F
0.005	1	0.095	0.999		1	0.837	0.866		3	0.326	0.980
0.009	1	0.171	0.998	0.052	3	0.080	0.998		2	0.733	0.896
0.011	1	0.208	0.997		2	0.312	0.986		1	0.998	0.526
0.013	1	0 246	0.996		1	0.858	0.851	0.100	5	0.038	0.999
0.014	1	0.265	0.995	0.054	3	0.088	0.998		4	0.131	0.995
0.015	1	0.283	0.994		2	0.332	0.984		3	0.363	0.975
0.016	1	0.302	0.993		1	0.876	0.835		2	0.773	0.878
0.017	1	0.321	0.992	0.056	3	0.097	0.997		1	0.999	0.500
0.018	2	0.048	0.999		2	0.352	0.982	0.110	5	0.055	0.998
	1	0.339	0.991		1	0.893	0.819		4	0.172	0.992
0.019	2	0.053	0.999	0.058	3	0.105	0.997		3	0.438	0.964
	1	0.358	0.990		2	0.372	0.980		2	0.842	0.837
0.020	2	0.058	0.999		1	0.908	0.802	0.120	6	0.022	0.999
	1	0.376	0.989	0.060	4	0.026	0.999		5	0.076	0.997
0.021	2	0.064	0.999		3	0.115	0.997		4	0.219	0.988
	1	0.394	0.987		2	0.392	0.978		3	0.514	0.950
0.022	2	0.070	0.999		1	0.922	0.785		2	0.895	0.793
	1	0.412	0.986	0.062	4	0.029	0.999	0.130	6	0.031	0.999
0.023	2	0.075	0.999		3	0.124	0.996		5	0.101	0.996
	1	0.431	0.984		2	0.413	0.975		4	0.271	0.983
0.024	2	0.082	0.999		1	0.934	0.768		3	0.589	0.933
	1	0.449	0.982	0.064	4	0.032	0.999		2	0.934	0.748
0.025	2	0.088	0.999		3	0.134	0.996	0.140	6	0.043	0.998
	1	0.466	0.980		2	0.433	0.972		5	0.131	0.994
0.026	2	0.094	0.998		1	0.944	0.751		4	0.328	0.976
	1	0.484	0.978	0.066	4	0.036	0.999		3	0.661	0.912
0.028	2	0.108	0.998		3	0.144	0.995		2	0.960	0.703
	1	0.519	0.973		2	0.454	0.969	0.150	7	0.017	0.999
0.030	2	0.122	0.998		1	0.953	0.733		6	0.059	0.998
	1	0.553	0.968	0.068	4	0.039	0.999		5	0.166	0.991
0.032	2	0.137	0.997		3	0.155	0.995		4	0.388	0.968
	1	0.587	0.962		2	0.474	0.966		3	0.728	0.887
0.034	2	0.152	0.996		1	0.961	0.716		2	0.976	0.661
	1	0.620	0.955	0.070	4	0.043	0.999	0.160	7	0.024	0.999
0.036	2	0.168	0.996		3	0.165	0.994		6	0.077	0.997
	1	0.651	0.947		2	0.495	0.962		5	0.205	0.988
0.038	3	0.036	0.999		1	0.967	0.699		4	0.450	0.957
	2	0.185	0.995	0.075	4	0.054	0.999		3	0.787	0.860
	1	0.682	0.938		3	0.194	0.992		2	0.987	0.622
0.040	3	0.041	0.999		2	0.545	0.953	0.180	7	0.044	0.998
	2	0.202	0.994		1	0.980	0.659		6	0.125	0.994
	1	0.712	0.929	0.080	4	0.066	0.998		5	0.295	0.978
0.042	3	0.047	0.999		3	0.225	0.990		4	0.575	0.930
	2	0.219	0.993		2	0.595	0.941		3	0.879	0.799
	1	0.740	0.918		1	0.988	0.621		2	0.996	0.555
0.044	3	0.053	0.999	0.085	4	0.080	0.997	0.200	8	0.025	0.999
	2	0.237	0.992		3	0.257	0.987		7	0.074	0.997
	1	0.767	0.906		2	0.643	0.928		6	0.187	0.988
0.046	3	0.059	0.999		1	0.993	0.586		5	0.397	0.963
	2	0.255	0.991	0.090	5	0.025	0.999		4	0.693	0.895
	1	0.792	0.894		4	0.095	0.997		3	0.938	0.736
0.048	3	0.066	0.999		3	0.291	0.984		2	0.999	0.500
	2	0.274	0.989		2	0.689	0.913	0.220	8	0.043	0.998
	1	0.815	0.881		1	0.996	0.554		7	0.115	0.994
0.050	3	0.073	0.998	0.095	5	0.031	0.999		6	0.263	0.980
	2	0.293	0.988		4	0.112	0.996		5	0.505	0.943

TABLE F.2   FINITE QUEUING FACTORS—POPULATION 20 (Continued)

X	M	D	F	X	M	D	F	X	M	D	F
	4	0.793	0.852		4	0.998	0.555	0.500	14	0.033	0.998
	3	0.971	0.677	0.380	12	0.024	0.999		13	0.088	0.995
0.240	9	0.024	0.999		11	0.067	0.996		12	0.194	0.985
	8	0.068	0.997		10	0.154	0.989		11	0.358	0.965
	7	0.168	0.989		9	0.305	0.973		10	0.563	0.929
	6	0.351	0.969		8	0.513	0.938		9	0.764	0.870
	5	0.613	0.917		7	0.739	0.874		8	0.908	0.791
	4	0.870	0.804		6	0.909	0.777		7	0.977	0.698
	3	0.988	0.623		5	0.984	0.656		6	0.997	0.600
0.260	9	0.039	0.998		4	0.999	0.526	0.540	15	0.023	0.999
	8	0.104	0.994	0.400	13	0.012	0.999		14	0.069	0.996
	7	0.233	0.983		12	0.037	0.998		13	0.161	0.988
	6	0.446	0.953		11	0.095	0.994		12	0.311	0.972
	5	0.712	0.884		10	0.205	0.984		11	0.509	0.941
	4	0.924	0.755		9	0.379	0.962		10	0.713	0.891
	3	0.995	0.576		8	0.598	0.918		9	0.873	0.821
0.280	10	0.021	0.999		7	0.807	0.845		8	0.961	0.738
	9	0.061	0.997		6	0.942	0.744		7	0.993	0.648
	8	0.149	0.990		5	0.992	0.624		6	0.999	0.556
	7	0.309	0.973	0.420	13	0.019	0.999	0.600	16	0.023	0.999
	6	0.544	0.932		12	0.055	0.997		15	0.072	0.996
	5	0.797	0.848		11	0.131	0.991		14	0.171	0.988
	4	0.958	0.708		10	0.265	0.977		13	0.331	0.970
	3	0.998	0.536		9	0.458	0.949		12	0.532	0.938
0.300	10	0.034	0.998		8	0.678	0.896		11	0.732	0.889
	9	0.091	0.995		7	0.863	0.815		10	0.882	0.824
	8	0.205	0.985		6	0.965	0.711		9	0.962	0.748
	7	0.394	0.961		5	0.996	0.595		8	0.992	0.666
	6	0.639	0.907	0.440	13	0.029	0.999		7	0.999	0.583
	5	0.865	0.808		12	0.078	0.995	0.700	17	0.047	0.998
	4	0.978	0.664		11	0.175	0.987		16	0.137	0.991
	3	0.999	0.500		10	0.333	0.969		15	0.295	0.976
0.320	11	0.018	0.999		9	0.540	0.933		14	0.503	0.948
	10	0.053	0.997		8	0.751	0.872		13	0.710	0.905
	9	0.130	0.992		7	0.907	0.785		12	0.866	0.849
	8	0.272	0.977		6	0.980	0.680		11	0.953	0.783
	7	0.483	0.944		5	0.998	0.568		10	0.988	0.714
	6	0.727	0.878	0.460	14	0.014	0.999		9	0.998	0.643
	5	0.915	0.768		13	0.043	0.998	0.800	19	0.014	0.999
	4	0.989	0.624		12	0.109	0.993		18	0.084	0.996
0.340	11	0.029	0.999		11	0.228	0.982		17	0.242	0.984
	10	0.079	0.996		10	0.407	0.958		16	0.470	0.959
	9	0.179	0.987		9	0.620	0.914		15	0.700	0.920
	8	0.347	0.967		8	0.815	0.846		14	0.867	0.869
	7	0.573	0.924		7	0.939	0.755		13	0.955	0.811
	6	0.802	0.846		6	0.989	0.651		12	0.989	0.750
	5	0.949	0.729		5	0.999	0.543		11	0.998	0.687
	4	0.995	0.588	0.480	14	0.022	0.999	0.900	19	0.135	0.994
0.360	12	0.015	0.999		13	0.063	0.996		18	0.425	0.972
	11	0.045	0.998		12	0.147	0.990		17	0.717	0.935
	10	0.112	0.993		11	0.289	0.974		16	0.898	0.886
	9	0.237	0.981		10	0.484	0.944		15	0.973	0.833
	8	0.429	0.954		9	0.695	0.893		14	0.995	0.778
	7	0.660	0.901		8	0.867	0.819		13	0.999	0.722
	6	0.863	0.812		7	0.962	0.726	0.950	19	0.377	0.981
	5	0.971	0.691		6	0.994	0.625		18	0.760	0.943

TABLE F.3 FINITE QUEUING FACTORS—POPULATION 30

X	M	D	F
0.004	1	0.116	0.999
0.007	1	0.203	0.998
0.009	1	0.260	0.997
0.010	1	0.289	0.996
0.011	1	0.317	0.995
0.012	1	0.346	0.994
0.013	1	0.374	0.993
0.014	2	0.067	0.999
	1	0.403	0.991
0.015	2	0.076	0.999
	1	0.431	0.989
0.016	2	0.085	0.999
	1	0.458	0.987
0.017	2	0.095	0.999
	1	0.486	0.985
0.018	2	0.105	0.999
	1	0.513	0.983
0.019	2	0.116	0.999
	1	0.541	0.980
0.020	2	0.127	0.998
	1	0.567	0.976
0.021	2	0.139	0.998
	1	0.594	0.973
0.022	2	0.151	0.998
	1	0.620	0.969
0.023	2	0.163	0.997
	1	0.645	0.965
0.024	2	0.175	0.997
	1	0.670	0.960
0.025	2	0.188	0.996
	1	0.694	0.954
0.026	2	0.201	0.996
	1	0.718	0.948
0.028	3	0.051	0.999
	2	0.229	0.995
	1	0.763	0.935
0.030	3	0.060	0.999
	2	0.257	0.994
	1	0.805	0.918
0.032	3	0.071	0.999
	2	0.286	0.992
	1	0.843	0.899
0.034	3	0.083	0.999
	2	0.316	0.990
	1	0.876	0.877
0.036	3	0.095	0.998
	2	0.347	0.988
	1	0.905	0.853
0.038	3	0.109	0.998
	2	0.378	0.986
	1	0.929	0.827
0.040	3	0.123	0.997
	2	0.410	0.983
	1	0.948	0.800
0.042	3	0.138	0.997
	2	0.442	0.980

X	M	D	F
	1	0.963	0.772
0.044	4	0.040	0.999
	3	0.154	0.996
	2	0.474	0.977
	1	0.974	0.744
0.046	4	0.046	0.999
	3	0.171	0.996
	2	0.506	0.972
	1	0.982	0.716
0.048	4	0.053	0.999
	3	0.189	0.995
	2	0.539	0.968
	1	0.988	0.689
0.050	4	0.060	0.999
	3	0.208	0.994
	2	0.571	0.963
	1	0.992	0.663
0.052	4	0.068	0.999
	3	0.227	0.993
	2	0.603	0.957
	1	0.995	0.639
0.054	4	0.077	0.998
	3	0.247	0.992
	2	0.634	0.951
	1	0.997	0.616
0.056	4	0.086	0.998
	3	0.267	0.991
	2	0.665	0.944
	1	0.998	0.595
0.058	4	0.096	0.998
	3	0.288	0.989
	2	0.695	0.936
	1	0.999	0.574
0.060	5	0.030	0.999
	4	0.106	0.997
	3	0.310	0.987
	2	0.723	0.927
	1	0.999	0.555
0.062	5	0.034	0.999
	4	0.117	0.997
	3	0.332	0.986
	2	0.751	0.918
0.064	5	0.038	0.999
	4	0.128	0.997
	3	0.355	0.984
	2	0.777	0.908
0.066	5	0.043	0.999
	4	0.140	0.996
	3	0.378	0.982
	2	0.802	0.897
0.068	5	0.048	0.999
	4	0.153	0.995
	3	0.402	0.979
	2	0.825	0.885
0.070	5	0.054	0.999
	4	0.166	0.995

X	M	D	F
	3	0.426	0.976
	2	0.847	0.873
0.075	5	0.069	0.998
	4	0.201	0.993
	3	0.486	0.969
	2	0.893	0.840
0.080	6	0.027	0.999
	5	0.088	0.998
	4	0.240	0.990
	3	0.547	0.959
	2	0.929	0.805
0.085	6	0.036	0.999
	5	0.108	0.997
	4	0.282	0.987
	3	0.607	0.948
	2	0.955	0.768
0.090	6	0.046	0.999
	5	0.132	0.996
	4	0.326	0.984
	3	0.665	0.934
	2	0.972	0.732
0.095	6	0.057	0.999
	5	0.158	0.994
	4	0.372	0.979
	3	0.720	0.918
	2	0.984	0.697
0.100	6	0.071	0.998
	5	0.187	0.993
	4	0.421	0.973
	3	0.771	0.899
	2	0.991	0.664
0.110	7	0.038	0.999
	6	0.105	0.997
	5	0.253	0.988
	4	0.520	0.959
	3	0.856	0.857
	2	0.997	0.605
0.120	7	0.057	0.998
	6	0.147	0.994
	5	0.327	0.981
	4	0.619	0.939
	3	0.918	0.808
	2	0.999	0.555
0.130	8	0.030	0.999
	7	0.083	0.997
	6	0.197	0.991
	5	0.409	0.972
	4	0.712	0.914
	3	0.957	0.758
0.140	8	0.045	0.999
	7	0.¹15	0.996
	6	0.256	0.987
	5	0.494	0.960
	4	0.793	0.884
	3	0.979	0.710
0.150	9	0.024	0.999

TABLE F.3   FINITE QUEUING FACTORS—POPULATION 30 (Continued)

X	M	D	F	X	M	D	F	X	M	D	F
	8	0.065	0.998		7	0.585	0.938		7	0.901	0.818
	7	0.155	0.993		6	0.816	0.868		6	0.981	0.712
	6	0.322	0.980		5	0.961	0.751		5	0.999	0.595
	5	0.580	0.944		4	0.998	0.606	0.290	14	0.023	0.999
	4	0.860	0.849	0.230	12	0.023	0.999		13	0.055	0.998
	3	0.991	0.665		11	0.056	0.998		12	0.117	0.994
0.160	9	0.036	0.999		10	0.123	0.994		11	0.223	0.986
	8	0.090	0.997		9	0.242	0.985		10	0.382	0.969
	7	0.201	0.990		8	0.423	0.965		9	0.582	0.937
	6	0.394	0.972		7	0.652	0.923		8	0.785	0.880
	5	0.663	0.924		6	0.864	0.842		7	0.929	0.795
	4	0.910	0.811		5	0.976	0.721		6	0.988	0.688
	3	0.996	0.624		4	0.999	0.580		5	0.999	0.575
0.170	10	0.019	0.999	0.240	12	0.031	0.999	0.300	14	0.031	0.999
	9	0.051	0.998		11	0.074	0.997		13	0.071	0.997
	8	0.121	0.995		10	0.155	0.992		12	0.145	0.992
	7	0.254	0.986		9	0.291	0.981		11	0.266	0.982
	6	0.469	0.961		8	0.487	0.955		10	0.437	0.962
	5	0.739	0.901		7	0.715	0.905		9	0.641	0.924
	4	0.946	0.773		6	0.902	0.816		8	0.830	0.861
	3	0.998	0.588		5	0.986	0.693		7	0.950	0.771
0.180	10	0.028	0.999		4	0.999	0.556		6	0.993	0.666
	9	0.070	0.997	0.250	13	0.017	0.999	0.320	15	0.023	0.999
	8	0.158	0.993		12	0.042	0.998		14	0.054	0.998
	7	0.313	0.980		11	0.095	0.996		13	0.113	0.994
	6	0.546	0.948		10	0.192	0.989		12	0.213	0.987
	5	0.806	0.874		9	0.345	0.975		11	0.362	0.971
	4	0.969	0.735		8	0.552	0.944		10	0.552	0.943
	3	0.999	0.555		7	0.773	0.885		9	0.748	0.893
0.190	10	0.039	0.999		6	0.932	0.789		8	0.901	0.820
	9	0.094	0.996		5	0.992	0.666		7	0.977	0.727
	8	0.200	0.990	0.260	13	0.023	0.999		6	0.997	0.625
	7	0.378	0.973		12	0.056	0.998	0.340	16	0.016	0.999
	6	0.621	0.932		11	0.121	0.994		15	0.040	0.998
	5	0.862	0.845		10	0.233	0.986		14	0.086	0.996
	4	0.983	0.699		9	0.402	0.967		13	0.169	0.990
0.200	11	0.021	0.999		8	0.616	0.930		12	0.296	0.979
	10	0.054	0.998		7	0.823	0.864		11	0.468	0.957
	9	0.123	0.995		6	0.954	0.763		10	0.663	0.918
	8	0.249	0.985		5	0.995	0.641		9	0.836	0.858
	7	0.446	0.963	0.270	13	0.032	0.999		8	0.947	0.778
	6	0.693	0.913		12	0.073	0.997		7	0.990	0.685
	5	0.905	0.814		11	0.151	0.992		6	0.999	0.588
	4	0.991	0.665		10	0.279	0.981	0.360	16	0.029	0.999
0.210	11	0.030	0.999		9	0.462	0.959		15	0.065	0.997
	10	0.073	0.997		8	0.676	0.915		14	0.132	0.993
	9	0.157	0.992		7	0.866	0.841		13	0.240	0.984
	8	0.303	0.980		6	0.970	0.737		12	0.392	0.967
	7	0.515	0.952		5	0.997	0 617		11	0.578	0.937
	6	0.758	0.892	0.280	14	0.017	0.999		10	0.762	0.889
	5	0.938	0.782		13	0.042	0.998		9	0.902	0.821
	4	0.995	0.634		12	0.093	0.996		8	0.974	0.738
0.220	11	0.041	0.999		11	0.185	0.989		7	0.996	0.648
	10	0.095	0.996		10	0.329	0.976	0.380	17	0.020	0.999
	9	0.197	0.989		9	0.522	0.949		16	0.048	0.998
	8	0.361	0.974		8	0.733	0.898		15	0.101	0.995

TABLE F.3   FINITE QUEUING FACTORS—POPULATION 30 (Continued)

X	M	D	F	X	M	D	F	X	M	D	F
	14	0.191	0.988		16	0.310	0.977		22	0.038	0.998
	13	0.324	0.975		15	0.470	0.957		21	0.085	0.996
	12	0.496	0.952		14	0.643	0.926		20	0.167	0.990
	11	0.682	0.914		13	0.799	0.881		19	0.288	0.980
	10	0.843	0.857		12	0.910	0.826		18	0.443	0.963
	9	0.945	0.784		11	0.970	0.762		17	0.612	0.936
	8	0.988	0.701		10	0.993	0.694		16	0.766	0.899
	7	0.999	0.614		9	0.999	0.625		15	0.883	0.854
0.400	17	0.035	0.999	0.500	20	0.032	0.999		14	0.953	0.802
	16	0.076	0.996		19	0.072	0.997		13	0.985	0.746
	15	0.150	0.992		18	0.143	0.992		12	0.997	0.690
	14	0.264	0.982		17	0.252	0.983		11	0.999	0.632
	13	0.420	0.964		16	0.398	0.967	0.600	23	0.024	0.999
	12	0.601	0.933		15	0.568	0.941		22	0.059	0.997
	11	0.775	0.886		14	0.733	0.904		21	0.125	0.993
	10	0.903	0.823		13	0.865	0.854		20	0.230	0.986
	9	0.972	0.748		12	0.947	0.796		19	0.372	0.972
	8	0.995	0.666		11	0.985	0.732		18	0.538	0.949
0.420	18	0.024	0.999		10	0.997	0.667		17	0.702	0.918
	17	0.056	0.997	0.520	21	0.021	0.999		16	0.837	0.877
	16	0.116	0.994		20	0.051	0.998		15	0.927	0.829
	15	0.212	0.986		19	0.108	0.994		14	0.974	0.776
	14	0.350	0.972		18	0.200	0.988		13	0.993	0.722
	13	0.521	0.948		17	0.331	0.975		12	0.999	0.667
	12	0.700	0.910		16	0.493	0.954	0.700	25	0.039	0.998
	11	0.850	0.856		15	0.663	0.923		24	0.096	0.995
	10	0.945	0.789		14	0.811	0.880		23	0.196	0.989
	9	0.986	0.713		13	0.915	0.827		22	0.339	0.977
	8	0.998	0.635		12	0.971	0.767		21	0.511	0.958
0.440	19	0.017	0.999		11	0.993	0.705		20	0.681	0.930
	18	0.041	0.998		10	0.999	0.641		19	0.821	0.894
	17	0.087	0.996	0.540	21	0.035	0.999		18	0.916	0.853
	16	0.167	0.990		20	0.079	0.996		17	0.967	0.808
	15	0.288	0.979		19	0.155	0.991		16	0.990	0.762
	14	0.446	0.960		18	0.270	0.981		15	0.997	0.714
	13	0.623	0.929		17	0.421	0.965	0.800	27	0.053	0.998
	12	0.787	0.883		16	0.590	0.938		26	0.143	0.993
	11	0.906	0.824		15	0.750	0.901		25	0.292	0.984
	10	0.970	0.755		14	0.874	0.854		24	0.481	0.966
	9	0.994	0.681		13	0.949	0.799		23	0.670	0.941
	8	0.999	0.606		12	0.985	0.740		22	0.822	0.909
0.460	19	0.028	0.999		11	0.997	0.679		21	0.919	0.872
	18	0.064	0.997		10	0.999	0.617		20	0.970	0.832
	17	0.129	0.993	0.560	22	0.023	0.999		19	0.991	0.791
	16	0.232	0.985		21	0.056	0.997		18	0.998	0.750
	15	0.375	0.970		20	0.117	0.994	0.900	29	0.047	0.999
	14	0.545	0.944		19	0.215	0.986		28	0.200	0.992
	13	0.717	0.906		18	0.352	0.973		27	0.441	0.977
	12	0.857	0.855		17	0.516	0.952		26	0.683	0.953
	11	0.945	0.793		16	0.683	0.920		25	0.856	0.923
	10	0.985	0.724		15	0.824	0.878		24	0.947	0.888
	9	0.997	0.652		14	0.920	0.828		23	0.985	0.852
0.480	20	0.019	0.999		13	0.972	0.772		22	0.996	0.815
	19	0.046	0.998		12	0.993	0.714		21	0.999	0.778
	18	0.098	0.995		11	0.999	0.655	0.950	29	0.226	0.993
	17	0.184	0.989	0.580	23	0.014	0.999		28	0.574	0.973

Index